BlackBerry Bold Made Simple

For the BlackBerry Bold 9700 and 9650 Series

■ ■ ■

Martin Trautschold
and
Gary Mazo

Apress®

BlackBerry Bold Made Simple

ISBN-13 (pbk): 978-1-4302-3117-2

ISBN-13 (electronic): 978-1-4302-3118-9

Printed and bound in the United States of America 9 8 7 6 5 4 3 2 1

Distributed to the book trade worldwide by Springer Science+Business Media, LLC., 233 Spring Street, 6th Floor, New York, NY 10013. Phone 1-800-SPRINGER, fax (201) 348-4505, e-mail `orders-ny@springer-sbm.com`, or visit `www.springeronline.com`.

For information on translations, please e-mail `rights@apress.com`, or visit `www.apress.com`.

Apress and friends of ED books may be purchased in bulk for academic, corporate, or promotional use. eBook versions and licenses are also available for most titles. For more information, reference our Special Bulk Sales–eBook Licensing web page at `www.apress.com/info/bulksales`.

This book is dedicated to our families—to our wives, Julie and Gloria, and to our kids, Sophie, Livvie and Cece, and Ari, Dan, Sara, Billy, Elise and Jonah.

Without their love, support, and understanding, we could never take on projects like this one. Now that the book is done, we will gladly share our BlackBerrys with them – for a little while!

Contents

About the Authors

Martin Trautschold is the founder and CEO of Made Simple Learning, a leading provider of Apple iPad, iPhone, iPod touch, BlackBerry, and Palm webOS books and video tutorials. He has been a successful entrepreneur in the mobile device training and software business since 2001. With Made Simple Learning, he helped to train thousands of BlackBerry Smartphone users with short, to-the-point video tutorials. Martin has now co-authored fifteen "Made Simple" guide books. He also co-founded, ran for 3 years, and then sold a mobile device software company. Prior to this, Martin spent 15 years in technology and business consulting in the US and Japan. He holds an engineering degree from Princeton University and an MBA from the Kellogg School at Northwestern University. Martin and his wife, Julia, have three daughters. He enjoys rowing and cycling. Martin can be reached at `martin@madesimplelearning.com`.

Gary Mazo is Vice President of Made Simple Learning and is a writer, a college professor, a gadget nut, and an ordained rabbi. Gary joined Made Simple Learning in 2007 and has co-authored the last thirteen books in the Made Simple series. Along with Martin, and Kevin Michaluk from CrackBerry.com, Gary co-wrote *CrackBerry: True Tales of BlackBerry Use and Abuse*—a book about BlackBerry addiction and how to get a grip on one's BlackBerry use. This book is being refreshed and reprinted by Apress and will be available this fall. Gary also teaches writing, philosophy, technical writing, and more at the University of Phoenix. He holds a BA in anthropology from Brandeis University. Gary earned his M.A.H.L (Masters in Hebrew Letters) as well as ordination as Rabbi from the Hebrew Union College-Jewish Institute of Religion in Cincinnati, Ohio. He has served congregations in Dayton, Ohio, Cherry Hill, New Jersey and Cape Cod, Massachusetts.

Gary is married to Gloria Schwartz Mazo; they have six children. Gary can be reached at: `gary@madesimplelearning.com`.

Acknowledgments

A book like this takes many people to put together. We would like to thank Apress for believing in us and our unique style of writing.

We would like to thank our Editors, Jim and Laurin, and the entire editorial team at Apress.

We would like to thank our families for their patience and support in allowing us to pursue projects such as this one.

Part I

Quick Start Guide

In your hands is one of the most capable devices to hit the market in quite some time: the BlackBerry Bold. This Quick Start Guide will help get you and your new Bold up and running in a hurry. You'll learn all about the buttons, switches, and ports, and how to use the responsive trackpad to help you get around. Our App Reference Tables introduce you to the apps on your Bold — and serve as a quick way to find out how to accomplish a task.

Getting Around Quickly

This Quick Start Guide is meant to be just that—a tool that can help you jump right in and find information in this book—and learn the basics of how to get around and enjoy your Bold right away.

We start with the nuts and bolts in our "Learning Your Way Around" section—what all the keys, buttons, switches, and symbols mean and do on your Bold. You will learn how to get inside the back of your Bold to remove and replace the battery, SIM card, and media card. Also learn how to use the **Menu** key, **Escape** key, **trackpad**, and other important buttons. We show you some great time-saving tips for menus and setting dates and times as well as how to multitask.

In "Working With the Wireless Network," we help you understand when the letters, numbers, and symbols at the top of your Bold screen tell you that you can make phone calls, send SMS text messages, send and receive email, or browse the Web. We also show you how to handle your Bold on an airplane when you might need to turn off the radios.

In "App Reference Tables," we've organized the app icons into general categories so you can quickly browse the icons and jump to a section in the book to learn more about the app a particular icon represents. Here are the tables:

- Getting Set Up (Table 2)
- Staying In Touch (Table 3)
- Staying Organized (Table 4)
- Being Productive (Table 5)
- Being Entertained (Table 6)
- Networking Socially (Table 7)
- Personalize Your Curve (Table 8)
- Add and Remove Software (Table 9)

Learning Your Way Around

To help you get comfortable with your Bold, we start with the basics–what the keys, buttons, ports, and trackpad do and how to open up the back cover to get at your battery, media, and SIM card. Then we move into how you start apps and navigate the menus. We end this section with a number of very useful time-saving tips and tricks about getting around the menus, setting dates and times, and many great ways to use the **Space** key.

Keys, Buttons, and Ports

Figure 1 shows all the things you can do with the buttons, keys, switches, and ports, on your Bold. Go ahead and try out a few things to see what happens. Try pressing the **Menu** key for two seconds, press and hold the **1** key, press the side **Convenience** keys, click on an icon, then press the **Escape** key. Have some fun getting acquainted with your device.

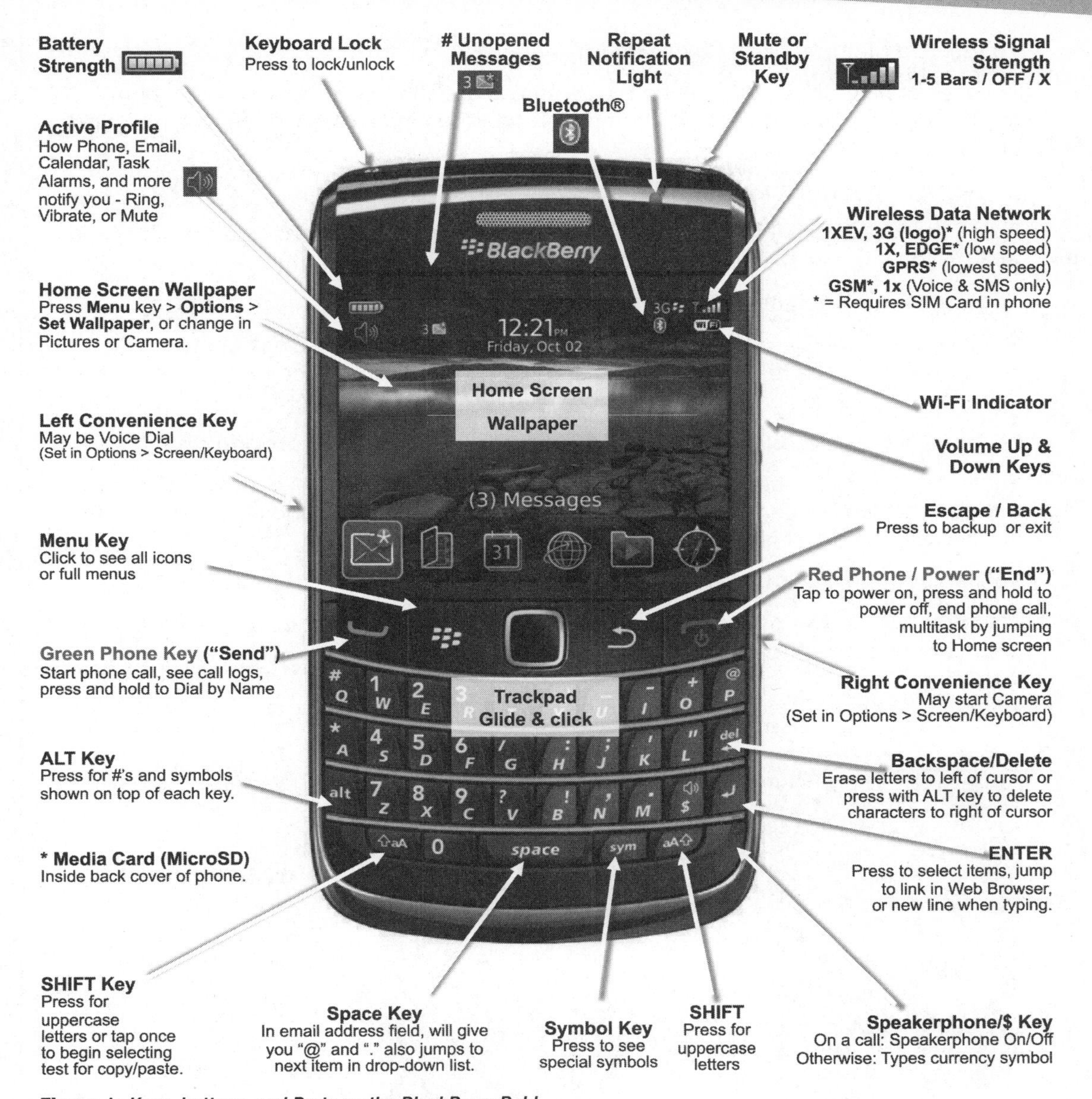

Figure 1. *Keys, buttons, and Ports on the BlackBerry Bold*

Inside Your Bold

You have to get inside your Bold to access your battery, SIM card slot, and media card slot. The following instructions and Figure 2 show you how.

To Remove the Back Cover:

Slide the cover down and off the bottom of the BlackBerry.

To Insert a Memory Card (MicroSD format):

You do not need to remove the battery. Gently place the media card with the metal contacts facing down and slide it completely into the media card slot shown in Figure 2 below.

To Remove the Memory Card:

For Bold 9700: Press it in with your fingernail until it pops out, then remove it. For Bold 9650: Slide it out with your fingernail.

To Remove or Replace the Battery:

Gently put your fingernail under the edge of the battery (with the black or gray semicircle—see Figure 2) and pry it up and out. To replace it, place it in the slot with the metal contacts touching and press it down firmly into the BlackBerry.

To Insert a SIM Card (Required to Connect to a GSM phone network):

For Bold 9700, you first need to remove the battery to see the SIM card slot. Place the SIM card on the left edge with the notch in the upper left corner and slide it completely into the SIM card slot (see Figure 2).

To Replace the Back Cover:

Place the back cover on the bottom half of the device and slide it up to the top until it clicks or locks into place.

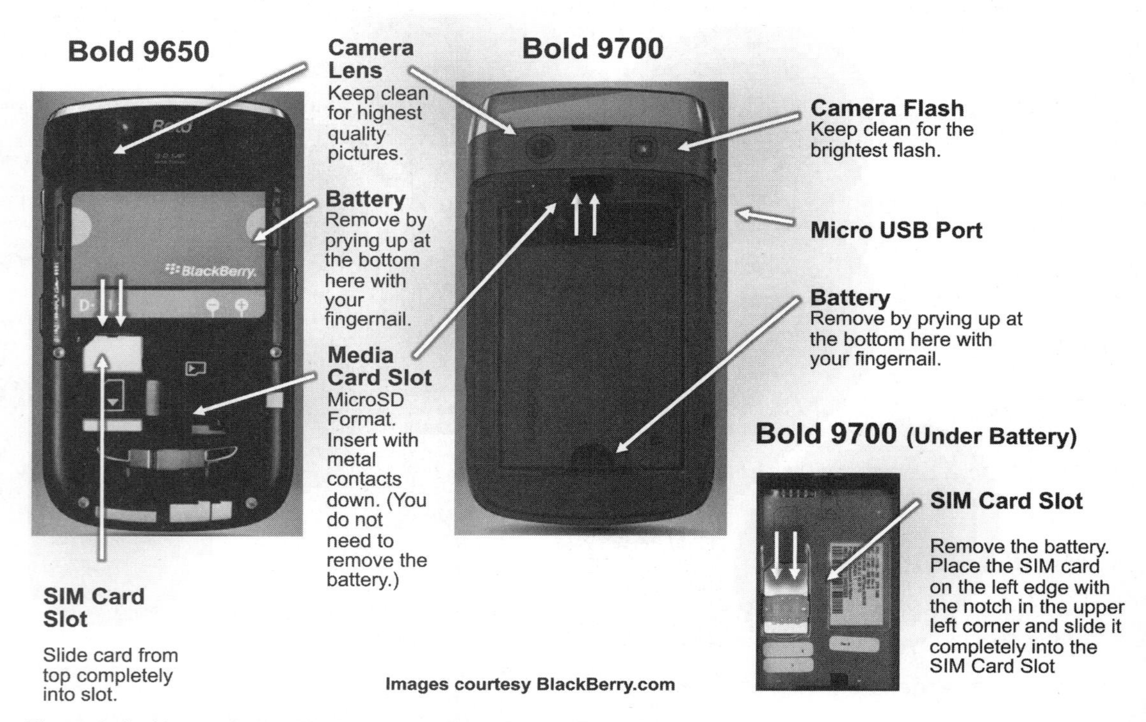

Figure 2. *Inside your Bold—The battery, media card, and SIM card slots*

Starting and Exiting Icons

You use the **trackpad**, **Menu** key, and **Escape** key to navigate around your BlackBerry, open folders, and select icons (Figure 3). The **Escape** key will get you back out one step at a time, while the **Red Phone** key will jump you all the way back to your **Home** screen. You can change the background image (wallpaper), also called the **Home** screen image, from your media player or camera. You can change the look and feel or **Theme** of your Bold by going into the **Options** icon and selecting **Theme** (see page 179.) Learn how to move icons (page 169) or hide icons (page 171).

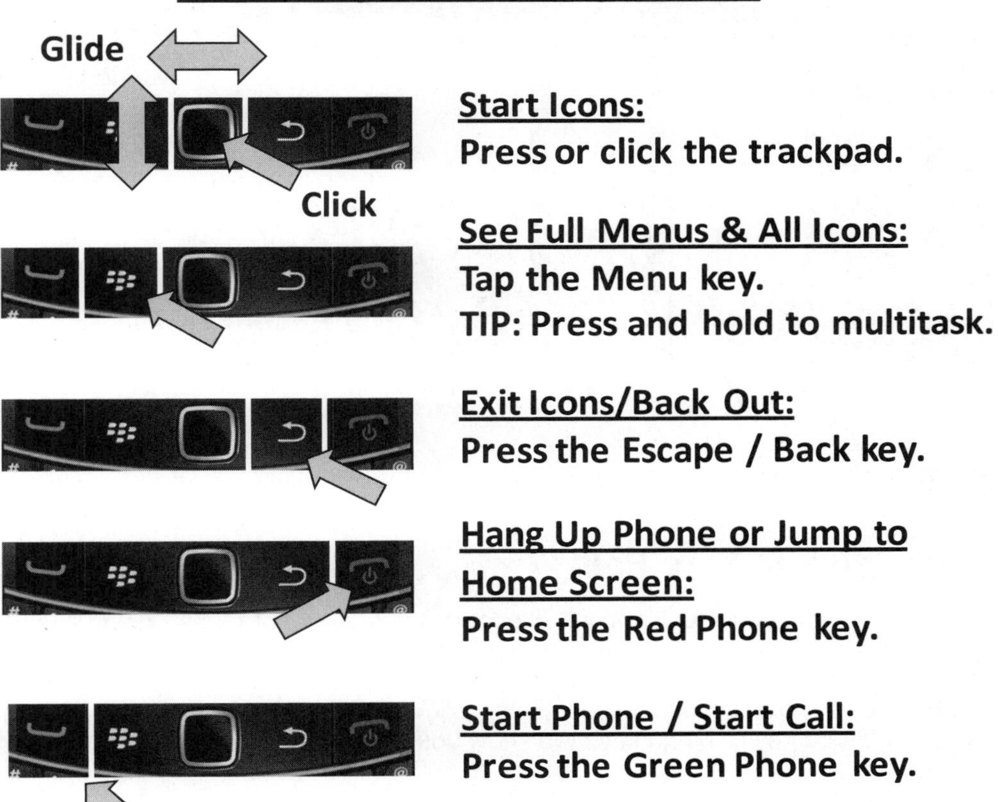

Figure 3. *How to start and exit icons (apps)*

Two Types of Menus (Full and Short)

One useful thing on your BlackBerry is the two types of menus: *full* and *short*. You see the full menu by pressing the **Menu** key and the short menu by pressing the **trackpad**. You will notice that the short menu will often times have exactly the thing you want to do highlighted, such as **Send** or **Forward**, saving you time with common tasks.

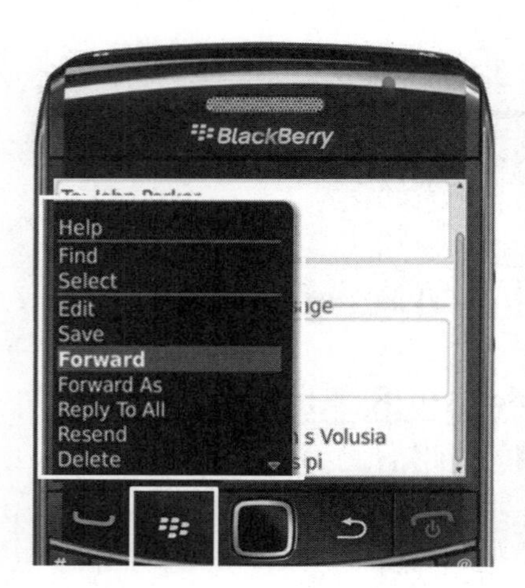

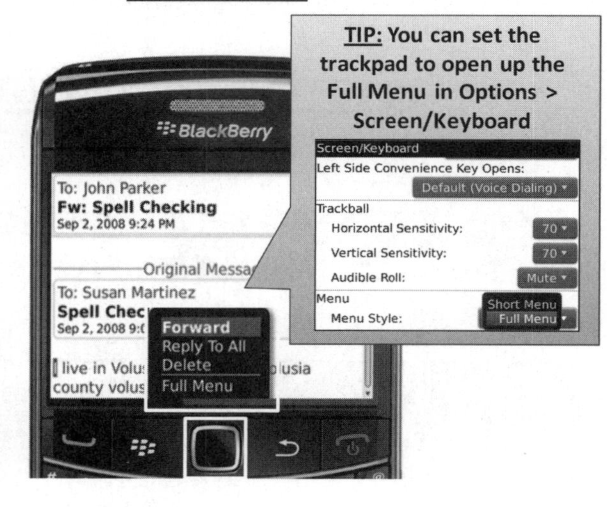

Figure 4. *Full and short menus*

Multitasking on Your Bold

Once you start using your Bold, you will almost immediately want to start doing a few things at the same time. Maybe you want to check your Calendar while listening to Pandora Internet radio, or cut, copy, and paste text from an email message into your Notepad. It is easy to multitask on your Bold. You actually have two choices: you can use the Red Phone key or Menu key. We show you how to use both next.

Multitask with the Red Phone Key

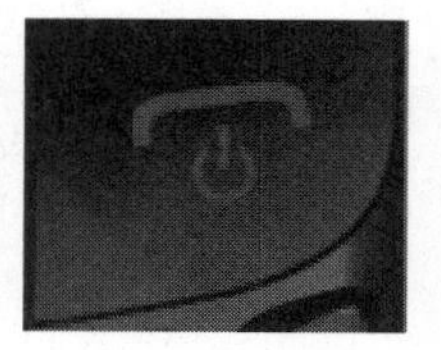

Press the Red Phone key (when not on a call) and you jump right to the Home screen. Or when on a call, press the Escape key, it does the same thing. Say you are writing an email and need to check the Calendar or want to schedule a new event.

1. Press the Red Phone key to jump to the Home screen.
2. Start the Calendar to check your schedule.
3. Press the Red Phone key again to return to the Home screen.
4. Click on the Messages icon to return exactly to where you left off composing your email message.

Remember, though, that if you always use the Red Phone key to jump out of icons and leave them running in the background, over time your BlackBerry will slow down.

Multitask with the Menu Key

You can also multitask using the Menu key. Just press and hold it to bring up the multitasking pop-up window.

1. Press and hold the Menu key to see the pop-up window of running apps.
2. Move or glide the trackpad to the app you want to start.
3. Click the trackpad on the Home icon, if you don't see the icon you want to start. Then click on the icon you want to start from the Home screen.
4. Repeat the procedure to return to the app you started in.

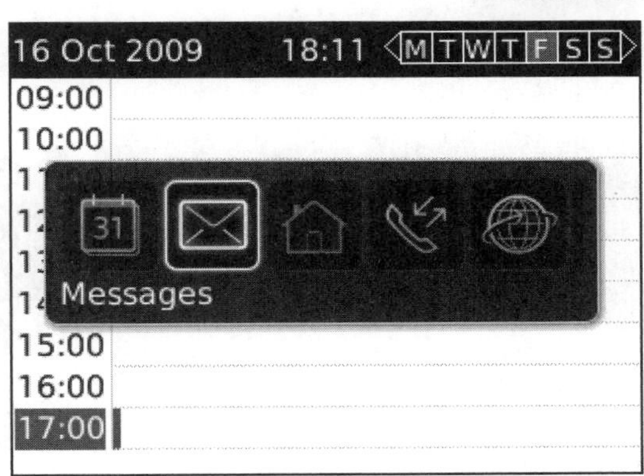

Too Much Multitasking May Slow You Down...

Is your Bold running slowly? We all like to multitask and doing so on your BlackBerry can help you be more productive! However, if you always use the **Red Phone** key or **Menu** key trick to jump out of icons and leave them running in the background, over time your BlackBerry will slow down. You need to make sure to close or exit using the **Escape** key to close unneeded icons. This will help speed your BlackBerry up. Or, you might want to do a soft reset as we show in our Troubleshooting chapter.

TIP: See page 521 to learn how to fix this slowness and other troubleshooting tips.

Jump to First Letter Trick for Menus and Drop-Downs

Getting around on your Bold, you will probably access at least 20 menus every day, sometimes many more. Since you use menus so often, saving a little time on every menu will result in a lot of overall time savings. The way to do this is by pressing the first letter of the menu item, list, or drop-down item on your keyboard to instantly jump down to it. See Figure 5.

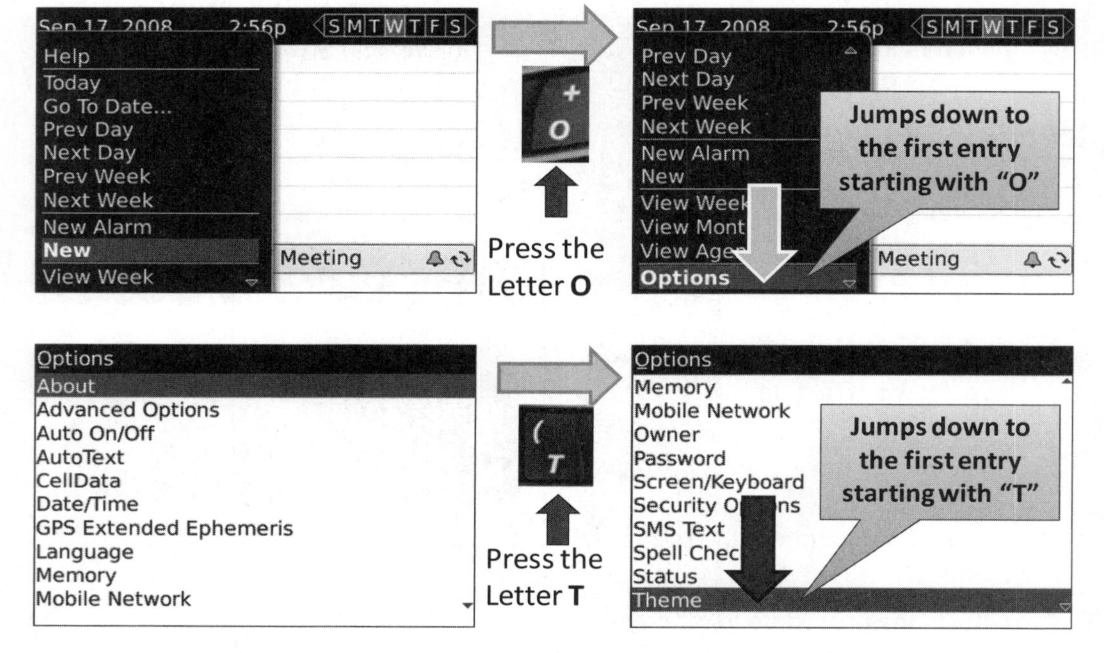

Figure 5. *Press the first letter to jump to a particular menu item or list item.*

Save Time with the Space Key

Pressing the **Space** key will save you time when you type email or web addresses. Also, pressing it twice at the end of a sentence will automatically insert a period and capitalize the first letter of the next sentence. See Figure 6.

Use the **Space** key when typing:

Email Addresses

- Type:
  - susan **Space** company **Space** com

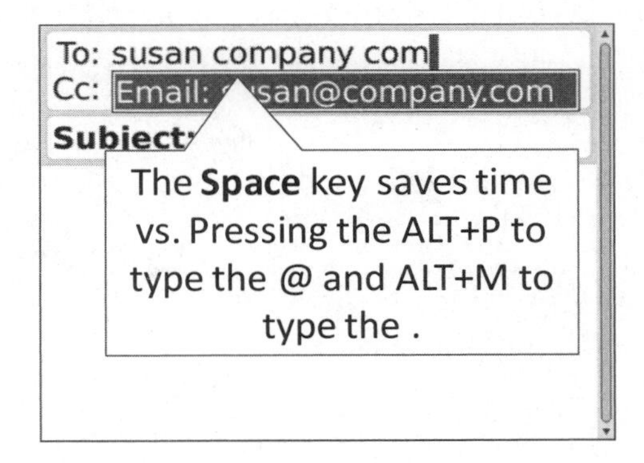

Web Addresses

- Get the dot "." in the address:
- Type:
  - www **Space** google **Space** com

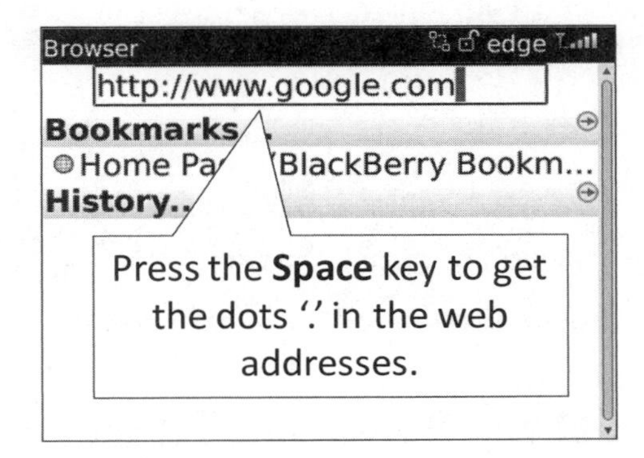

TIP: Save time typing emails by pressing the **Space** key twice at the end of each sentence. You will get an automatic period and the next letter will be uppercase.

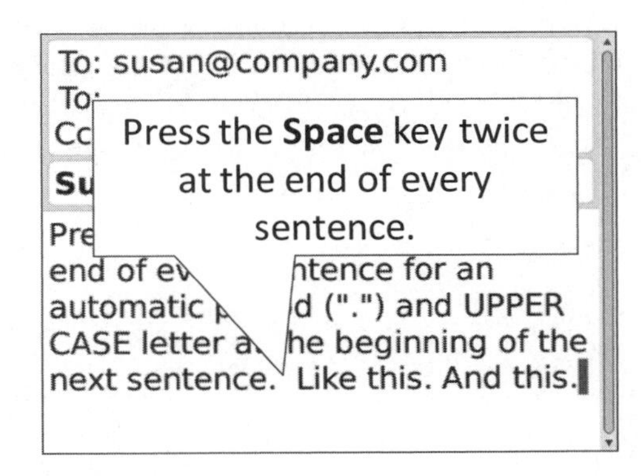

Figure 6. *Use the Space key to quickly type email and web addresses*

Setting Dates and Times with the Space Key

When you are setting dates or times, try using the **Space** key to jump to the next month, day, hour, or 15 minute interval (see Figure 7). It can be a great time saver if you want to schedule something just a month, day, hour, or 15 minutes ahead.

The **Space** key will jump to the next item in a date, time, or drop-down list field. Will jump to next 15 minutes in minute field.

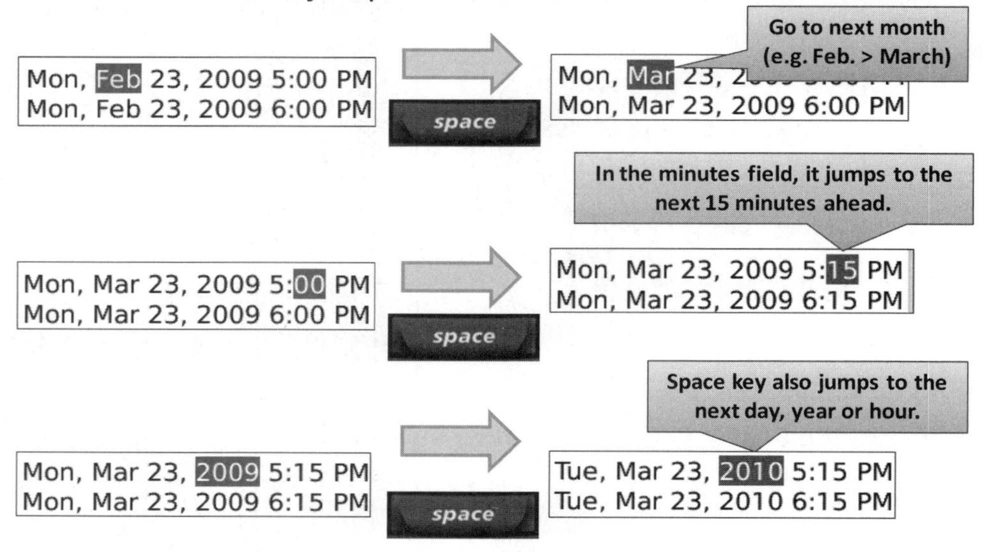

Figure 7. *Use space key to advance a month, day, year, or 15 minutes.*

Setting Dates and Times with Number Keys

When you are setting dates or times, you can save time rolling and clicking by using the **Number** keys to enter an exact date or time. For example, if you want to go to the 19th, type **1** and **9** and you're done (see Figure 8).

The **NUMBER KEYS** will allow you to type exact numbers in date/time fields. Like “2” “5” for the 25th of the month.

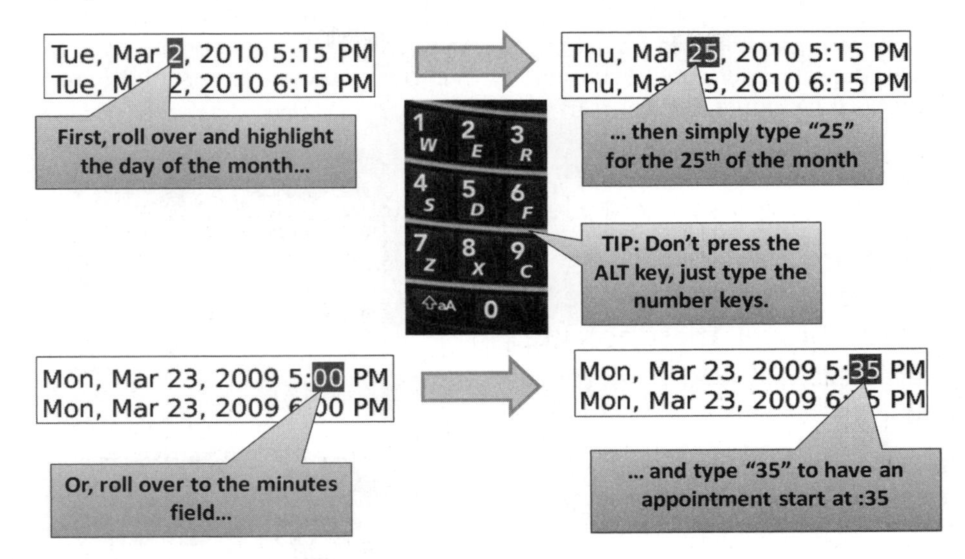

Figure 8. *Use number keys to set dates and times*

Space Key as Page Down

Save your thumb from the pain caused by constantly moving or gliding down to read more in an email, web page, or other item. Tap the **Space** key instead which works as your **Page Down** key.

- The **Space** key also doubles as a **Page Down** key.
- This can be very useful when reading:
  - Email
  - Web pages
  - Email attachments

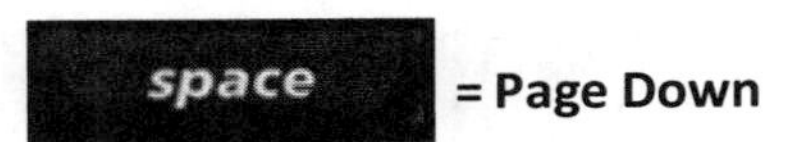

Quickly Typing Accented or Special Characters

The fastest way to type accented or special characters on your BlackBerry is to press and hold the underlying letter key, such as **a**, and then roll the **trackpad** until you see the character you need. Using this trick you can even quickly type special characters such as a trademark or copyright symbol. Remember, you can also use the **Sym** key for basic symbols (see Figure 9).

- To type an accented letter or special character, press and hold a letter key and roll the trackpad.
- Example: Press and hold the letter A to scroll through these accented characters: À Á Â Ã Ä Å Æ (both upper and lower case).
- Letters this trick works on are: E, R, T, Y, U, I, O, P, A, S, D, K, C, V, B, N, M
- Some other common characters:
  - V key for ¿, T key for ™, C key for ©, R key for ®

Figure 9. *How to quickly type accented letters or special characters*

Working with the Wireless Network

Since most of the functions on your Bold work only when you are connected to the Internet (email, Web, **App World**, **Facebook**, and so forth), you need to know when you're connected. Understanding how to read the status bar can save you time and frustration. You will also want to know how to quickly turn off your wireless radio or other radios (e.g. Bluetooth, Wi-Fi) when you get on an airplane.

Reading Your Wireless Network Status

In Table 1, we show you how to read the status icons on the top of your Bold screen so you can save time and keep you connected.

Table 1. *Reading Your Wireless Network Status*

In the upper right corner, If you see letters & symbols...	Email & Web	Phone Calls	SMS Text	Speed of Data Connection
3G (3G with logo)	✓	✓	✓	High
1XEV or EDGE	✓	✓	✓	Medium
1X or GPRS	✓	✓	✓	Slow
WiFi With any letters shown	✓	✓	✓	High
WiFi Without letters	✓	✗	✗	High
1x, edge, gprs, GSM or 3G	✗	✓	✓	None
OFF / X	✗	✗	✗	None

The following list shows the various wireless signal strength icons that will appear (signal strength varies between one and five bars):

Strong signal:

Weak signal:

No signal:

Radio off:

Traveling with Your Bold—Airplane Mode

When you travel, on most airlines you can simply turn off all your wireless connections (Mobile Network radio and Bluetooth) and continue using your BlackBerry.

Here's how to turn off all your wireless connections:

1. Click the **Manage Connections** icon.

2. Then click **Turn All Connections Off** until you see the word **Off** next to your wireless signal strength indicator.

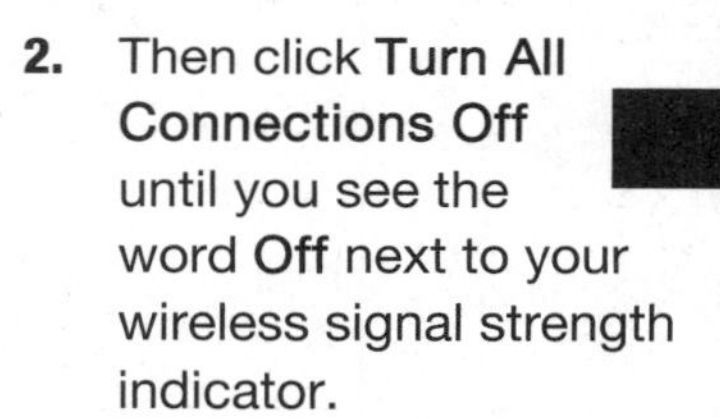

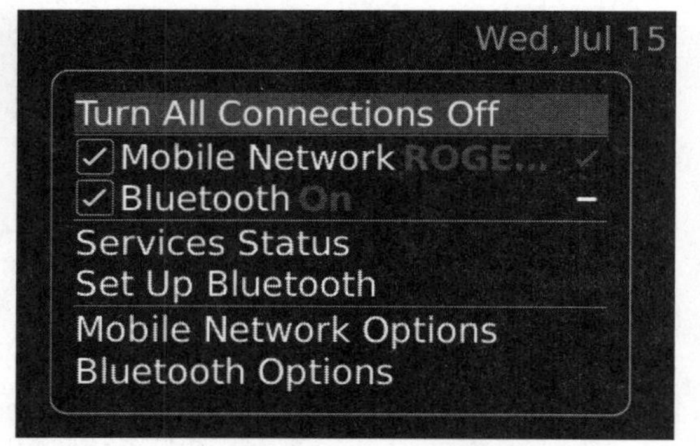

When you land and want to turn your connections on again, go back into **Manage Connections** and click the top option: **Restore Connections**.

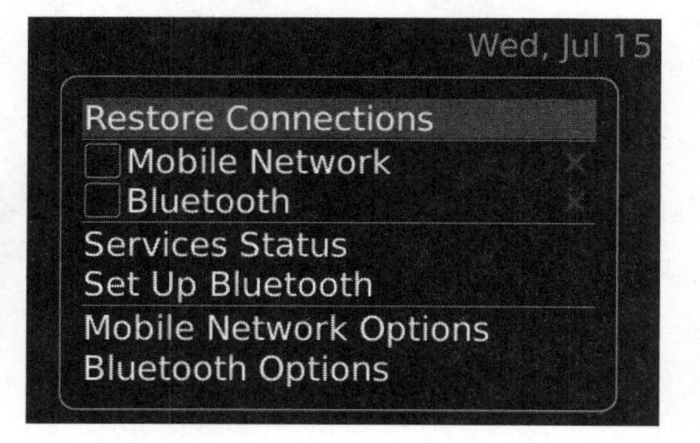

> **TIP:** To check/uncheck any option, just highlight and hit the **Space** key.

Bluetooth On, Network Off: Some airlines will allow you to keep your Stereo Bluetooth connection turned on while in flight. This allows you to listen to your music or watch a video using a Stereo Bluetooth headset. In order to keep Bluetooth on, but turn off your radio, uncheck **Mobile Network** but leave **Bluetooth** checked as shown.

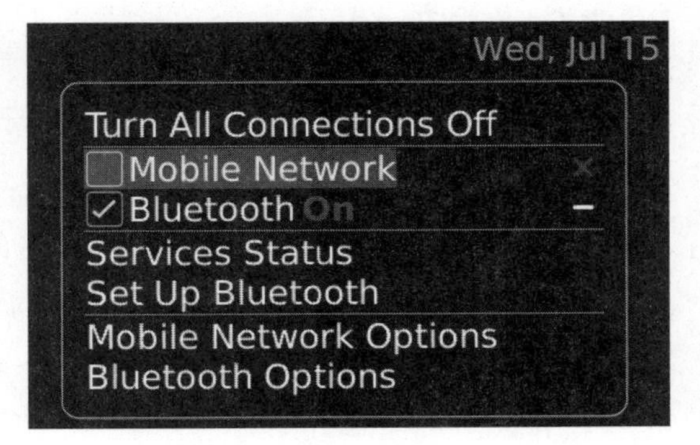

App Reference Tables

This section gives you a number of tables that group the apps on your Bold, as well as other apps you can download, into handy reference tables. Each table gives you a brief description of the app and tells where to find more information in this book.

> **NOTE:** The hotkeys listed in the following tables only work after you turn them on (see page 541 for instruction on this).

Getting Set Up

Table 2 includes some icons and quick links to help you get your email, Bluetooth, contacts, calendar, and more loaded on to your Bold. You can even boost the memory by adding a memory card, which is important to install if you like to have more pictures, videos, and music on your Bold.

Table 2. *Getting Set Up*

To Do This...	Use This...(Hotkey)	Where to Learn More
Find Your Setup Icons	**Setup folder**	Click on this icon to see your other Setup icons.
Email Setup, Date/Time, Fonts, and more.	**Setup Wizard**	Page 35.
Set up Your Internet Email	**Personal Email Setup**	Page 36.
Set up your Bluetooth headset	**Setup Bluetooth**	Page 441.
Share Addresses, Calendar, Tasks, and Notes with Your Computer	**BlackBerry Desktop Manager (for Windows)** **BlackBerry Desktop Manager (for Apple Mac)**	Page 65. See Videos at www.MadeSimpleLearning.com Page 125.
Add Memory to Store your Music, Videos, and Pictures	© SanDisk Corp. **Media Card**	Page 359.
Load up Your Music, Pictures, and Videos	Windows or Mac **Mass Storage Mode**	Page 144.
Fine-Tune Your Internet Email Setup & More	**Your Wireless Carrier web site**	See list of web sites on page 55.
Turn Off	**Power Off**	Click this icon to turn off your BlackBerry.
Lock Your Keyboard (Avoid speed dial in pocket or purse)	**Keyboard Lock** *(K)*	Click this icon, or tap the lock key on the top of your BlackBerry to lock it.

Staying In Touch

Getting familiar with the following icons and buttons in Table 3 will help you stay in touch with your friends and colleagues. Whether you prefer calling, emailing, texting, or using instant messaging, your Bold has many options. Use the Browser to stay up-to-date with the latest happenings on the Web.

Table 3. *Staying In Touch*

To Do This...	Use This... *(Hotkey)*	Where to Learn More
Read & Reply to Email	Messages *(M)*	Email—page 231. PIN Messaging—page 342. Attachments—page 243.
Send & Read SMS Text and MMS Messages	SMS & MMS	See page 328.
View All Your Saved Messages	Saved Messages *(V)*	Click to see all messages you have saved in your Messages inbox.
Get on the Internet/Browse the Web	Browser *(B / W)*	Page 461.
Call Voicemail	Press & Hold 1	Page 202.
Start a Call Dial by Name View Call Logs	Green Phone Phone & Call Logs	Phone – page 202. Call Logs - page 213. Conference Call – page 227.
Dial by Voice	Voice Dialing	Page 219.
Send an Instant Message to Another BlackBerry	BlackBerry Messenger *(N)*	BBM - Page 345.

To Do This...	Use This... *(Hotkey)*	Where to Learn More
Use Your Favorite Instant Messengers	AOL Google ICQ Yahoo MSN Messenger	Page 358.
Flying on an Airplane, Need to Turn Off the Radio	**Manage Connections**	Airplane – Page 16. Troubleshooting—Page 526.

Staying Organized

From organizing and finding your contacts to managing your calendar, taking written or voice notes, and calculating a tip using your built-in calculator, your Bold can help you do it all. Table 4 provides information for helping you to stay organized.

Table 4. *Staying Organized*

To Do This...	Use This... *(Hotkey)*	Where to Learn More
Manage Your Contact Names & Numbers	**Contacts** *(C)* **Address Book** *(A)*	Basics—Page 269. Add New—Page 272.
Manage your Calendar	**Calendar** *(L)*	Page 289. Sync to PC—Page 65. Sync to Mac—Page 125.
Set a Wakeup Alarm, Use a Countdown Timer or Stopwatch	**Clock**	Page 504.
Store All Your Important Passwords	**Password Keeper**	Page 507.
Find Lost Names, Email, Calendar Entries, and More	**Search** *(S)*	Page 509.

Being Productive

Sometimes you need to get work done on your Bold. Use the icons in Table 5 to get things done on your BlackBerry.

Table 5. *Being Productive*

To Do This...	Use This... (Hotkey)	Where to Learn More
Manage Your To-Do List	**Tasks** *(T)*	Page 311. Sync to PC—Page 65. Sync to Mac—Page 125.
Find Things, Get Directions, See Traffic	**Google Maps**	Page 491.
Take Notes, Store Your Grocery List, and More	**MemoPad** *(D)*	Page 317. Sync to PC—Page 65. Sync to Mac—Page 125.
Cannot Type a Note? Leave Yourself a Quick Voice Note	**Voice Note Recorder**	Page 506.
View & Edit Microsoft Office Word, Excel, and PowerPoint	**Word to Go, Sheet to Go, Slideshow to Go**	Page 246.
Calculate your MPG, a Meal Tip, and Convert units	**Calculator** *(U)*	Page 503.

Being Entertained

Use these icons and hotkeys listed in Table 6 for entertainment on your BlackBerry.

Table 6. *Being Entertained*

To Do This...	Use This... *(Hotkey)*	Where to Learn More
Listen to Free Music – with customized music stations	**Pandora**	Page 376.
Listen to Free Music – with customized music stations	**Slacker**	Page 378.
Play Music, Videos, and See Pictures	**Media**	Music—Page 363. Videos—Page 397. Ring Tones—Pages 203 and 221. Pictures—Page 392. Voice Notes—Page 506.
Snap Pictures of Anything, Anytime	**Camera**	Page 381.
Capture Video of Anything, Anytime	**Video Camera**	Page 397.
Take a Break and Play a Game	**Games Folder**	Page 478.
Use the Built-In Text-Based Help *(What, you can't find it in this book?)*	**Help** *(H)*	Page 159.

Networking Socially

Connect and stay up-to-date with friends, colleagues, and professional networks using the social networking tools on your Bold shown in Table 7.

Table 7. *Networking Socially*

To Do This...	Use This...*(Hotkey)*	Where to Learn More
Connect with Facebook Friends	Facebook	Page 406.
Follow people and Tweet using Twitter	Twitter	Page 409.
Connect with Colleagues	Linked In	Page 416.

Personalize Your Bold

Use the icons and hotkeys shown in Table 8 to personalize the look and feel of your Bold.

Table 8. *Personalize Your Bold*

To Do This...	Use This...*(Hotkey)*	Where to Learn More
Change Your Phone Ringer, if Emails Vibrate or Buzz, and more	Sounds	Page 191.
Change Your Wallpaper or Home Screen Picture	Media/Pictures	From Pictures - Page 184. From Home Screen – Page 178.
Change Your Theme	Options *(O)*	Theme—Page 179.

To Do This...	Use This...*(Hotkey)*	Where to Learn More
Change Your Font Size	Options *(O)*	Screen/Keyboard—Page 176.
Change Your Programmable Convenience Keys	Options *(O)*	Screen/Keyboard—Page 187.

Add or Remove Apps

Use the apps shown in Table 9 to add software and capabilities to your BlackBerry.

Table 9. *Add and Remove Software*

To Do This...	Use This...	Where to Learn More
Add New Software (also remove it)	**App World**	Add Icons—Page 431.
Add New Apps	**Browser** *(B)*	Add Icons—Page 475.
Find the Icons You Download	**Downloads folder**	Glide to and click on this folder to see your downloaded icons.
Remove Icons and Programs	**Options** *(O)*	Page 481.

Part I

Introduction

Welcome to your new Bold. In this section, we will introduce you to how the book is organized and where to find useful information. Inside the front and back cover, check out **Day in the Life** section, where we give you some scenarios to describe how you can use your Bold for work and play. We even show you how to find some great tips and tricks sent right to your Bold.

Congratulations on Your BlackBerry Bold!

In your hands is one of the most powerful smartphones available—the BlackBerry Bold 9650 or Bold 9700 Series.

Key Features on the BlackBerry Bold

Your BlackBerry Bold has many shared features with the BlackBerry family and some unique features as well.

Multimedia & Network Features

- Camera (3.2 mega pixel)
- Media Player (Pictures, Video, and Audio)
- Video Recording
- GPS Built-in
- 3G Network Capable

Media Card—Expansion Memory Card

Enhance the usable memory on your BlackBerry by a tremendous amount with an 8 or 16 gigabyte (GB) or even higher-capacity MicroSD memory card.

Getting the Most Out of *BlackBerry Bold Made Simple*

You can read this book cover-to-cover, but you can also peruse it in a modular fashion, by chapter or topic. Maybe you just want to check out BlackBerry **App World**, try the Web Browser, get set up with your email or contacts, or you might just want to load up your music. You can do all this and more with our book.

You will soon realize that your Bold is a very powerful device. There are, however, many secrets locked inside that we help you unlock throughout this book.

Take your time—this book can help you on your way to learning how to best use, work, and have fun with your new Bold. Think back to when you tried to use your first Windows or Mac computer. It took a little while to get familiar with how to do things. It's the same with the Bold. Use this book to help you get up-to-speed and learn all the best tips and tricks more quickly.

Remember that devices this powerful are not always easy to grasp—at first.

You will get the most out of your Bold if you can read a section and then try out what you read. We all know that reading and then doing an activity gives us a much higher retention rate than simply reading alone.

So, in order to learn and remember what you learn, we recommend to:

Read a little, try a little on your Bold, and repeat!

How This Book Is Organized

Knowing how this book is organized will help you more quickly locate things important to you. Here we show you the main organization of this book. Remember to take advantage of our abridged table of contents, detailed table of contents, and our comprehensive index to help you quickly pinpoint items of interest.

Day in the Life of a Bold User

Located inside the front and back cover, this is an excellent piece of information full of easy-to-access cross-reference page numbers. So if you see something you want to learn, simply thumb to that page and learn it—all in just a few minutes.

Part 1: Quick Start Guide

Learning Your Way Around: Learn about the keys, buttons, and ports on your Bold as well as how to get inside and change the battery, SIM card, or memory card. Then learn many time-saving tips about getting around quickly as well as how to multitask.

Working with the Wireless Network: Learn how to read letters, numbers, and symbols at the top of your Bold screen so you know when you can make phone calls, send SMS text messages, send and receive email, or browse the Web. Also learn how to handle your Bold on an airplane.

App Reference Tables: Quickly peruse the icons or apps grouped by category. Get a thumbnail of what all the apps do on your Bold and quick page numbers to jump right to the details of how to get the most out of each app in this book.

Part 2: Introduction

You are here now . . . .

Part 3: You and Your BlackBerry Bold

This is the meat of the book, organized in 36 easy-to-understand chapters packed with loads of pictures to guide you every step of the way.

Part 4: Hotkey Shortcuts

You will want to check out this section for a complete listing of all the hotkey shortcuts to get things done quickly on your Bold. Save yourself countless hours of rolling, searching, and clicking by using Home screen hotkeys, Email hotkeys, Calendar hotkeys, Web Browser hotkeys, and Media Player hotkeys. These hotkeys will make you much faster with your Bold—giving you a more enjoyable user experience.

Quickly Locating Tips, Cautions, and Notes

If you flip through this book, you can instantly see these items based on their formatting. For example, if you want to find all the Calendar tips, you would flip to the Calendar chapter and quickly find them.

TIPS, CAUTIONS, and NOTES are all formatted like this, with a gray background, to help you see them more quickly.

Free BlackBerry Email Tips and Free Videos

Check out the author's web site at `www.madesimplelearning.com` for a series of very useful bite-sized chunks of BlackBerry tips and tricks. We have taken a selection of great tips out of this book and even added a few new ones. Click on the **Free Tips** section and register for your tips in order to receive a tip right in your Bold inbox about once a week. Learning in small chunks is a great way to master your Bold.

The authors also offer some videos showing you tips and tricks about your BlackBerry Bold on their web site at `www.madesimplelearning.com`. In addition, there are over 100 video tutorials for purchase to show you how to use your Bold that you can watch on your computer and your BlackBerry.

Part 

You and Your BlackBerry Bold

This is the heart of *BlackBerry Bold Made Simple*. In this part of the book, you'll find clearly labeled chapters—each explaining the key features of your Bold. You'll see that most chapters focus on an individual app or a specific type of application. Many of the chapters discuss applications that come with your Bold, but we also include some fun and useful apps you can download from BlackBerry **App World**. Sure, the Bold is for fun, but it's for a whole lot more as well, so you'll learn how to be productive with **Word To Go**, **Sheet To Go**, and **Slideshow To Go** in this section, too. We finish with some handy troubleshooting tips that can help if your Bold isn't working quite right.

Chapter 1

Email Set Up

This chapter focuses on getting your email set up as well as wireless syncing of your contacts if you use Google Mail (Gmail) or have a BlackBerry Enterprise Server (BES). Sometimes email setup can cause problems, so we show you how to fix some common errors and make sure you receive all your email on both your computer and your BlackBerry. While you can adjust email setup from your BlackBerry, you may also want to adjust your email settings from your computer; we cover that in this chapter, as well.

If your BlackBerry is tied to a BES, we show you how to get connected or'*activated* on the server. And, more recently, Research In Motion, BlackBerry's maker, announced free BlackBerry Enterprise Server Express software. This is a fantasic value if your organization uses Microsoft Exchange and would like to take advantage of the BES features without additional software licensing and, in most cases, without additional hardware costs.

Learn Your BlackBerry and Getting Around

Much of the basics about the keys, buttons, and using your Bold as well as great tips and tricks are contained in our **Quick Start Guide** at the beginning of this book. In the Quick Start Guide, we also show you how to save time with many **Space key**, **Number** key, and **Menu** item shortcuts, and how to multitask.

Getting around Your Home Screen, Folders, and Icons

Depending on the particular phone company that supplied your BlackBerry, you may see more or fewer icons on your **Home** screen–similar to your computer's desktop. You will see a different background picture than the one shown in Figure 1-1. The background wallpaper in that figure was taken with the BlackBerry Bold Camera.

TIP: Check out Home Screen Hotkeys in Part 4 of this book on page 541 to learn how to enable single key startup of all your most-used icons, such as Messages (in Figure 1-1), Contacts, Calendar, and more.

1. Press the **Menu** key to go from your limited set of icons to see all your icons.
2. Scroll the **trackpad** down to see any icons that might be off the bottom of the screen.
3. Click on any folders, such as **Media**, to see more icons inside the folders
4. Press the **Escape** key to back out of a folder or return to **your** Home screen with just a few icons and your picture.
5. Your BlackBerry is fully customizable, so you can change the look and feel (called the **Theme**) and even the picture you see as the background.

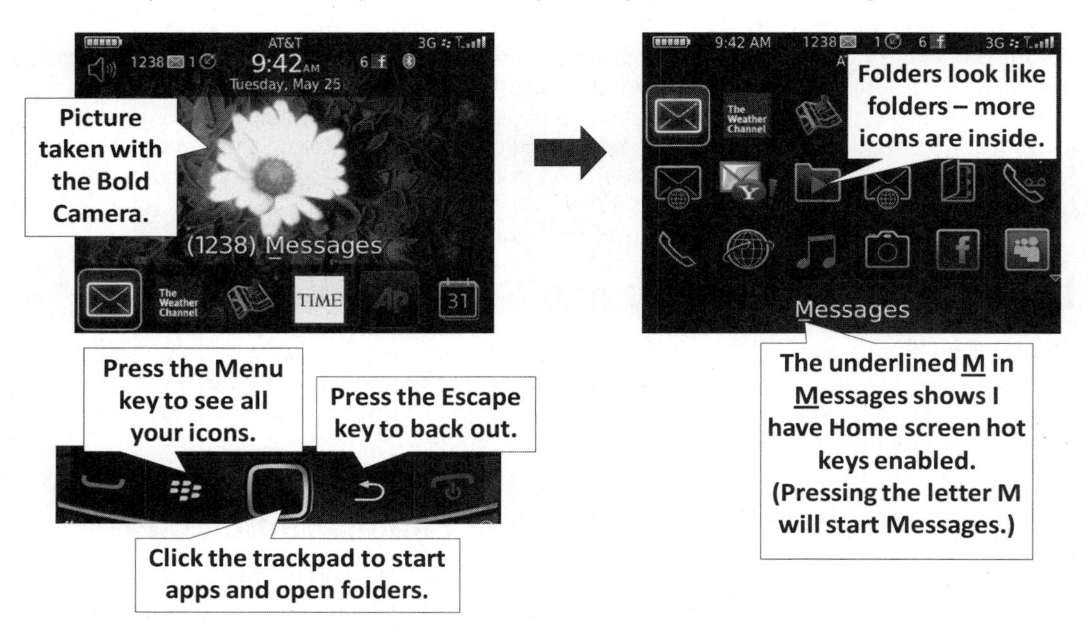

Figure 1-1. *Use Your **Menu** key, **trackpad**, and **Escape** keys to navigate around your bold*

TIP: To change your background wallpaper, from your **Home** screen press the **Menu** key twice until you see a menu appear. Select **Options** from the menu. Scroll down to **Set Wallpaper** and click on **Change Wallpaper**. Then, select from a number of pre-loaded background images, or you can even snap a picture and set it as your wallpaper.

The Setup Wizard

When you first turn on your BlackBerry, you will likely be presented with the Setup Wizard. If you ignored or closed it, you can get back to it by locating and clicking on the **Setup Wizard** icon. You may need to click on the **Setup** folder in order to find the **Setup Wizard** icon.

You will then be presented with a number of screens, which are not shown in this book because they are self-explanatory. Go ahead and follow the steps suggested; they will give you a good jump start on getting your BlackBerry set up and learning some of the basics.

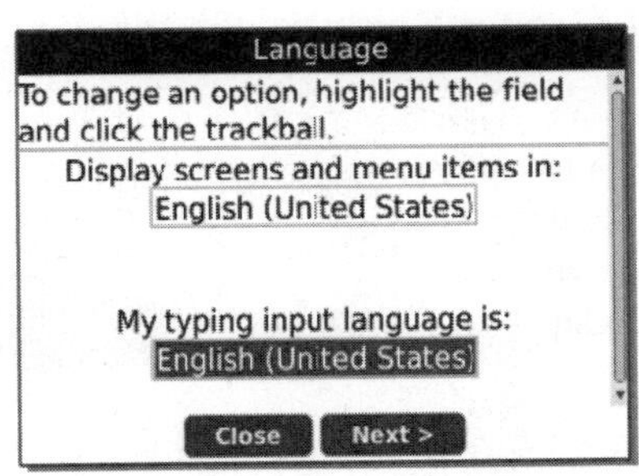

Just click on any field, such as date, time, or time zone to make an adjustment and then glide down and click **Next**.

Remove the unused languages to save space.

Next, you will see a few screens giving you some tips about how to use the basic keys on your Bold. (The screens you see in the Setup Wizard vary depending on which wireless carrier you have, so we have not included them in the book.)

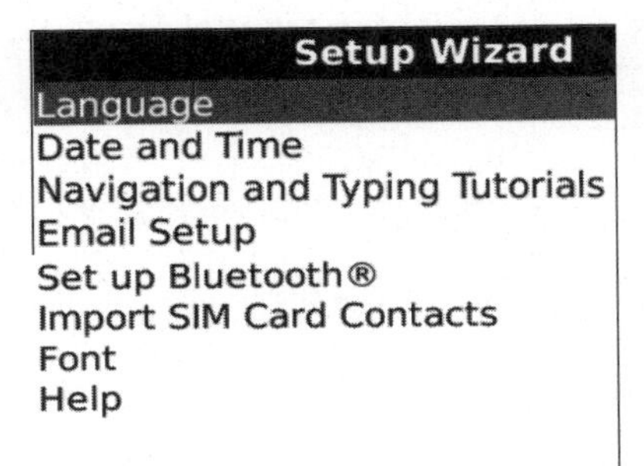

Finally, you will see a screen similar to this one to the right with the following items (you may also see a few different items listed):

Language – see page 255.

Date and Time—see page 175.

Navigation and Typing Tutorials – see our Quick Start guide and Chapter 6 "Typing, Spelling and Help.

Email setup—continue reading this chapter.

Set up Bluetooth— see page 441

Import SIM Card Contacts—see page 269.

Font—see page 175.

Help—see page 159.

Setting Up Email for the First Time

Your BlackBerry is designed to retrieve your email from up to ten different email accounts and, if you are connected to a BlackBerry Enterprise Server, one corporate email account. When your BlackBerry receives your email, all your messages will be displayed in your **Messages** inbox.

You can set up your basic email and even change things like your email signature right from your BlackBerry.

> **NOTE:** If you want to set up your BlackBerry to work with email coming from a BlackBerry Enterprise Server, skip to the "Setting Up Your Corporate Email" section on page 39.

Personal or Internet Email Setup

You can set up your Personal or Internet Email from two places:

The **Setup Wizard** icon and the **Personal Email Set Up** icon , both of which are usually inside the **Setup** folder, but may be on your Home screen of icons.

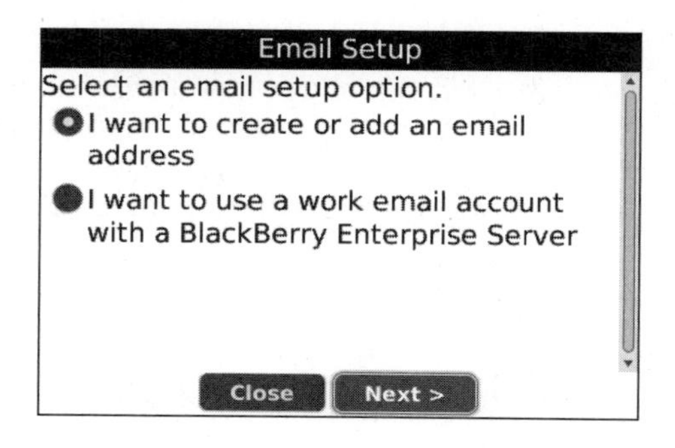

1. If you already happen to be inside in the **Setup Wizard**, then click **Next** to get to the **Personal Email Setup** screen.
2. Or, if you see a list of items, then click on **Personal Email Setup**.
3. Select **I want to create or add an email address** and click **Next** to get to the login screen shown to the right.

NOTE: You may need to create a new account if this is the first time you are logging into **Email Setup** on your BlackBerry.

If so, be sure to click the checkbox next to **Remember me on this device** and you won't have to keep entering your username/password every time.

4. Click the **Log In** button. Any accounts you have already set up will be listed at the top, as shown in the screen on the right.
5. To add an account, click the **Add** button.

6. Now, select from the various types of email accounts. In this case, we'll choose **Gmail**.
7. If you have another type of account (such as **Yahoo!**, **AOL**, and so forth), click on that entry.
8. If you don't see your email account type, then click **Other**.

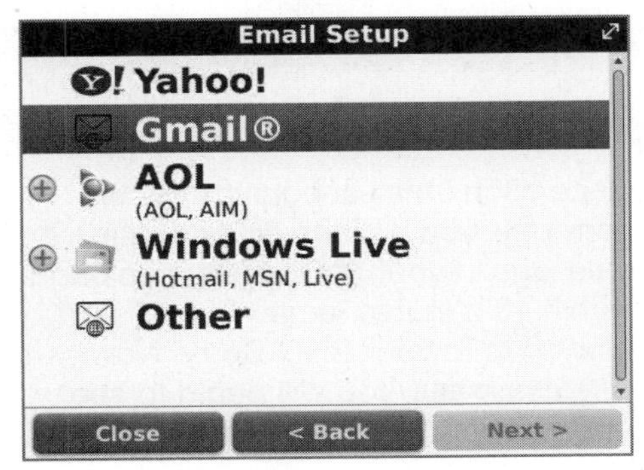

9. Next, type your email address, using your full email address (name@gmail.com) and password, then click **Next** to attempt the login.

> **TIP:** Click the Show Password checkbox to see the letters in your password—this can be helpful to prevent mistyping it.

If everything is correct, then you will see a screen similar to Figure 1-2.

Figure 1-2. *The final setup screen on **Email Setup** and the **Activation Complete** pop-up window*

Once each email account is set up correctly, you will see an **Activation** email message in your **Messages** inbox and within 15 minutes, email should start flowing in to your BlackBerry. Now, with full wireless email, you will begin to see why so many are addicted to their BlackBerry Smartphones.

From: Activation Server
Email account information
Dec 10, 2009 12:24 PM

Congratulations, you have successfully set up martintrautschold@yahoo.com with your BlackBerry(R) device. You should begin receiving new messages in approximately 20 minutes.

Setting Up Your Corporate Email

Your Help Desk or Information Technology department will typically set up your corporate email (that is, Enterprise Activation). If not, all you need is your activation password and you can set this up on your own—right from the BlackBerry, using the **Setup Wizard**.

> **NOTE:** If you have not received your Activation password, then you need to ask your Help Desk or Technology Support department for that password before you may complete this process.

Setup Corporate Email Using the Setup Wizard

1. Start your **Setup Wizard** icon and select **Email Setup**, and then choose I want to use a work email account with a BlackBerry Enterprise Server from the menu.

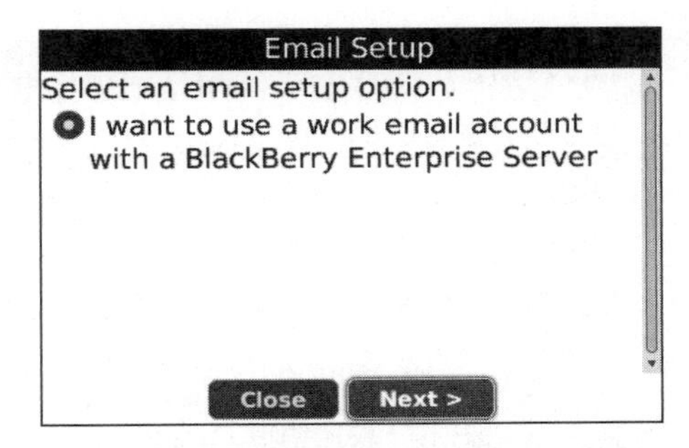

2. Verify that you have your Enterprise Activation password.

3. Type in the email address along with the activation password you received from your Help Desk or System Administrator.

4. Press the **trackpad** and select **Activate**.

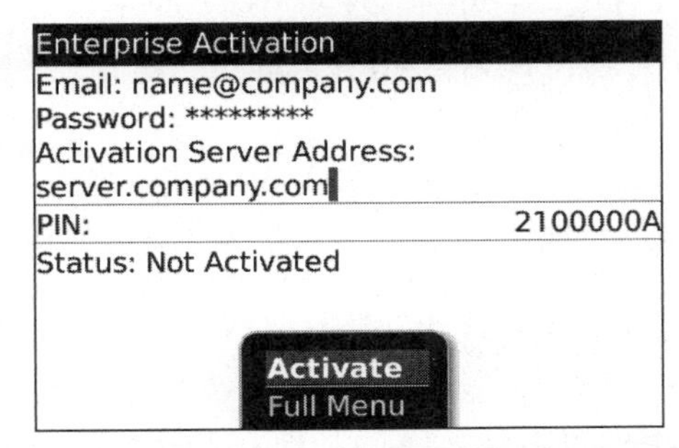

5. You will then see many messages on the screen, such as Establishing Secure Connection, Downloading Contacts, Downloading Calendar. The entire process may take 15 minutes or more, depending on how much data is being sent as well as the strength of your wireless connection.

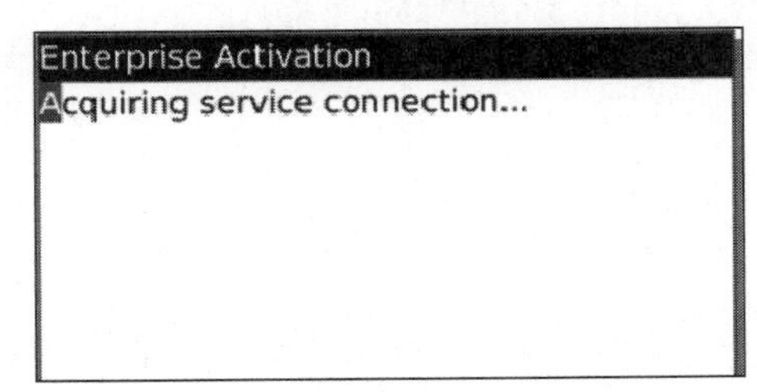

6. If you see an error message, then verify your password, that your wireless radio is turned on, and that you are in a strong coverage area. If you still have problems, then please contact your Technology Support group.

7. Once you have seen all the Enterprise Activation messages go by, and you see a completion message, then try email and check your **Address Book** and **Calendar** to see if it looks like everything was loaded correctly. If not, then contact your Help Desk for support.

How Can I Tell If I Am Activated on the BlackBerry Server?

A simple rule of thumb is that if you can send and receive email and you have names in your BlackBerry address book (**Contacts** icon), then it has been successfully set up (the Enterprise Activation is complete.)

To verify if your BlackBerry successfully configured with your Server:

1. Start your BlackBerry **Contact List** / **Address Book**.
2. Click the **Menu** key to the left of the **trackpad**.
3. Glide the **trackpad** up or down to find a specific menu item called **Lookup**.
4. If you see **Lookup**, then you are connected. (This **Lookup** command allows you to do a **Global Address List** (**GAL**) lookup from your BlackBerry Enterprise Server to find anyone in your organization—and then add them to your personal **Address Book** on your BlackBerry.)

Maintaining Your Email Accounts

You may need to add, edit, or delete email accounts. You also might want to fine-tune your email signature (Sent from my BlackBerry) that gets attached to the bottom of each email you send from your Bold.

Adding More Email Personal Addresses

You can add up to ten email addresses to your Bold. Only one can be an Enterprise or Corporate type email address, but the rest can all be personal addresses, such as POP3 or IMAP addresses.

1. Click on your **Personal Email Setup** icon (may be in your **Setup** folder).
2. Enter your username and password, if requested, and click **Log In**.
3. Scroll down to the bottom of the list of email addresses and click the **Add** button as shown.
4. Follow the steps to add the new account as we described previously.

Hiding Extra Email Account Icons

As you set up each email account, you will notice that a new icon appears on your **Home** screen tied to that particular account. If you like having individual icons, you can leave them alone. However, since all your email goes into your main **Messages** icon, you can hide these icons to clean up your **Home** screen. Follow these steps to hide these extra icons:

1. Highlight the icon you want to hide.
2. Press the **Menu** key.
3. Select **Hide**.

If you want to get the icon back, then follow the steps shown in the "Hiding and Un-hide Icons" section on page 169 in Chapter 8.

Edit or Delete an Email Account (Signature and Advanced Settings)

To change your email account name, password, signature, advanced settings, or synchronization options, you need to edit your email account settings. If you don't use a certain email account anymore, you may want to delete that particular address.

1. Click on the **Personal Email Setup** icon.
2. Log in if requested.

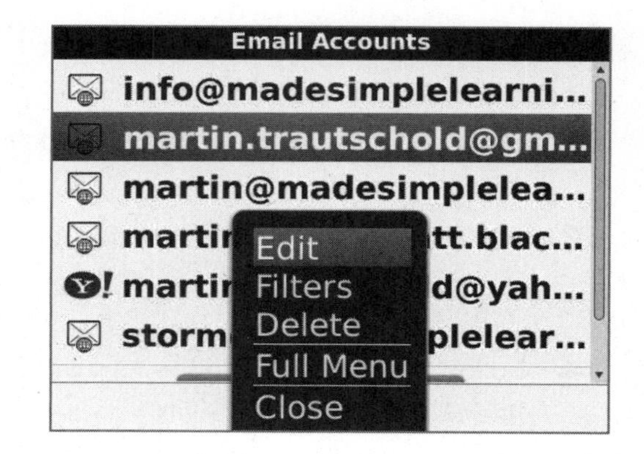

3. Click the **trackpad** on a particular Email account you want to edit or delete.

4. To delete the account, select **Delete** from the short menu and confirm your selection. If you are done, press the **Escape** key to exit to your **Home** screen.

5. To edit the account, select **Edit** from the short menu.

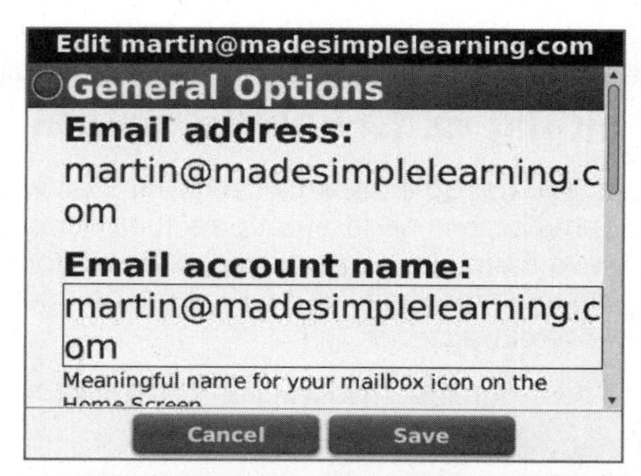

6. Now, you can adjust the following items:

   General Options:

   - **Email account name** (the name for the email account icon).
   - **Your Name** (the name that appears instead of the email address when you send messages, also known as the **Friendly Name**).
   - **Signature** (the signature attached to each email you send from your BlackBerry). Usually, you will want to change this from **Sent from my (wireless carrier) BlackBerry Device** to something more personal, such as your name and phone number or your company name.

7. Scroll down to click on **Login Information** to be able to change your password for this account.

8. Click on **Delivery Options** to be able to add an email account to send **Auto Blind Carbon Copies** of every email you send from your BlackBerry.

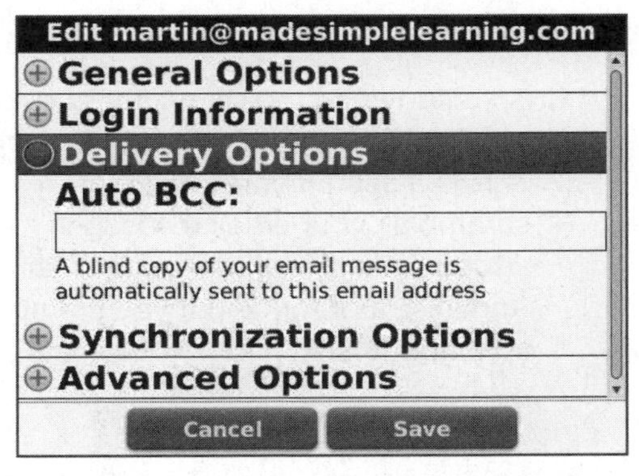

9. Click on **Synchronization Options** to adjust whether or not email messages you delete on your BlackBerry are synced wirelessly to also be deleted from your main email inbox. To check or uncheck the box, highlight it and press the **Space key**.

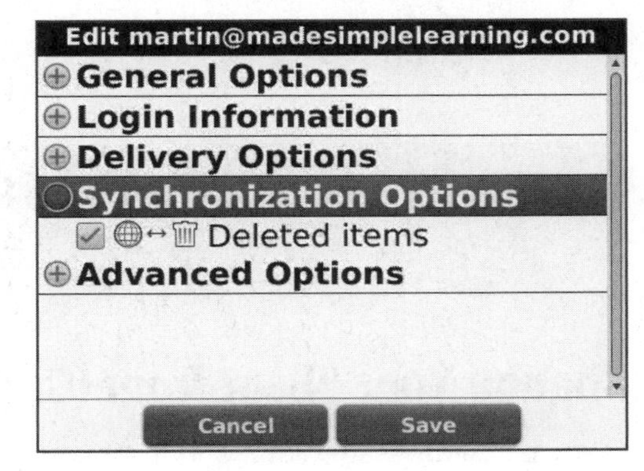

10. Scroll down to **Advanced Options** to adjust the **Email Server** name. Contact your Email Administrator for help if you are not sure about the **Email Server** name.

11. Scroll down to select whether or not you use **SSL** (Secure Socket Layer). This is a security protocol that scrambles your email messages for added safety. Contact your Email Administrator for help if you are not sure about SSL.

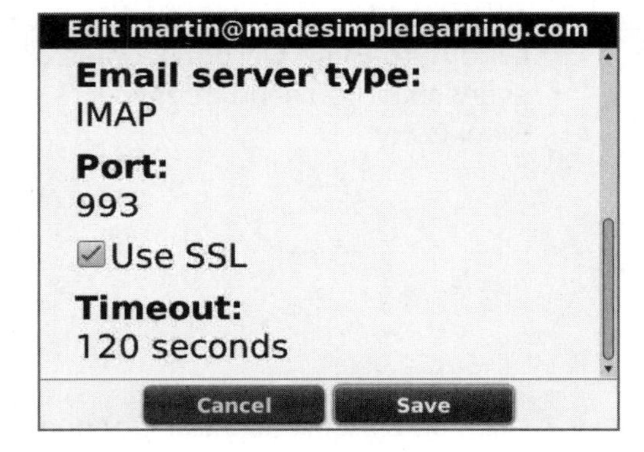

TIP: Press the **Space key** to get the "at" (@) and "dot" (.) symbols whenever typing an email. For example, for sara@company.com, type, **sara SPACE company SPACE com.**

Change Your "Sent from (Carrier) Wireless" Signature

Every email message you send from your BlackBerry will have an auto signature attached to the bottom of it. Usually it will have your carrier name on it. For example, if you use AT&T, then your default auto-signature may say "Sent from my AT&T BlackBerry." To change this signature, you need to follow the steps shown previously in the "Edit or Delete Email Account" section on page 41.

NOTE: If you are not sure of your advanced email settings, open up the mail client on your computer. If using MS Outlook or similar, click **Accounts** or **Email Accounts** from the menu. Then, review the advanced settings for that account. If you use web mail exclusively, then contact your service provider to ask for assistance with getting your BlackBerry set up.

If you are still having trouble with email setup, then please contact your email service provider or your BlackBerry wireless carrier's (phone company) technical support.

Wireless Email Reconciliation

The BlackBerry allows you to turn on or off **wireless reconciliation,** which is the feature that synchronizes deletion of email between your regular mailbox and your BlackBerry. In other words, if you delete an email on your BlackBerry, you could set it up so that the same email message is also automatically deleted from your regular email account. Usually, this is turned on by default, but you can disable it.

NOTE: If you work at an organization that supplied your BlackBerry to you, this feature may be controlled centrally by your Administrator and may not be adjustable.

Some wireless carriers (phone companies) do not support this feature (or don't support it fully) unless your BlackBerry is tied to a BlackBerry Enterprise Server.

1. Start your **Messages** icon.
2. Hit the **Menu** key and select **Options.** Or, press the letter **O** to jump down to **Options**.
3. Select **Email Reconciliation**.
4. Set **Delete On** to **Handheld**.
5. Set **Wireless Reconcile** to **Off**.

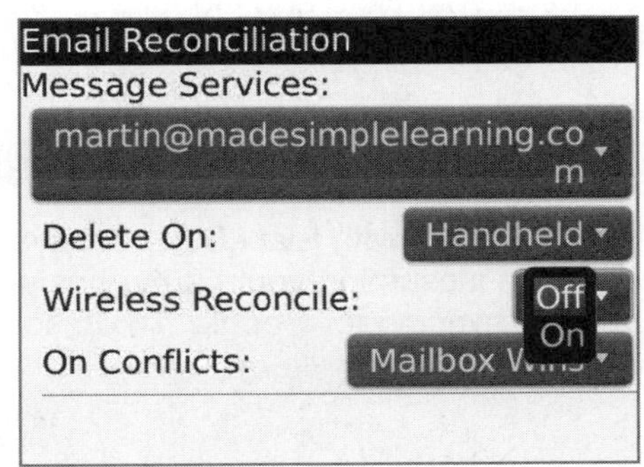

If you want to turn Email Reconciliation back on, set **Delete On** to **Mailbox & Handheld**, and set **Wireless Reconcile** to **On**.

If you have turned on the **Wireless Reconcile** and want to get rid of old email that you deleted from either your main email inbox or from your BlackBerry, then press the **Menu** key and select **Purge Deleted Items,** and then select the email address.

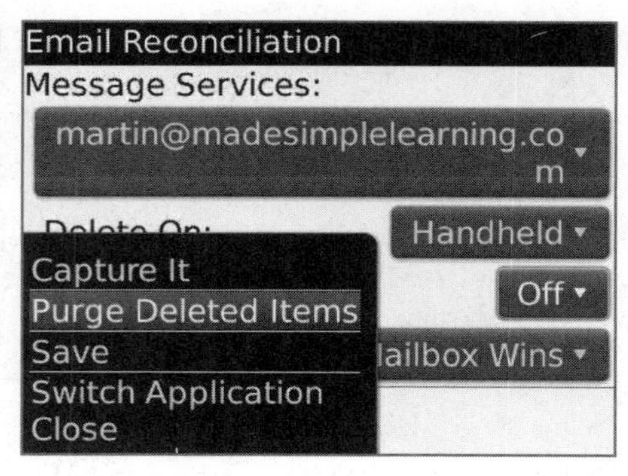

TIP: Clean Up Your Main Email Inbox On the Road

If you like to keep your main email inbox (which you view on your computer) cleaned up, make sure to keep **Wireless Reconcile** turned **On**. Just use downtime during the day to clean up your BlackBerry email inbox. This way, any email messages you delete on your BlackBerry (such as spam or other unnecessary emails), will not appear in your computer's email inbox. It is quite refreshing to come back to a very clean inbox after a day away from your desk.

Sync Google Contacts Using Email Setup

If you use Google for email and to manage your contacts, then you can set up a wireless sync for your **Google Contacts**. This is a great feature because you no longer need to connect your Bold with a sync cable to your computer to update contacts changes between your Bold and Google. The contact updates, like email, all happen automatically and wirelessly.

Setting Up Google Wireless Contacts Syncing

A new feature with BlackBerry 5.0 system software and BlackBerry Internet Service software (assuming your carrier has upgraded) is that you now can select wireless contact syncing for Google. To do so, follow these steps.

CAUTION: You can also sync your **Google Contacts** using the **Google Sync** app (see page 305). Do not try to sync **Google Contacts** with both methods—you are asking for trouble. Instead, if you choose to use this method for contacts syncing, then use **Google Sync** for only your **Google Calendar**.

TIP: If you forgot to check this sync box on the first-time setup, you can get to it in the Edit email account area shown on page 41.

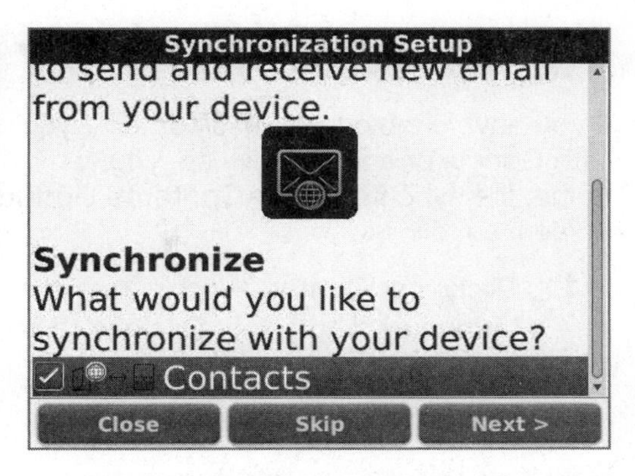

1. Highlight your email account from the main list of accounts, click the **trackpad**, and select **Edit**.
2. Roll down and click **Synchronization Options** as shown here, and check/uncheck your boxes.

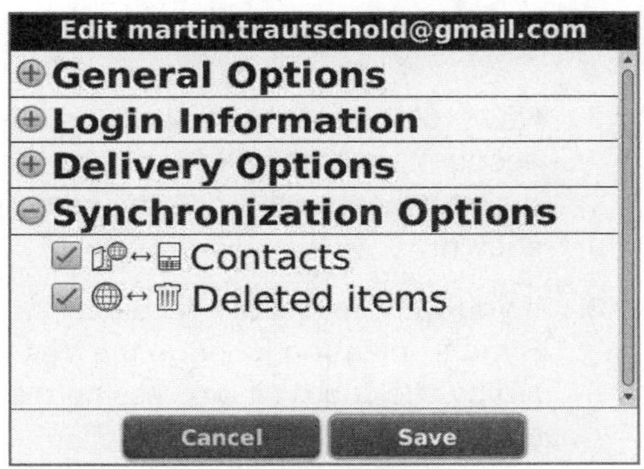

The first time you enable this contact sync feature, you will see a screen showing you the progress of the initial sync. If you have many contacts, (for example, 1,000 or more) it could require over 10 minutes to finish the first-time sync.

You may also have to try it a few times if it fails the first time.

As of publishing time, this only worked with Google/Gmail accounts, but we believe more account types could be added in the future.

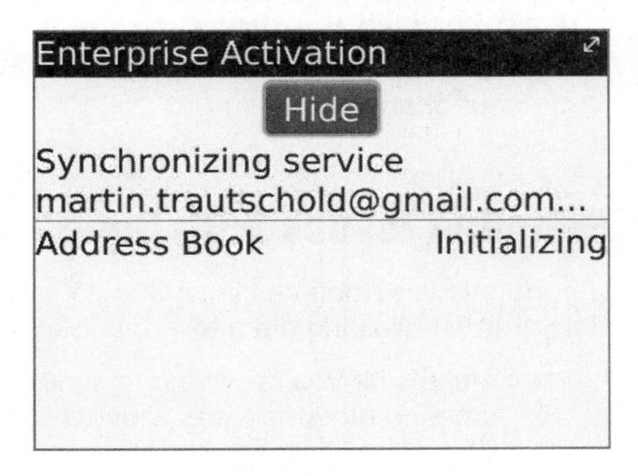

Turning On or Off Wireless Contact Syncing

If you have enabled wireless contact syncing with Google or another service, you can adjust the sync from your **Contacts Options** screen, as follows.

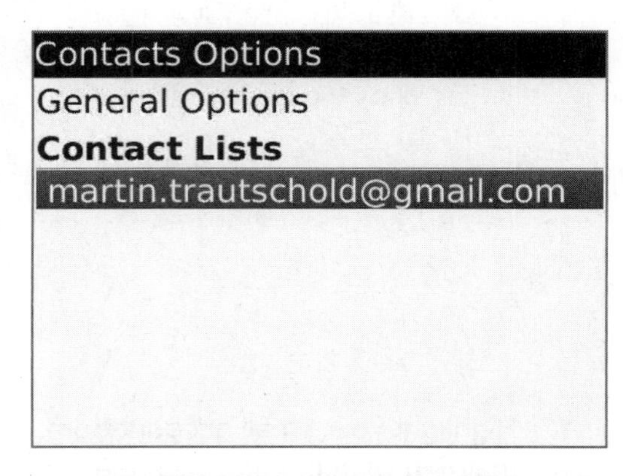

1. Click your **Contacts** icon, press the **Menu** key, and select **Options** to see this screen.

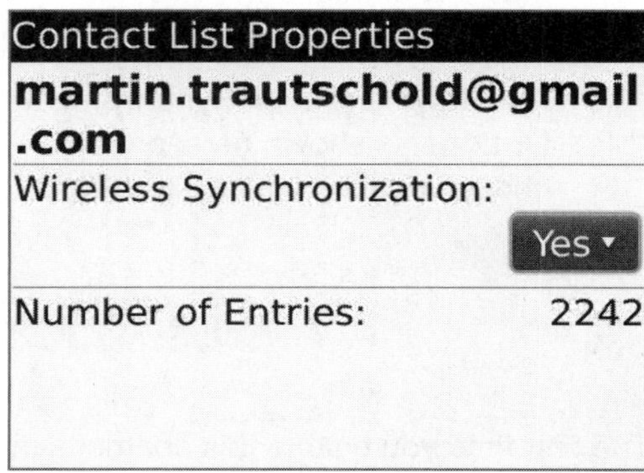

2. Now, roll down and click on the email account under the **Contact List** to make adjustments using the screen shown.
3. If you want to disable **Wireless Synchronization**, change the **Yes** to a **No** by clicking on it or pressing the **Space key**.
4. This screen also shows you how many contact entries are being synced. Save your changes.

Handling Issues with Google Contacts Wireless Sync

Unfortunately, sometimes your Google contacts do not sync correctly with your BlackBerry. Following are a few tips to try to help you get up and running:

- Make sure you are trying to sync a Google Mail account; at publishing time, no other services allowed syncing of contacts.
- As of publishing time, you could not sync Google Mail contact groups.
- Make sure you have enabled **Wireless Synchronization** on your BlackBerry smartphone. To verify this, start the **Contacts** application, press the **Menu** key, and select **Options**. Scroll down to and click your Gmail address under **Contact Lists**. The setting next to **Wireless Synchronization** should be set to **Yes** (see preceding steps and images).

Troubleshooting Your Email Accounts

Sometimes, your email accounts just don't work quite right. Here are a few tips to handle some of the more common errors.

Unfortunately, sometimes email will not be as easy to set up as shown in the previous section.

Verify Usernames and Passwords

Your first action, which usually handles about 80% of the problems, is to simply re-type your email address and password. The reason could be as simple as a wrong character typed in your email address or password. Watch the password very carefully, especially if you have numbers or other characters. Always try retyping them a few times before doing anything else.

Verify Your Email Server Is POP3 or IMAP

Some email servers cannot be accessed by the BlackBerry Internet Service, so you cannot use these types of email accounts on your Bold. Contact your email service provider and tell them you are trying to access your email from a BlackBerry smartphone and verify the server is of a type called **POP3** or **IMAP**.

Verify Your Email Server Settings (Advanced Settings)

Another setup issue might be that your server uses SSL (Secure Socket Layer) security or might have a non-standard email server name.

Contact your service provider to find out about these settings. To change these settings on your Bold, you need to follow the steps shown in the "Edit or Delete Email Account" section on page 41.

Solving a Gmail Enable IMAP Error Message

If you receive an error message in your email inbox telling you to turn on IMAP settings in Gmail, you need to log in to your Gmail account from your computer and follow these steps.

1. Click the **Settings** link (usually in the very top right corner).
2. Click the Forwarding and POP/IMAP tab.
3. Make sure that **IMAP Access** is set to **Enable IMAP**, as shown in Figure 1-3.
4. Then click the **Save Changes** button.

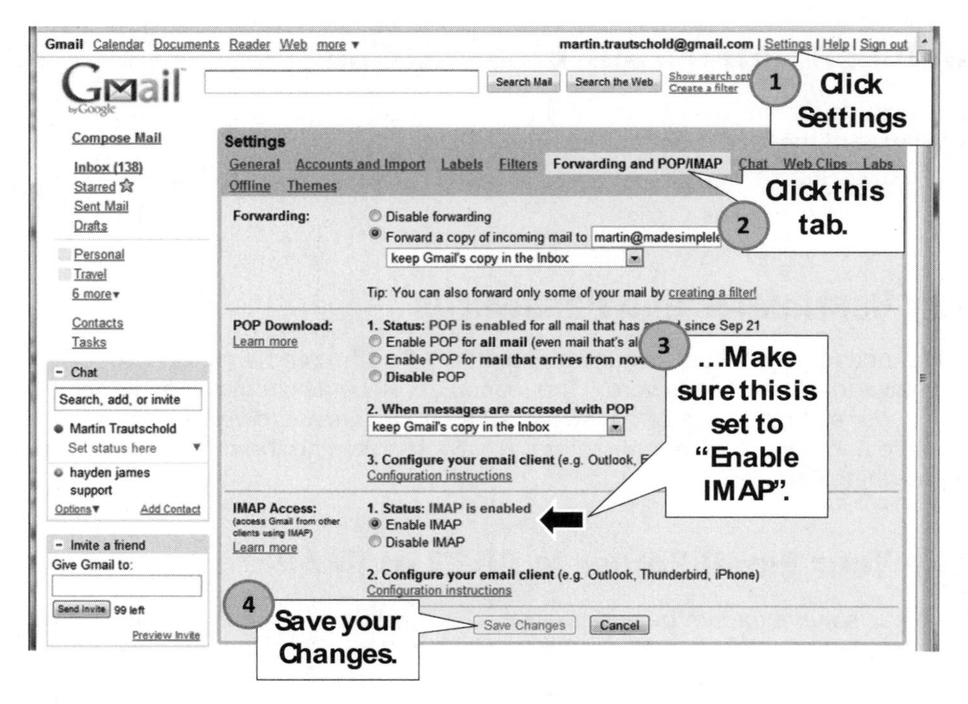

Figure 1-3. *Gmail Settings screen*

5. After you make this change, you will need to go back into your BlackBerry **Personal Email Setup** icon.

6. Highlight the invalid account (in this case Gmail) and click the **trackpad** or **Menu** key to select **Validate**.

7. On the next screen, you need to re-enter your password and click **Next**.
8. If everything is OK, then you will see this pop-up saying **Your password has been successfully validated**.

Why Is Some Email Missing?

If you download your email messages to your computer using an email program such as Microsoft Outlook, Outlook Express, or similar, and you use BlackBerry Internet Service on your BlackBerry for Email, then you need to turn on a specific setting in your computer's email program. If you do not leave a copy of messages on the server from your computer's email, you may end up receiving all email on your computer, but only a limited set of email on your BlackBerry.

Why Does This Happen?

By default, most email programs pull down or retrieve email from the server every one to five minutes and then erase the retrieved messages from the server. By default, the BlackBerry Internet Service usually pulls down email every 15 minutes or so. So, if your computer has pulled down the email every five minutes and erased it from the server, your BlackBerry will only receive a very limited set of messages (those that haven't yet been pulled down by your computer).

How to Fix This?

The answer is to set your computer's email program to keep your messages on the server. This way, the BlackBerry will always receive every email message. Here's how:

1. In the email program on your computer (such as Microsoft Outlook), find the location where you can configure or change your email accounts. For example, Tools, Configure Accounts, Account Settings, or something similar (Figure 1-4).
2. Select or change the appropriate email account. (Sometimes you just double-click the account to edit it.)

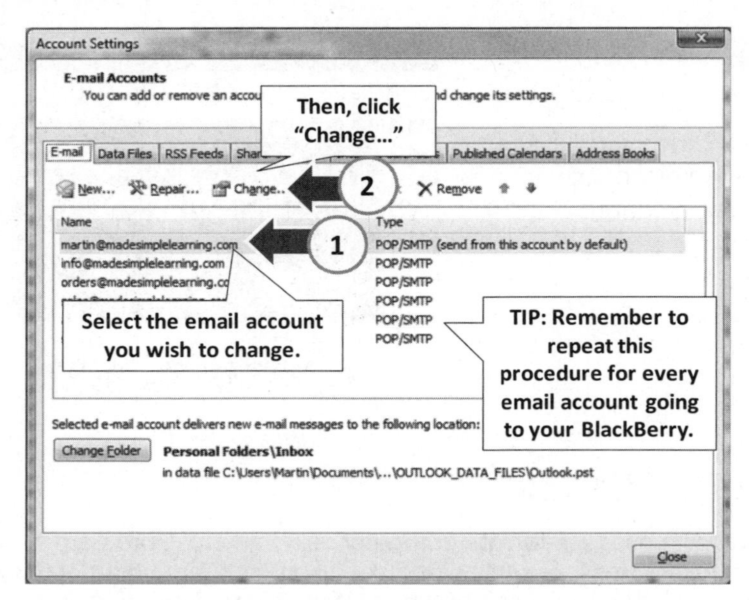

Figure 1-4. *Email Account Settings screen in Microsoft Outlook*

3. You will then usually go to an **Advanced Settings** area to make changes to **Leave a copy of the message on the server**. In Microsoft Outlook, click the **More Settings** tab and then the **Advanced** tab (Figure 1-5).

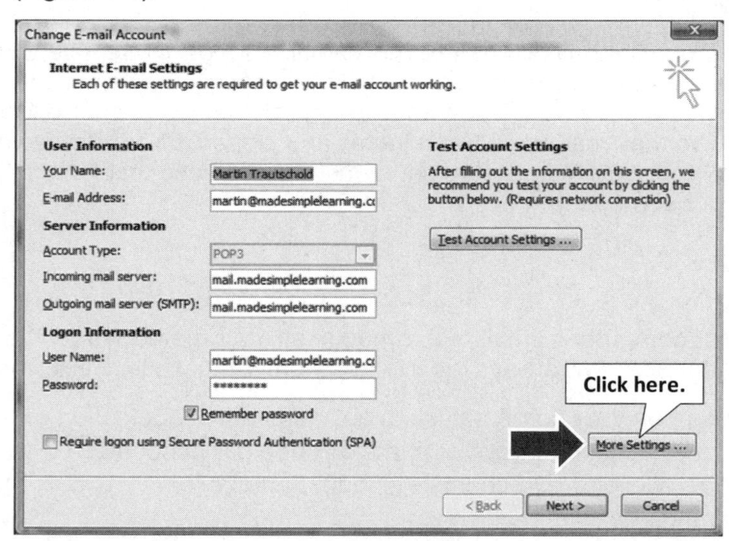

Figure 1-5. *Internet Email Settings in Microsoft Outlook*

4. At the bottom of the screen, under **Delivery**, check the **Leave a copy of messages on the server** box.

5. Then check the **Remove from server after X days** box. We suggest changing the number to about10 days. This allows you time to make sure that the message reaches both your BlackBerry and your PC—but doesn't clutter the server for too many days. If you make the number of days too high, you may end up with a Mail Box Full error and have your incoming email messages bounced back to the senders.

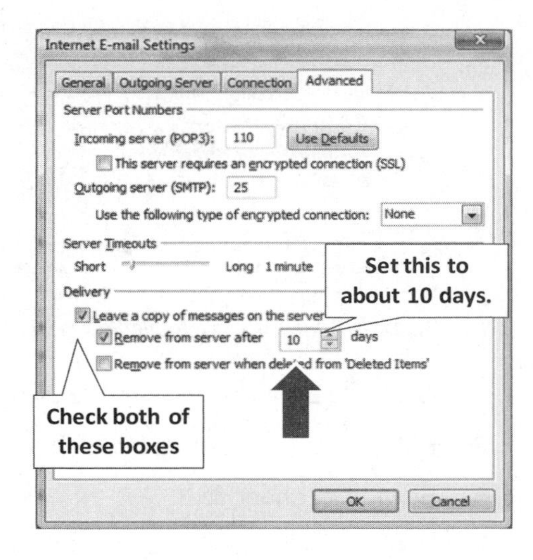

NOTE: Remember to repeat this process for every email account that you have going to both your computer and your BlackBerry.

Correcting the "Invalid Account. Please Validate" Message

From time to time, you may see an invalid email account message, such as the one shown in Figure 1-6, either inside your BlackBerry **Email Setup** icon or when you log in to your wireless carrier's web site. This may happen if you have changed your email account password, or sometimes it just happens if the system encounters an unforeseen error—through no fault of your own! The following sections provide the steps to correct the problem.

Using Your Computer

You can log in to the BlackBerry Internet Service site from your computer to correct the invalid account. See Figure 1-6.

1. Log in to your BlackBerry Wireless Carrier's web client (see page 55 for list of sites).

Figure 1-6. *BlackBerry Internet Service showing invalid email Accounts*

2. Click the **Edit** icon next to the invalid account.
3. Enter your information, including your password.
4. Save your changes.
5. You will then see a message that says **Your email account has been successfully validated.** The invalid account icon will change to a check mark in the **Valid** column, as shown in Figure 1-7.

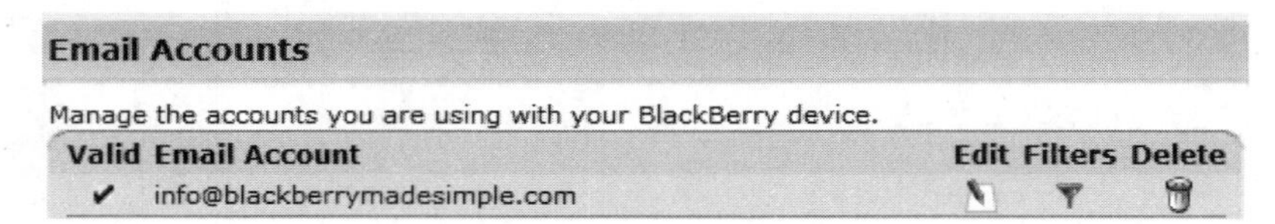

Figure 1-7. *Email account successfully validated in BlackBerry Internet Service*

Using Your BlackBerry

You can also correct invalid email accounts right on your BlackBerry.

1. Log in to your **Email Setup** icon (usually in the **Setup** folder).
2. After logging in, you will see a screen similar to the image shown to the right.
3. To correct the **Invalid Account** errors, highlight the account and click the **trackpad** to select **Validate.**

4. Now, re-type your password and click **Next** to validate the account.
5. You will see a message similar to the one shown to the right for each validated account.
6. Repeat the process for any other accounts shown as invalid.

Set Up or Adjust Email Accounts from Your Computer

You can also set up your personal email accounts from your computer using your carrier's web site. You also might notice that when you send an email from the BlackBerry, the signature is something very basic, such as **Sent from BlackBerry Device via T-Mobile**, or **Sent from AT&T Wireless**, depending on your carrier. You can easily change your email signature and perform other tasks from your carrier's web site, but you'll first need to create an account. If your account was already set up when you activated your phone, just log in with your username and password and skip ahead to the "Changing Your Email Auto Signature" section.

Setting Up BlackBerry Email from Your Computer

From your computer, find your way to your carrier's web site in Internet Explorer, Firefox, or Opera (a partial list of sites is listed below). Once there, log in to your personal account page.

NOTE: These web sites change frequently! Some carriers imbed or include the BlackBerry Email setup pages within the main carrier web site. Please check with your wireless carrier if any of the following links are incorrect, or you don't see your carrier listed. You may also want to check for updated sites at the bottom of this web page: `http://na.blackberry.com/eng/support/blackberry101/setup.jsp#tab_tab_email`

Alltel (USA): http://www.alltel.blackberry.com

AT&T/Cingular (USA): http://www.att.blackberry.com/

Bell / Solo Mobile (Canada): https://bis.na.blackberry.com/html?brand=bell

Cellular South (USA): https://bis.na.blackberry.com/html?brand=csouth1

Rogers Wireless (Canada): https://bis.na.blackberry.com/html?brand=rogers

Sprint/Nextel (USA): https://bis.na.blackberry.com/html?brand=sprint

T-Mobile (USA) – (Login to main site): http://www.t-mobile.com/bis/

T-Mobile (Germany): http://www.instantemail.t-mobile.de/

Telus Mobility (Canada): https://bis.na.blackberry.com/html?brand=telus

Verizon Wireless (USA): https://bis.na.blackberry.com/html?brand=vzw

Virgin Mobile (Canada): https://bis.na.blackberry.com/html?brand=virginmobile

Vodafone (UK): https://bis.eu.blackberry.com/html?brand=vodauk

If you cannot find a link directly to your phone company's BlackBerry Internet Service site from this list or from http://na.blackberry.com/eng/support/blackberry101/setup.jsp#tab_tab_email, then you should log in to your own phone company's web site and look for a button, tab or link (varies by website) that might say **Phone & Accessories**, **Device**, **Handheld** or **Support** from your home page and then **Setup BlackBerry Email** or **Setup Handset Email**. If you still cannot get to your BlackBerry Internet email setup, then please contact your phone company.

On many of the sites listed, however, you will first need to create your BlackBerry Internet Service account on a screen similar to this one.

Create a New Account

To create a new account, follow these steps:

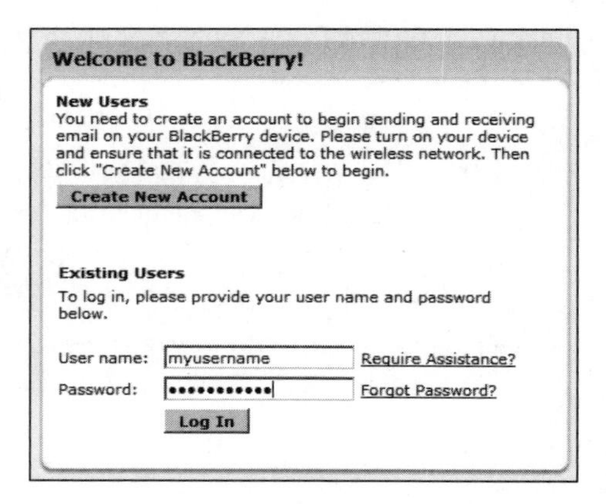

1. Click the **Create New Account** button at your provider's site.
2. You will then see a Legal Agreement. In order to continue, check the **I have read this agreement** box and click the **I Agree** button. After accepting the legal agreement, you should now see a screen like that shown in Figure 1-8.

Account Setup

To begin creating your BlackBerry Service account, type your device details below.

Device PIN:

Device IMEI:

Cancel Continue

To find your PIN perform one of the following actions:

- In the BlackBerry device options or settings, click **Status**.
- Look for the PIN and IMEI information on the outside of the box that your BlackBerry device or BlackBerry-enabled device came in.
- Turn the BlackBerry device off and remove the battery. Look for the sticker on the BlackBerry device with the PIN information where the battery is usually located.

Copyright 2006-2009 Research In Motion Limited. All rights reserved. Legal Information.

Figure 1-8. *Set up BlackBerry Internet Service Account—requesting PIN and IMEI*

3. Both of the numbers you need are located in your **Options** icon on your BlackBerry, so refer back to your Bold. Click the **Options** icon from your BlackBerry Home screen.

4. Press the letter **S** on your keyboard a few times until you get to the **Status** item and click it.

5. Once in **Status**, look at the lines marked **PIN** and **IMEI**, as shown in Figure 1-9.

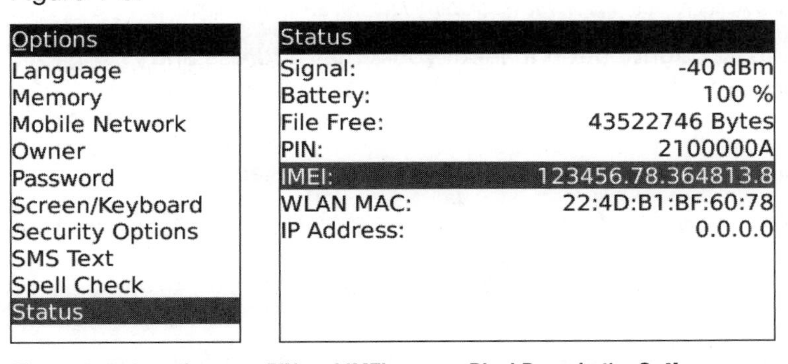

Figure 1-9. *Locating your PIN and IMEI on your BlackBerry in the* ***Options*** *app*

6. Type in the **PIN** and **IMEI** in the fields on the web site screen—but remove any spaces or dots. You should then come to a series of screens on the Main BlackBerry Internet Service web page (see Figures 1-10, 1-11, 1-12).

7. Log in or click the link from your carrier's web site, to access a screen similar to the one in Figure 1-10.

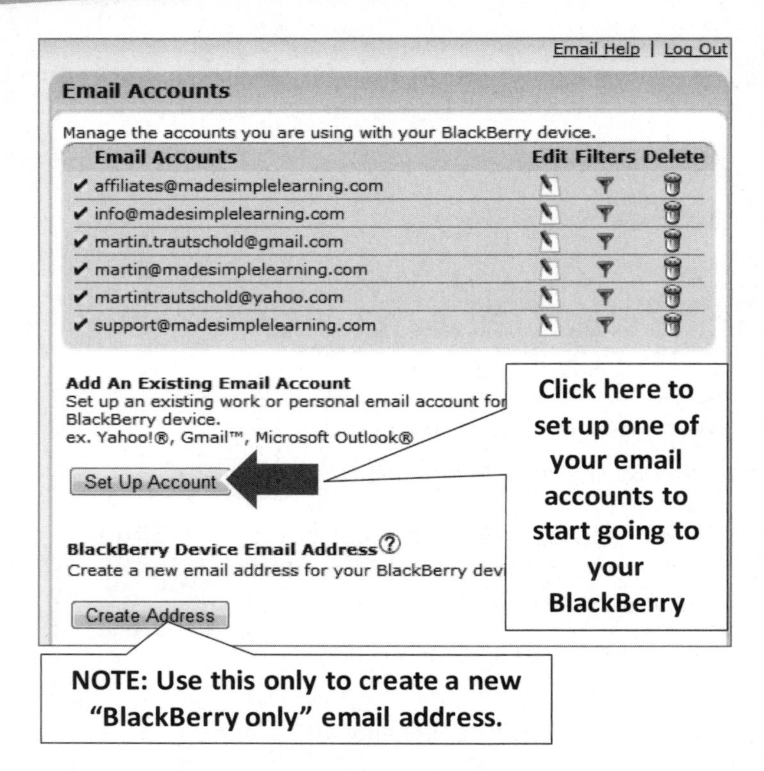

Figure 1-10. *Email Account Settings on the BlackBerry Internet Service web site*

8. Click the **Setup Account** button to input your email address and password (Figure 1-11). Then click **Next** and your account will be set up.

   Shortly thereafter, your first email will come in on the BlackBerry.

Set Up An Existing Email Account

Set up the BlackBerry Internet Service to deliver email messages from your personal or work email account to your BlackBerry device. Type your email address and the password you use to access the account. Open help to determine which password to type.

Email address: martintrautschold@yahoo.cor

Password: ••••••••••••

Confirm password: ••••••••••••

Cancel Next

10 Research In Motion Limited. All rights reserved. End User Agreement. Legal Information.

Figure 1-11. *Adding an existing email account to the BlackBerry Internet Service*

After the email account is successfully set up, you will then receive confirmation email on your BlackBerry, usually titled **Activation**.

10:30a Activation... Congratulatio...
10:29a Activation... Congratulatio...
10:29a Activation... Congratulatio...
10:29a Activation... Congratulatio...

Repeat step 8 for each of your email accounts to set them up.

Once you have all your email accounts configured, you will see them listed as shown in Figure 1-12. You can then customize (**Edit**), filter email (**Filter**) or remove them (**Delete**) by selecting the icons on the right side, as shown in Figure 1-12.

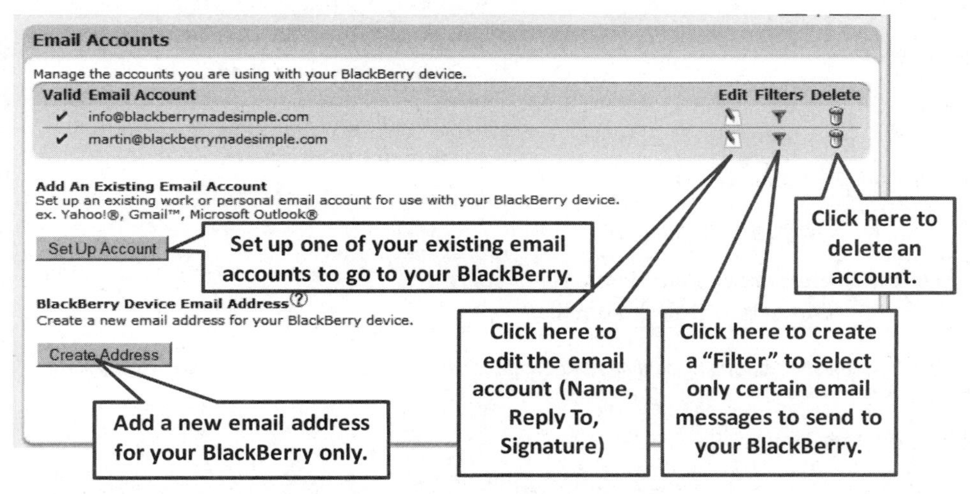

Figure 1-12. *Editing or adding email accounts in the BlackBerry Internet Service*

Changing Your Email Auto Signature

On your email accounts page, you should see an icon for editing each of your accounts that you have set up. You can then add a unique signature for every email account that you have set up.

> **TIP:** You can also change this signature directly from your BlackBerry—see page 41.

1. Select the **Edit** icon next to the email account you wish to work with (Figure 1-13).
2. Make any changes in the fields provided to you (Figure 1-13).
3. In the signature box, simply type in the new signature you wish to appear at the bottom of that particular email account's messages.

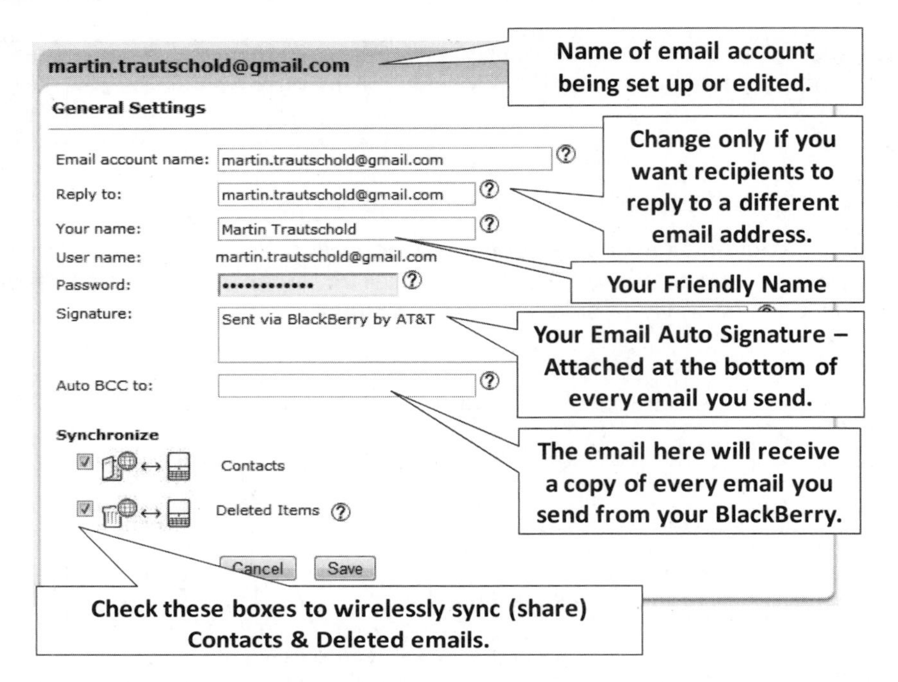

Figure 1-13. *Adjusting settings including Auto-Signature and Sync Settings from BlackBerry Internet Service*

4. Click the **Save** button.

> **NOTE:** As of publishing time, only Gmail had the ability to wirelessly sync or share Contacts. We expect more services to be added in the future (for example, Yahoo!.)

We recommend that you test your new settings by sending an email from your Blackberry to yourself or another email account and verify that the new signature is included.

You can also add signatures that you can select "on the fly" from your BlackBerry while typing emails, using the **AutoText** feature (see page **163**).

Adjusting Advanced Email Account Settings

From the email settings screen, you can also adjust advanced email settings which you will need if your email is not working correctly on your BlackBerry.

1. Click on the **Advanced Settings** link (sometimes it does not look clickable) at the top of the email **Edit** screen to see the screen shown in Figure 1-14.
2. Enter or edit your **email server name**.

3. Enter or edit your email server port.
4. Check the box if your email server uses SSL (Secure Socket Layer) encryption.
5. Press Save to save your changes..

martin@madesimplelearning.com

General Settings | **Advanced Settings**

Email server: mail.madesimplelearning.com
Email server type: IMAP
Port: 993
Timeout: 120 seconds
SSL: ☑

Cancel Save

10 Research In Motion Limited. All rights reserved. End User Agreement. Legal Information.

Figure 1-14. *Advanced Settings screen on the BlackBerry Internet Service web site*

BlackBerry Enterprise Server Express (Free Software)

If you work at an organization that uses a Microsoft Exchange or Microsoft Windows Small Business Server, then you can now acquire the BlackBerry Enterprise Server (BES) Express software for free. (See the advantages of having a BES later in this section.)

Image courtesy www.blackberry.com

A BES is a server that typically sits behind your organization's firewall and securely connects your BlackBerry to corporate email and data, as well as wirelessly synchronizes (shares) contacts, calendar, tasks, and memo items between your corporate computer and your BlackBerry.

BES allows your organization to gain all the benefits of a BlackBerry Server with no additional software costs. This saves thousands of dollars over the old pricing model by Research In Motion. You might want to let your IT group know about this great new deal if you use BlackBerry devices at your work place.

You should be able to support up to 75 BlackBerry users on the same box as your email server. You can support up to 2,000 users by putting the BES Express software on a separate server.

How to Get the BES Express software?

To acquire BES Express, go to `www.blackberry.com`, and follow these steps:

1. Click the **Software** link at the top.
2. Click **BlackBerry Enterprise Server Express** under **Business Software** in the left column to see a screen similar to the one shown in Figure 1-15. (You might be able to get straight to the page by typing this link: `http://na.blackberry.com/eng/services/business/server/express/`.)
3. Then follow the on-screen steps to start the free download.

Figure 1-15. *BlackBerry Enterprise Server Express Software web site*

4. Once the software is downloaded, the administrator uses a web-based console to set up and administer all users. The full setup instructions for BES Express are beyond the scope of this book, but please follow the on-screen help and tutorials found at `www.blackberry.com`. See Figure 1-16.

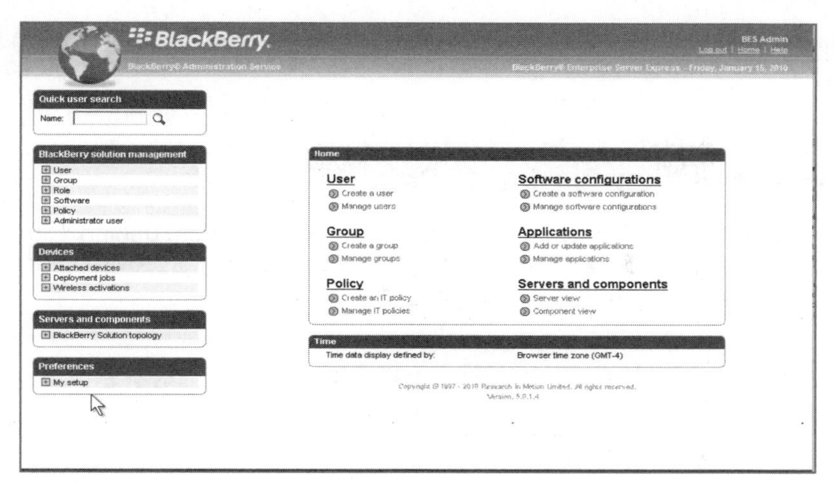

Figure 1-16. *Setting up the BES Express Software*

Benefits of Being Connected to a BlackBerry Server

Individuals and small office BlackBerry users can get easy access to a BES. Just perform a web search for "hosted BES" or "hosted blackberry enterprise server" to locate a number of providers.

> **TIP:** See the previous BES Express section for information on free BES software.

Connecting your BlackBerry to a BES (v4.0 and higher) will give you all of the following benefits.

Strong Encryption of Email within Your Organization

All email sent from your BlackBerry to other users in your organization will be fully encrypted with military-grade Triple-DES encryption provided by the server.

Full Two-Way Wireless Synchronization

Full two-way wireless updates between your BlackBerry and corporate desktop account for:

- Address Book
- Calendar
- Task List
- MemoPad (Notes)

This means you will be able to add new information or make changes to anything in your Address Book, Calendar, Task List, or MemoPad on your BlackBerry and in minutes it will appear on your desktop computer.

Securely Connect to Your Corporate Data

The BES server provides a highly secure connection to corporate data behind your firewall. This can provide productivity benefits to your organization by providing access to view and update important information in the field.

Push Applications Wirelessly to All Your BlackBerry Users

The BES administrator can create software configurations for groups of users and push them out wirelessly to any or all BlackBerry users. This saves on IT support costs.

Global Address List Lookup

Using GAL Lookup (**Lookup** for short), you can immediately look up anyone in your organization from your BlackBerry, even if there thousands of people at your organization.

Out-of-Office Auto Reply and Email Signature

Your Out-of-Office reply and Email Signatures are things you can adjust right from your BlackBerry if you are connected to a BlackBerry Server. Go to **Messages** (**Email**), press the **Menu** key, and select **Email Settings** to adjust these items.

Meeting Invitations

Just like on your desktop computer, you may check people's availability; you can invite attendees to meetings you schedule right on your BlackBerry. And, just like on your desktop, you may accept, decline, or tentatively accept meeting invitations you receive on your BlackBerry. Learn more on page 304.

> **NOTE:** Sending and responding to meeting invitations is a feature available even without having your BlackBerry tied to a BlackBerry Enterprise Server.

Chapter 2

Windows PC Setup

This chapter shows you how to install Desktop Manager Software on your Windows computer. Then we show you how to synchronize your contacts, calendar, tasks and memos, backup and restore, and more.

If you want to transfer files and media, then check out our chapter on transferring files for Windows users found on page 105 (you may need some of the instructions in this chapter on how to install Desktop Manager if you want to use it as your method to transfer files).

Have an **Apple Mac** computer? Please go to the "Apple Mac Setup" chapter on page 125.

Unless you work at an organization that provides you access to a BlackBerry Enterprise Server, if you are a Windows computer user, you will need to use BlackBerry Desktop Manager Software to do a number of things:

- Transfer or **synchronize** your personal information (addresses, calendar, tasks, notes) between your computer and your BlackBerry (see page 71)
- Back up and restore your BlackBerry data (see page 95)
- Install or remove application icons (see page 89)
- Transfer or sync your media (songs, videos & pictures) to your BlackBerry (see the next chapter on page 105)

CAUTION: Do not sync your BlackBerry with several computers, or sync separately to an online service such as Google. You could end up corrupting your BlackBerry and/or other databases—ending up with duplicates or worse, deleted items.

Download Desktop Manager for Windows

Each new version of RIM's Desktop Manager Program has come with more functionality and more versatility than the previous versions. So, it is always a good idea to keep up-to-date with the latest version of the Desktop Manager software.

The Disk from the BlackBerry Box

It is fairly likely that the disk that arrived with your brand new BlackBerry has a version of Desktop Manager that is already out-of-date. This is because many times, they produced the CDs months ago, and in the meantime a new version has been released. So we recommend downloading the latest version from the Internet directly from `www.blackberry.com`, as shown in the following sections.

Check Your Current Version

If you have already installed Desktop Manager, you should check which version you currently have. The easiest way to do that is start up your **Desktop Manager** program, go to **Help**, and then to **About Desktop Manager**. You will see right here that the version number of your particular version is shown. If you don't have version 5.0.1 or higher—it is time to upgrade.

To Get the Latest Version of Desktop Manager:

Try typing (or clicking—if you are reading this in e-book format) this link: `http://na.blackberry.com/eng/services/desktop/`

Otherwise, perform a web search for **BlackBerry Desktop Software download** and pick the search results that look correct to go to the `blackberry.com` site.

This should bring you to the BlackBerry web site. Now, follow the links to get to the download screen shown in Figure 2-1.

Then, click the **Download** button at the bottom to get started.

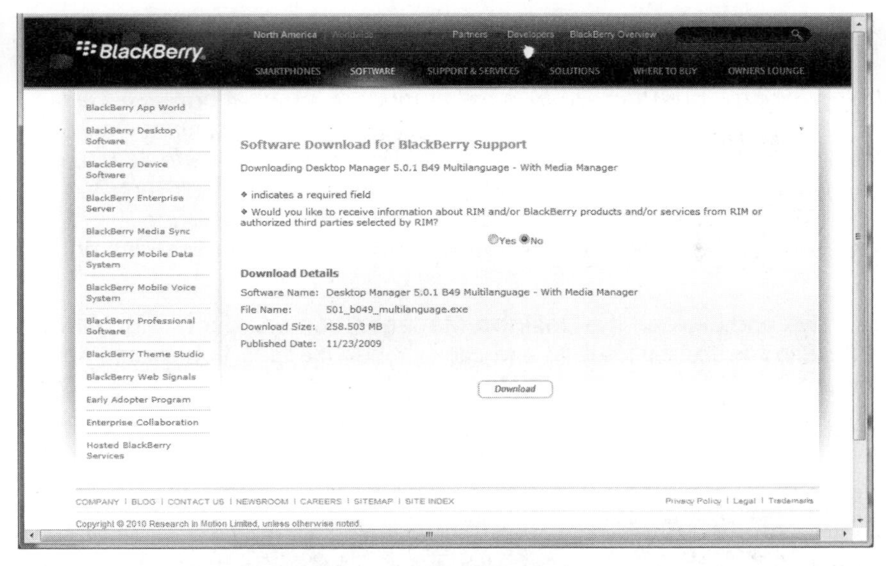

Figure 2-1. *BlackBerry Desktop Manager Download web site*

Save the file to a place where you will remember it. This is a large file, so it may take some time to download.

Install Desktop Manager

Locate and double-click the installation file that you downloaded. It will usually be in your **Downloads** folder, unless you changed the default, and will probably look something like Figure 2-2.

Figure 2-2. *Locating the downloaded installation file on your computer*

What the file looks like will depend on your View in Windows Explorer (for example, Small Icons, Large Icons, List, or Details). The first few numbers in the file name correspond to the version of Desktop Manager; in this figure, it is 501 for version 5.0.1.

After you double-click the install file, then follow the directions to complete the installation. Choose **Integrate with a personal email account** unless your BlackBerry is tied to a BlackBerry Enterprise.

Overview of BlackBerry Desktop Manager

One of the great things about your BlackBerry is the amount of information, entertainment, and fun that you can carry in your pocket at all times. But, what would happen if you lost your BlackBerry or lost some of your information? How would you get it back? What if you wanted to put music from your computer on your BlackBerry? Fortunately, your BlackBerry comes with a program called BlackBerry Desktop Manager, which can back up, synchronize, add media, and load new applications on your BlackBerry.

To get started, go ahead and click on the **Desktop Manager** icon on your computer–or go to **Programs ➤ BlackBerry ➤ Desktop Manager** and click. When it starts, you should see a screen similar to Figure 2-3.

Make sure your BlackBerry is plugged into the USB cable provided and attached to the computer.

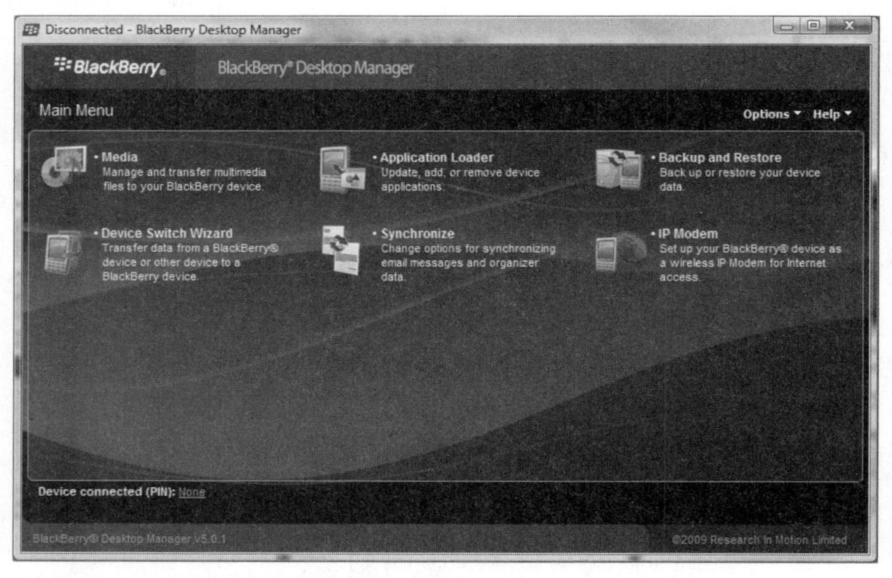

Figure 2-3. *BlackBerry Desktop Manager main screen*

> **NOTE:** Version 5.0.1 is shown in Figure 2-3; your version may be higher.

You will see the following five or six icons (the **IP Modem** icon may not show up for you, we tell you why below) in **Desktop Manager**:

The **Media** icon is for transferring media (songs, videos, pictures, and ringtones) between your computer and your BlackBerry. We discuss the **Media** icon (**Media Manager** and **Media Sync** on page 105).

The **Application Loader** is for installing or removing BlackBerry icons and upgrading your BlackBerry System Software version.

The **Backup and Restore** icon is for making a full backup of all your data on your BlackBerry and restoring (or selected databases) at a later time.

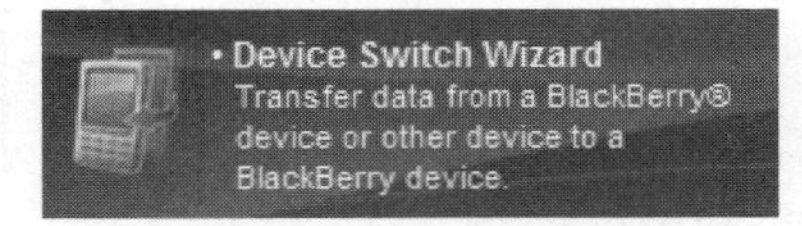

Along the second row is the **Device Switch Wizard.** This **Wizard** can be very helpful for moving your data from an old device (BlackBerry or non-BlackBerry) to a new BlackBerry.

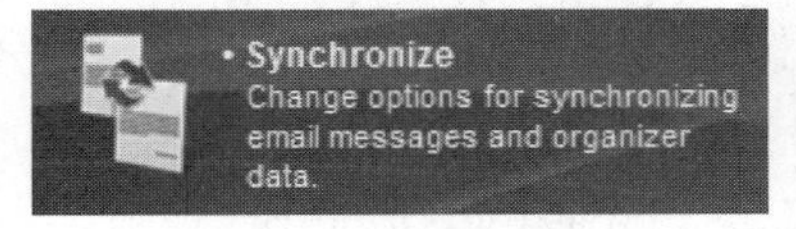

Next is the **Synchronize** icon that controls the settings for synchronizing your data including your address book, calendar, tasks, memos and more to keep your computer and your BlackBerry up-to-date with one another.

The **IP Modem** icon allows you to connect your Laptop computer to the Internet using your BlackBerry. You may not see the **IP Modem** icon. Some wireless carriers disable this feature within Desktop Manager causing the icon to disappear once you connect your BlackBerry to your computer. Learn more about this **IP Modem** feature on page 454.

TIP: You can also install new software icons wirelessly right on your BlackBerry; see page 475.

NOTE: Most wireless carriers require you purchase a separate "BlackBerry as a Modem" or "Tethering" data plan in order to use this feature.

Entering Your Device Password

If you have enabled password security on your BlackBerry, you will have to enter your password on your computer right after you connect your BlackBerry, as shown in Figure 2-4.

Figure 2-4. *BlackBerry Desktop Manager Device Password screen*

Device Switch Wizard

If you are upgrading from another BlackBerry, a Palm device or a Windows Mobile device, you will want to use the **Device Switch Wizard** in Desktop Manager.

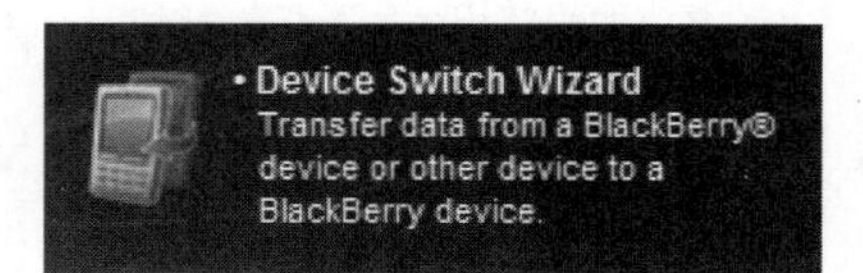

After clicking the **Device Switch Wizard**, you will see that you have two options depending on the type of device you were using before your BlackBerry Bold, as shown in Figure 2-5.

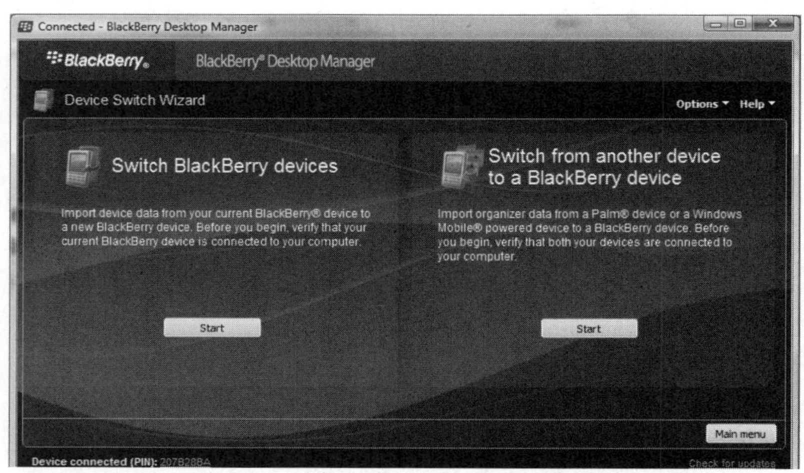

Figure 2-5. *Device Switch Wizard main screen in Desktop Manager*

Moving from Another BlackBerry

You'll first need to connect that old (called *current*) BlackBerry to your computer in order to get started. Then click the **Start** button and follow the on-screen instructions. Your current BlackBerry data will be backed up to your computer. Then, you connect your new Bold to the computer so all data (including email setup and other settings) can be restored.

NOTE: You will not be able to copy all third-party icons (applications) you added to your old BlackBerry to your new BlackBerry. The Bold will have a newer operating system and may have a different screen size, so some third-party applications that worked fine on your old BlackBerry will not work on the Bold.

Moving from a Palm or Windows Mobile Device

You will need to connect both your Palm/Windows Mobile handheld and your new BlackBerry to your computer at the same time and click the **Start** button to see Figure 2-6. Follow the on-screen directions. *Because the devices use different operating systems, not all the information from the Palm or Windows Mobile device will be copied to your BlackBerry.*

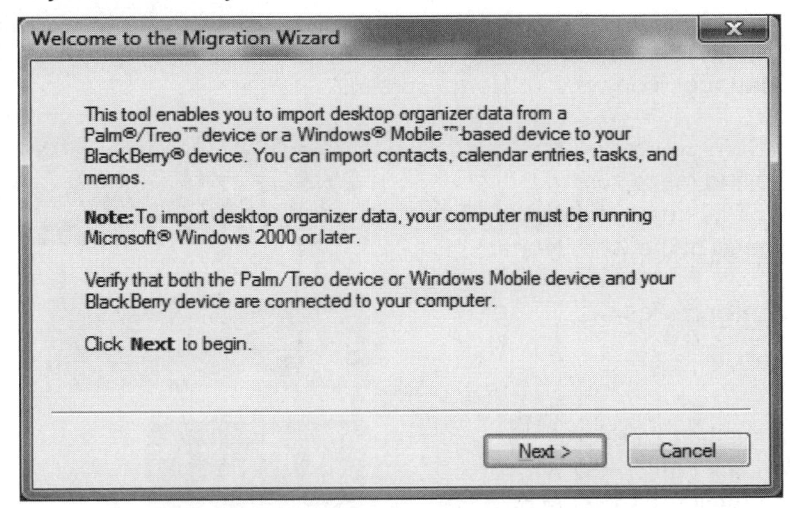

Figure 2-6. *BlackBerry Desktop Manager Migration Wizard*

Note You will not be able to copy any icons or software from your Palm/Windows Mobile device to your BlackBerry. If you have a favorite application from Palm/Windows Mobile, then check out BlackBerry App World (see page 431) or the software vendor's web site to see if they have a version compatible with your BlackBerry Bold.

Synchronize Your BlackBerry

You have probably come to rely on your BlackBerry more and more as you get comfortable using it. Think about how much information you have stored in there.

TIP: If you use Google for your contacts, you may already have set up synchronization when you setup your Gmail account (see the "Sync Google Contacts" section of Chapter 1: Email Setup, on page 46).

Now ask yourself: "Is all my information safely stored on my computer?"

Then ask: "Is all my BlackBerry information synchronized or shared with the information in my computer software?"

Synchronizing your BlackBerry with Desktop Manager is very important. Your data will be safe and backed up or shared with the correct program on your computer–making things like your Calendar, Address Book, and Tasks more useful.

Set Up the Sync

The first thing to do is to open your Desktop Manager software as you usually do by clicking on the **Desktop Manager** icon on your **Home** screen.

1. Connect your BlackBerry to your computer using the USB cable and make sure you see your **BlackBerry PIN number** in the lower left corner instead of the word **None**.

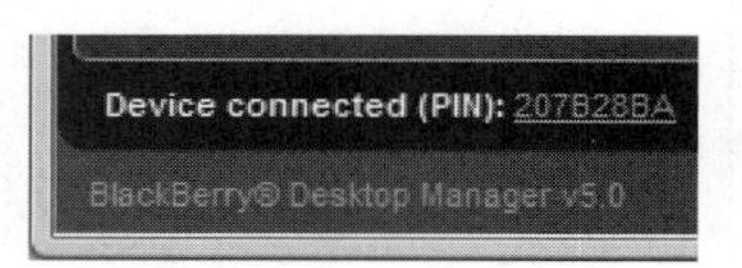

2. Then, click the **Synchronize** icon.

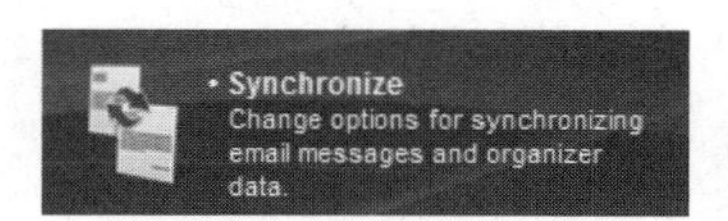

3. Before you sync for the first time, click on the **Synchronization** link right under where it says **Configure** on the left-hand side of the screen.

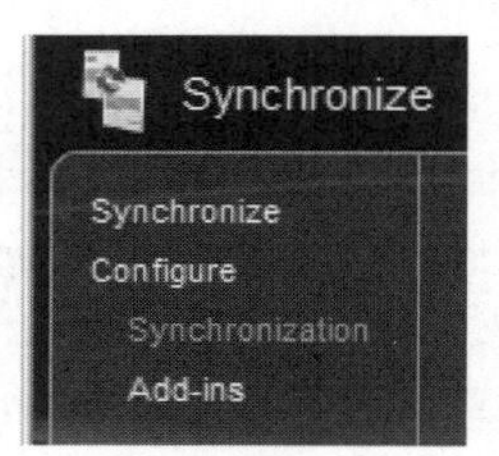

4. Click on the **Synchronization** button in the right side next to **Configure synchronization settings for my desktop program.**

NOTE: If the **Synchronization** button is grayed out and not clickable, please make sure your BlackBerry is connected to your computer. If you see **None** instead of your PIN number in the lower left corner, your BlackBerry is not connected to your computer.

5. Now you will see the main **IntelliSync** program window shown in Figure 2-7.

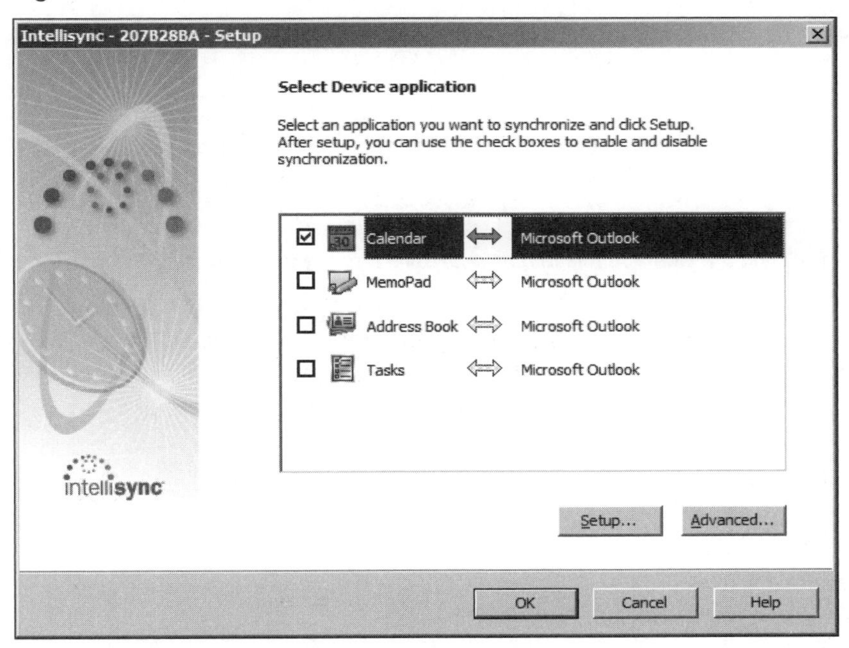

Figure 2-7. *Intellisync main Setup screen in Desktop Manager*

6. To get started, just check the box next to the icon you want to sync or on the name of the icon, or click the check box and then click the **Setup** button at the bottom. For example, clicking **Calendar** and **Setup** will bring you to a few screens with details for how to sync your computer's Calendar to your BlackBerry.

7. Select your desktop application from the list in Figure 2-8 and click **Next**.

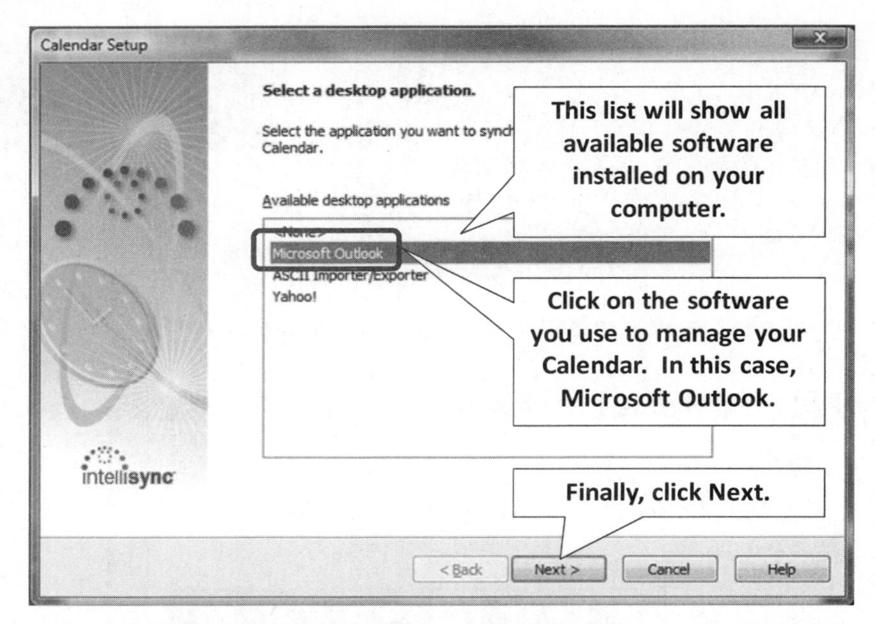

Figure 2-8. ***Select a Desktop Application*** *to sync using* ***Intellisync*** *in Desktop Manager.*

8. Now you will see options for **Two way** or **One way sync**. The **Two way sync** means any changes you make on your computer or BlackBerry will be synchronized to the other device. This is what you usually will want. Under special circumstances, you might require or want **One way sync** see Figure 2-9

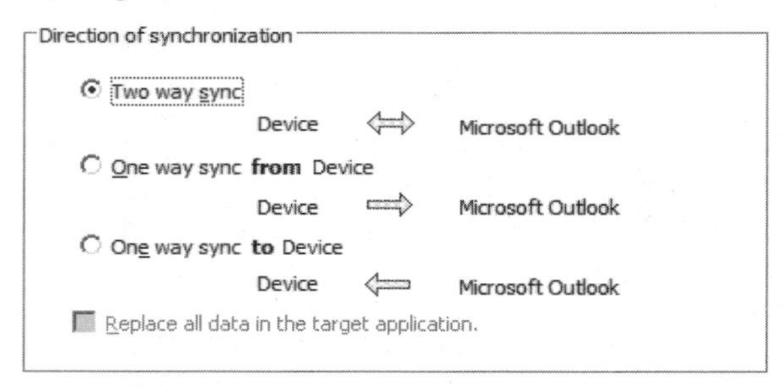

Figure 2-9. *Select One way sync or Two way sync for your information*

9. Click **Next** to see an advanced screen with more options. This one shows options for the Calendar. Address Book, Tasks, and MemoPad may have different options. We recommend settings as shown in Figure 2-10 to help make sure you never miss out on any data you enter on your BlackBerry if you forget to sync every day. These settings will sync Calendar events up to 30 days old from your BlackBerry and 180 days into the future.

Microsoft Outlook options for Calendar

These settings are used during data exchange operations involving Microsoft Outlook.

Outlook user profile

Outlook

Calendar date range

Transfer all scheduled items

Transfer only future items

Transfer items within a range of days

30 Days prior to today

180 Days after today

Alarm settings

Remove alarm for past items

Figure 2-10. *Select additional options for your Calendar sync*

10. Repeat the procedure for all the applications you want synced. Click **Next**, then click **Finish** on the next screen.

You will see similar screens for all four applications, with some minor variations (see Figure 2-11).

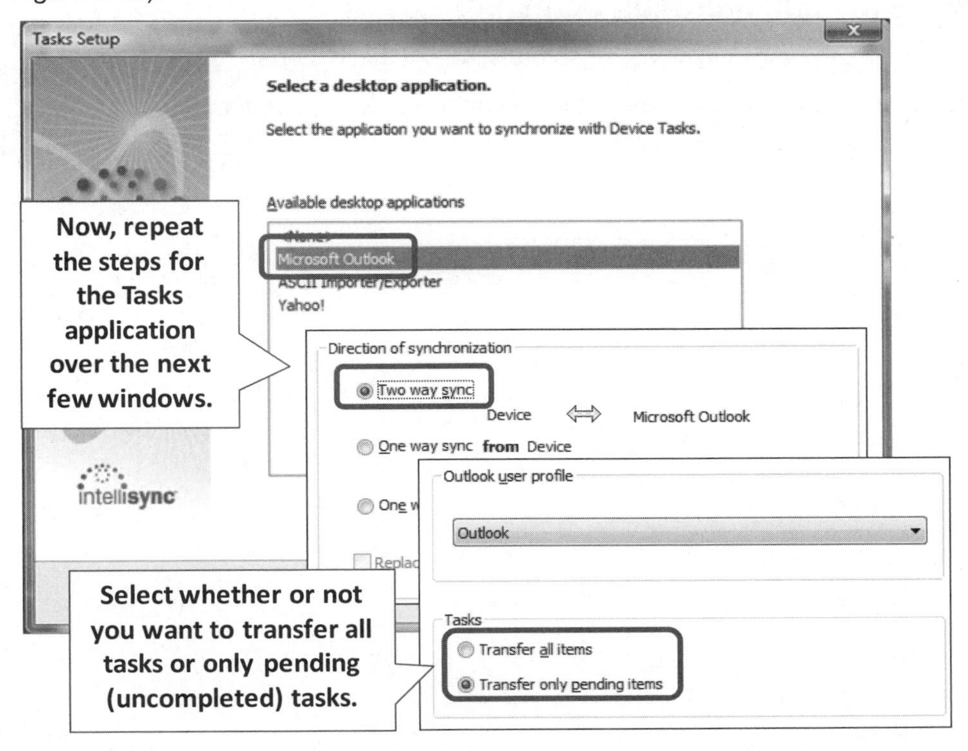

Figure 2-11. *Task sync options in Desktop Manager*

Once the setup is complete for two-way sync for all four applications, your screen should look similar to the one below.

If you are syncing all four applications, it will look like this:

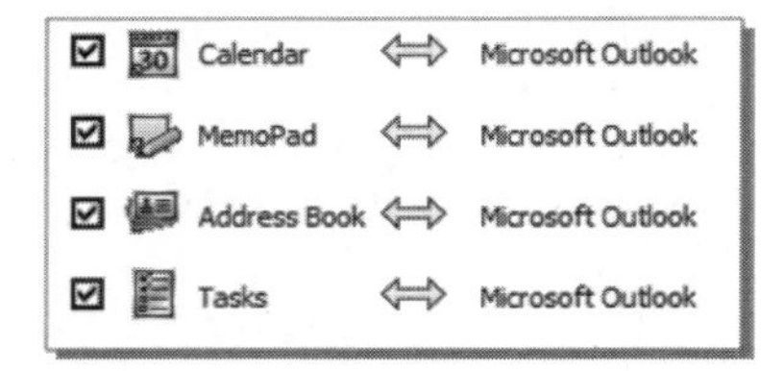

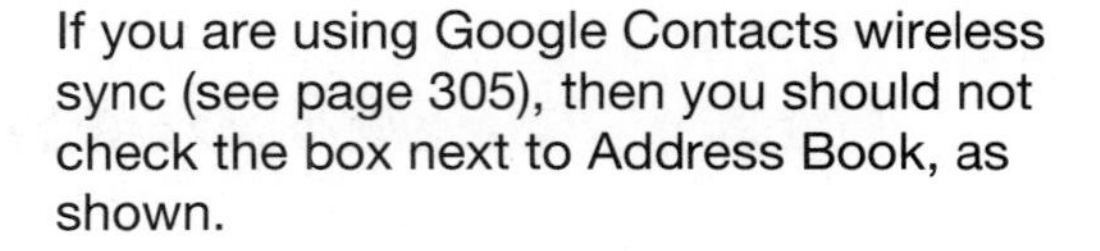
If you are using Google Contacts wireless sync (see page 305), then you should not check the box next to Address Book, as shown.

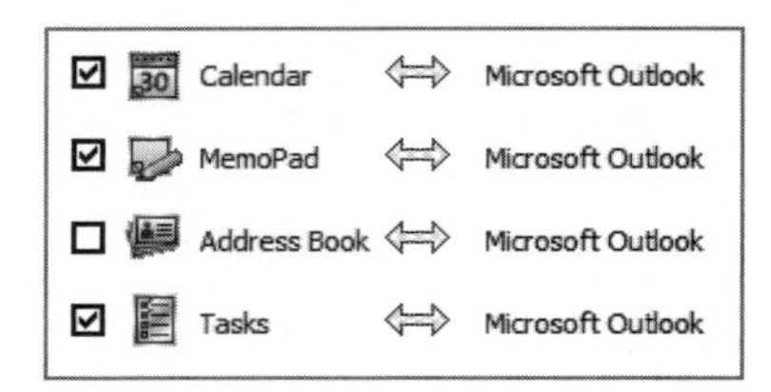

After the configuration is set, go back to the main **Synchronization** screen and put a check mark in the **Synchronize Automatically** box if you want the Desktop Manager to automatically synchronize as soon as you connect your BlackBerry to your computer.

Finally, close out all the **Sync** setup windows to save your changes.

Advanced Sync Configuration Screens

In order to see the advanced sync setup screens, follow these steps:

1. Get into the main **Setup** screen, as shown in the previous section.
2. Click the **Advanced** button on the screen shown in Figure 2-12.

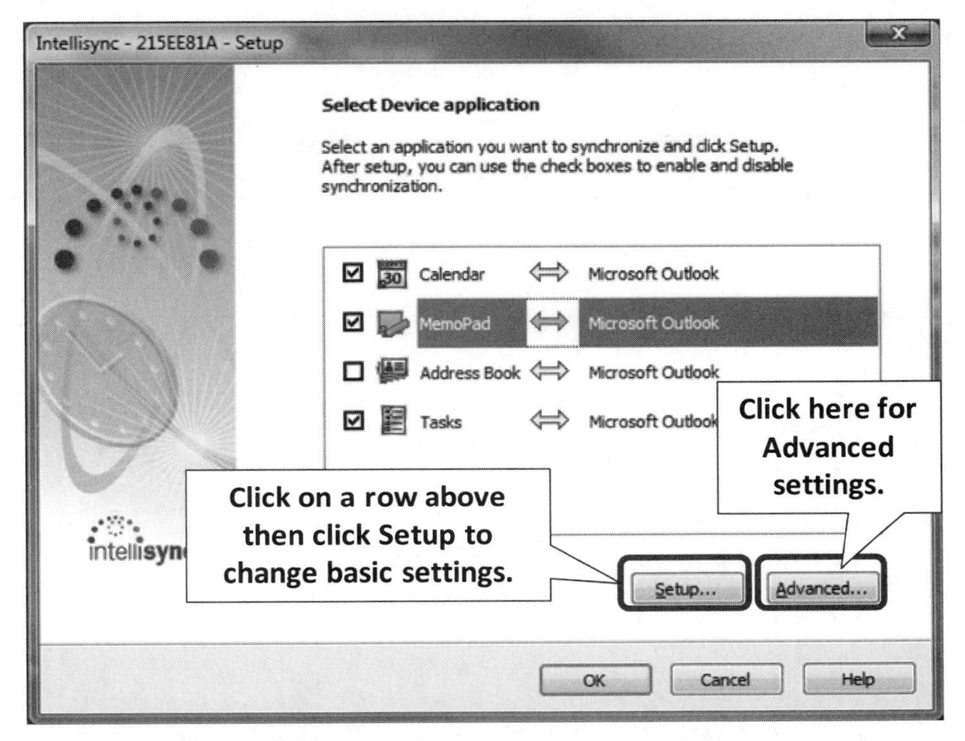

Figure 2-12. *Getting to basic and advanced settings for your sync configuration*

3. After clicking **Advanced**, you will see a screen similar to the one in Figure 2-13.

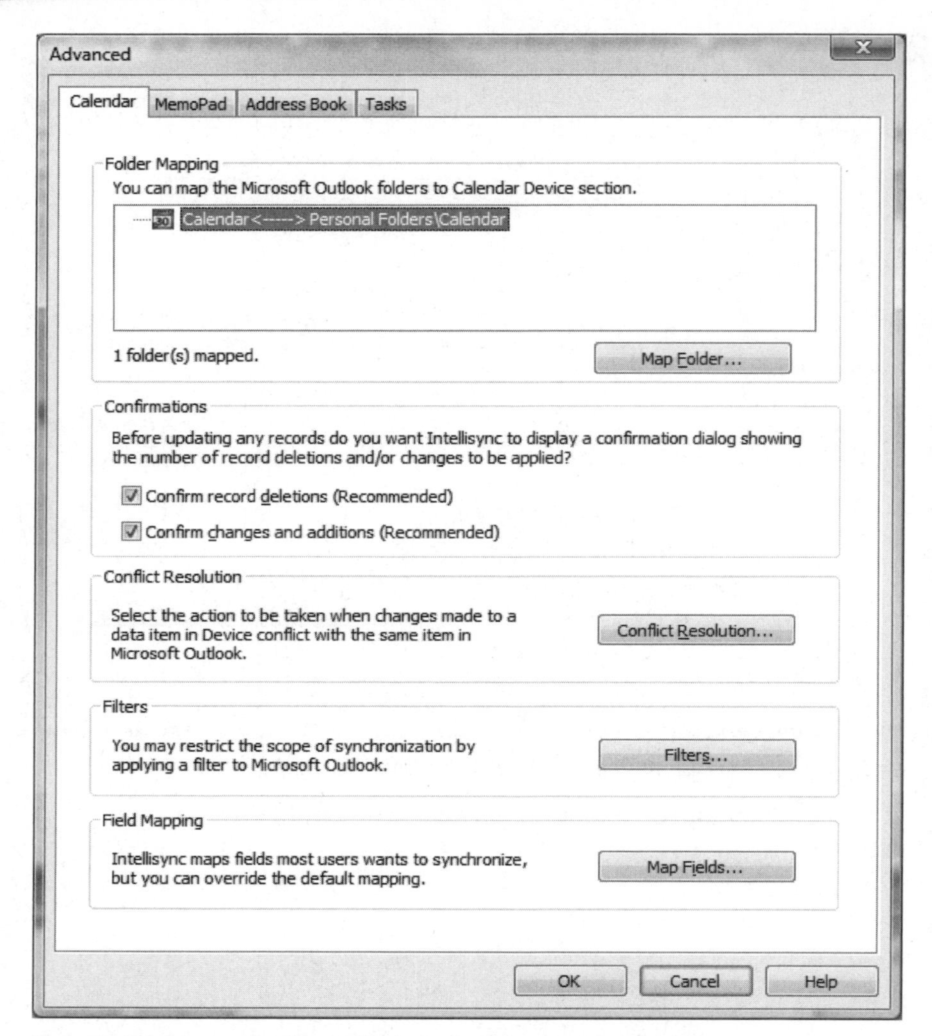

Figure 2-13. *Advanced settings tabbed screen for your sync configuration*

From this **Advanced** settings screen, you can perform any of the following actions:

- Map Folder
- Conflict Resolution
- Filters
- Field Mapping

Map Folder

The **Map Folder** function allows you to select one or several folders to map from your desktop application to sync your BlackBerry. In Figure 2-14, notice there are several Outlook folders from which to choose to map to sync the Calendar to the BlackBerry.

1. Use the plus signs (+) in the right column to expand or collapse the views (see Figure 2-14).
2. Select individual items (such as **Calendar**) from the left by clicking on them.
3. Click the **Add** button in the middle to add this item to the sync.
4. To remove a selected item from the sync, click on it in the right column, then click the **Remove** button in the middle.

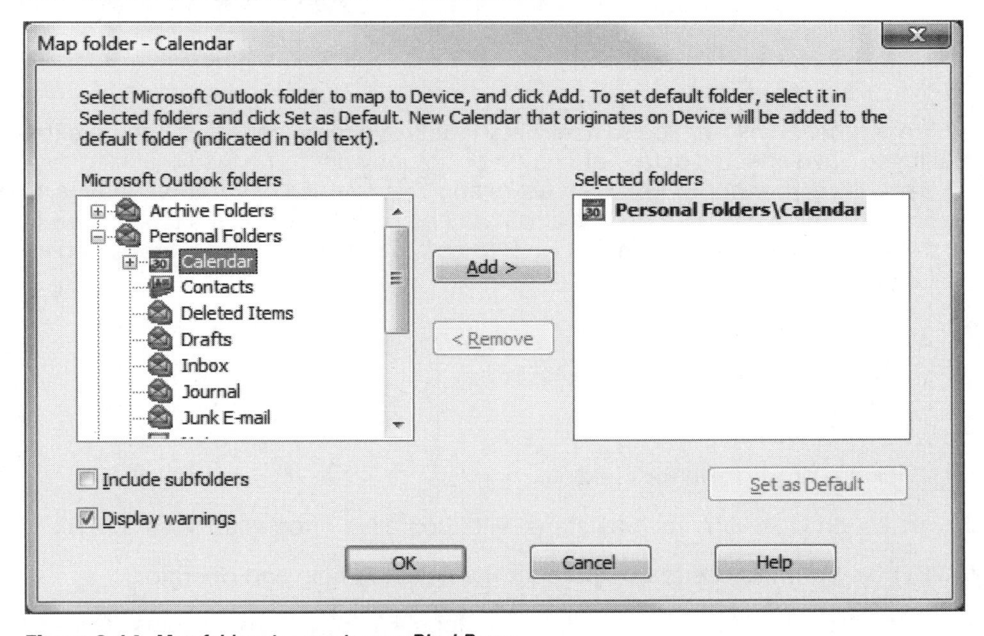

Figure 2-14. *Map folders to sync to your BlackBerry*

Conflict Resolution

The **Conflict Resolution** function allows you to determine if you want to review each sync change and determine whether the handheld or your computer will "win" should any conflicts arise (or you should be asked each time). Being asked each time is the default and recommended setting (see Figure 2-15).

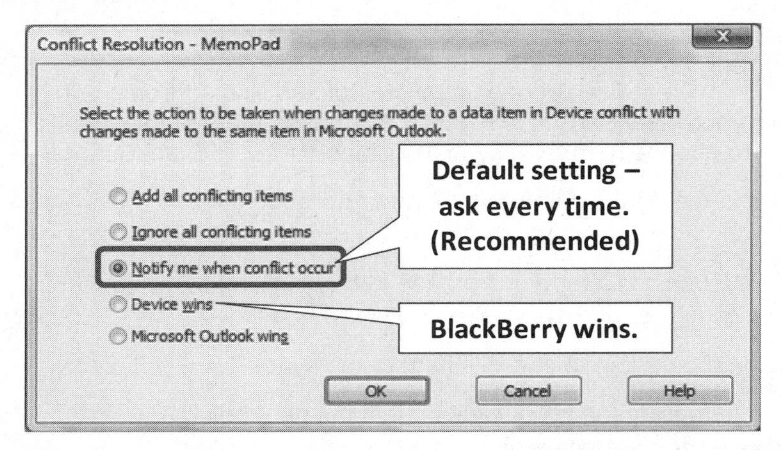

Figure 2-15. *Conflict Resolution choices window from Advanced sync settings (for MemoPad application)*

Filters

The **Filters** function allows you to filter data that is synchronized. This can be extremely useful if you have specific data that you do or do not want to be synchronized to your BlackBerry from your computer. With filters and creative use of information typed in your desktop application, you can do just about anything you want. To set up a new filter, Click the **Filters** button shown in Figure 2-13 and follow the following steps and in Figure 2-16.

1. Click **New.**
2. Enter a name for your **Filter.**
3. Click **OK.**
4. Click the drop-down for **Field.**
5. Select an Operator, such as starts with, contains, or equals.
6. Type a **Value** to use to compare for the selected field and operator.
7. Click the **Add to List** button. Repeat steps 4-7 for additional fields, if desired.
8. Click the **Rules** tab at the top.
9. Select one of the two conditions: **All Conditions** or **One or more.**
10. Click **OK** to return to the screen shown in the upper right corner of Figure 2-16.
11. Check or uncheck the box that says **Delete from device any data that does not match the filter** as desired. Then click **OK** again.
12. Finally, click OK to save your changes.

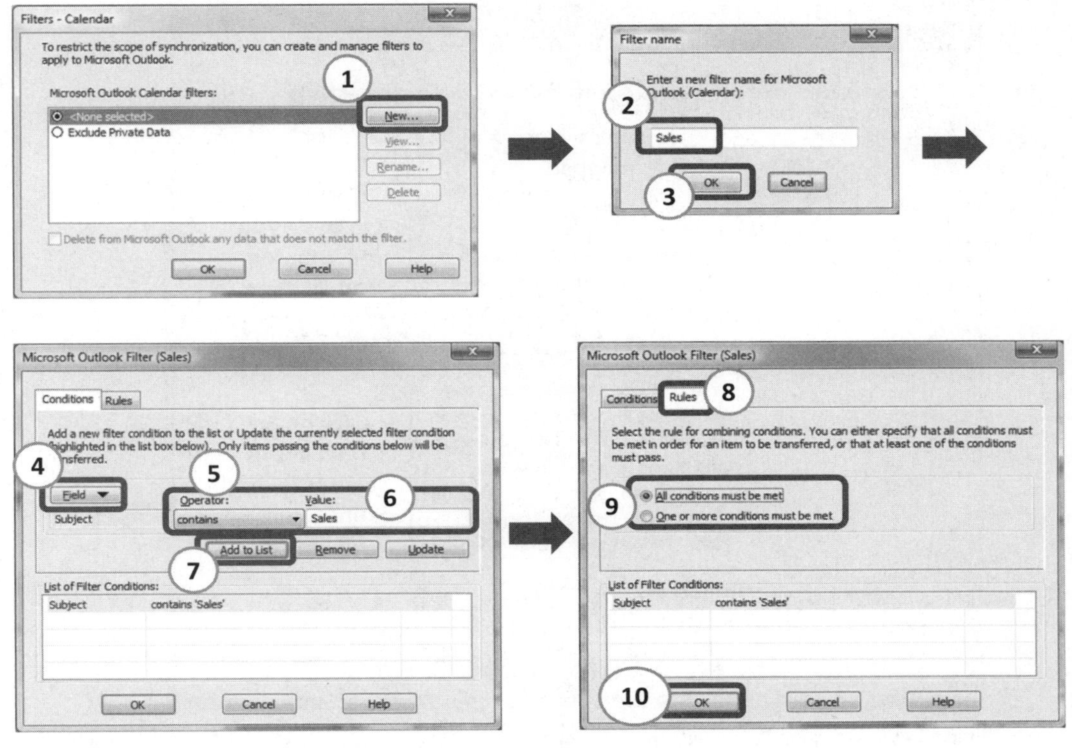

Figure 2-16. *Creating a new sync Filter in Desktop Manager*

Field Mapping

The **Field Mapping** function allows you to map individual fields from your computer application into your BlackBerry. This can be useful if you need to fine-tune the information that is put onto your BlackBerry (see Figure 2-17). You can do the following actions:

- To **map** a field, click between the left and right columns until you see the double-arrow.
- To **unmap** a field, click between the left and right columns until it is blank (the double-arrow goes away).
- To **change which field is mapped** on the right column, drag it up or down and drop it.
- You may need to scroll down the list to see all the possible fields to be mapped.

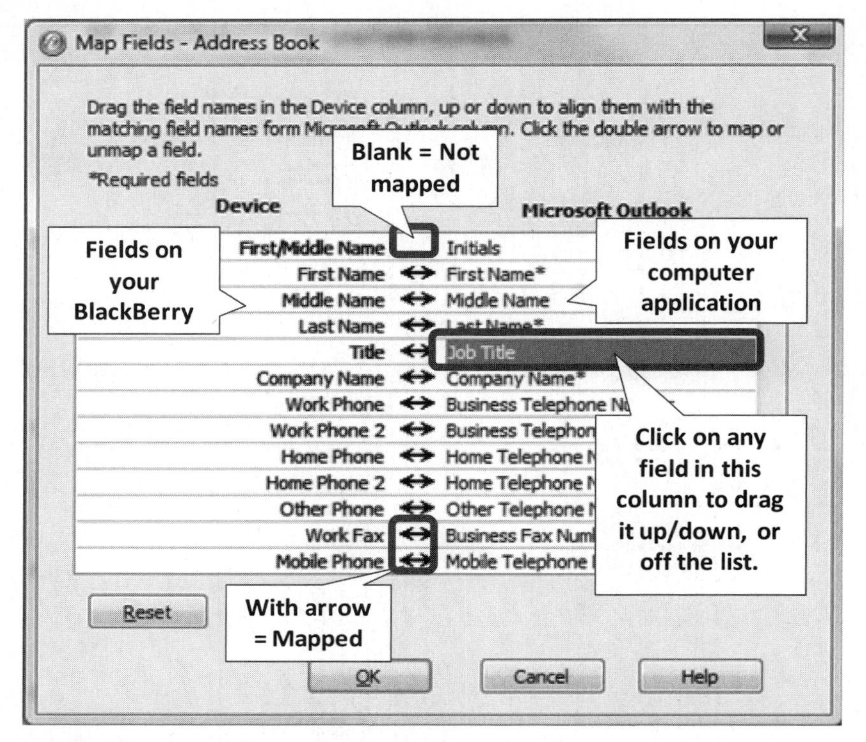

Figure 2-17. *Adjusting field mapping for the sync*

To return to the main Synchronize screen, click **OK** or **Save.**

Running the Sync

To get the sync started the first time, you need to do the following steps. You can automate the sync so you don't have to do these steps every time. We show you how to automate the sync in the next section.

1. Click the **Synchronize** link in the very left-hand column (see Figure 2-18).
2. Make sure the box next to **Synchronize organizer data** is checked.
3. Click the **Synchronize** button in the middle of the window to start your sync.

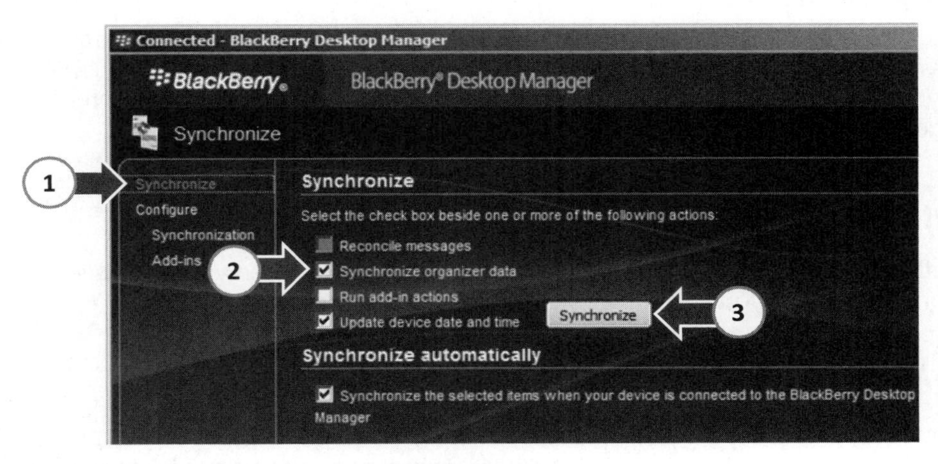

Figure 2-18. *Steps to start the sync in Desktop Manager*

After starting the sync, you will see a small window pop up showing you status of the current sync.

Automating the Synchronization

If you click the checkbox at the bottom of the Synchronize screen shown in Figure 2-18, your BlackBerry will automatically sync every time you connect it to your computer.

Accepting or Rejecting Sync Changes

During the sync, if there are additions or deletions found in either the BlackBerry or the computer application, a dialogue box will appear giving you the option to **Accept** or **Reject** the changes. Click **Details** if you want to see more about the specific changes found (see Figure 2-19).

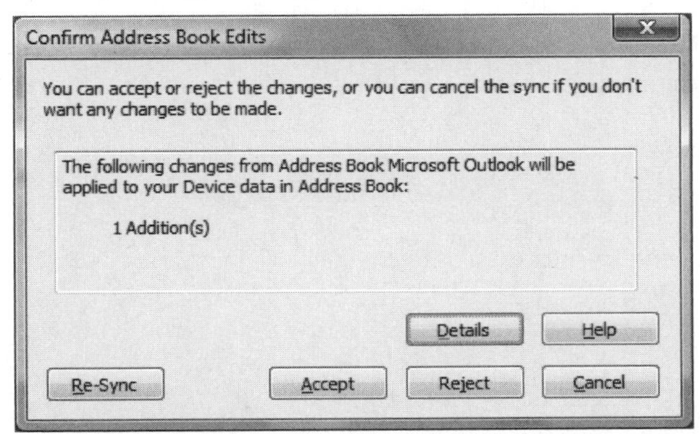

Figure 2-19. *Accepting or rejecting changes found during the sync*

Usually, we recommend to **Accept** the changes unless something looks strange. Finally, the Synchronization process will come to an end and your data will be transferred to both your BlackBerry and your computer. If you want more details on what has changed on your BlackBerry and computer, click the **Details** button.

Troubleshooting Your Sync

Sometimes you will encounter errors or **caution** messages when you try to sync. In this section, we try help you through some of the more common issues.

Default Calendar Service Has Changed Message

Sometimes you may see a message similar to Figure 2-20. What usually happens when you add a new email address is that it takes precedence as the default email address or service for all new Calendar entries you add on your BlackBerry.

Click **Cancel** on the screen shown in Figure 2-20 and follow the steps that follow to verify that everything is ok before you sync again.

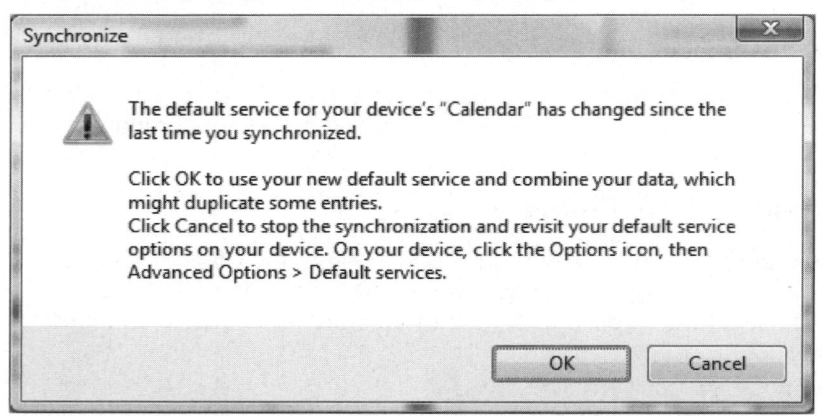

Figure 2-20. *Default calendar service has changed message*

After pressing the **Cancel** button, follow these steps:

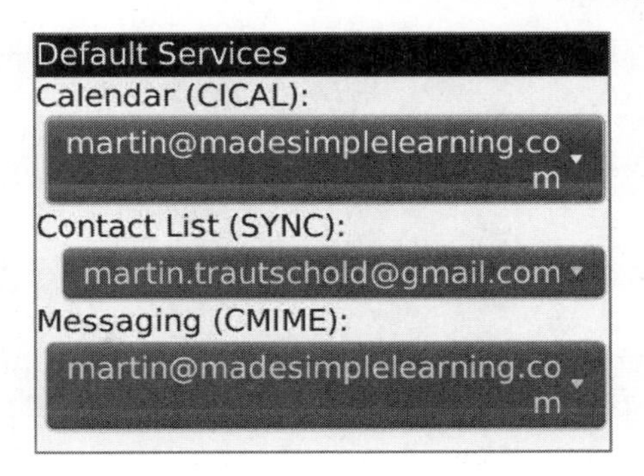

1. Click on the **Options** icon on your Bold.
2. Click **Advanced Options**.
3. Click **Default Services**.
4. You will see a screen similar to the one shown. Verify that the email address under the **Calendar** item at the top is set correctly. If not, click and adjust it.
5. Press the **Menu** key and select **Save**.
6. Re-sync using **Desktop Manager** and if you see the same error, ignore it by clicking **OK**.

Close and Restart Desktop Manager

Try closing down **Desktop Manager** and restarting it, sometimes this can help.

Remove and Reconnect Your BlackBerry

Sometimes a simple disconnect and reconnect can also help. Give it a try.

Fixing Specific Errors with Calendar, Address Book, or MemoPad or Tasks Sync

Try the sync again after it fails and watch it closely, you should note where it fails–on the Calendar, Address Book, MemoPad, or Tasks–by watching the status screen.

Once you figure out where the sync fails, then you can try one thing to get it running again: clearing out or deleting the problem database from your BlackBerry and starting the sync again.

> **CAUTION:** Doing this process will force you to lose any changes you have made on your BlackBerry since your last successful sync.

1. From the main **Desktop Manager** screen, click on **Backup & Restore.**
2. Click **Backup.** Make a note of the file name and location; you may need to use it later to restore data if this troubleshooting does not work. In Figure 2-21, the backup file name is **Backup-(2008-12-12)-1.ipd**.

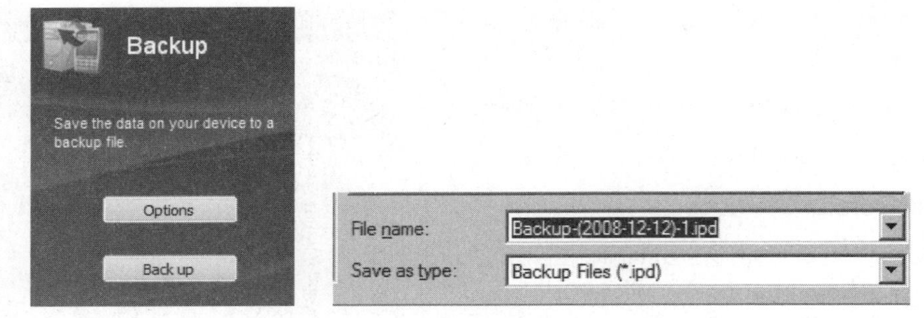

Figure 2-21. *Perform a full backup and note the file name*

Once your full backup is completed, click the Advanced button from the Backup and Restore screen shown in Figure 2-22.

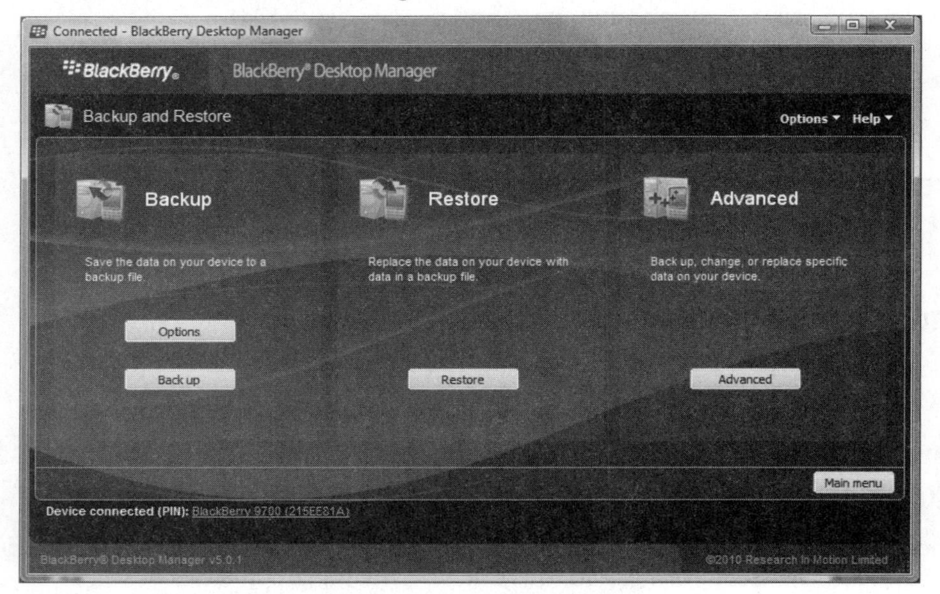

Figure 2-22. *Main Backup and Restore screen in Desktop Manager*

3. Now, you will be on the screen shown in Figure 2-22. Locate the problem database in the right-hand window (BlackBerry) and click the **Clear** button at the bottom. In Figure 2-23, we are getting ready to clear out the **Address Book** and **Address Book – All** from the BlackBerry (press the **Ctrl** key to click on and select more than one.)

4. Both are selected in the right-hand window **Device Databases**, then we click the **Clear** button.

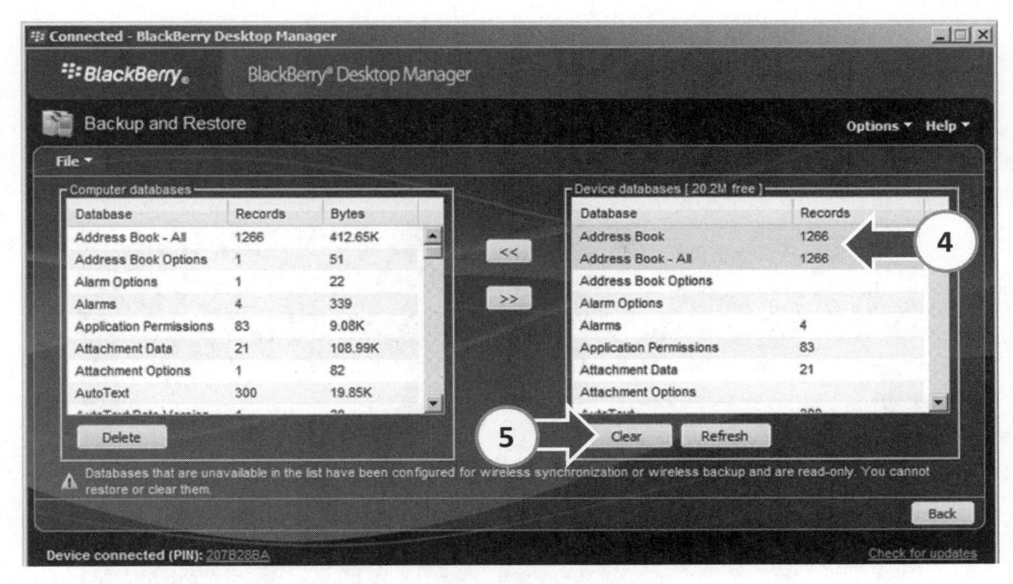

Figure 2-23. *Clear or erase a specific database from the BlackBerry*

5. Once they have been cleared out, then click the **Back** button.

6. Now try to sync again. Hopefully, this will correct the sync problem.

If the problem has not been corrected, then you can restore the Address Book by doing the following steps:

1. Return to the advanced **Backup and Restore** window as shown earlier in this section.

2. In the upper right corner, click **File**, then **Open** the full backup file you just created, as shown in Figure 2-24.

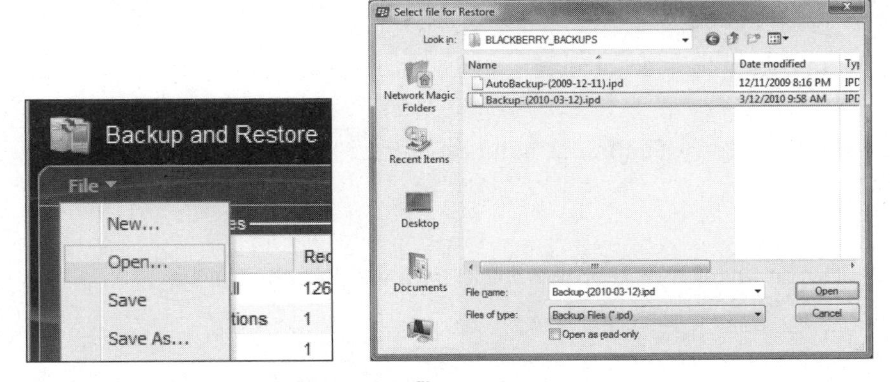

Figure 2-24. *Opening a previous backup file to restore*

3. Now you can highlight the correct databases from the full backup in the left-hand window (**Computer databases**). In this example, we have clicked on the **Address Book – All** database (see Figure 2-25).

4. To restore the selected database to your BlackBerry, click the >> button in the middle of the screen.

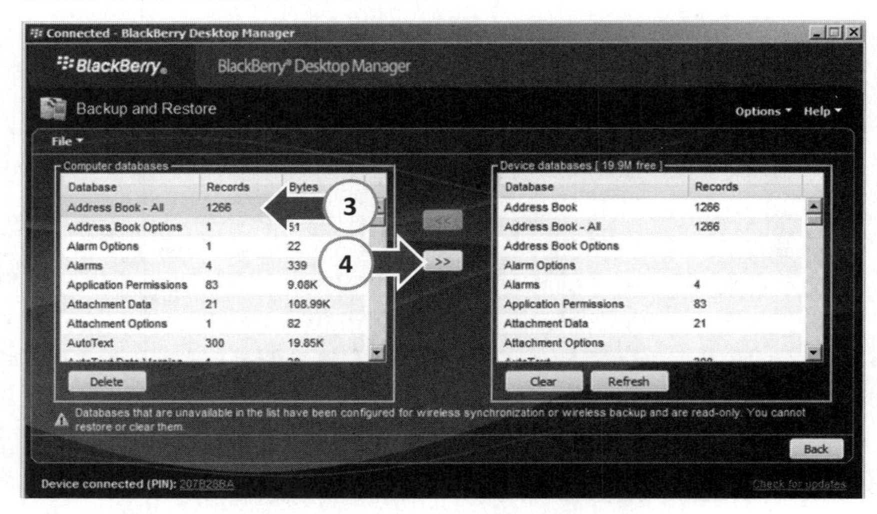

Figure 2-25. *Restoring a particular database to your BlackBerry*

Getting More Help for Desktop Manager Issues

Some of the Desktop Manager sync issues can be particularly tricky. Before you pull out too much hair, you should go try the BlackBerry Technical Knowledgebase. You should also try some of the more popular BlackBerry online discussion forums to see if others have experienced and solved similar issues.

BlackBerry Knowledgebase:

1. From your computer web browser, go to: `http://na.blackberry.com/eng/support/`.

2. Click the **BlackBerry Technical Solution Center** link (or similar) in the left column.

Web Search:

- Pull up your favorite web browser and try a web search for the particular issue you are facing.

BlackBerry Forums:

- www.crackberry.com
- www.blackberryforums.com
- www.pinstack.com
- www.blackberrycool.com

Application Loader

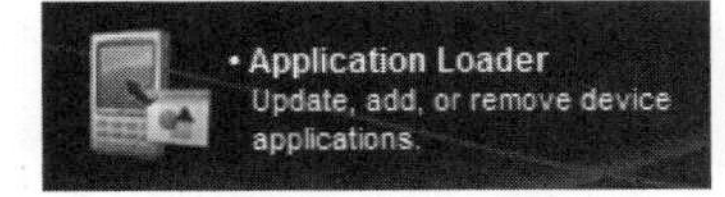

Use this icon to add or remove software from your BlackBerry. You can also use this to update your BlackBerry Operating System or System Software. There are easier ways to load or remove software from your device; see the App World chapter on page 431 and the Add / Remove Software chapter on page 475.

First, make sure your BlackBerry is connected to your computer and it is shown in the lower left corner of the **Desktop Manager** screen next to **Device connected (PIN):**. If you see **none**, then you will need to try to get it connected. Some of the easier things to do are unplug/replug the USB cable, try plugging into another USB port, shutting down and restarting Desktop Manager, or restarting your computer.

Now, click the **Application Loader** icon to see the screen in Figure 2-26.

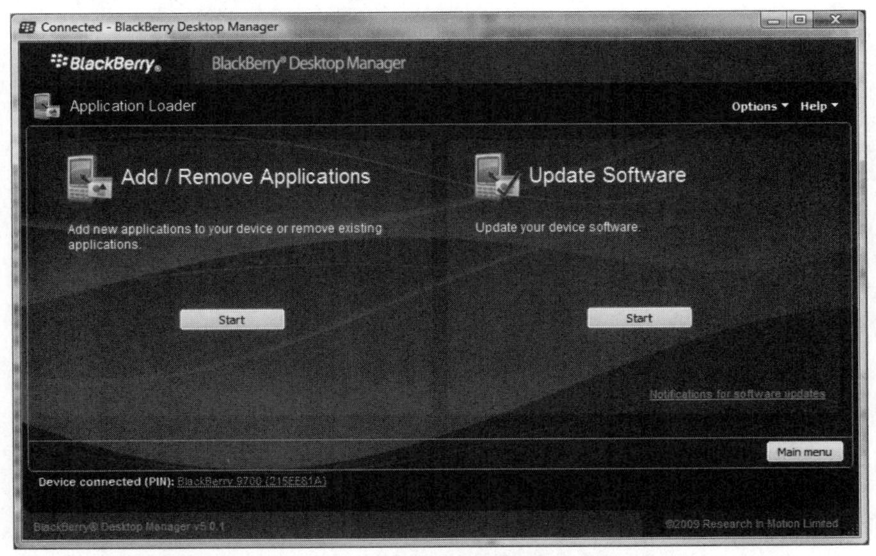

Figure 2-26. *Application Loader main window in Desktop Manager*

Add/Remove Applications

To add or remove applications, which could be third-party or portions of the main system software and core applications (such as language files), follow these steps:

1. Click the **Start** button under **Add/Remove Applications** on the screen in Figure 2-26.

> **NOTE:** You will first see a **Task In Progress** window showing you that the software is reading your current BlackBerry configuration and installed software.

2. Listed at the top of the screen shown in Figure 2-27 is your current BlackBerry System Software. This device is running 5.0.0; your device may show a different version.

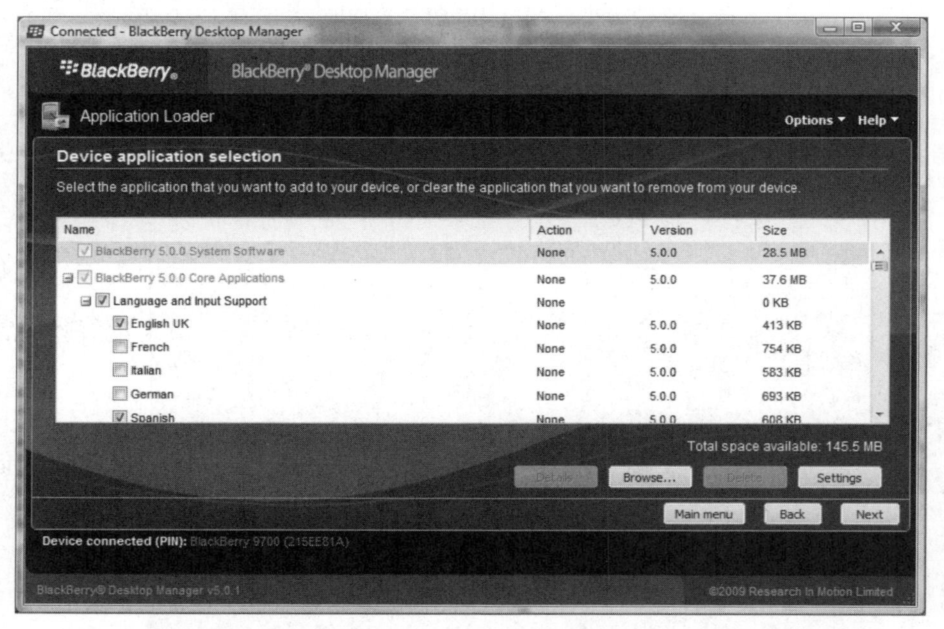

Figure 2-27. *Add/Remove Software screen in Application Loader of Desktop Manager*

3. Notice that you can add or remove languages using this tool.
   a. To add an item, place a **check** in the check box next to it.
   b. To remove an item, **uncheck** a check box next to it.
4. Scroll down using the scroll bar on the right edge to see more language options. Part of the way down, you will notice **Supplemental SureType Wordlists**, as shown in Figure 2-28.

5. If you work in the finance profession, legal profession, or medical profession, then you may want to add some of these customized dictionaries. These will help when you use SureType or the spell checker guesses what you are trying to type. In Figure 2-28, we want to add **English Financial** and **Medical** terms, so we checked both boxes.

6. Notice the **Action** column shows **Install** as a status.

Name	Action	Version	Size
Supplemental SureType(TM) Wordlists	None		0 KB
English	None		0 KB
Financial Terms	Install	5.0.0	33 KB
Legal Terms	None	5.0.0	34 KB
Medical Terms	Install	5.0.0	264 KB
French	None		0 KB
Financial Terms	None	5.0.0	31 KB
Legal Terms	None	5.0.0	36 KB

Figure 2-28. *Supplemental word lists for Financial, Legal' and Medical terms*

7. Finally, to see all your installed third-party applications, scroll down to the bottom of the list, as shown in Figure 2-29. You can see that we have a number of apps installed and we have decided to remove or uncheck two of them: **U2 Mobile Album** and **BlackBerry Developer Conference Mobile Guide.**

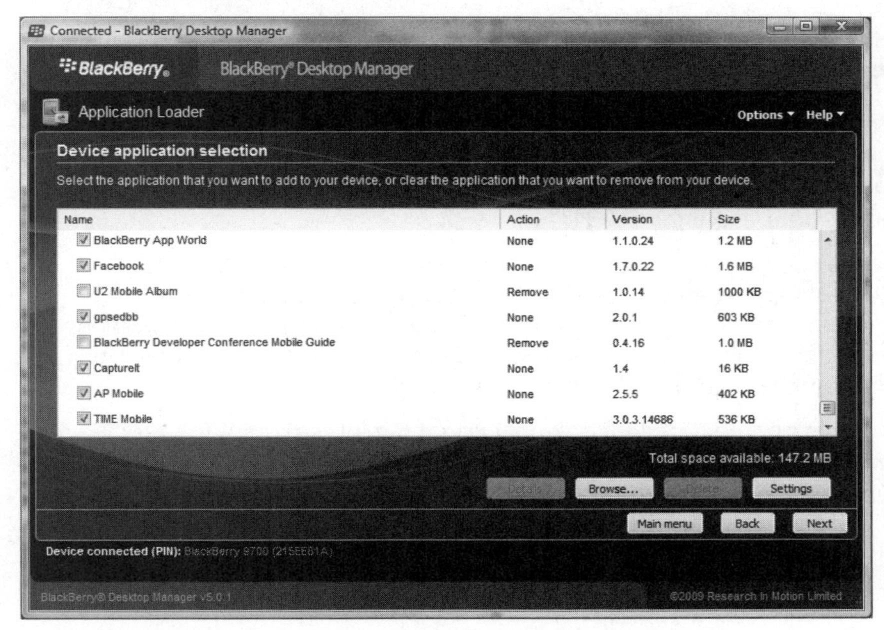

Figure 2-29. *Third-party applications are shown at the bottom of the list*

8. Finally, to complete adding or removing applications, click the **Next** button in the lower right corner to see the **Summary** screen shown in Figure 2-30.

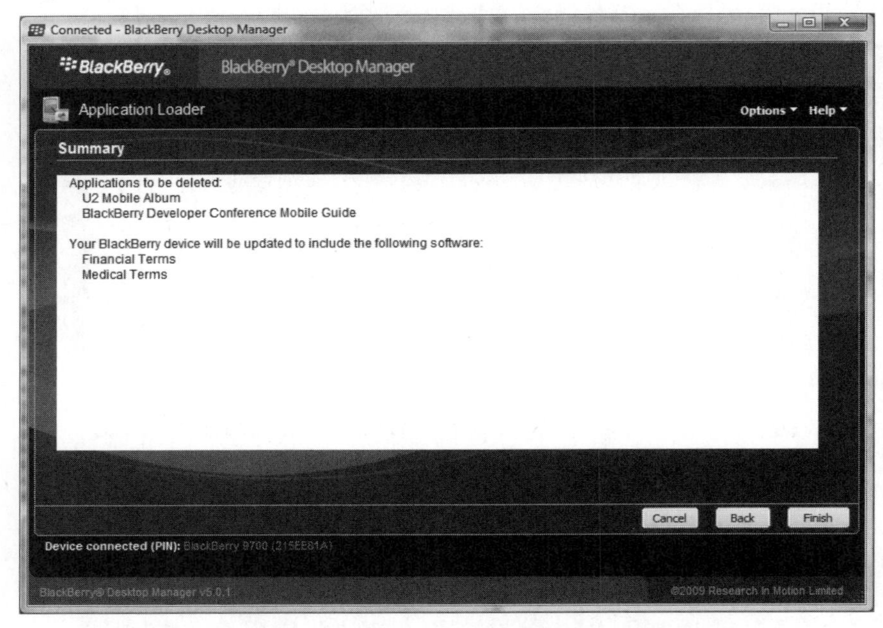

Figure 2-30. *Application Loader Summary screen showing your selections*

9. If you see that you have made a mistake, then click **Back** to return to the previous screen; otherwise click the **Finish** button to execute the listed actions. While the software is working, you will see a status window similar to Figure 2-31.

   In some cases, this process is very fast, taking just a minute or so. However, in other cases, especially if you are updating the System Software or any part of it (such as adding or removing core dictionaries like we did in this example), you will see a message indicating that **This task might take up to 30 minutes to complete.**

   In our testing, the process took only about 6 minutes, but it felt like 30 minutes!

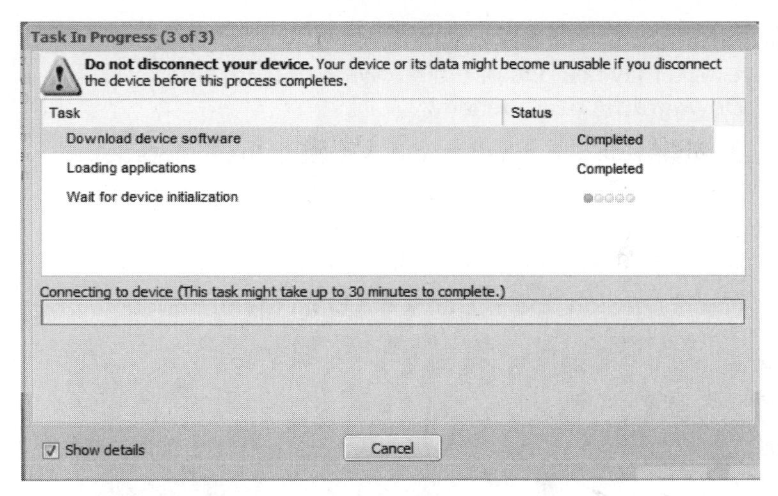

Figure 2-31. *Application Loader status screen*

Have patience while this is happening, because if you disconnect your BlackBerry from your computer while this reboot is happening, your BlackBerry might become unusable.

10. Finally, when the process is finished, you should see a small status message in the upper left corner, as shown in Figure 2-32.

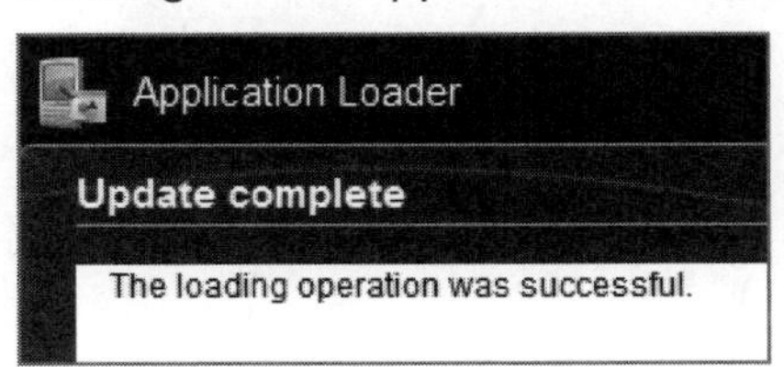

Figure 2-32. *Application Loader successful completion message*

Update Device Software

Also in Application Loader, you can update the device software which is the System Software running on your BlackBerry smartphone. You can actually upgrade and downgrade using this feature.

1. Click the **Start** button under the **Update Software** section of the **Application Loader** main screen (Figure 2-26).

2. The software will then check your BlackBerry and the Internet to see if there are updates available.

3. Finally, you will see a screen similar to the one shown in Figure 2-33.

4. Depending on what version you have installed on your BlackBerry and what is available, you may see one or more rows with **(Current), (Upgrade),** or **(Downgrade)** next to them.

   a. To **upgrade** your software, check the box next to the **(Upgrade)** item and click the **Next** button.

   b. To **downgrade** your software, check the box next to the **(Downgrade)** item and click the **Next** button.

   c. To make no changes do nothing, click the **Main Menu** button.

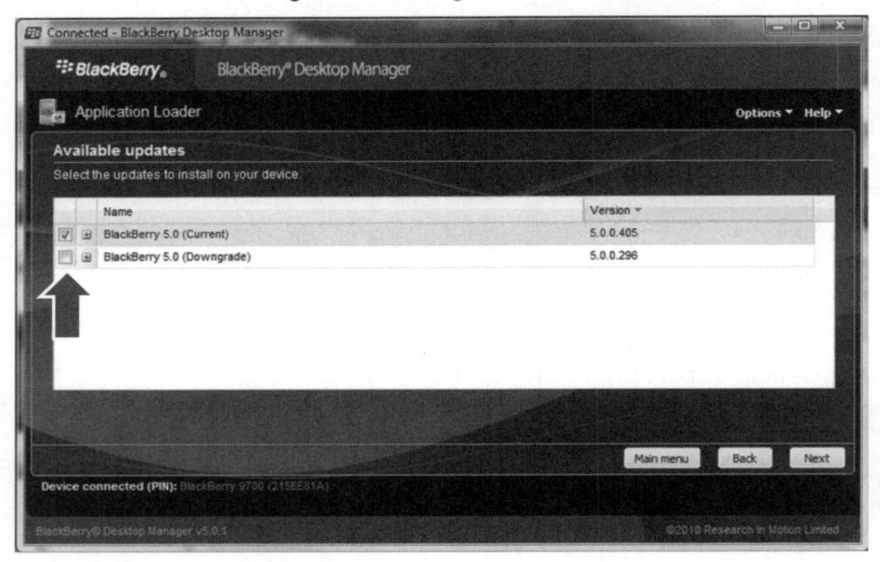

Figure 2-33. *Update System Software screen in Application Loader*

5. If you are upgrading or downgrading software, then you will see some screens telling you it is backing up your BlackBerry, then erasing it and reinstalling software, then restoring your data. This process could take more than 10 minutes.

6. Finally, you will see an **Update summary** screen similar to Figure 2-34.

7. Click the **Main Menu** button to finish the process and return to the main Desktop Manager window.

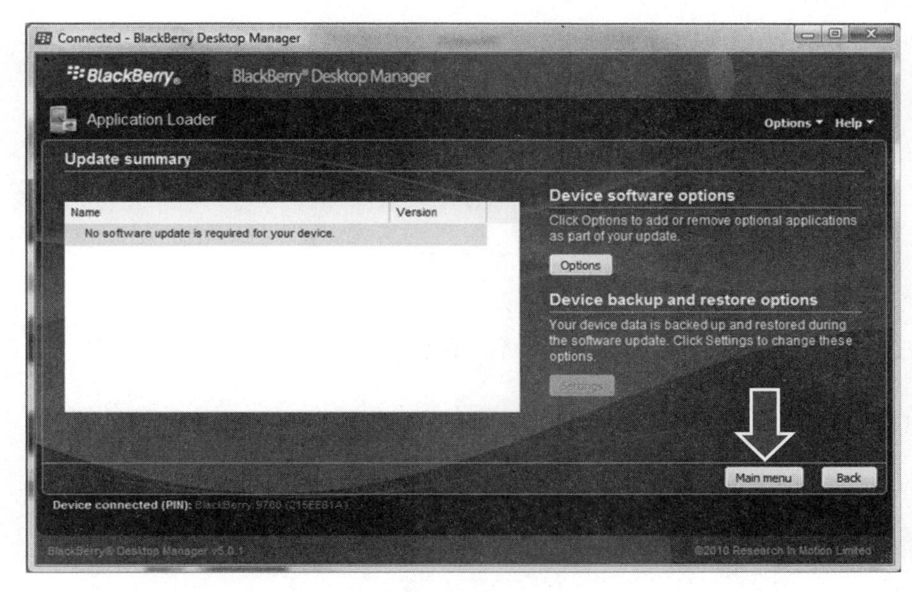

Figure 2-34. *Update summary screen in Application Loader*

Backup and Restore

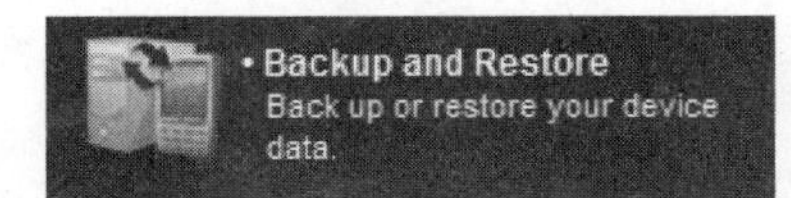

Use this feature to protect the important data on your BlackBerry. You can also use some of the advanced features to help with troubleshooting your Desktop Manager sync (see page 84).

Click the **Backup and Restore** icon to see the screen in Figure 2-35.

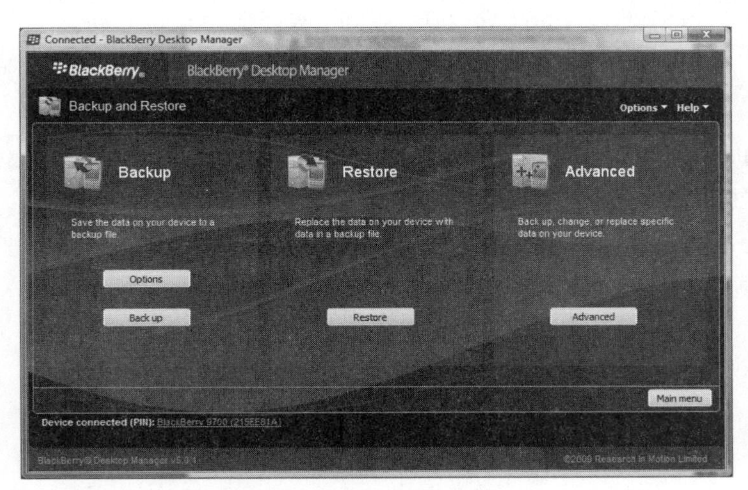

Figure 2-35. *Backup and Restore main window*

Backup and Restore Options screen: Click the **Options** button under the **Backup** heading on the left side of the page to see the **Backup Options** screen shown in Figure 2-36.

Backup Options

Additional device memory

☐ Back up on-board device memory

Encryption

☐ Encrypt backup file

Automatic backup settings

☑ Back up my device automatically every 7 day(s)

When backing up my device automatically

◉ Back up all device application data

○ Back up all device application data except for

☐ Email messages

☐ Application data that is synchronized with an organizer application on my computer

OK | Cancel | Help

Figure 2-36. *Backup Options screen*

If you want to back up data stored in your main BlackBerry memory (such as pictures, video recordings, voice notes, or other information) then you would check the box next to **Back up on-board device memory**.

If you have especially confidential information and want to encrypt your backup file for added security, then check the box next to **Encrypt backup file**.

The default for **Automatic Backup Settings** are every 7 days and to back up all the device data. You can adjust this to be any number of days; try fewer days if you are more worried about your data being lost.

The backup can take several minutes or more depending on how much information you have stored on your device. The easiest way to speed up the backup is to back up less information. Since your email is on your computer or email server anyway, you might want to check **Email Messages**. This would skip backing up email. If you sync regularly with your computer, then you could check the box next to **Application data that is synchronized with an organizer application on my computer**.

Click **OK** to save your options settings.

Back Up Your BlackBerry

In order to back up your BlackBerry, follow these steps:

1. Connect your BlackBerry to your computer.
2. If you are not already in the **Backup and Restore** menu, from the main **Desktop Manager** screen, click the **Backup and Restore** icon.
3. Next, click the **Back up** button in the image shown to the left to start your backup process.
4. Now you will see a dialog box pop up asking you to select a folder in which to store your full backup file (see Figure 2-37).

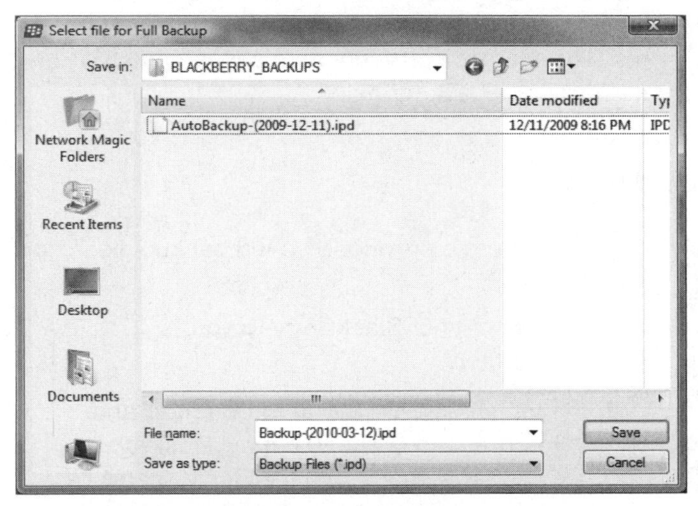

Figure 2-37. *Choose a folder to store the backup file*

5. Use the drop-down list at the top or one of the icons to the left to select your backup file location.
6. Notice the format of the file name **Backup-(2010-03-12).ipd** is in YEAR-MONTH-DATE format so you can easily see the date of your backup.

IMPORTANT: The authors have both experienced computer hard disk failures. We highly recommend storing the backup on at least one external location. This could be a USB thumb drive, an external USB drive, another computer on your network, or any location off of your hard disk.

7. Once you have selected the location of your backup file, click the **Save** button to start the backup.

8. You will then see a status window similar to Figure 2-38.

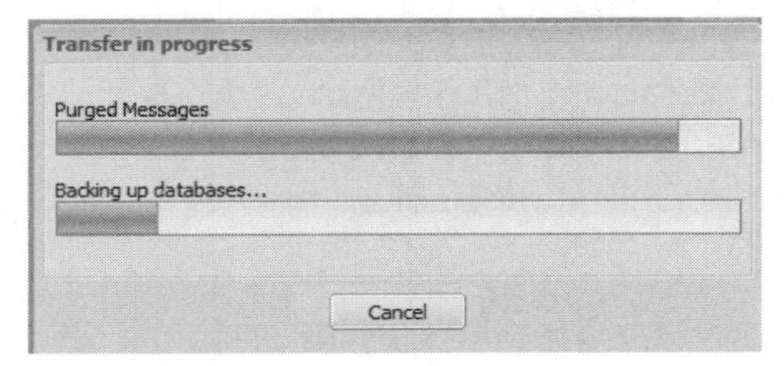

Figure 2-38. *Backup status window*

9. Finally, you will see a **Backup Successfully Created** message at the end. If you see any error message, try looking at any help available on the screen. Sometimes, just re-doing the backup will solve the issue. If that does not help, try the BlackBerry discussion forums (see page 537) or BlackBerry Technical Knowledge base for help.

Restore Your BlackBerry

To restore from a previously saved backup file, follow these steps:

1. Connect your BlackBerry to your computer.

2. If you are not already in the **Backup and Restore** menu, from the main **Desktop Manager** screen, click the **Backup and Restore** icon.

3. Next, click the **Restore** button as shown in the image to the left.

4. You will see a dialog box pop up asking you to select a folder in which to store your full backup file see Figure 2-39).

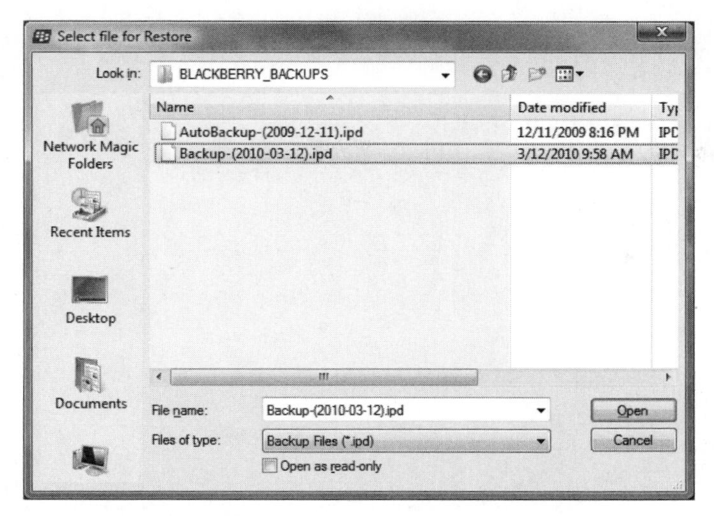

Figure 2-39. *Select folder and backup file to use for the restore*

5. Use the drop-down list at the top or one of the icons to the left to locate your backup file to restore from. Notice the format of the file name **Backup-(2010-03-12).ipd** is in YEAR-MONTH-DATE format so you can easily see the date of your backup.

6. Once you have located the file to use to restore data to your BlackBerry, click the **Open** button.

7. Next, you will see a list of details of the information contained in the file you just opened. This allows you to confirm you want to use this restore file. Scroll down to see things such as total number of **Contacts** (**Address Book**) and **Calendar** entries; make sure they seem reasonable.

8. Click **Yes** to start the restore process (see Figure 2-40).

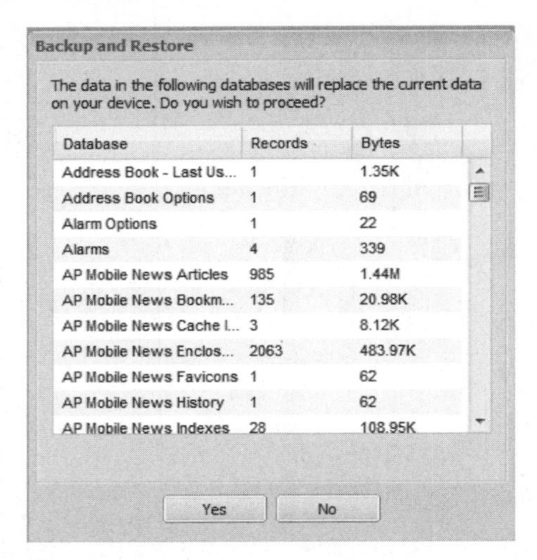

Figure 2-40. *Confirm restore screen*

NOTE: In the screen shown in Figure 2-40, you do not see a number of Address Book entries, only **Address Book Options** and **Address Book - Last Used**; this is because we happen to be using a wireless sync with Google Contacts. With any wireless sync, you cannot back up or restore those databases (Addresses, Calendar, and so forth) using Desktop Manager. All these items are essentially backed up all day long with the wireless sync process.

Advanced Backup & Restore

You would use the **Advanced** feature in **Backup and Restore** in order to selectively back up, restore, or erase individual databases (Addresses, Calendar, MemoPad, and so forth) on your BlackBerry.

NOTE: We showed how to erase only your Address Book and restore it to help with troubleshooting the Desktop Manager sync in the "Fixing Specific Errors" section on page 85.

1. Connect your BlackBerry to your computer.
2. If you are not already in the **Backup and Restore** menu, from the main **Desktop Manager screen**, click the **Backup and Restore** icon.
3. Next, click the **Advanced** button in the right portion of the screen to see the screen shown in Figure 2-41.

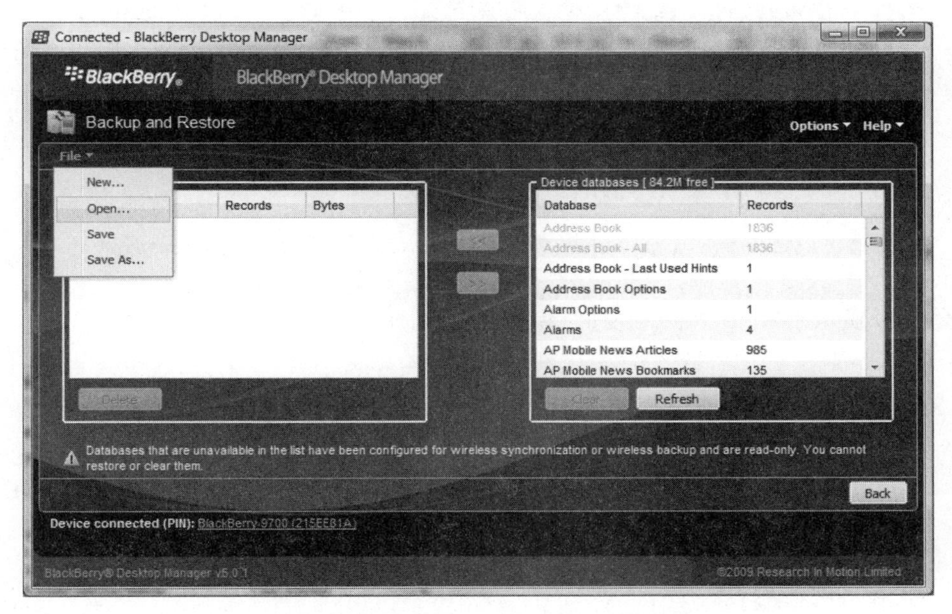

Figure 2-41. ***Opening a backup file in the advanced Backup and Restore screen***

4. The left side of the screen shows a backup file from your computer. If the left side is blank, then you will need to open up a backup file to use. Select **File ➤ Open** as shown in Figure 2-41.
5. Navigate to a specific folder and backup file to open and then click **Open**.
6. When the file is open, you will see the left-hand window fill up with the contents of that backup file (see Figure 2-42).

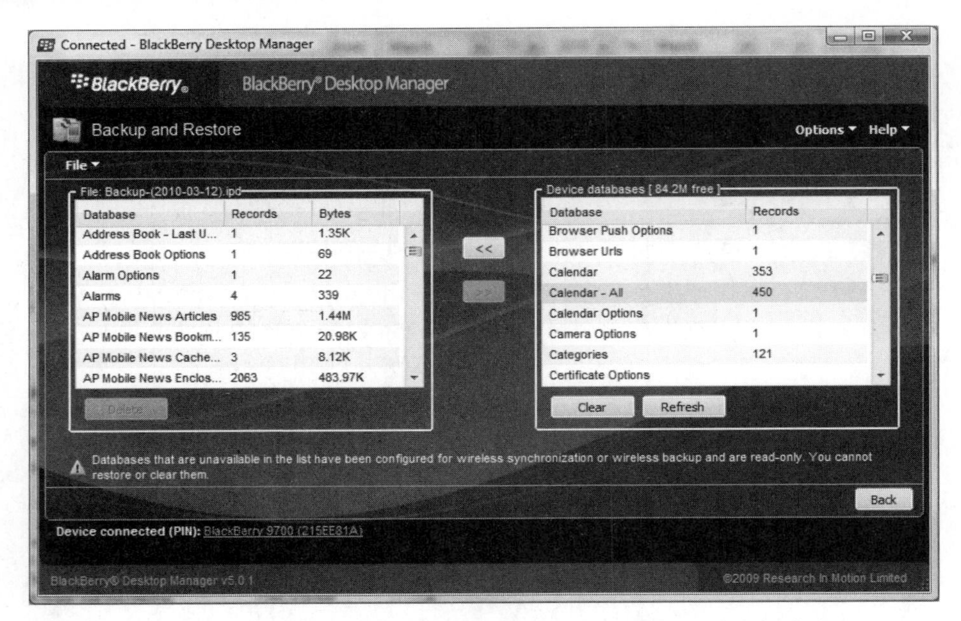

Figure 2-42. *Advanced Backup and Restore screen with backup file open in left window*

Now, you can selectively back up, erase, or restore individual databases on your BlackBerry by following the steps in the next sections.

Back Up Specific Databases

In order to selectively back up a single databases or selected databases, for example the Calendar, Address Book, or NotePad, follow these steps and refer to Figure 2-41.

1. Click in the right-hand window (**Device Databases**) to highlight and select a database or hold the **Control** key on your keyboard to select several databases.
2. Click the << button in the middle to copy that information to the backup file on the left.

Erase or Clear Specific Databases

In order to selectively erase or clear a single database or several databases from your BlackBerry, follow these steps and refer to Figure 2-42.

1. Click in the right-hand window (**Device Databases**) to highlight and select a database or hold the **Control** key on your keyboard to select several databases.
2. Click the **Clear** button under the right-hand window.

3. You will then see the screen in Figure 2-43; click **Yes** to confirm and delete the listed databases.

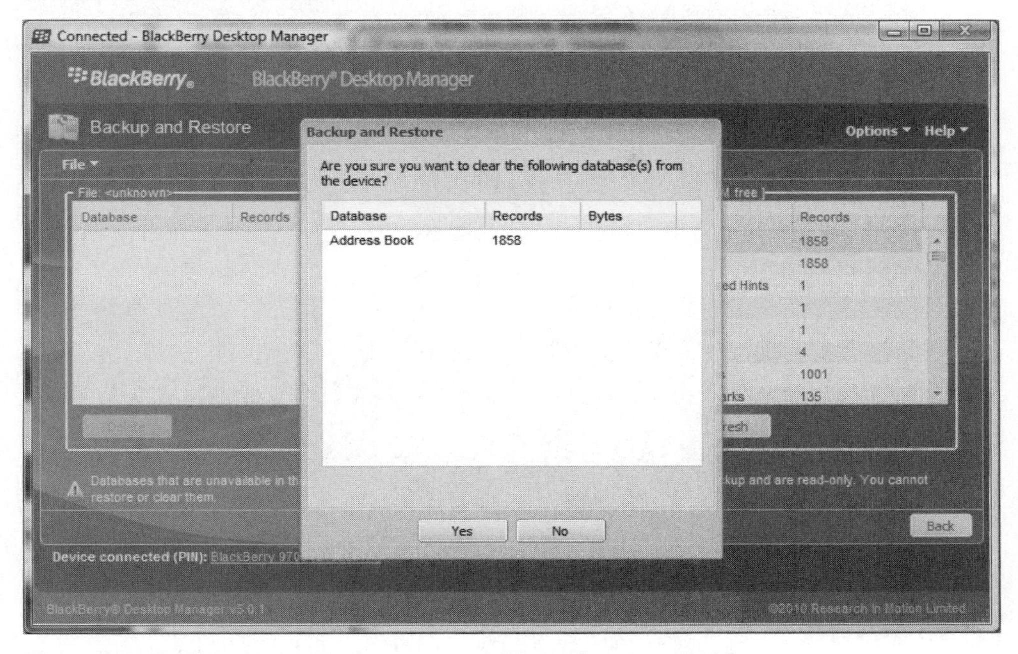

Figure 2-43: *Confirm you want to clear or erase databases from your BlackBerry*

How to Restore Specific Databases

In order to restore a file to your BlackBerry, you need to follow these steps and refer to Figure 2-44.

1. You need to have a backup file open in the left window. If the left window is blank, you need to click **File ➤ Open** in the upper left corner to open a backup file.

2. Click to select one or more database files in the left-hand window under **File: Backup-(year-month-day).ipd**.

3. Click the >> button in the middle to copy that information to your BlackBerry on the right side. In Figure 2-44, we have selected the **Calendar** database with 450 entries from the backup and are getting ready to copy it to the BlackBerry.

4. Start the restore process by clicking**Yes** on the next screen, which will look very similar to Figure 2-43.

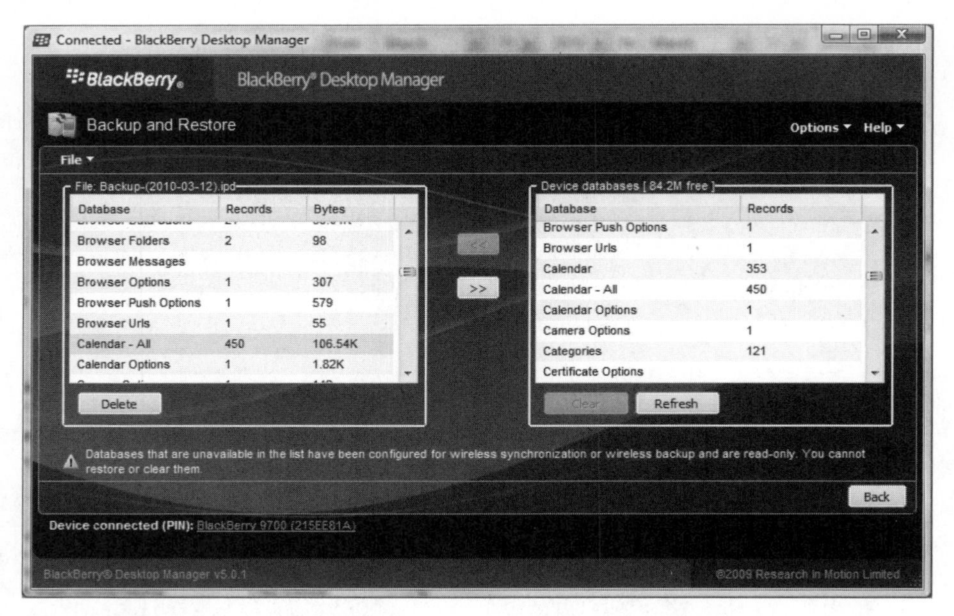

Figure 2-44. *Restoring specific databases from a backup file to your BlackBerry*

IP Modem

You may not see this icon in Desktop Manager because your wireless carrier may have disabled it or may provide separate software.

See the "Modem" chapter starting on page 451 to learn how to set up your BlackBerry as an IP Modem or Dial-up Internet connection for your laptop computer.

Chapter 3

Windows PC Media and File Transfer

In this chapter we will help you get your important files and media from your Windows computer to your Bold. Your Bold is quite a capable media player on which you can enjoy music, pictures, and videos. Like your computer, your BlackBerry can even edit Microsoft Office documents.

You have a variety of choices about how to transfer documents and media and we explore all the most popular ones in this chapter. You will quickly see that some methods such as Mass Storage Mode transfer work well for large numbers of files, whereas you will want to use the Media Sync program to transfer your music playlists. There are a few ways to load up media (music, videos, pictures) and Microsoft Office documents (for use with Documents to Go) onto your BlackBerry:

- Desktop Manager Media Manager (page 106)
- Desktop Manager BlackBerry Media Sync (page 114)
- Mass Storage Mode Transfer (page 144)
- Email the files to yourself as attachments (*if they are small enough)*

The **Mass Storage Mode** transfer method allows you to directly copy or drag-and-drop any file types to your BlackBerry when it looks like another disk drive to your computer. (you will see a drive C: normally, but when you connect your BlackBerry, you may see a new drive letter such as E:, F:, G:). We recommend this Mass Storage Mode transfer to copy your Microsoft Office documents (for use with Documents to Go) onto your BlackBerry into the **Documents** folder. See page 144.

More options are popping up all the time and will vary depending on who supplied your BlackBerry (for example, Verizon's Rhapsody MediaSync)

Before we begin, we strongly recommend that you install a Micro SD memory card into your BlackBerry handheld. With the price of memory coming way down, you can get a 2 GB, 4 GB, or even an 8GB Micro SD card for little money. Obviously, the bigger the card, the more media files you can store on your device. (See our chapter on Media Cards on page 359 to learn how to insert a media card)

Using the Media Manager (in Desktop Manager)

NOTE: Remember, you should use BlackBerry Media Sync instead of Media Manager if you want to sync iTunes or Windows Media playlists (see page 114).

To use the **Media Manager**, follow these steps.

1. Start BlackBerry Desktop Manager.
2. Plug in your BlackBerry Device with the USB Cable.
3. Click the **Media** icon.

4. Click the **Launch** button under the Media Manager icon.
5. You may see a license agreement that you need to accept before you can continue.

Scanning Your Computer for Media Files

When you start the Media Manager for the first time, it may ask if you want it to scan your computer for all music, video, and picture files that you could use on your BlackBerry device. This takes a while to do, but it is worthwhile if you have lots of pictures, music, and videos scattered over your computer. So click **Yes**

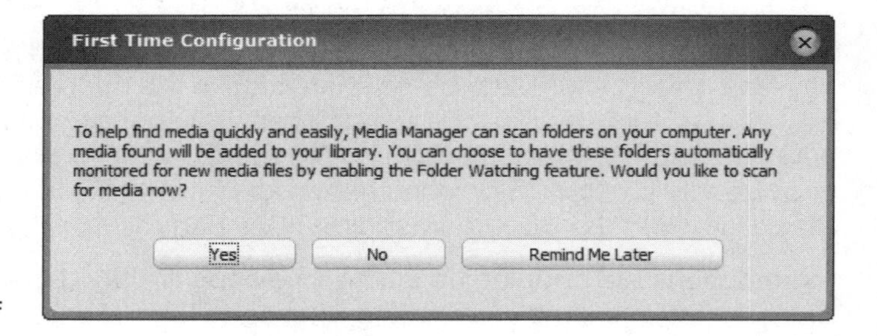

CAUTION: If you have a lot of media on your computer, this scanning process could take more than 10 minutes to complete.

After you do this, you can see that the **Media Manager** tells you exactly how many of each kind of file it contains. Under the icon for each type of media, you can click on **Manage Media** to rename, regroup, or organize your media.

Watched Folder Settings

You may see another window for Watched Folder Settings. These are folders that are scanned by the Media Manager to see if any changes have occurred (new songs, videos, pictures, and so forth) that should be synchronized with your BlackBerry.

Once you click **OK**, you will see the software scanning the selected Watched Folders and see a status window. You can **Pause** or **Cancel** the process if it takes too long.

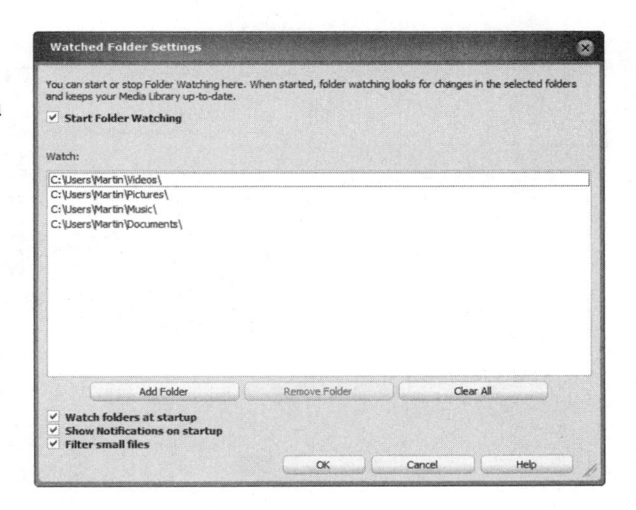

Enter Password

If you have a device password set on your BlackBerry, you will need to enter it before Media Manager can see the files stored on your BlackBerry.

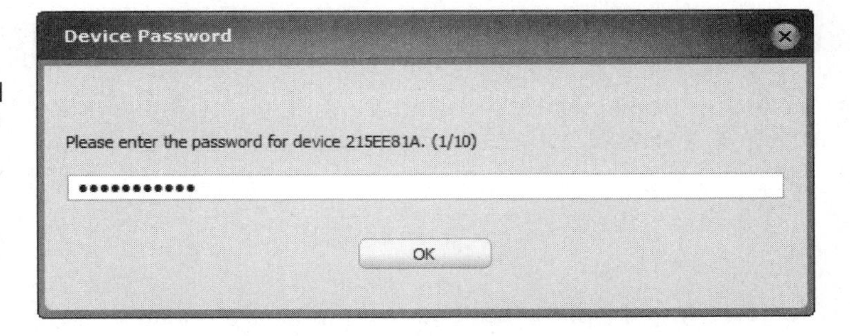

The Main Media Manager Window

Once the program loads, you will see a window similar to Figure 3-1.

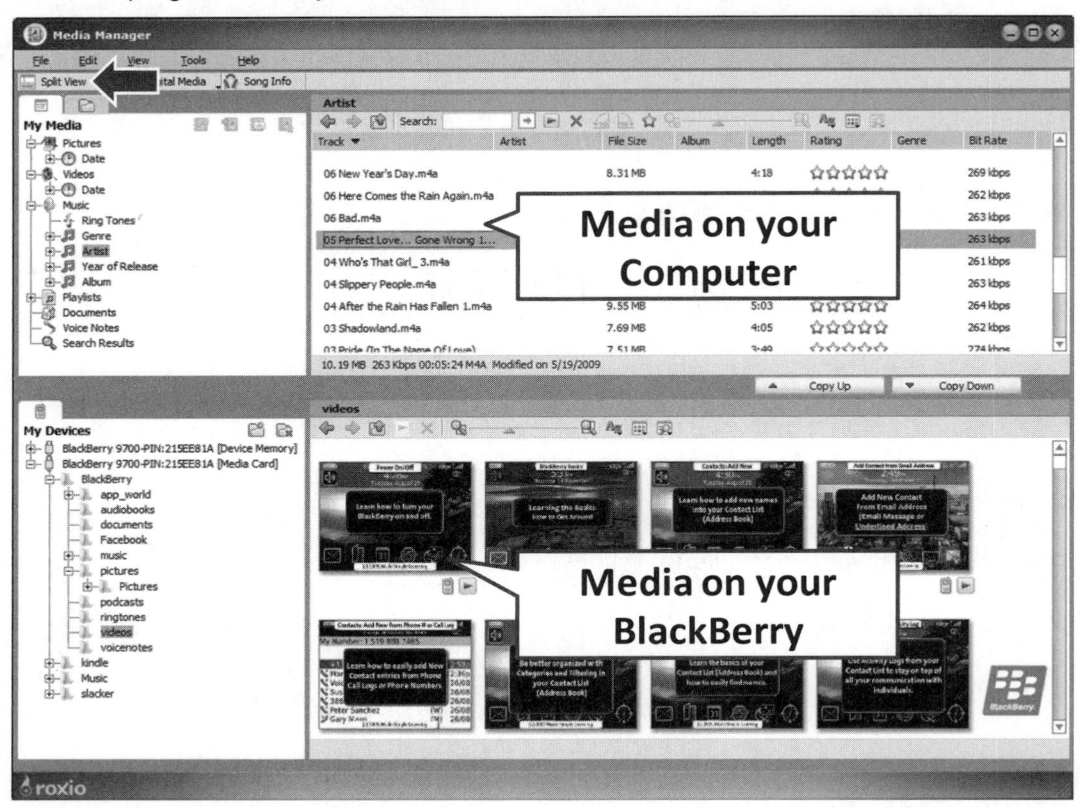

Figure 3-1. *Media Manager Main Window*

The **Media Manager** window show you the media files that are stored on your computer in the top half and on your BlackBerry and on your media card in the bottom half.

Make sure you have clicked **Split View** in the top left in order to see this view of both your computer and your BlackBerry.

You can drag the slider bar above the pictures/media to increase or decrease their size.

Locating Media on Your Comptuer

Use the top left portion of the Media Manager window to look for media you want to copy to your computer.

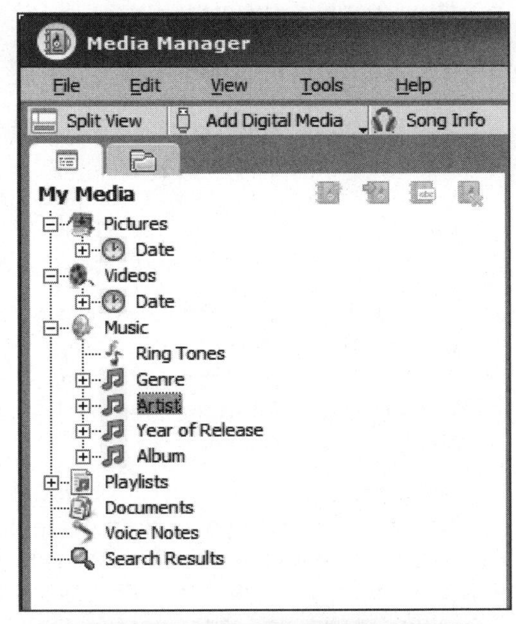

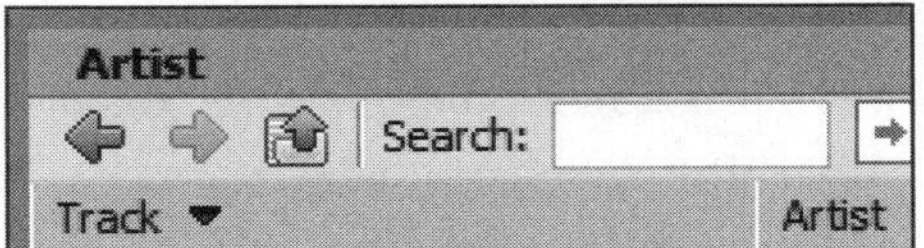

Click on the plus sign (+) next to any item to expand the view. You can use any of the following items to help you find media.

- Click on **Music** and then under that **Genre**, **Artist**, **Year of Release**, **Album**, or **Playlists** to further refine your view.
- Click on **Pictures** to view pictures and click on the **Date** item to narrow views by dates.
- Click on **Videos** to view your videos.
- Click on **Documents** to view documents compatible with your BlackBerry.
- Click on **VoiceNotes** to view voice notes.
- Type in a **Search** string in the top row to search for particular media.

Copy Music to Your BlackBerry

1. In the top half of the **Media Manager** screen, click on a folder to view your music or playlist.

2. In the bottom left corner of the **Media Manager** screen, click the plus sign (+) next to the BlackBerry with the **[Media Card]** at the end of it to see all the folders stored on your media card.

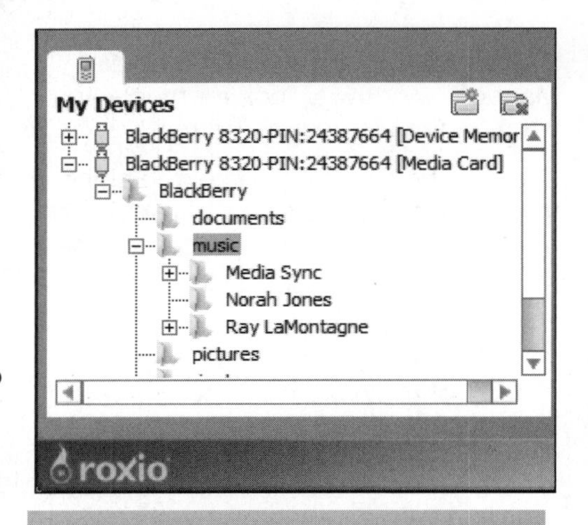

3. Click on the folder to which you want to copy your media–music, videos, ringtones, and so forth. In this case, because we want to transfer music, we click on the **music** folder.

4. Now, you should see the music you want to copy from your computer in the top half of the screen and the music folder from your BlackBerry media card in the bottom half of the screen.

NOTE: On your **Media Manager**, your will see *your* BlackBerry model number instead of the one shown.

5. Now, click to highlight songs, playlists, videos, or any other media in the top window.

6. Click the **Copy Down** button in the middle of the screen.

7. Repeat the procedure for more songs or any other type of media.

8. You may see a window asking you if you want the Media Manager program to copy the song and convert it for optimal playback on the BlackBerry–you can select either copy with conversion, without conversion, or look at advanced conversion options.

   We generally recommend letting the Media Manager convert your media for optimal playback on your BlackBerry. (However, this conversion may not work with videos, which are much more challenging to convert than music and are beyond the scope of this book.)

9. Select **OK** and the song (or songs) will now be copied onto your BlackBerry media card. Verify the copy by looking on the lower window and seeing the song on your *media card.*

Copying Pictures

The only difference between copying pictures and copying music is that on the **Media Manager** screen, in the top window, under **My Media**, just select **Pictures** and your pictures will be displayed in the top window. Make sure that down below, you collapse the **Music** menu and open up your **pictures** folder on your media card to ensure that your files will be copied to that folder.

Select your pictures (if you want more than one, just hold down the **Ctrl** key on your keyboard and then press and click each picture you want; they will all highlight.) Then, click the **Copy Down** button and let them be **converted** and they will go right on your media card (see Figure 3-2).

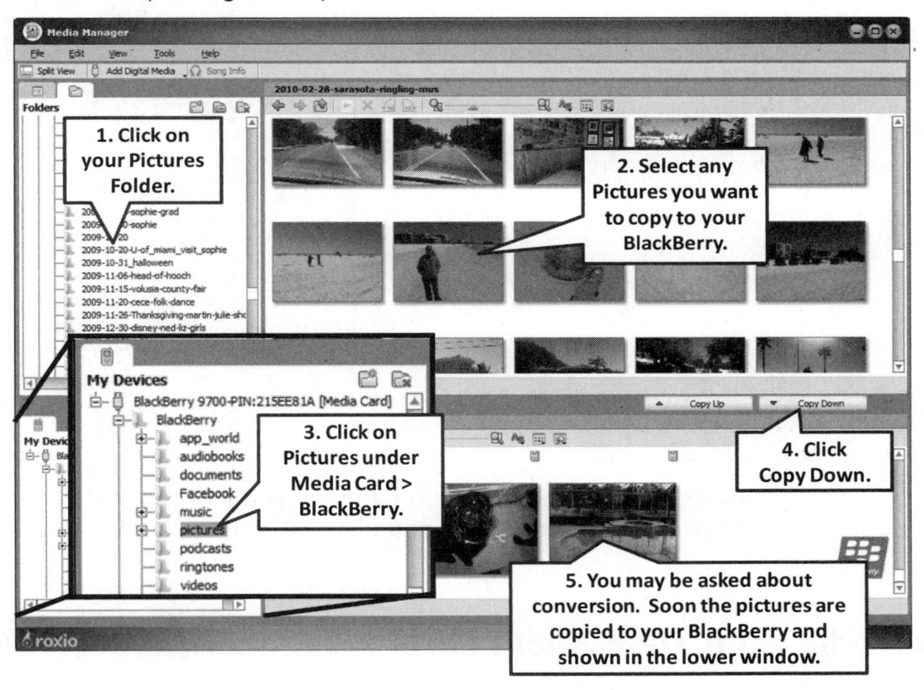

Figure 3-2. *Copying pictures and other items to your BlackBerry using Media Manager*

Copying (Microsoft Office) Documents

Repeat the steps for copying pictures and music, but in the top window under **My Media,** just select **Documents** and your documents folders on your computer will be displayed in the top window. Navigate to the correct folder for your particular documents.

Also, make sure that down below, you open up your **documents** folder on your media card to ensure that your files will be copied to that directory.

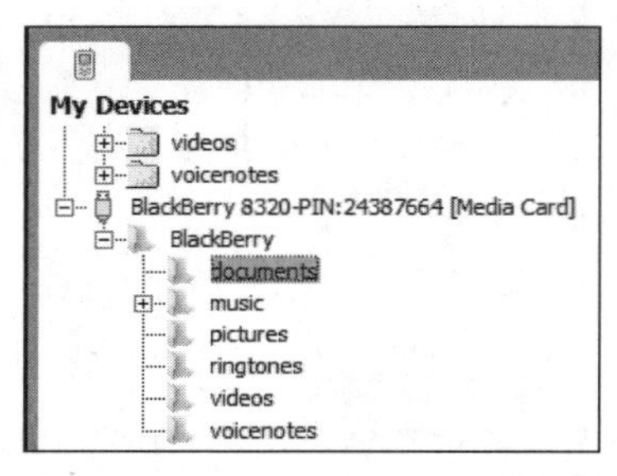

Just select your files in the top window. Then draw a box around the files, or click on one and press **Ctrl** + **A** (Windows) to select them all. Or hold the **Ctrl** key (Windows) down and click on individual files to select them.

Then, just click the **Copy Down** button and for pictures let them be converted (for Documents, let them be Copied with no Conversion) and they will go on to your media card.

Deleting Media from Your Media Card

You can use Media Manager to free up space on your media card. We first recommend copying or backing up the items you will delete. Use the drag–and-drop methods described previously to copy items from your media card to your computer (see Figure 3-3).

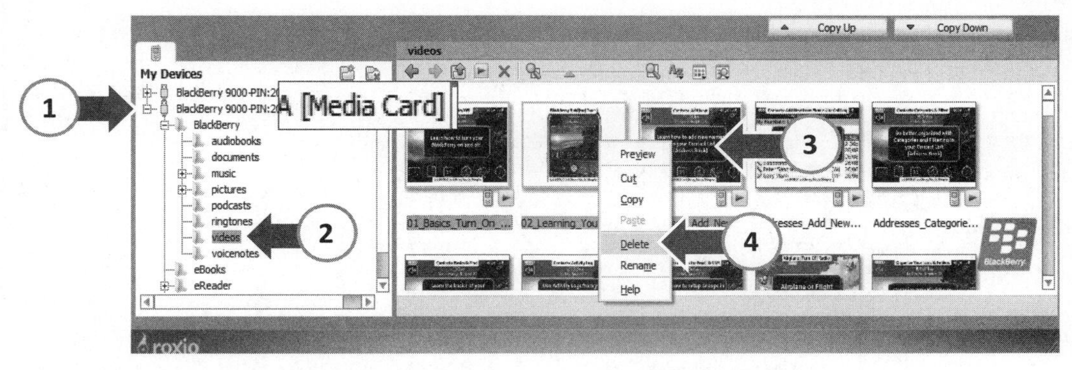

Figure 3-3. *Selecting and deleting media from your BlackBerry with Media Manager*

1. Click the + sign next to the **BlackBerry nnnn-PIN:xxxxxxx [Media Card]** in the lower left window to see all the folders on the media card.
2. Then click on a folder from which you wish to delete media on the media card.
3. Select the items, using **Ctrl + click** or **Shift + click** to select a list of items.
4. Once the items to delete are selected, press the **Delete** key on your keyboard, or right-click and select **Delete**.

TIP: Videos and songs will usually be the largest items on your BlackBerry–deleting these items will free up more space than deleting individual pictures.

Troubleshooting Media Manager

Sometimes when you are previewing a file or performing some other function, Media Manager might crash and stop responding. If this is the case, you can stop the program by following these steps:

1. On your Windows computer, press three keys simultaneously: **Ctrl + Alt + Del**.
2. If you are given a choice, then select **Start Task Manager**.
3. From the **Window Task Manager** (Figure 3-4), click on the **Processes** tab.
4. Scroll down and highlight the Image Name of **MediaManager**, as shown.
5. Click the **End Process** button at the bottom.

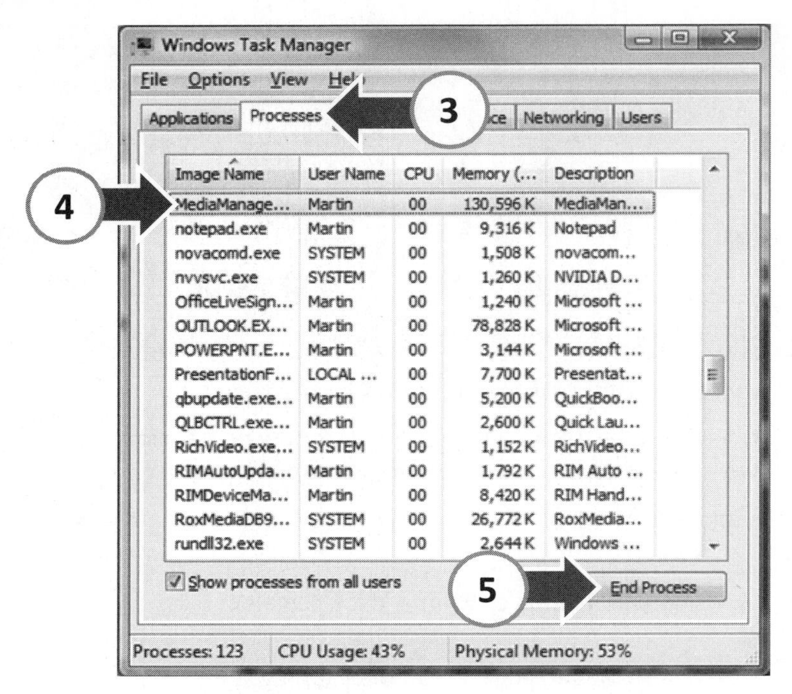

Figure 3-4. *Windows Task Manager*

6. On the next screen that says **Do you want to end this process?**, click **End process** to stop the program.

7. Now, you can restart the program and try again.

BlackBerry Media Sync

NOTE: You should use the BlackBerry Media Manager instead of the Media Sync if you want to sync non-iTunes media and you need to convert music and video to be viewable on your BlackBerry (see page 106).

Perhaps the easiest way to get music into playlists (and now your album art) is using the BlackBerry Media Sync program. If you are an iTunes user and you have playlists already in your iTunes program, the Media Sync program allows you to transfer those playlists directly to your BlackBerry.

TIP: Running Media Sync Separately from Desktop Manager

Normally, you just launch Media Sync from inside Desktop Manager; however, you can download and run Media Sync separately from BlackBerry Desktop Manager. To do this:

Open up a web browser on your computer and go to: `www.blackberry.com/mediasync`

Click the **Download for PC** link.

Once you have the file downloaded, just run the installation program. A window will appear letting you know the application has been installed properly.

To launch the application, just go to **Start→All Programs→BlackBerry Media Sync** and click on the icon. Make sure that your BlackBerry is connected via the USB cable to your computer–but **don't** have Desktop Manager running when you do this.

Start Media Sync from Desktop Manager

To use the **Media Sync** from within Desktop Manager, follow these steps.

1. Start BlackBerry Desktop Manager.
2. Plug in your BlackBerry Device with the USB Cable.
3. Click the **Media** icon.

4. Click the **Launch** button under the BlackBerry Media Sync icon. You may need to accept a license agreement to continue.

5. After clicking **Launch**, you may see a window telling you an update is available. Click the **Download** button and follow the steps to install the updated software.

NOTE: If a song is in iTunes and is DRM-protected (see page 121), then it is NOT possible to sync it to your BlackBerry.

Enter BlackBerry Password for Media Sync

If you have set a password to protect your BlackBerry, then you will need to enter your password to continue.

Figure 3-5. *Media Sync Password screen.*

Media Sync Setup

When you first start Media Sync, it may show you a setup screen similar to Figure 3-6.

1. Change the name of your device, if you like.
2. Select where your media should be stored; leave this on your media card (see page 359 to learn about media cards).

3. Use the slider bar to keep more or less space free after the sync. The default is 10% and should be fine.
4. Click on **iTunes** or **Windows Media Player** for where you store your music.
5. Click **OK** to continue.

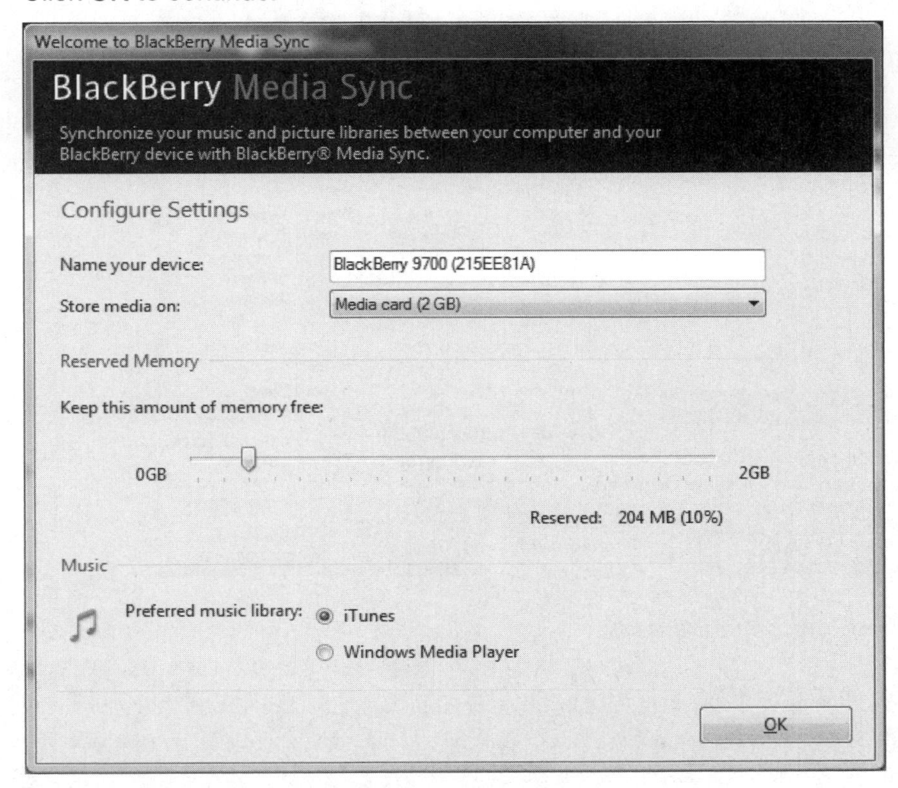

Figure 3-6. *Media Sync options (device).*

After your initial **Configure Settings** screen is complete, you should then see a screen similar to the one shown in Figure 3-7.

Media Sync—Syncing Music

You are now ready to set up your Music Sync. To do so, follow the next set of steps.

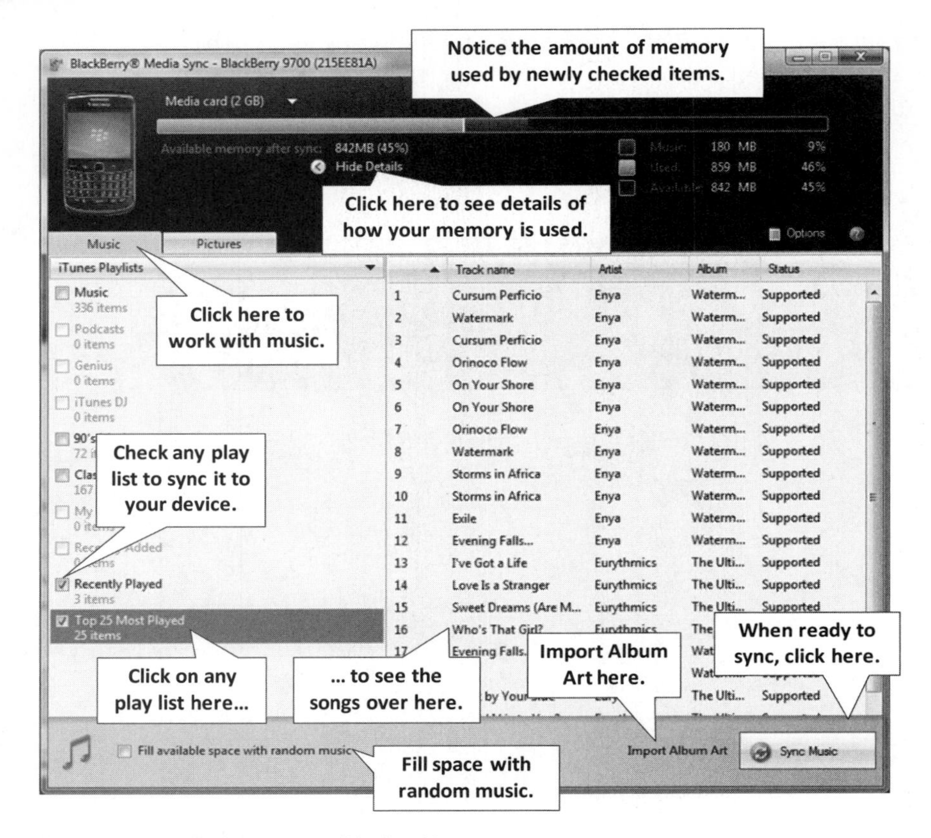

Figure 3-7. *Media Sync screen (Music tab).*

NOTE: At the top of Figure 3-7, you see a Bold 9700 model with a 2GB media card; you will not see this on your computer; instead, you will see the name and model number of *your* BlackBerry as well as the size of *your* media card.

Make sure to click on the **Music** tab in the upper left corner as shown in Figure 3-7 to configure your music sync.

TIP: To get back to the **Configure Settings** screen we just completed, just click the **Options** button in the upper right corner of Figure 3-7 and click on the **Device** tab at the top (see Figure 3-5).

To import your Album Art, just click the **Import Album Art** link next to the Sync Music button in the lower right corner of Figure 3-3. If you are importing from iTunes, iTunes will automatically start automatically.

NOTE: If iTunes has a dialog box open when it automatically starts, you will have to close out the dialog box and re-try the **Import Album Art** button.

Figure 3-8. *Status screen for importing Album Art in BlackBerry Media Sync*

Once the import is done, you will see a little pop-up window saying it is **Finished importing Album Art**.

To select a playlist to sync to your device, just check the box next to the playlist. When you check it, watch the **Memory Bar** at the top to make sure that you have not **Exceeded Available Memory** with your selections. If you have, just uncheck the playlists until you get back under 100% of memory usage.

If you want to fill the available space with random music, then place a check next to the box in the lower left corner that says **Fill available space with random music**.

In order to see the details of what is occupying the space on your memory card, click the **Show Details** button underneath the Available memory after sync number in the middle upper part of the window.

Once you are done with your selection of playlists, click the **Sync Music** button in the lower right corner. You will see the sync status in the upper portion of the window.

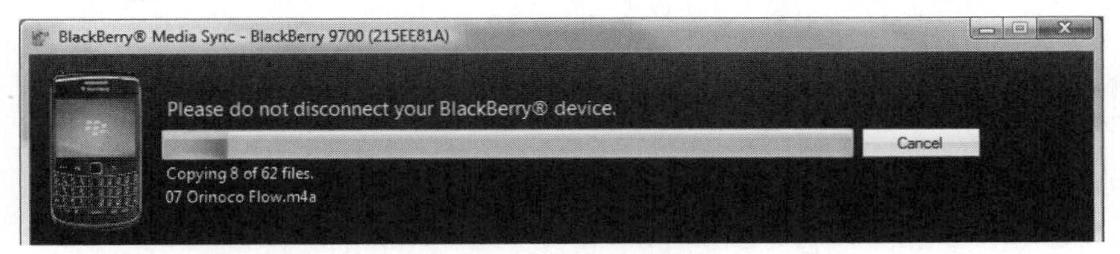

Figure 3-9. *Status screen for syncing files to the BlackBerry*

Media Sync—Syncing Pictures

If you would like to transfer or sync pictures between your BlackBerry and your computer, click the **Pictures** tab in the upper left corner, as shown in Figure 3-4.

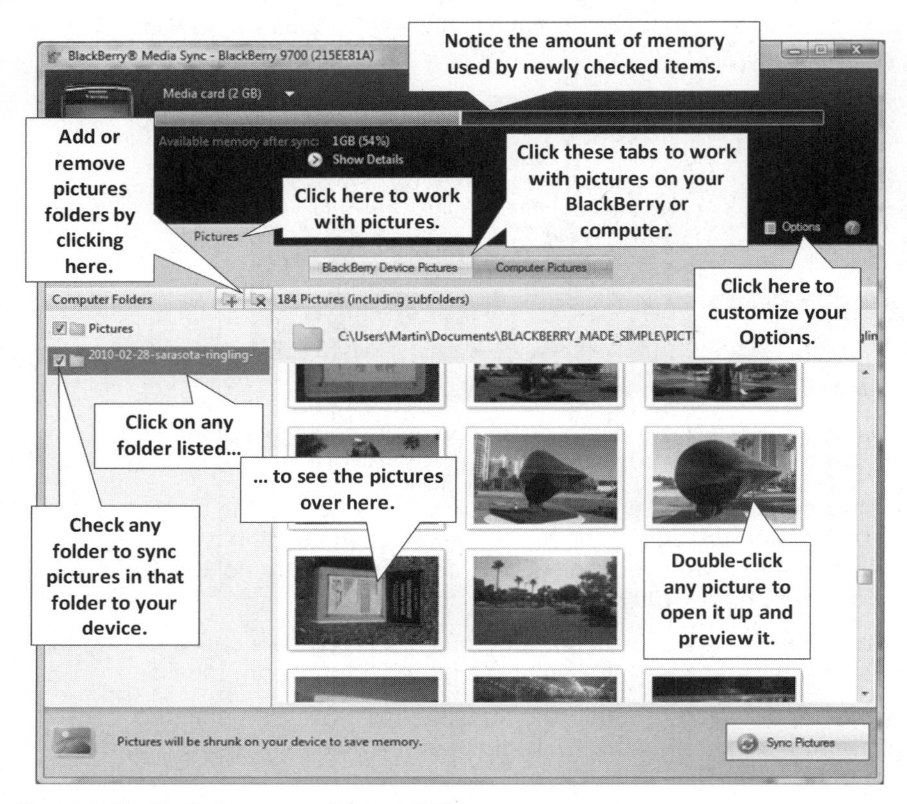

Figure 3-10. *Media Sync screen (Pictures tab)*

To switch between viewing pictures on your computer and on your BlackBerry, click the buttons in the middle upper section of the screen (Figure 3-4).

To add or remove picture folders on your computer, click the folder + and – icons at the top of the folder column.

Once you have folders listed, then click on a folder to display all pictures in that folder.

Check the box next to the folder to sync all the pictures in that folder to your device. If you have previously synced a folder, you can uncheck it to remove those pictures from your device to save space.

When you check each pictures folder, watch the **Memory Bar** at the top to make sure that you have not **Exceeded Available Memory** with your selections. If you have, just uncheck the folders until you get back under 100% of memory usage.

In the lower left corner of Figure 3-4, you see the text **Pictures will be shrunk on your device to save memory**. That is a setting you can change by clicking on the **Options** link in the upper right corner. See Figure 3-5 and related descriptions for help.

In order to see the details of what is occupying the space on your memory card, click the **Show Details** button underneath the Available Memory after sync number in the middle upper part of the window.

Once you are done with your selection of pictures, click the Sync Pictures button in the lower right corner. Your sync status will be shown in the top portion of the window.

Some Songs Could Not Be Synchronized

Look carefully at the **Synchronization complete** message (Figure 3-11). If some songs are protected, then they will not be synced to your BlackBerry. In this example, 35 songs were protected.

Figure 3-11. *Synchronization complete message*

To see the list, click the **Click to view a list of those songs** link. Then you should see a window similar to Figure 3-12. Click the plus sign (+) next to **Protected** to see all the protected songs.

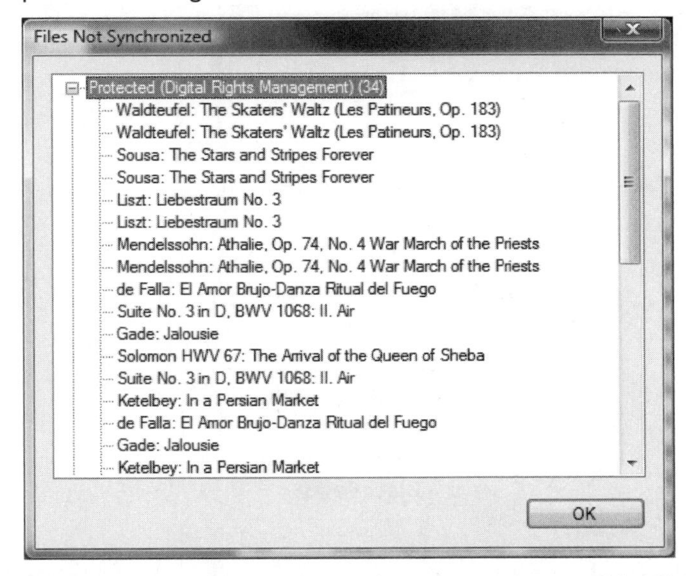

Figure 3-12. *Viewing songs that could not be synced due to Digital Rights Management protection*

DRM Protected–Digital Rights Management

Now, it is important to remember that some music that is purchased on iTunes contains **DRM (Digital Rights Management)** restrictions–that means that most iTunes music can only be played on iPods and through iTunes. Most likely, these are older songs purchased before the middle of 2009.

Any other music you might have put in your iTunes library–such as CDs you loaded into your computer or music that does not have DRM restrictions—will transfer into the appropriate playlist. Make sure that you don't disconnect your BlackBerry while the music is transferring.

Once the Sync is done, close out the **Media Sync** window. Now jump to page 363 to learn how to use music on your BlackBerry.

Media Sync Options Screen

To see the **Options** screen (Figure 3-13), click the **Options** button in the upper right corner of the main **Media Sync** window (Figure 3-10). Notice there are two tabs at the top: **General** and **Device**.

General Tab

Click the **General** tab at the top of the **Options** screen to see the screen shown in Figure 3-13.

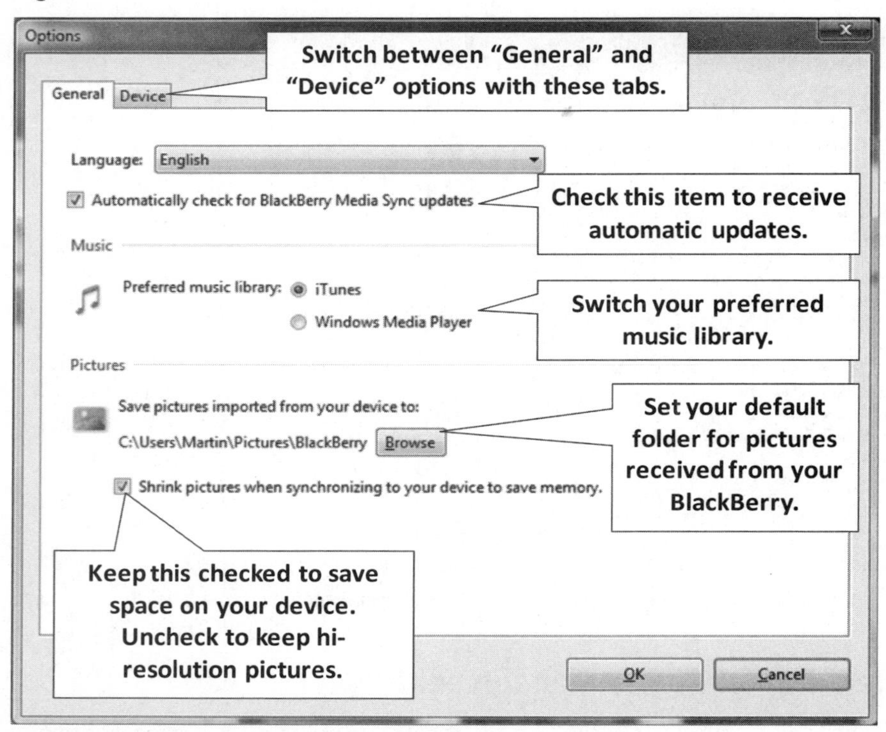

Figure 3-13. *Media Sync Options (General tab)*

On the **General** tab, you can do the following:

- Select your **language** from the drop-down list.
- Select **Automatically check for Media Sync Software updates** by checking the box (it is checked by default).
- Select your **Preferred music library**. If you have changed from **iTunes** to **Windows Media Player** or vice-versa, you can select either here.
- Select the folder to store pictures that are transferred from your BlackBerry to your computer by clicking the **Browse** button.
- You can also decide how high the resolution should be for pictures you sync to your BlackBerry. The default is to **Shrink pictures when synchronizing to your device to save memory.** Uncheck it if you want higher-resolution pictures and aren't worried about the extra storage space required.

Device Tab

Click the **Device** tab at the top to see settings related to your device, as shown in Figure 3-14 below.

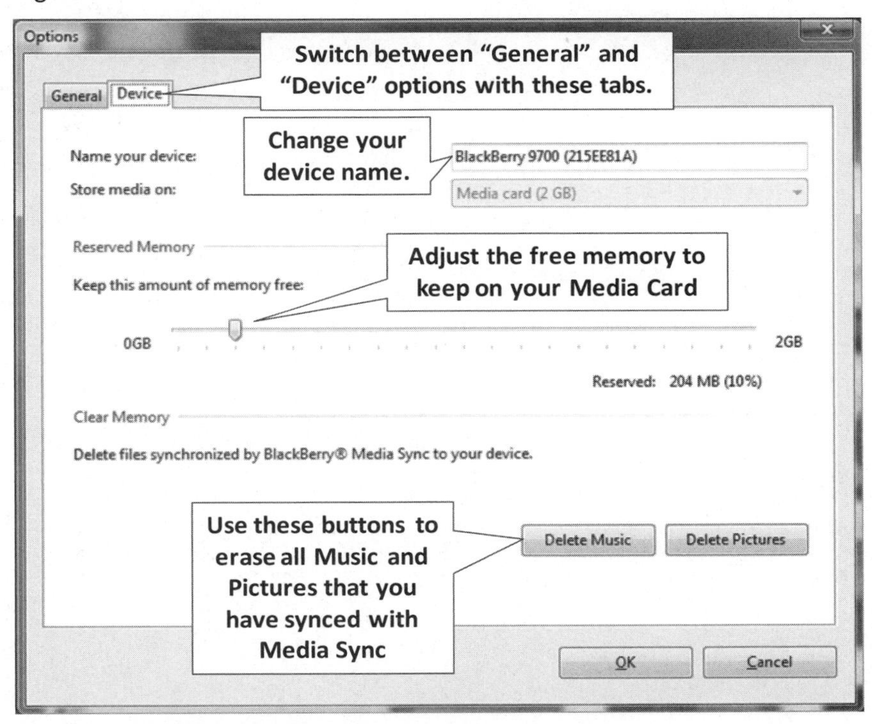

Figure 3-14. *Media Sync Options (Device tab)*

On the Device tab, you can do the following.

- Change your **Device Name** for your BlackBerry.
- Change the amount of **Reserved Memory** on your device after the sync. Use the **slider bar**.
- Erase all the music and pictures synced to your BlackBerry by clicking the **Delete Music** or **Delete Pictures** buttons.

Chapter 4

Apple Mac Setup

This chapter shows you how to install the new BlackBerry Desktop Manager Software on your Apple Mac computer and do the basics of synchronizing your contacts, calendar, tasks and memos, backup and restore, and more. If you want to transfer files and media with your Mac, then check out our chapter on transferring files for Mac users found on page 141. (You may need some of the instructions in this chapter on how to install Desktop Manager if you want to use it as your method to transfer files.)

Have a **Microsoft Windows** computer? Please go to page 65.

Do you use more than one computer (such as at work, home, and so forth)?

> **CAUTION:** Make sure to check the **sync with other computers** option if you plan on syncing your BlackBerry with multiple computers using Desktop Manager. Otherwise, you could end up corrupting your BlackBerry and/or computer databases!

Do you want a wireless, two-way automated sync?

Try using Google Contacts, Google Calendar, and Google Sync for your BlackBerry. All of these are free applications and give you a full two-way wireless sync.

BlackBerry Desktop Manager for Mac

For years, Windows users have enjoyed the seamless synchronization of their contacts, calendar, notes, and tasks with their PC via the BlackBerry Desktop Manager Software. Now, for the first time, this "piece of mind" that comes with knowing your data is fully backed up is available to the Mac user.

If you are a Windows user, this will seem familiar, yet very much streamlined for the Mac. If you have never used BlackBerry Desktop Manager, you will now be able to not only synchronize your data, but you will be able to back up, restore, sync your iTunes playlists, and more.

Download and Install Desktop Manager for Mac

Desktop Manager for Mac software is available for free from BlackBerry.com.

1. Open up your web browser and go to the download page (Figure 4-1): `http://na.blackberry.com/eng/services/desktop/mac.jsp`
2. Fill out the required information on the download page and then click to download the software.

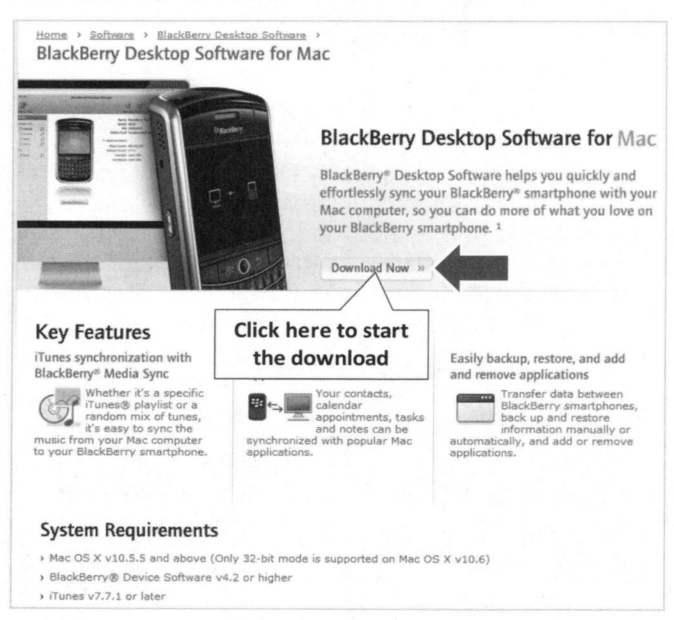

Figure 4-1. *Locating the download file on BlackBerry.com web site*

3. Once the file is downloaded, you will be presented with the screen shown in Figure 4-2.

Figure 4-2. *Starting the installation*

4. Double click the **BlackBerry Desktop Manager.mpkg** file to open the installation program.

5. Your Mac will display a **caution** message similar to the one shown in Figure 4-3.

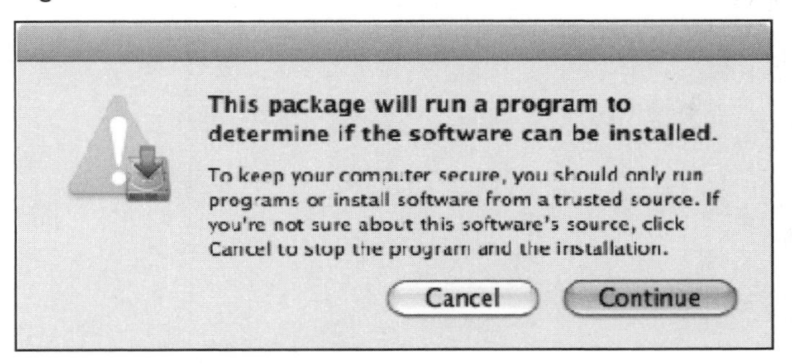

Figure 4-3. *Mac installation* ***caution*** *screen*

6. Select **Continue** to move forward with the installation process.

NOTE: If you have been using either Pocket Mac or The Missing Sync to synchronize your BlackBerry with the Mac, you will receive another **caution** note letting you know that in order to proceed, the connection between your BlackBerry and the third-party synchronization software will need to be discontinued.

7. If you already has some other software installed, you will see another **caution** message similar to Figure 4-4. Click **Install Anyway** to move forward with the installation process.

Figure 4-4. *Additional warning screen about additional BlackBerry software*

8. Follow the on-screen prompts as your Mac installs the new Desktop Manager Software (see Figure 4-5).

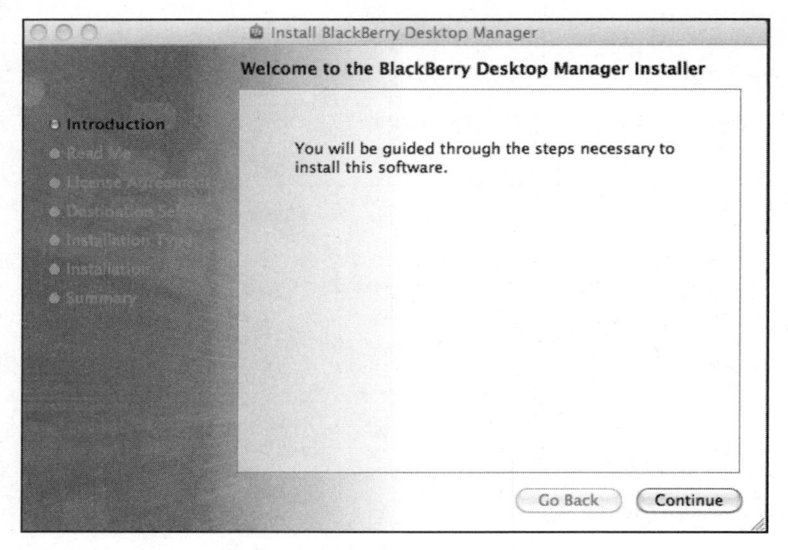

Figure 4-5. *First installation screen*

9. Click **Read License** to read the software license or click **Agree** to proceed (see Figure 4-6).

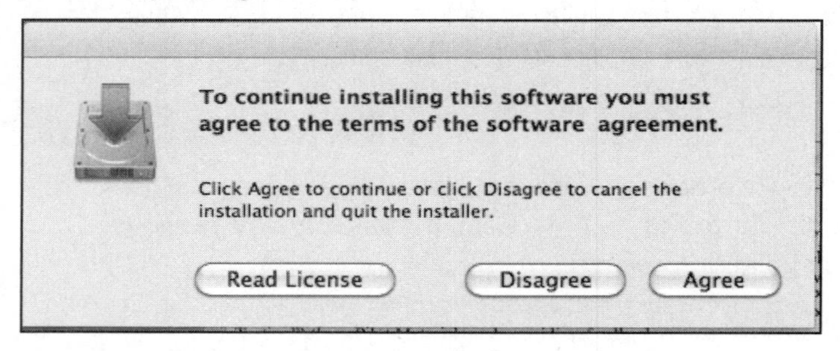

Figure 4-6. *Installation license agreement pop-up window*

10. For most Mac users, only one drive will be shown (see Figure 4-7), but it is possible that you might have more than one possible location for the install. Choose the correct drive and click **Continue**.

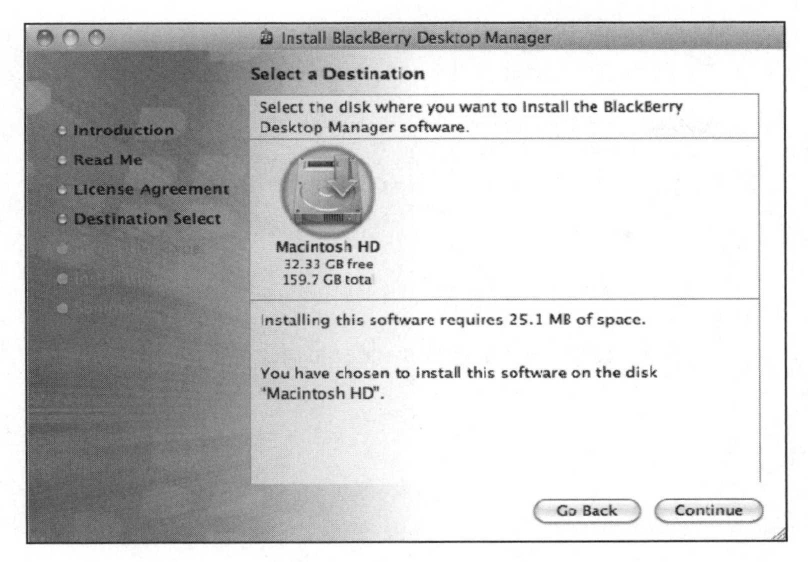

Figure 4-7. *Choose location of the installation*

11. If you have a **Password** set on your Mac, you will be prompted to enter it at this time in order to proceed with the installation (see Figure 4-8).

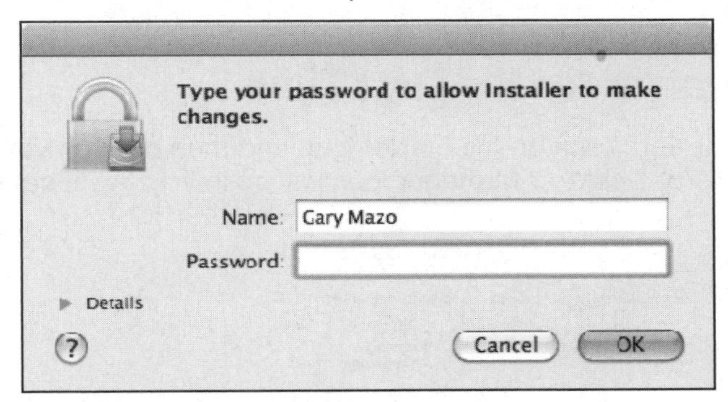

Figure 4-8. *Password required to complete installation*

12. You will be asked to restart your computer when the installation is complete–agree to this by clicking **Continue Installation** (see Figure 4-9).

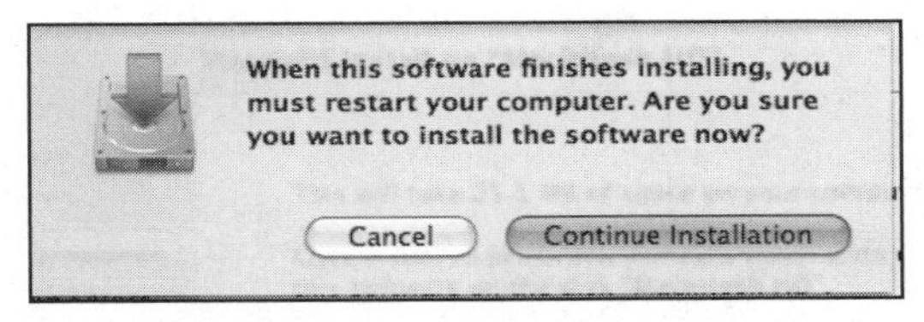

Figure 4-9. *Request to restart after installation.*

13. When the installation is complete, before the restart, you should see the screen shown in Figure 4-10 indicating that the software was installed successfully.

Figure 4-10. *Successful software installation screen*

Starting Desktop Manager for the First Time

To locate the **Desktop Manager** app, click on the **Finder** icon and then click on your **Applications** icon. The **BlackBerry Desktop Manager** icon will be in your **Applications** directory.

Figure 4-11. *Locating BlackBerry Desktop Manager for Mac*

Double-click the **BlackBerry Desktop Manager** icon and the **Welcome** screen will appear, showing you information about your particular BlackBerry (see Figure 4-12).

On this **Welcome** screen, you can adjust your device options.

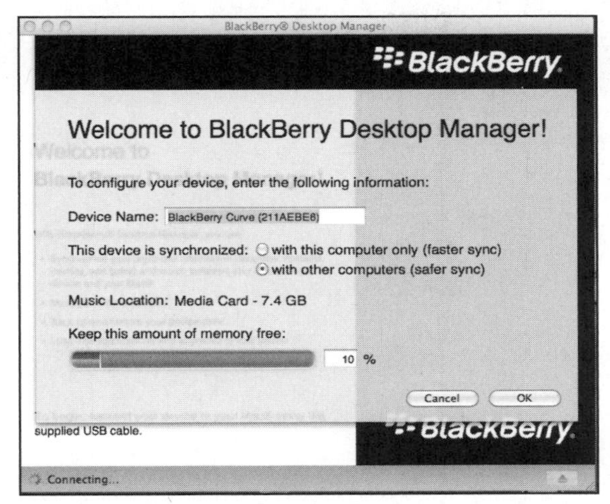

Figure 4-12. *BlackBerry Desktop Manager for Mac Welcome screen*

On this screen, you should select how to synchronize your BlackBerry in the **This Device is Synchronized** field.

If you synchronize your BlackBerry with other computers, a network server, or Google Sync for Calendar or Contacts, select **with other computers (safer sync)**.

CAUTION: If you want to sync with more than one computer, then you will need to be sure to check this option:

with other computers (safer sync)

OTHERWISE, YOU COULD END UP CORRUPTING YOUR BLACKBERRY AND/OR COMPUTER DATABASES!

If you are only planning on syncing your BlackBerry with this one Mac, you can choose **with this computer only (faster sync)**.

Main View in Desktop Manager

Desktop Manager will show you a picture of your BlackBerry device and a clean interface displaying information along the left-hand bar and commands along to top bar (see Figure 4-13).

NOTE: This picture shows a BlackBerry 8900; on your screen, you will see your own BlackBerry device.

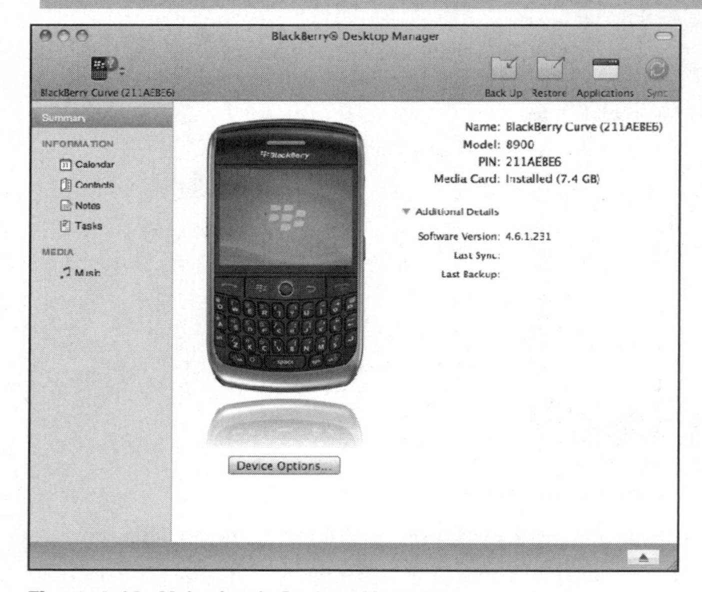

Figure 4-13. *Main view in Desktop Manager*

Using Desktop Manager for Mac

One of the first things you will notice is the **Device Options** button below the picture of your BlackBerry (see Figure 4-13).

Clicking on this brings you to the **Device Options** screen, the same screen you saw when you first started Desktop Manager (see Figure 4-14).

Figure 4-14. *Device Options in Desktop Manager for Mac*

Device Options

You can rename your device if you like (the default is simply your model and PIN number.) If you want to automatically sync your device each time you connect, just check the first check box. If you want to see a desktop icon showing your BlackBerry as an external disk, just check the **Show as disk on the desktop** box. Once you do this, you can easily transfer files between your Mac and your BlackBerry using the drag-and-drop method you have used on your Mac to copy between folders and disks.

Device Is Synchronized with One or Several Computers

Here you have the option of choosing whether you BlackBerry syncs only with this Mac or with other computers. If your BlackBerry syncs with your PC and with your Google account, you should select the **with other computers (safer sync)** option to avoid duplicating entries in your contacts and calendars.

Backup Options

Click the **Backup icon** along the tabs on the top of the **Device Options** screen (Figure 4-15).

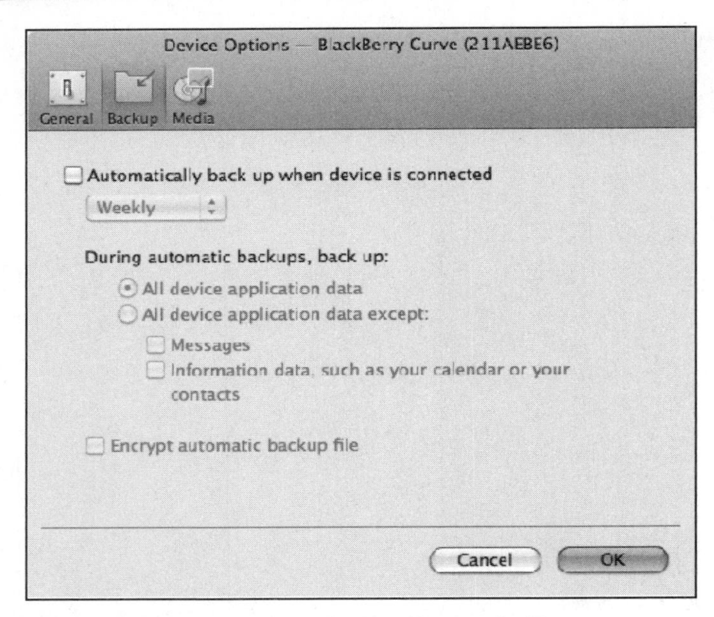

Figure 4-15. *Backup options in Desktop Manager for Mac*

If you want to create a backup each time you connect your device, just check the **Automatically back up when device is connected** box. You can then specify exactly what you wish to be backed up. We cover the backup options a little later.

Set Up Your Sync Options

Click **OK** or **Cancel** to return to the main screen of Desktop Manager (Figure 4-13). Look at the left column under where it says **INFORMATION** (see Figure 4-16).

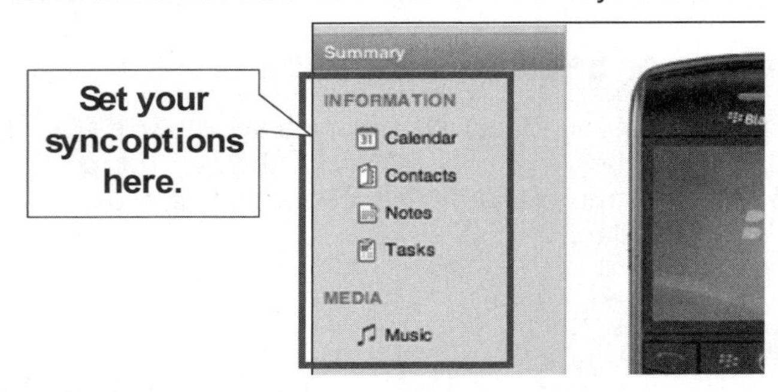

Figure 4-16. *Sync options in Desktop Manager for Mac*

This is where you set the sync options for the Calendar, Contacts, Notes, and Tasks.

1. Click any of the items below INFORMATION. In this case, we will start with the **Calendar** to be taken to the sync setup screen.
2. This screen (Figure 4-17) has a similar look and feel as the calendar sync screen within iTunes for those who are familiar with syncing an iPhone or iPod touch and a Mac.

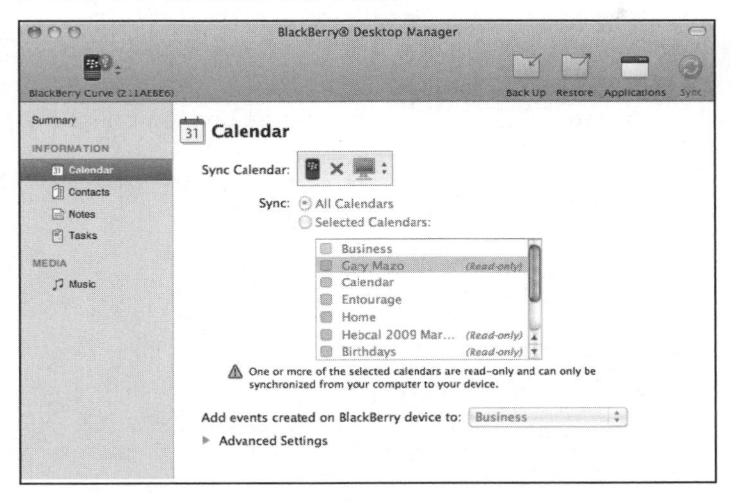

Figure 4-17. *Calendar sync options screen*

3. Desktop Manager will notice all the calendars you have on your BlackBerry. In this example, we use Google Calendar and have many different calendars, all set to a unique color. You see this in the list of calendars.
4. Click on the box with the **Sync Calendar** option. The red **X** on this icon shows that the BlackBerry calendar will not be synced with the Mac calendars.

5. If you click on the box and select the next item in the drop-down, the picture will change to show that now you desire a two-way sync between the Mac Calendar and the BlackBerry calendar.

6. You can also select into which calendar you want events to go that we create on your BlackBerry. The default on this one is my Business calendar, but you can change that to any calendar that you have set up on the device.

Add events created on BlackBerry device to: Business

Advanced Settings

Click on the **Advanced Settings** tab [Advanced Settings] at the bottom and the options shown in Figure 4-18 are revealed.

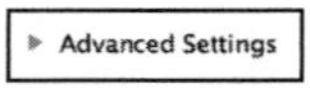

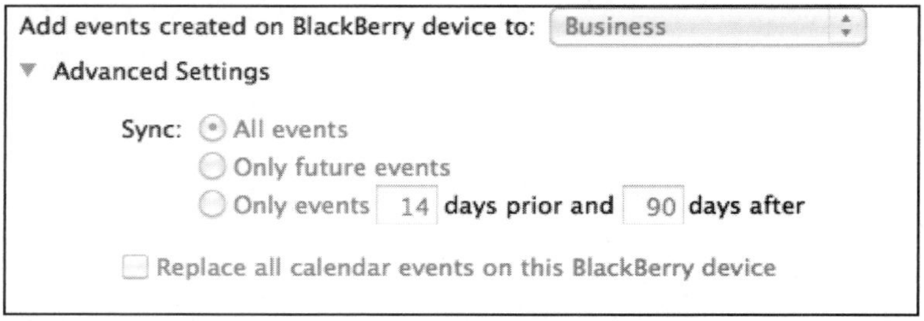

Add events created on BlackBerry device to: Business
Advanced Settings
Sync: All events
Only future events
Only events 14 days prior and 90 days after
Replace all calendar events on this BlackBerry device

Figure 4-18. *Advanced sync settings in Desktop Manager for Mac*

Like Desktop Manager for the PC, you can specify as to whether you want to sync **All events**, **Only future events**, or you can set individual parameters for synchronization.

> **TIP:** We recommend not selecting **Only future events** unless you have a strong reason to do so. Say you made some notes to an event that was held yesterday in the BlackBerry Calendar **Notes** field. If you select **Only future events** here, these important notes would not be transferred to your Mac.

To replace all calendar events on the BlackBerry with events from your Mac's calendar, just click the check box at the bottom of the screen.

Syncing Contacts, Calendar, Notes, and Tasks

The procedure for setting the sync options for the **Contacts**, **Tasks**, and **Notes** is identical to what we just did. The only settings that change are the groups or events to choose within each category (Figure 4-19).

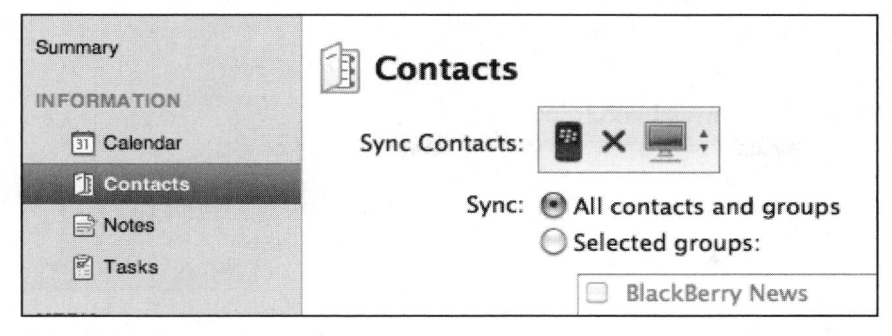

Figure 4-19. *Sync Contacts setup screen*

In this screen, you can see that you can click next to **Sync Contacts** just as we did previously and choose either to not sync with the Mac or to perform a two-way sync.

You can then select to sync either **All contacts or groups** or only **Selected groups** from your Address Book.

Automating the Synchronization

In order to have your BlackBerry sync every time you connect it to your Mac, you will need change a setting on the **Device Options** screen.

1. Click the **Device Options** button below the picture of your BlackBerry on the main screen (see Figure 4-20).
2. Then check the check box next to **Automatically sync when device is connected**.

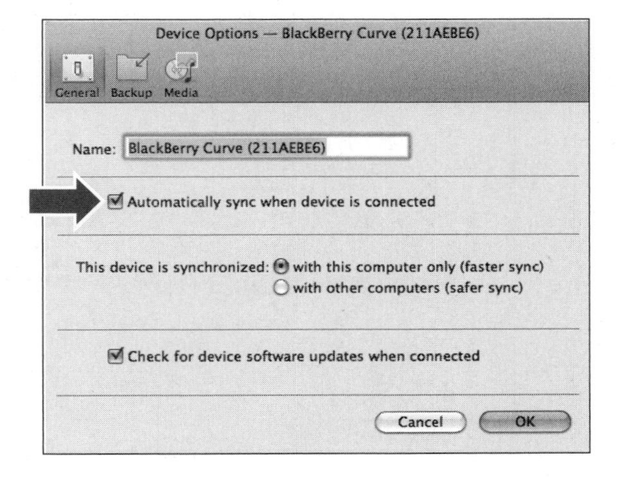

Backup and Restore

One of the great features now available to Mac users is the ability to **Backup** and **Restore** either your entire BlackBerry contents or just selected information on your Mac. **Backup** and **Restore** begins with the two icons at the top of the main screen in Desktop Manager.

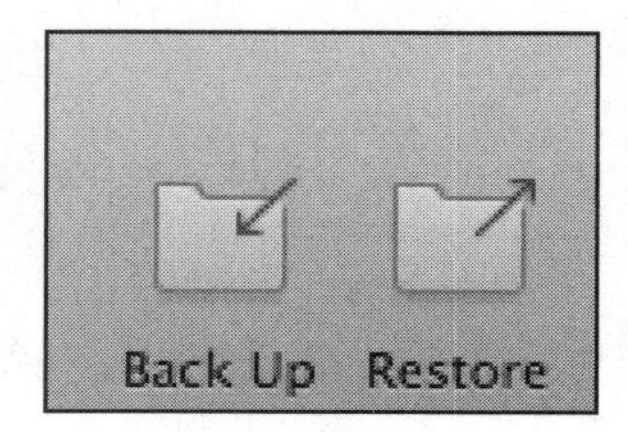

Using Backup

1. Click on the **Backup** icon and you will be taken to the screen shown in Figure 4-20, where you specify exactly what information you wish to be backed up on your Mac.

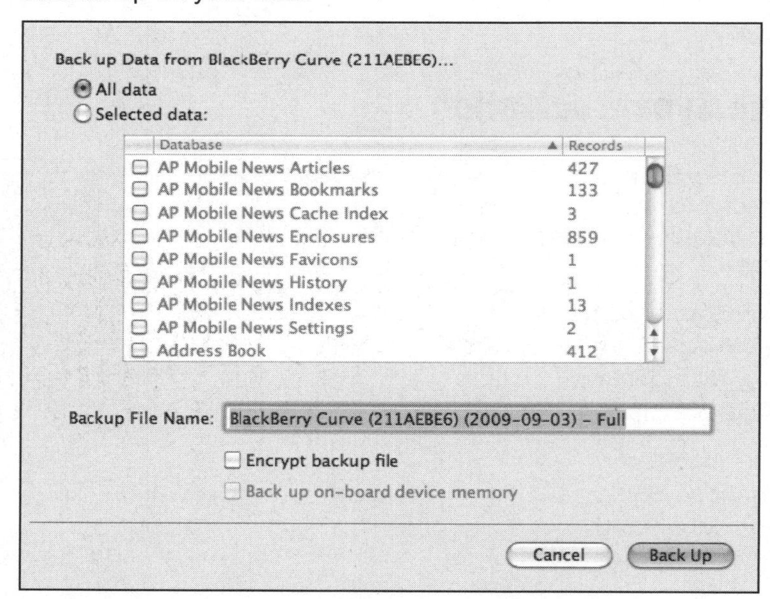

Figure 4-20. *Backup screen in Desktop Manager for Mac*

2. Select either **All data** or **Selected data** and then choose exactly which items you wish to back up.
3. Let's say that you are only really concerned with backing up your **Contacts**, **Calendar**, and **Notes**. Just check off each of those boxes and your backup will complete much faster.
4. You can specify the name of your backup for easy retrieval in the future.
5. Once you have made all your backup selections, click the **Back Up** button and the progress of the backup will be displayed in a dialog box.

Restoring from Backup

We all know that sometimes, unexplained things happen and we lose information in our BlackBerry. Maybe we try to update the OS and make a mistake or maybe we sync with other computers and the information gets corrupted. Now, Mac users have a reliable and safe way to restore data on their devices.

1. Click the **Restore** icon along the top row of the main screen in Desktop Manager. You will then be taken to the **Restore** options screen (Figure 4-21).

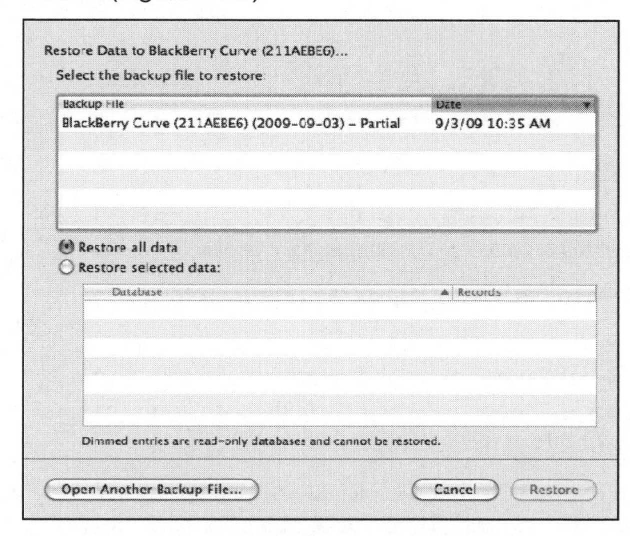

Figure 4-21. *Restore screen in Desktop Manager for Mac*

2. If you have made a backup file on your Mac (which is required so you have a file you can restore), it will be shown in the top box under **Backup File**. If you have multiple backup files, they will all be listed here.

3. Select the file from which you wish to restore information (or, if you did a selective backup, selected data will be displayed in Figure 4-22).

4. Click the **Restore** button and your BlackBerry will be restored just as it was when you made the backup file.

Add or Remove Applications

For the first time, Mac users are now able to add or remove applications on their BlackBerry from the Desktop Manager environment.

1. Click on the **Applications** icon (in the upper right corner of the main screen) and you will be taken to the **Applications** screen in Desktop Manager (Figure 4-22).

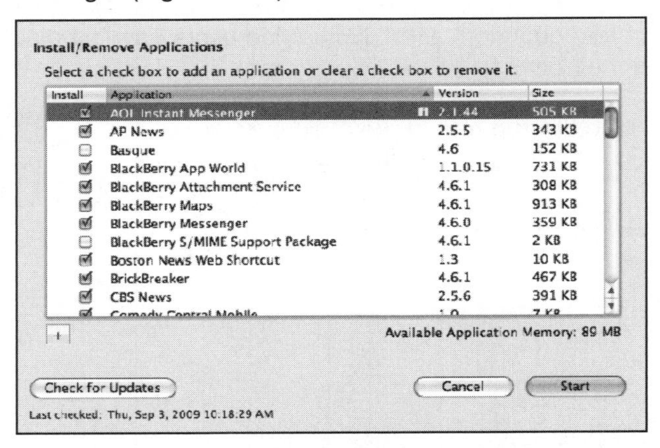

Figure 4-22. *Install/Remove Applications window in Desktop Manager for Mac*

2. To check for updates, click the **Check for Updates** button in the lower left-hand corner.

3. Place a **check mark** in any application that isn't checked already, and it will be installed on your device. Conversely, uncheck any box, and that application will be removed from your BlackBerry.

4. Click the **Start** button and the selected or deselected applications will be either installed or uninstalled, depending on your selections.

TIP: For peace of mind, it is always a good idea to perform a backup both before and after you add or delete applications from the device.

Update Your BlackBerry Device Software

When BlackBerry has issued a new version of device software for your BlackBerry smartphone, you will see an alert window in Desktop Manager when you connect your device to your Mac. This window will tell you that a new version of device system software is available and ask you if you would like to update your BlackBerry. To update your device, click OK or Update depending on the screen shown.

Setting Up the BlackBerry as a Modem for Your Mac

See our Tethered Modem chapter starting on page 451 for help with setting up your BlackBerry to connect your Mac to the Internet as a dial-up modem.

Chapter 5

Apple Mac Media and File Transfer

Your BlackBerry can be a great media player. In order to get all your songs, videos, and other media on your BlackBerry, you'll need to learn some of the information in this chapter.

In this chapter we show you how to load up media (music, videos, pictures) and Microsoft Office documents (for use with Documents to Go) onto your BlackBerry:

- Use BlackBerry Desktop Manager for Mac software
- Use Mass Storage Mode transfer

Syncing Media with Desktop Manager for Mac

Start up Desktop Manager for Mac as shown on page 451 and make sure your BlackBerry is connected to your Mac with a USB cable. Click on the third icon along the top called **Media**. (Figure 5-1).

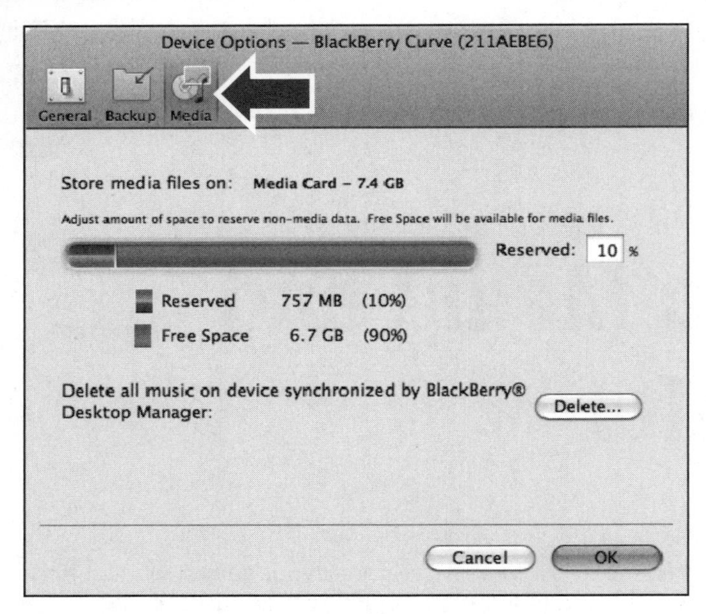

Figure 5-1. *Media sync in Desktop Manager for Mac*

By default, Desktop Manager will reserve 10% of the space on your media card for non-media data. You can adjust this amount in the **Reserved** box. The smaller the number you input, the more space you will have for media files on the media card.

Delete All Music

Click on the **Delete** button under the **Delete all music on device...** statement and you can remove any or all music that is stored on your device.

Why would you want to do that? Let's say that you have been dragging and dropping music on your BlackBerry (which was one of the only options for Mac users unless you were using Pocket Mac or the Missing Sync). Or, let's say you were using a program like the Missing Sync (which would sync iTunes playlists, but did not bring in the album art) .

You now have the option of syncing your iTunes playlists complete with album art. So, you might want to start fresh and get rid of the other music on your BlackBerry.

Syncing Music

BlackBerry Desktop Manager allows you to sync your iTunes playlists right onto the media card of your BlackBerry. Just click the **Music** icon in the **Media** section along the left-hand column of the main screen.

You will then be taken to the **Music** sync screen. There are some very nice options, as shown in Figure 5-2.

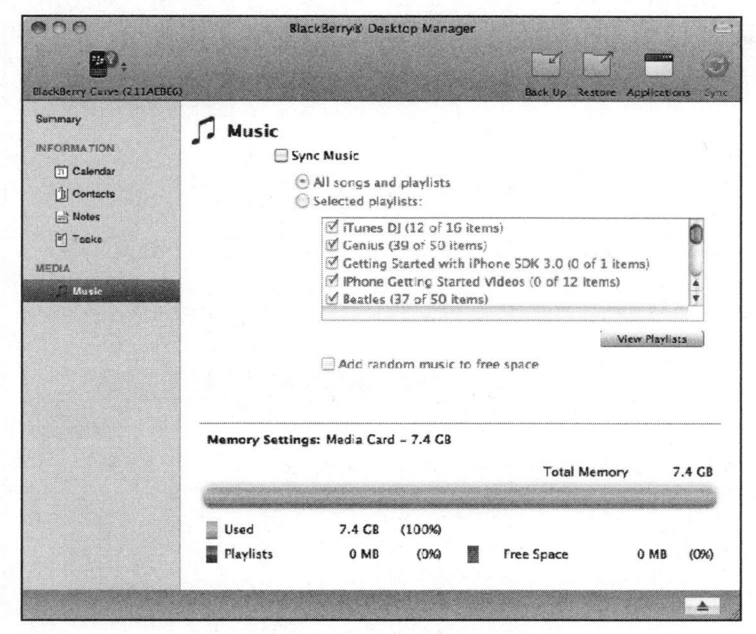

Figure 5-2. *Sync Music screen in Desktop Manager for Mac*

Like iTunes, you can choose to sync **All songs and playlists** or **Selected playlists**. Just place a check mark in the **Sync Music** box at the top of the screen and you can then select which playlists you wish to sync between your Mac and your BlackBerry.

Place a check mark in the **Add random music to free space** box and additional songs will randomly be placed on to the media card.

On the screen shown in Figure 5-3, you can see that I just selected four playlists that I wanted on my BlackBerry by placing check marks in the appropriate boxes.

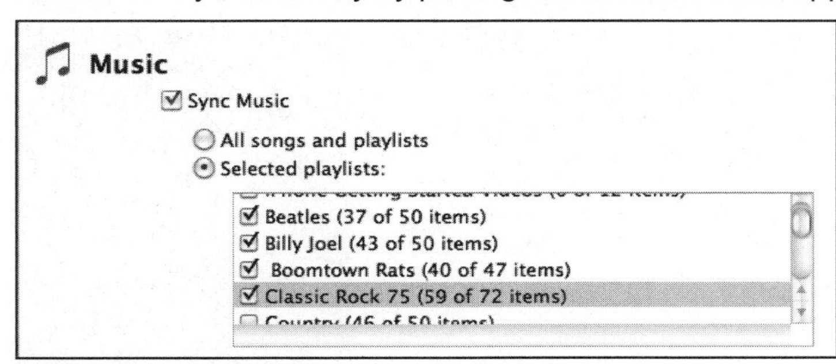

Figure 5-3. *Sync specific playlists in Desktop Manager for Mac*

Click the **Sync** icon at the top right of the screen to perform the music sync to your BlackBerry.

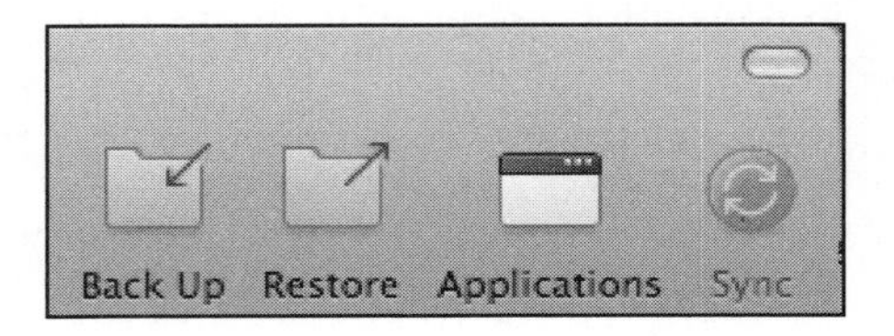

Mass Storage Mode Transfer for Your Media Card

This works whether you have a Windows or a Mac computer. We will show images for the Mac computer process, and it will be fairly similar for a Windows PC. This transfer method assumes you have stored your media on a MicroSD media card in your BlackBerry.

To get to this screen, click your **Options** icon, then scroll down and click on **Memory.**

Make sure your media card **Mass Storage Mode Support** is **On** and other settings are as shown.

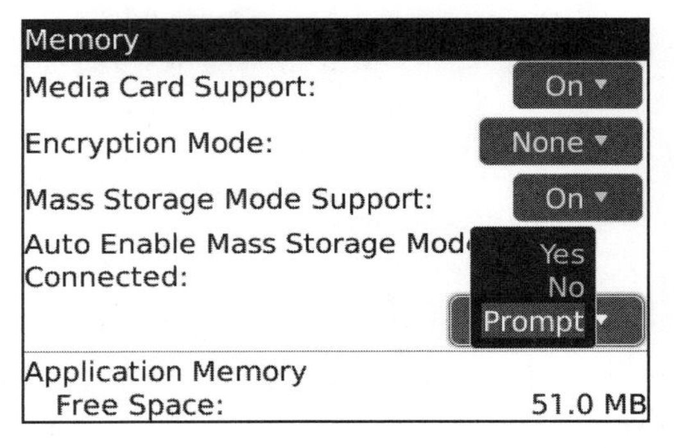

Now, connect your BlackBerry to your computer with the USB cable. If you selected **Prompt** for **Auto Enable Mass Storage Mode**, you will see a question similar to this: **Turn on Mass Storage Mode?.**

Answer **Yes** (you should probably check the box that says **Don't Ask Me Again**). When you answer Yes, then your media card looks just like another hard disk to your computer (similar to a USB Flash Drive).

TIP: If you set the **Auto Enable Mass Storage Mode** setting to **Yes**, then you won't be asked this question; the media card on the BlackBerry will automatically look like a mass storage device (disk drive letter).

Using Your BlackBerry in Mass Storage Mode

NOTE: You will need to install Desktop Manager for Mac in order to be able to use this **Mass Storage** option. This is because there are drivers required to connect your BlackBerry to your Mac.

Once connected, your Mac will see your BlackBerry as a mass storage device and mount it as an external drive. The Mac will be able to identify what kind of BlackBerry you have and put the name of the BlackBerry in the icon on your desktop.

The BlackBerry will also be visible if you click on your **Finder** icon in the dock. It will be listed under **Devices**.

Explore the Drive

Right-click on the icon for the BlackBerry and choose **Open**, or double-click on the desktop icon and open the drive (see Figure 5-4).

Now, you can explore your BlackBerry as you would any drive. You can copy pictures, music, or video files by just dragging and dropping to the correct folder, or you can delete files from your BlackBerry by clicking on the appropriate folder, selecting a file or files, and dragging them to the trash.

NOTE: Your music, video, ringtone, and picture files are located in the folder called BlackBerry.

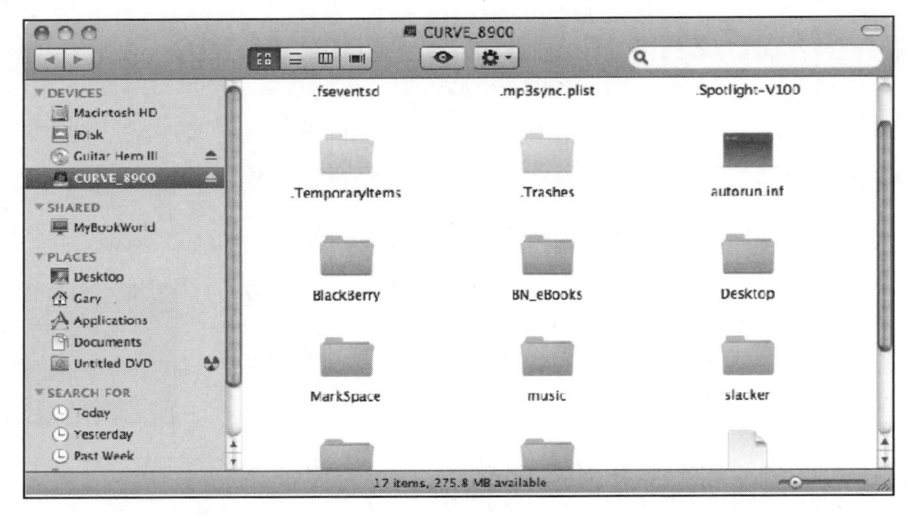

Figure 5-4. *Finding media on the Bold*

Copying Files Using Mass Storage Mode

After your BlackBerry is connected and in Mass Storage mode, just open up your computer's file management software. On your Mac, start your **Finder**. Look for another hard disk or BlackBerry (model number) that has been added. On Windows, open up **Windows Explorer**.

> **NOTE:** You will see your own BlackBerry model number, such as BOLD_9700 or BOLD_9650.

When you plug your BlackBerry into your Mac, it will identify the main memory and the contents of the Micro SD card as two separate drives and place them right on your desktop for easy navigation.

On your Mac, click on the **Finder** icon in the lower left-hand corner of the Dock. On Windows, click on the disk drive letter that is your BlackBerry media card.

You will see your **DEVICES** (including both BlackBerry drives) on the top and your **PLACES** (where you can copy and paste media) on the bottom.

To copy pictures (or other items) from your BlackBerry, follow these steps:

1. Select the pictures from the **BlackBerry/pictures** folder using one of the following methods:
   a. Draw a box around some pictures to select them.
   b. Click on one picture to select it.
   c. Press **Command+A** (Mac) **or Ctrl+A** (Windows) to select them all.
   d. Press the **Command** key (Mac) or **Ctrl** key (Windows) and click to select individual pictures.
2. Once selected, **Ctrl+click** (Mac) or **right-click** (Windows) on one of the selected pictures and select **Cut** (to move) or **Copy** (to copy). A status bar with then open, as shown in Figure 5-5.

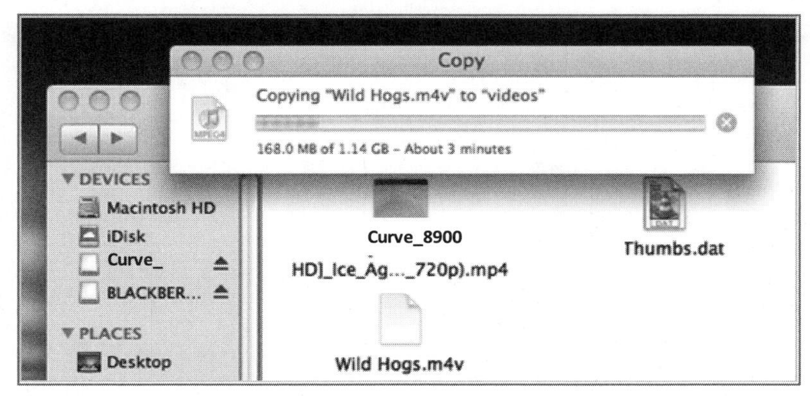

Figure 5-5. *Status of Copying Files to your BlackBerry.*

3. Press and click on any other disk/folder on your computer, such as like **My Documents**, and navigate to where you want to move/copy the files.
4. Once there, **right-click** (or **Ctrl+click**) again in the right window where all the files are listed and select **Paste**.

You can also delete all the pictures/media/songs from your BlackBerry in a similar manner. Navigate to the BlackBerry/(media type) folder, such as **BlackBerry/videos.** Press the key combination shown earlier on your computer keyboard to select all the files then press the **Delete** key on your keyboard to delete all the files.

You can also copy files from your computer to your BlackBerry using a similar method. Just go to the files you want to copy and select them. **Right-click** (or **Ctrl+click**) and click **Copy** and then paste them into the correct BlackBerry/(media type) folder.

> **Important:** Not all media (videos), pictures (images), or songs will be playable or viewable on your BlackBerry. Use Desktop Manager for Mac to transfer the files and most files will be automatically converted for you.

Chapter 6

Typing, Spelling, and Help

In this chapter we look at the **trackpad,** help you type as fast and accurately as you can on your small BlackBerry keyboard, and show you where to get help on your Bold.

> **TIP:** If you haven't already done so, please check out our "Quick Start Guide" in the front section of this book (page 3) for a picture of what every key does on your BlackBerry—and some great navigation tips and tricks.

The Trackpad

One great navigation feature on your BlackBerry is the front **trackpad**, which can glide in any direction as well as be clicked. Though this may take some getting used to for seasoned BlackBerry users (who are accustomed to the familiar trackball or side trackwheel), you'll quickly see that the **trackpad** gives you lots of freedom to scroll up and down and left to right using your thumb. It is incredibly intuitive to use.

Furthermore, clicking in the **trackpad** brings up an innovative **Short Menu** that's context-sensitive. Sometimes, it is so sensitive it almost seems as if it's reading your thoughts. Try it: Type an email, then click the **trackpad**. Notice that **Send** is highlighted so you can simply click the **trackpad** again to send your email.

You'll also find some great **trackpad** features you can adjust to suit your own preferences.

Trackpad Sound and Sensitivity

We're all different in how quickly we like to navigate the **Home** screen and how sensitive we want the **trackpad** to be. Here's how you can adjust **trackpad** sensitivity.

1. Click on the **Options** icon.
2. Scroll down and click on **Screen/Keyboard**.
3. Scroll down until you see **trackpad**, with three adjustable fields underneath.
4. Click on the **Horizontal Sensitivity** or **Vertical Sensitivity** number. The default is 70; higher is more sensitive, lower is less.
5. Click next to **Audible Roll** to set **Click** or **Mute. Click** produces an audible sound when you move the **trackpad**, **Mute** makes **trackpad** movement silent.
6. Press the **Short Menu** key and select **Save**.

TIP: You may find that your trackpad moves too quickly up or down lines. If so, then reduce the **Vertical Sensitivity** to 50 or lower.

Typing Tips for Your Bold

Now let's look at a number of tips that can help you become more proficient with your Bold. Be sure to check out the Quick Start Guide at the beginning of this book for even more tips and tricks, as well as many images to help understand some of the topics covered in this chapter.

Press and Hold for Automatic Capitalization

Here's an easy way to get uppercase letters as you're typing: simply press and hold the letter to capitalize it.

Caps Lock and Num/Alt Lock

If you want to lock the **Caps** key to TYPE ALL UPPERCASE, press the **Alt** key, then press the right **Shift** key. Tap the either **Shift** key to turn off **Caps** lock.

To type only numbers or the symbols shown on the top of the keys, turn on **Num/Alt** lock by pressing the **Alt** key and then the left **Shift** key. Tap the either **Shift** key to turn off **Num/Alt** lock.

Automatic Period and Cap at End of Sentence

At the end of a sentence, just press the space key twice to get a "." (period). The next letter you type will be automatically capitalized.

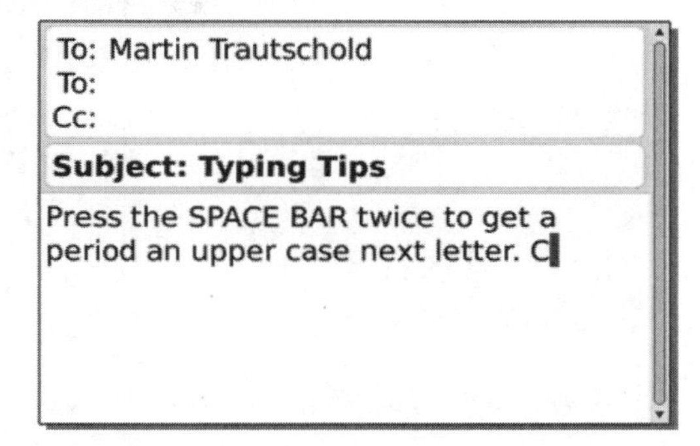

Typing Symbols: Using the Alt and Sym Keys

There are two ways to enter symbols on your BlackBerry. Several symbols are shown on the top of each key, and you can access these by holding the **Alt** key while pressing the key with the symbol you want. The other set of symbols—not shown on the keyboard—are accessed by pressing the **Sym** (Symbol) key. Both keys allow you to quickly add symbols to your text.

Suppose you want to use a **#**, for example. You just press and hold the **Alt** key while pressing the **Q** key, as shown in Figure 6-1.

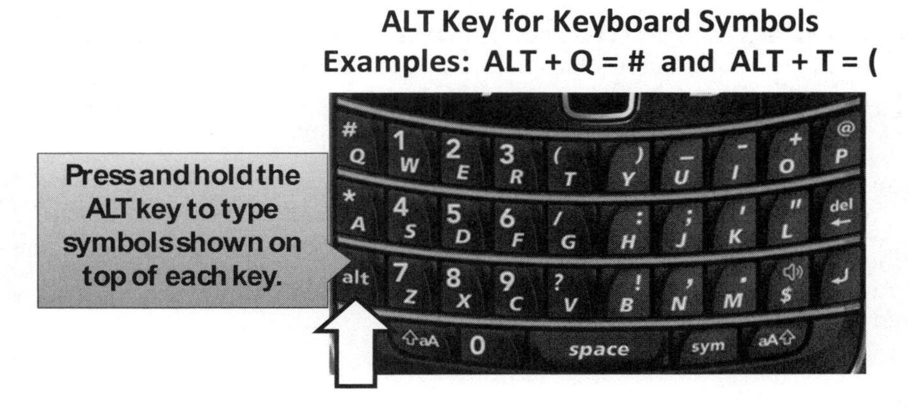

Figure 6-1. *Use the Alt key to access the symbols shown on the top of each key.*

Sometimes you need a symbol that's not shown on the keyboard. For these, you can press the **Sym** key to see a list of other symbols available to you. Press the **Sym** key to the right of the space key.

This key brings up the symbol menu shown in Figure 6-2.

Symbol Key:
For Symbols Not on Keyboard

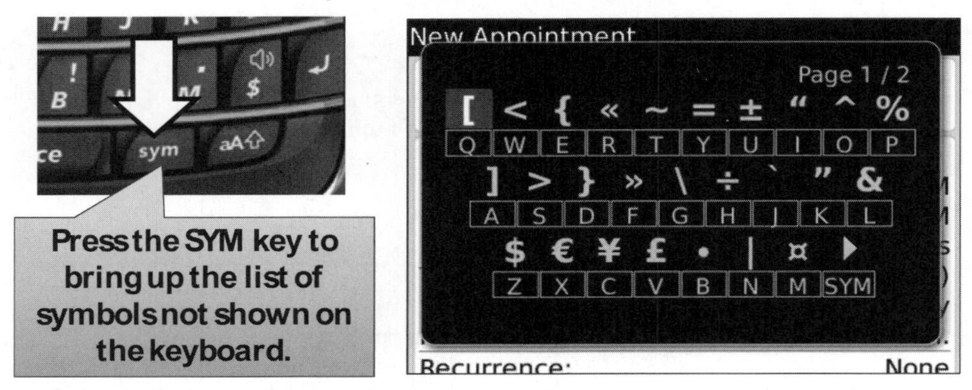

Figure 6-2. *Accessing symbols not shown on the keyboard*

Select the symbol by pressing the associated letter or gliding and clicking the **trackpad**. In the image above, if you press the letter **C** on your keyboard, you'd get the symbol for the Japanese Yen currency.

Quickly Typing Accented Letters and Other Symbols

You can easily type standard accented characters such as Á by pressing and holding a letter on the keyboard and gliding the **trackpad**. (See page 3 of the **Quick Start Guide**.)

Try this: Press and hold the letter **A** to scroll through these accented characters related to A: **À Á Â Ã Ä Å Æ** (both upper and lower case).

- Letters this trick works on are: E, R, T, Y, U, I, O, P, A, S, D, K, C, V, B, N, M.
- Letters that give you some other common characters: V key for ¿, T key for trademark symbol, C key for copyright symbol, R key for registered trademark symbol.

Editing Text

Making changes to your text is easy with the BlackBerry, as Figure 6-3 shows.

1. Glide the **trackpad** left or right to position the cursor.
2. Edit text using **Del** or **Alt+Del**.
   a. Use the **Del** key to erase characters to the left of the cursor.
   b. Hold the **Alt** Key and press **Del** to delete characters under the cursor.

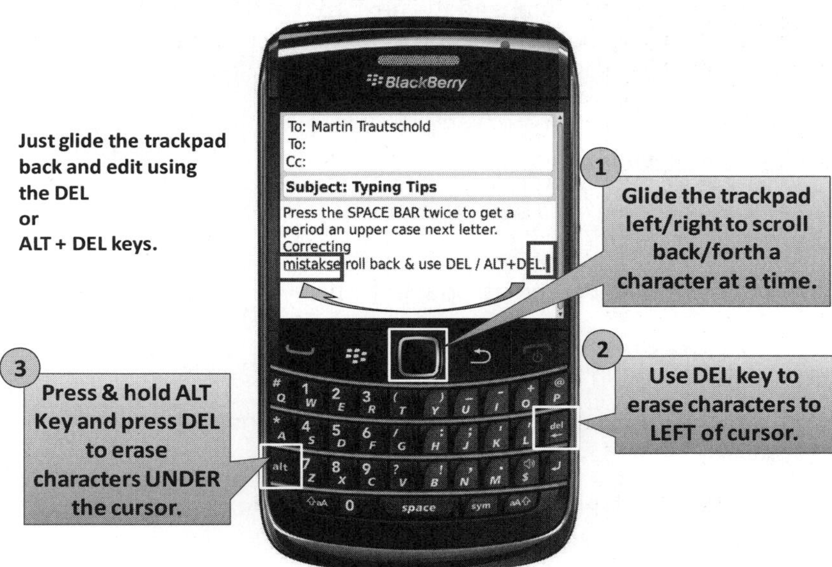

Figure 6-3. *How to edit text on your BlackBerry Bold*

The Mighty Space key

Like many of the keys on the BlackBerry, the space key can do some very handy things for you as you type.

Using the Space Key While Typing an Email Address

On most handhelds, when you want to enter the **@** or a period for an email address, you need a complicated series of commands—usually a **SHIFT** or **ALT** plus another key.

On the BlackBerry, you don't need to take those extra steps. When you're typing the email address, just press the space key once after the user name (martin, for instance) and the BlackBerry will automatically insert the **@**—"martin@".

Type in the domain name and press the space key again and, presto—the BlackBerry automatically puts in the period—"martin@madesimplelearning." No additional keystrokes necessary. Just finish the email address by adding the "com"—"martin@madesimplelearning.com".

Quickly Navigating Drop-Down Lists

The space key is also great at moving you down to the next item in a list. In a minute field, for example, pressing **Space** jumps to the next 15 minutes. In an hour field, you jump to the next hour, and in a month field, you jump to the next month. In any other type of field, pressing the space key jumps you to the next entry.

Give it a try.

1. Click on your **Calendar** icon.
2. Open up a new calendar event by clicking the **trackpad** anywhere in **Day** view.
3. Glide down to the month and press **Space.** Notice you advance one month.
4. Glide over to the hour and press **Space.** You advance one hour.
5. Finally, glide to the minutes and press **Space**. Notice that you move 15 minutes forward.

These great tricks let you quickly reschedule calendar events.

Using Letter Keys to Select Items in Lists and Menus

You can use the letter keys on your keyboard to instantly jump to the first item matching either letter on the key (if there are two letters), or to jump to a matching menu item, or to a matching item in a list (like the long list that appears when you press the Options icon).

Using Number Keys to Type Dates and Times

You can use the number keys on your keyboard to instantly type a new date or time or to select an entry in a drop-down list with that number.

Examples include:

- Typing 40 in the minute field to set the minutes to 40;
- Typing 9 in the hour field to get to 9 AM or PM.

This also works in fields where drop-down list items start with numbers, as in the Reminder field in calendar or tasks. Typing a 9 immediately jumps you to the **9 Hours** setting.

Using the Spell Checker

Your BlackBerry comes with a built-in spell checker. Normally, the spell checker is turned on to check everything you type—you see the little dots underlining questionable entries as you type on your BlackBerry. The underlining goes away when the spell checker matches your words with those in the dictionary. However, spell check is not on by default for outgoing email messages, so you'll have to turn it on if you want them checked.

If your mistakes are not auto-corrected with the **AutoText** feature (see page 163), the spell checker can help you quickly correct many typing or spelling errors. When your BlackBerry finds what it thinks is a misspelled word, it underlines it as shown in Figure 6-4.

Figure 6-4. *The Blackberry's spell checker underlines words it thinks are misspelled.*

To correct one of these words, glide the cursor back into the word with the **trackpad** and click. As Figure 6-5 shows, you'll see a list of suggestions and you can just glide to and click on the correct word. If the word is spelled correctly and you want to add the word to your custom dictionary, press the **Menu** key and select **Add to Dictionary**. You'll see more on the custom dictionary in Figure 6-6.

Selecting Corrections

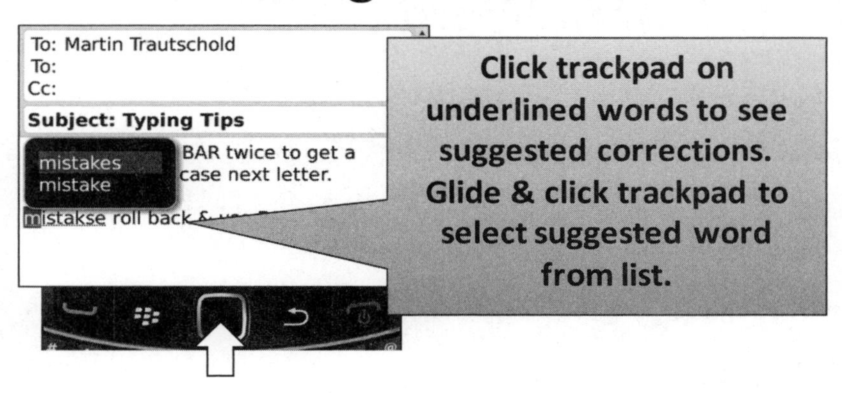

Figure 6-5. *The spell checker can help you correct your mistakes.*

Custom Dictionary

If you use a lot of special words in your emails, such as local place names, that aren't in the standard dictionary, you may want to add them to your own unique custom dictionary. The advantage of this is that you'll never again be asked to replace that word with something suggested. And if you misspell this custom word, you'll be prompted with the correct spelling. Once you realize how useful it can be, you'll often want to add words to your custom dictionary.

Adding Words to the Custom Dictionary

To add a new word, follow these steps as shown in Figure 6-6.

1. Type in a word you use frequently that's not in the standard dictionary. In this example, we are using **Flagler** County, a county in Florida that's not in the standard dictionary.
2. During the spell-checking process, the spell checker will notice the word Flagler, which it believes is misspelled.
3. You'll see that the spell checker suggests options for replacing the word, none of which are correct.
4. Press the **Menu** key.
5. Click **Add to Dictionary** to add the word to your own custom dictionary.

> **TIP:** You may also see **Cancel Spell Check**. You can use this to end the spell check without verifying the rest of the document.

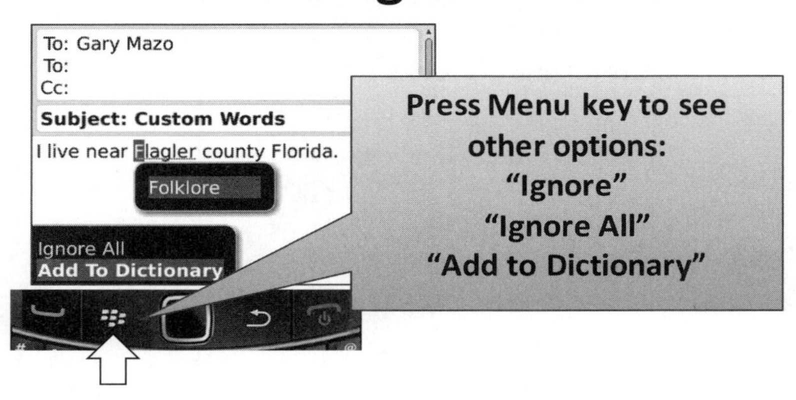

Figure 6-6. *Adding new words to your custom dictionary*

Next time we type **Flagler**, it will not be shown as misspelled. What's even better is that the next time we misspell Flagler (e.g., Flaglr), the spell checker will find it and give us the correct spelling.

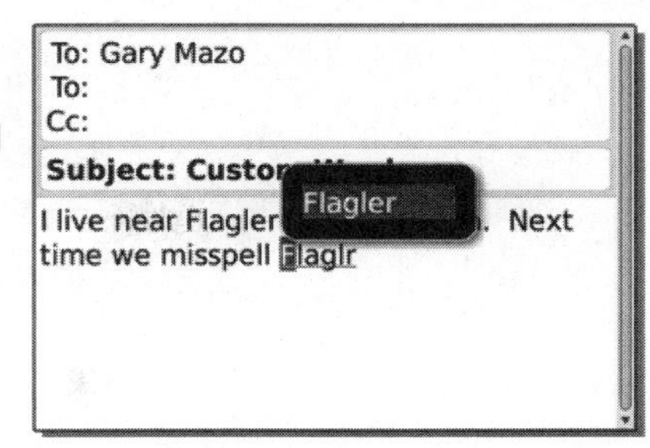

Editing or Deleting Words from the Custom Dictionary

Mistakes will happen, and it's fairly easy to click the wrong menu item and inadvertently add wrong words to the custom dictionary. We have done this plenty of times!

1. Get to the **Spell Check** options screen via the **Messages** icon > **Menu** key > **Options** > **Spell Check** or **Options** icon > **Spell Check**

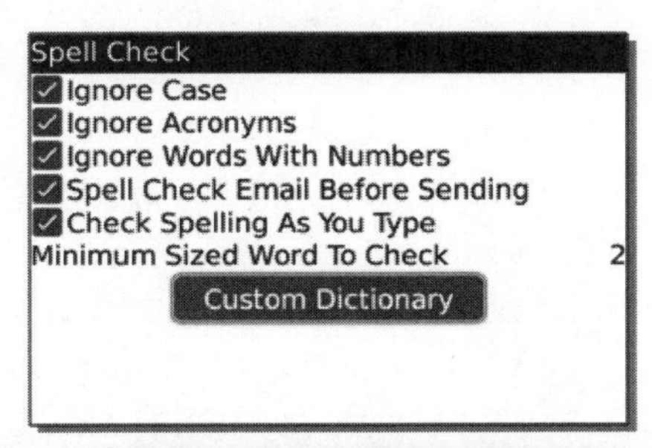

2. Click the **Custom Dictionary** button at the bottom.
3. Now you'll see a list of every word in your custom dictionary.
4. Scroll down or start typing a few letters to **Find** the word. In this case, we know we want to remove the word **misspellg** from the dictionary, so we type the letter **m** to instantly jump to those entries that start with m.

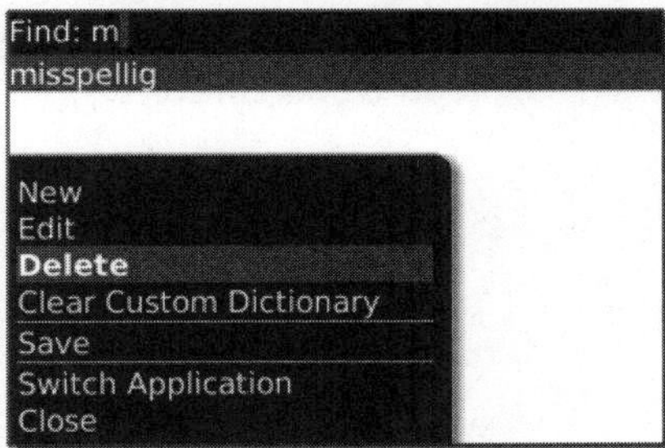

5. Highlight the word you want to edit or delete.
6. Press the **Menu** key and select **Delete** to get rid of the word or **Edit** to enter any changes.
7. Press the **Menu** key or **Escape** key and **Save** your changes.

Custom Dictionary: Edit
misspelling

TIP: You can force a spell check before sending email messages . By default, most BlackBerry Smartphones will not do a spell check before sending email. If you prefer, you can ignore all misspelled (underlined) words and send. Below we show you how to enable spell check for outgoing email.

Enabling Spell Check for Outbound Email

One of the great features of your new BlackBerry is that you can automatically check the spelling of your emails before you send them. However, this feature must be enabled; it is not turned on when you take your BlackBerry out of the box. Like the spell checker on your computer, you can even add words to the dictionary. Spell check will save you embarrassing misspellings in your communications, which is especially important with such a small keyboard.

1. Click your **Messages** (Email) icon.
2. Press the **Menu** key and select **Options.**

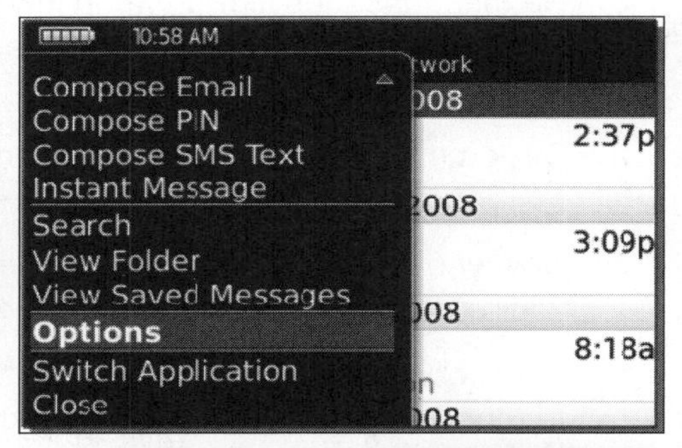

TIP: You can also start with the **Options** icon and then select **Spell Check**.

3. Click on **Spell Check**

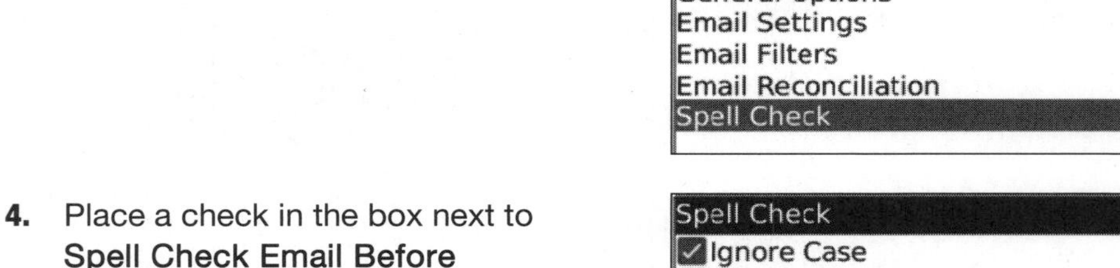

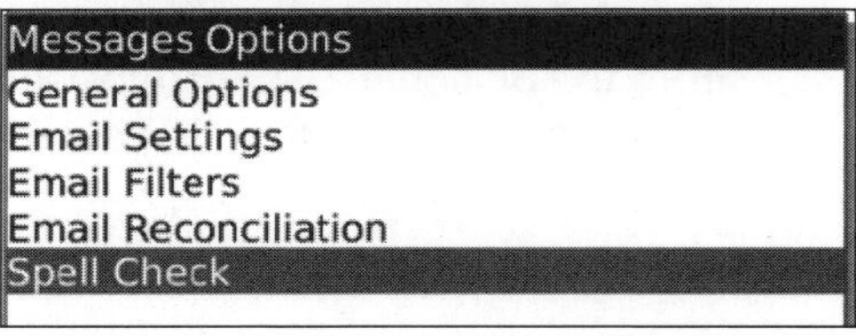

4. Place a check in the box next to **Spell Check Email Before Sending** to enable spell checking on outbound email. Click on it with the **trackpad** or press the space key to check it.
5. Press the **Menu** key or **Escape** key and **Save** your changes.

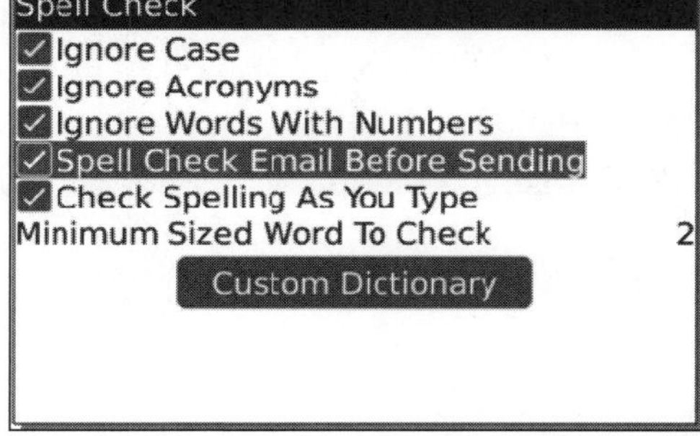

Using BlackBerry Text-Based Help

There may be times when you don't have this book or our video tutorials handy and you need to find out how to do something on your BlackBerry right away.

You can get to the help menu from the **Help** icon—and almost every application on the BlackBerry has a built in contextual help menu that can answer some of your basic questions.

Using the Help Menus Inside Applications

You can access **Help** from virtually any application. Let's take a look at the **Help** menu built into the **Calendar**.

1. Click your **Calendar** icon, or press the **Home** screen hotkey **L** (see page 541).

2. In most applications on your BlackBerry, **Help** is the top menu item, or near the top. Press the **Menu** key, scroll up to and click on **Help**. You'll see a number of topics about the **Calendar**.

3. To select any of these options, just glide the **trackpad** to highlight the item (such as **Calendar basics**, as shown) and click the **trackpad**.

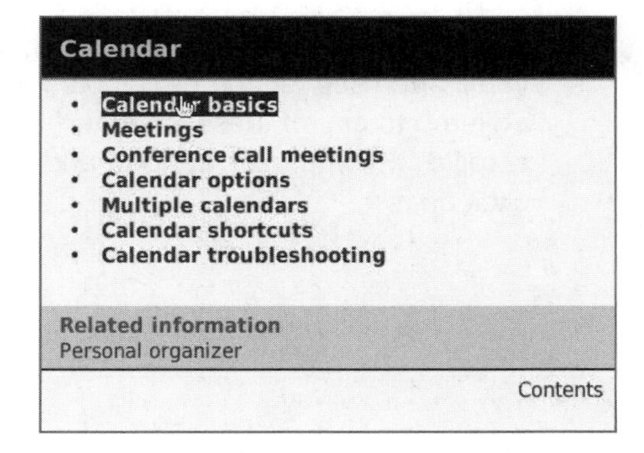

4. Continue to glide and click the **trackpad** on any topic you'd like to learn about. Press the **Escape** key to back up one level in the help menus.

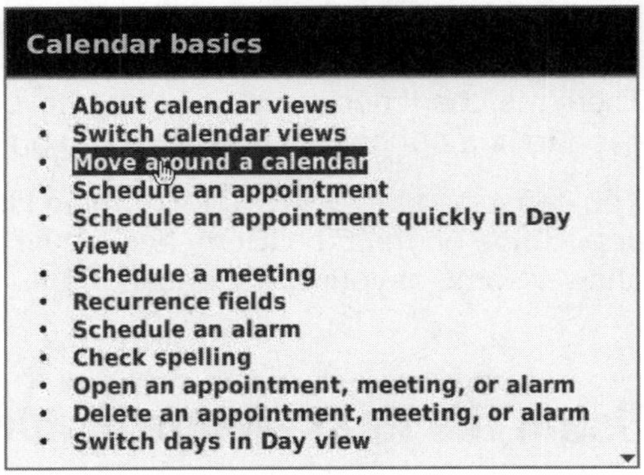

Calendar basics

- About calendar views
- Switch calendar views
- Move around a calendar
- Schedule an appointment
- Schedule an appointment quickly in Day view
- Schedule a meeting
- Recurrence fields
- Schedule an alarm
- Check spelling
- Open an appointment, meeting, or alarm
- Delete an appointment, meeting, or alarm
- Switch days in Day view

5. Within each topic, you'll see a screen like this one showing you the actual steps to follow to complete the task.

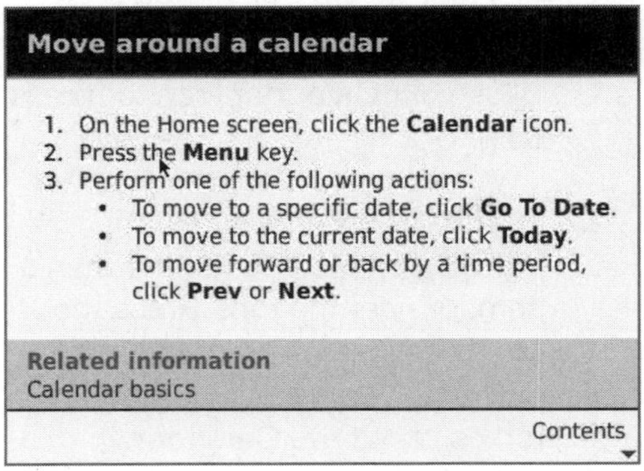

Move around a calendar

1. On the Home screen, click the **Calendar** icon.
2. Press the **Menu** key.
3. Perform one of the following actions:
   - To move to a specific date, click **Go To Date**.
   - To move to the current date, click **Today**.
   - To move forward or back by a time period, click **Prev** or **Next**.

Related information
Calendar basics

Contents

6. In the gray bar at the bottom, you'll often see related help topics. To jump to any of these, simply scroll down with the **trackpad** and click on it.

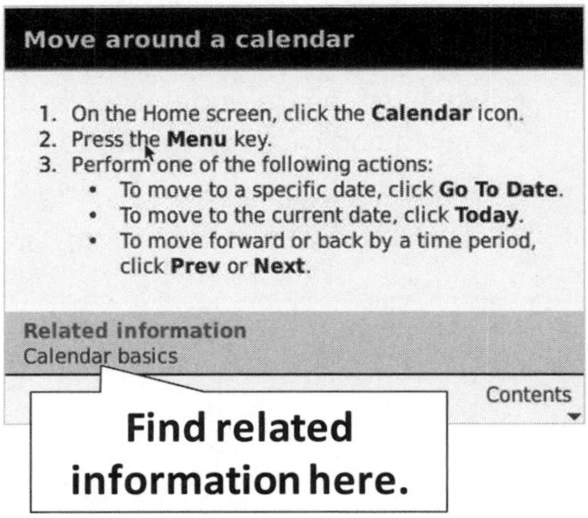

Move around a calendar

1. On the Home screen, click the **Calendar** icon.
2. Press the **Menu** key.
3. Perform one of the following actions:
   - To move to a specific date, click **Go To Date**.
   - To move to the current date, click **Today**.
   - To move forward or back by a time period, click **Prev** or **Next**.

Related information
Calendar basics

Contents

Find related information here.

Overall Help Contents and Finding Text

To jump back to a list of all the main help topics, click on **Contents** in the lower right corner.

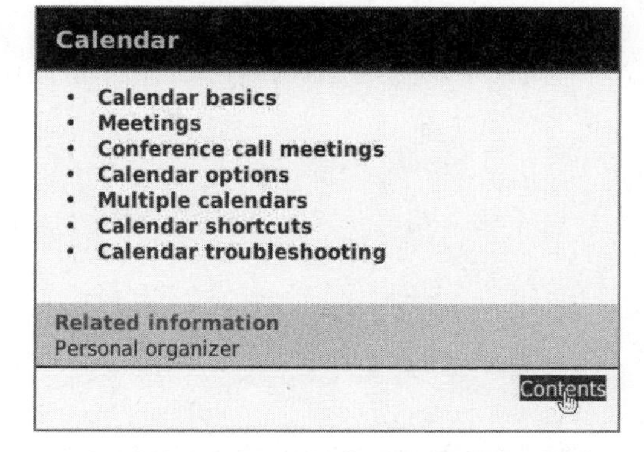

To find text on the ***currently displayed Help page***, press the **Menu** key and select **Find.** You can also get to this **Find** menu item if you glide the cursor to the top of the screen and click the **trackpad**.

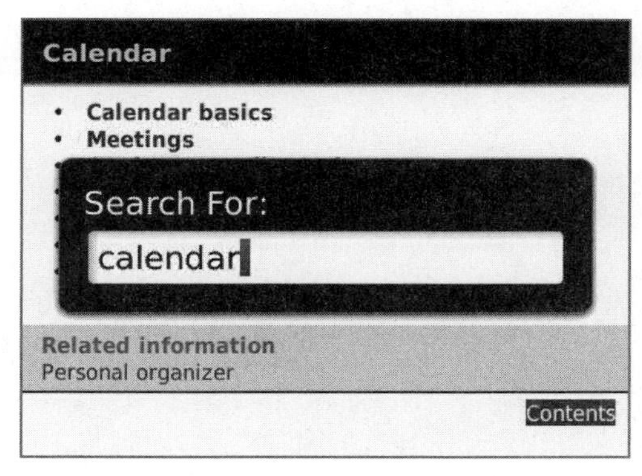

Chapter 7

Speed Your Typing with AutoText

In this chapter we show you some great tips for using **AutoText** to save time and increase accuracy for things you type many times a day. **AutoText** automatically corrects certain typing errors, and you can also customize it for, say, directions or by creating a date and time stamp with a two-letter shortcut. This is great if you are taking notes and don't want to bother typing out the current date and time—two letters of AutoText will do the trick!

Saving Time with Auto-Correcting AutoText

Sometimes, typing on the little BlackBerry keyboard produces less than desirable results. Fortunately, the preloaded **AutoText** automatically corrects many common typing mistakes, like leaving out an apostrophe in the word **aren't** or transposing the letters in **the** (see Figure 7-1). For other words you misspell or mistype often, you can create an AutoText entry to solve the problem.

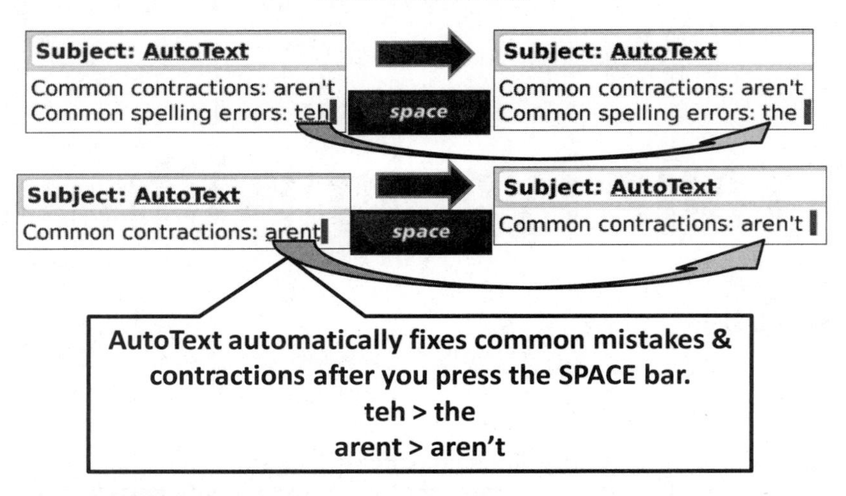

Figure 7-1. *AutoText fixes common spelling and typing mistakes so you save time.*

> **TIP:** Knowing **AutoText** is there to help you get things right allows you to type with greater abandon on your BlackBerry. Take a few minutes to browse the preloaded AutoText entries, especially the contractions, so you can learn to type them without ever using the apostrophe.

You can also use **AutoText** for more advanced tasks, like automatically typing an email signature (Page 163), driving directions, a "canned" email, routine text describing your products or services, legal disclaimer text, anything!

Creating a New Custom AutoText Entry

You can get into the **AutoText** list from the **Edit AutoText** menu item when you are typing an email (by pressing the **Menu** key) or from the main **Options** icon. To create a new entry, follow these steps.

1. Click on your **Options** icon.
2. Click on **AutoText** to see the list of entries.

TIP: Skip typing apostrophes! Learn the **AutoText** contractions to save time as you type future emails.

AutoText
acn (can) SmartCase
acn (can) SmartCase
adn (and) SmartCase
adn (and) SmartCase
agian (again) SmartCase
agian (again) SmartCase
ahd (had) SmartCase
ahd (had) SmartCase
ahppen (happen) SmartCase
ahppen (happen) SmartCase

3. Under **Replace:** type in the new entry you want to add, in this case **dirh** for **Directions to home**, making sure you have no entries that exactly match.

Find: dirh
* No Phrases *

* No Phrases * means no matches.

4. Hit the **Enter** key to start adding your new entry.
5. Glide down to under **With:** and enter the text you want to appear when you type your new AutoText word **dirh**.
6. Press the **Menu** key and select **Save**.

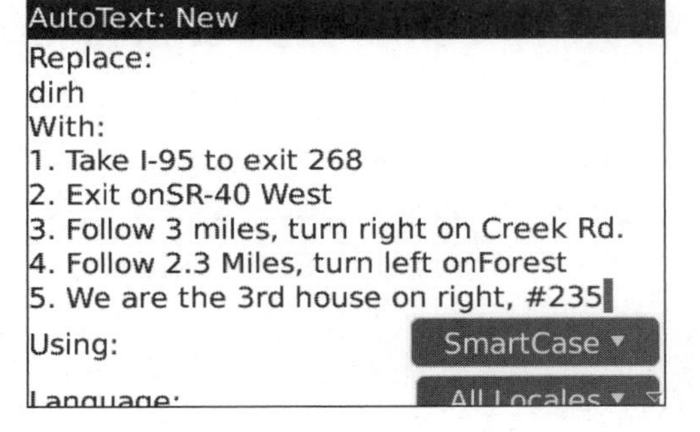

TIP: Type the directions on your computer and email them to yourself, then copy and paste them into AutoText from the email.

Using New AutoText Entries

Now when someone wants directions to your house, all you have to do is type **dirh**.

To: susan@company.com
To:
Cc:
Subject: Directions to house
Here's how to get to my house:
Dirh

When you press the space key, the full directions appear from your new AutoText entry.

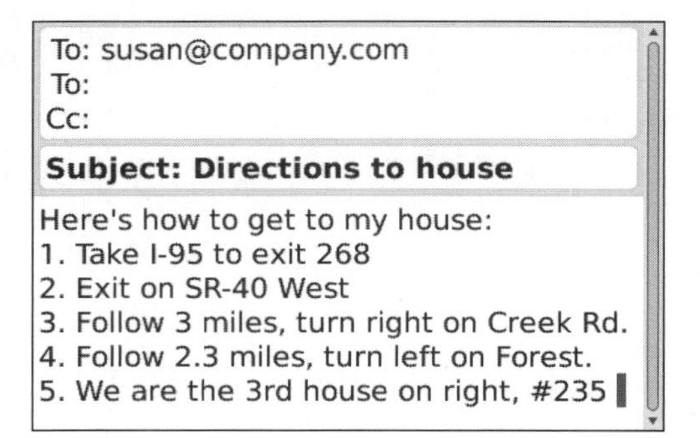

Advanced AutoText Features—Macros

Macros are shortcuts for other functions, such as displaying the current time and date, your PIN number or owner information, or even pressing the backspace or delete keys, and AutoText is the key to using them.

Creating a New Entry with Macros

Let's create a new entry called "ts" ("Time Stamp") that will instantly show the current time and date.

1. Start creating a new entry as you did above and use the letters **ts** for time stamp.
2. Press the **Menu** key and select Insert Macro.

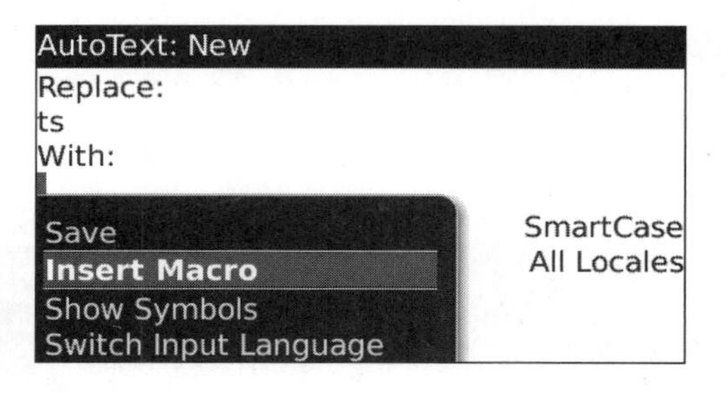

TIP: All macros start with the percent sign (%) and you can simply type them instead of selecting them from the menu.

3. Scroll up or down and select the macro you want.

   In this case, we want a Short Date (%d), which is mm/dd/yy format.

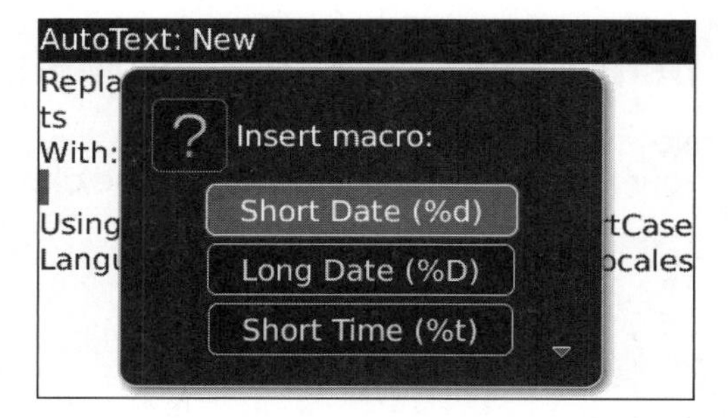

4. Type a space, a hyphen, and another space.
5. Now press the Menu key and select Short Time (%t).

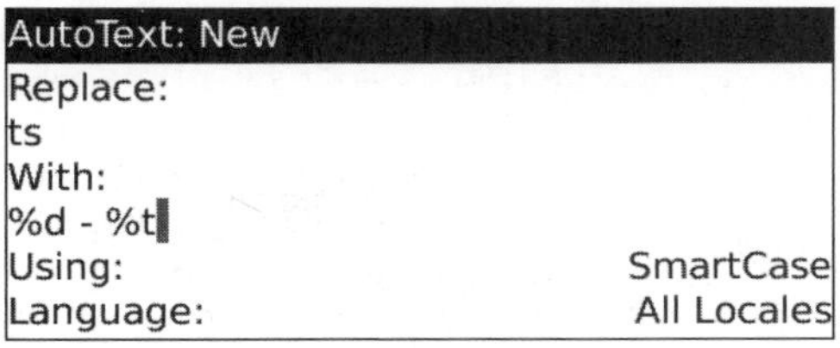

AutoText: New
Replace:
ts
With:
%d - %t
Using: SmartCase
Language: All Locales

Using Your New AutoText Entry

Now whenever you want to put the current date and time, just type **ts** and press **Space**.

Press the Space key to see the date/time.

Title: Meeting ts

Title: Meeting 1/31/2009 - 9:25a

Here's a list of the standard **AutoText Macros.** Here's what they look like:

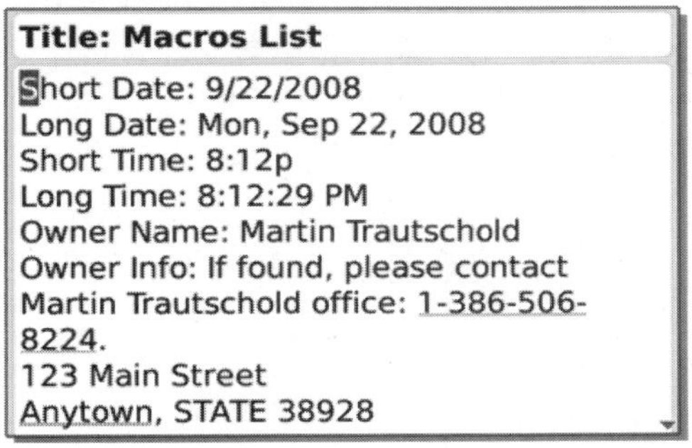

%d Short Date

%D Long Date

%t Short Time

%T Long Time

%o Owner Name

%O Owner Information

%p Your Phone Number

%P Your PIN number

%b Backspace

%B Delete

%% Percent

Title: Macros List
Short Date: 9/22/2008
Long Date: Mon, Sep 22, 2008
Short Time: 8:12p
Long Time: 8:12:29 PM
Owner Name: Martin Trautschold
Owner Info: If found, please contact Martin Trautschold office: 1-386-506-8224.
123 Main Street
Anytown, STATE 38928

> **TIP:** Instead of pressing the **Menu** key and selecting the macro, just type the letters like **%t** for short time

Edit or Delete an AutoText Entry

Sometimes you need to edit or remove an **AutoText** entry. This isn't very different from creating a new entry.

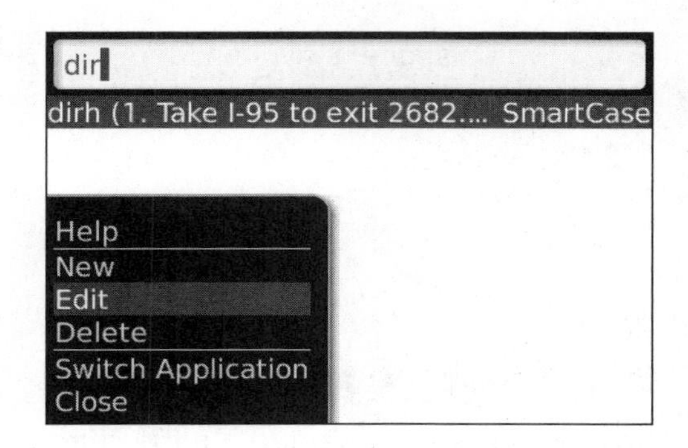

1. You can edit **AutoText** in a couple of places:
   - If you are editing an email message, press the **Menu** key and select **Edit AutoText.**
   - From your **Home** screen, click on the **Options** icon and select **AutoText**.
2. Type a few letters to find the **AutoText** entry.
3. Press the **Menu** key select **Edit** or **Delete**.
4. Once you are done editing your **AutoText** entry, press the **Menu** key and select **Save**.

Chapter 8

Personalize Your BlackBerry

In this chapter you'll learn some great ways to personalize your BlackBerry. You can move or hide your icons, organize them using folders, set convenience keys, and change your Theme or the look and feel of your BlackBerry by adjusting font sizes and types.

> **TIP:** You can change your **Home** screen image (your wallpaper). If you press the **Media** icon, you can select from a number of preloaded background images, or you can even snap a picture and immediately set it as your wallpaper. (See exactly how to do this on page 184.)

Moving, Hiding, and Deleting Icons

You may not need to see every single icon on your **Home** screen, or you may want to move your most frequently used icons up to the top row for easy access. We show you how to work with icons when you are in one of the standard Themes on your BlackBerry (see page 179). However, the steps may vary a little, depending on which Theme is active your BlackBerry.

Moving Your Icons within a Folder

Press the **Menu** key to see an array of all your icons. If the icon you want to move is inside a particular folder, like Downloads or Applications, glide to and click on that folder.

1. Glide over to highlight the icon you want to move using the **trackpad**. In this case, we have highlighted the **Maps** icon so we can move it.
2. Press the **Menu** key (to the left of the **trackpad**) to bring up the **Move** menu item.

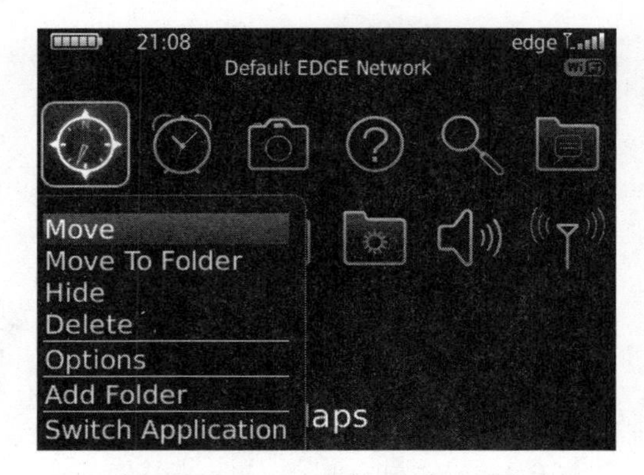

3. Once you select **Move,** you'll see arrows pointing around the icon (as shown). Start moving it wherever you want by rolling the **trackpad**.
4. Finally, click the **trackpad** to set the moved icon at the new location.

How Do I Know When I'm in a Folder?

When you are in a folder, you see a little icon at the top of your screen with a folder tab image and the name of the folder. In the image below, you can see that you're in the **Applications** folder.

Moving Your Icons Between Folders

Sometimes you want to move icons to your **Home** folder to make them more easily accessible. Or, you might want to move some of the icons you seldom use from your **Home** folder to another folder to clean up your **Home** screen.

Let's say we want to move the **Docs to Go** icon from our **Applications** folder to our **Home** folder, so it's more easily accessible.

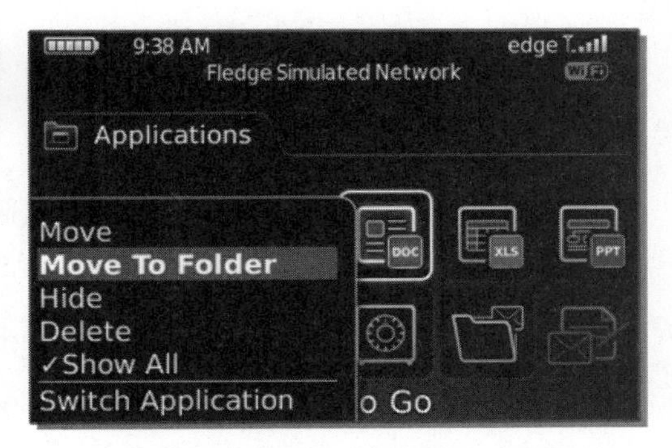

1. Highlight the **Docs to Go** icon as shown.
2. Press the **Menu** key to the left of the **trackpad** and select the **Move to Folder** item.
3. Now, we want to move this icon out of our **Applications** folder and into the **Home** folder, so we click on **Home** at the top of the list.

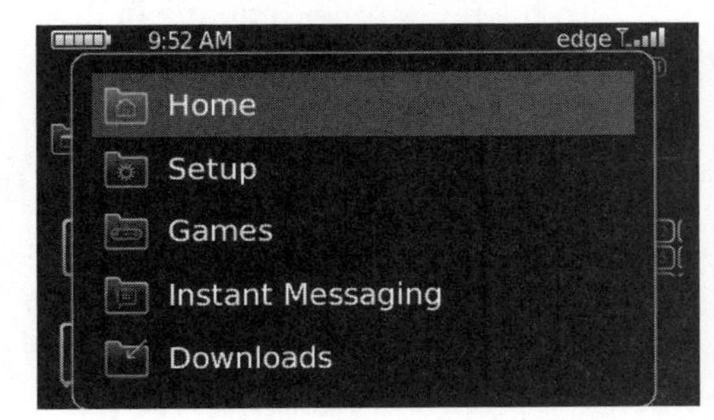

Next we press the **Escape** key to exit from the **Applications** folder back to the **Home** folder to locate our newly moved **Docs to Go** icon; in this case it is near the bottom of the list of icons.

Hiding and Un-Hiding Icons

At times you may want to hide an unused icon to make your BlackBerry easier to use. Hiding and un-hiding icons is easy; the steps are similar to those you took to move icons.

1. Press the **Menu** key to see all your icons on the **Home** screen.
2. If the icon you want to hide is inside a folder, click on the folder so you can see the icon.
3. Highlight the icon you want to hide.
4. Press the **Menu** key.
5. Select **Hide**.

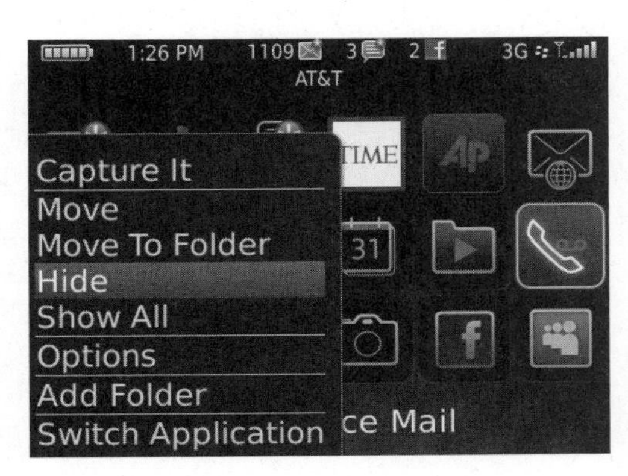

To **un-hide** an icon, follow these steps.

1. From your **Home** screen, press the **Menu** key until the menu appears.
2. Select **Show All**.
3. Highlight the grayed out (hidden) icon.
4. Press the **Menu** key
5. Select **Hide** to turn off the checkmark next to it.
6. This will make your icon visible again.
7. To get rid of other grayed out (hidden) icons, press the **Menu** key again and select **Show All**.

Setting Your Top Icons on Your First Home Screen

Many BlackBerry **Themes** show only a limited set of your top icons on your First **Home** screen (see the image to the right).

When you press the **Menu** key you see the Second **Home** screen with the rest of icons.

To make your BlackBerry faster to use, you want to move those icons that you use most often to this First **Home** screen.

*The First **Home** Screen with a limited set of icons.*

In order to move your icons to the First **Home** screen, you need to move them up to the top row of icons on the Second **Home** screen. In this example, let's move the **Word to Go** icon (see image below) into one of the top spots.

1. Press the **Menu** key to see your Second **Home** screen with all the icons.
2. Highlight the icon for **Word to Go** as shown.
3. Press the **Menu** key and select **Move.**
4. Move it up into one of the top six spots with the trackpad.
5. Click the **trackpad** to set it into place.

6. Now we see **Word to Go** in the limited set of icons on the First **Home** screen.

Using Folders to Organize Your Icons

On your BlackBerry you can create or delete folders to better organize your icons. Typically, you'll find a few folders already on your BlackBerry—Applications, Settings, Downloads, and Games—that are created by default. You can arrange your icons to suit your own needs by adding new folders and moving icons into them .

Creating a New Folder

NOTE: At the time of publication, you could only create folders one level deep. In other words, you can only create new folders from the **Home** folder, not when you are already inside another folder. This may change with new software versions.

Here's how to create a new folder.

1. From your **Home** screen, press the **Menu** key until you see a menu appear.
2. Select Add Folder.

NOTE: If you don't see the **Add Folder** menu item, press the **Escape** key to get back to your **Home** folder.

3. After you select the **Add Folder** menu item, you'll see this screen to the right.
4. Enter a **Name** for your folder.
5. You can click on the folder icon and glide left and right to check out all the different colors and styles available for your folder icon.

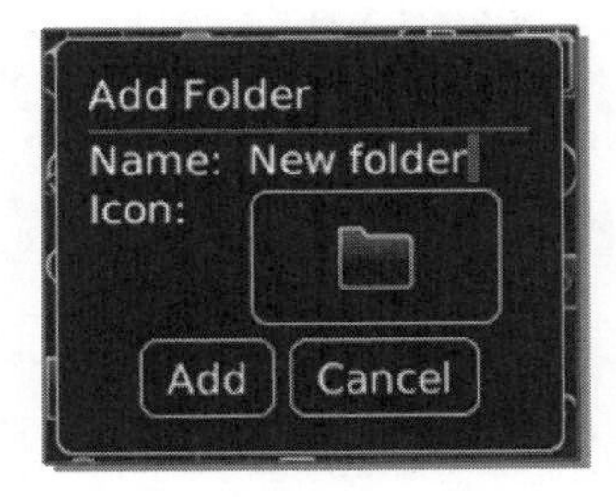

6. When you're done selecting the folder icon style, click on it and glide down to click on the **Add** button to finish creating your folder.

7. Now you'll see your new folder.

Moving Icons between Folders

Once you create your new folder, you'll want to move icons into it to organize them, and also so your **Home** screen is not too crowded. Please see our instructions on page 170 on how to move icons.

Editing a Folder

You can edit a folder by highlighting it, pressing the **Menu** key, and selecting **Edit Folder**.

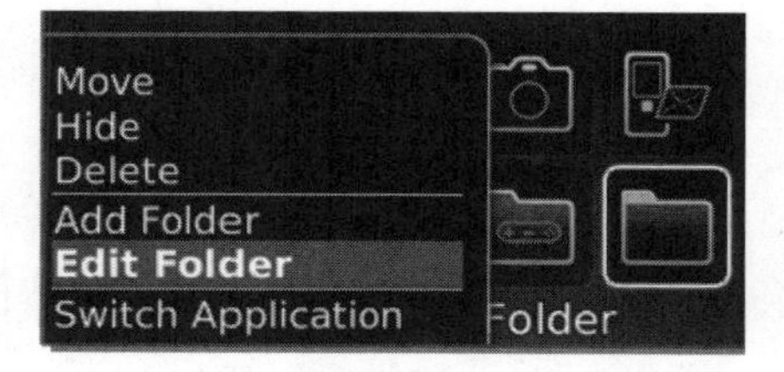

Then you can change the folder name, icon, and color and save your changes.

Deleting a Folder

Whenever you want to get rid of a folder, just highlight it, press the **Menu** key, and select **Delete**.

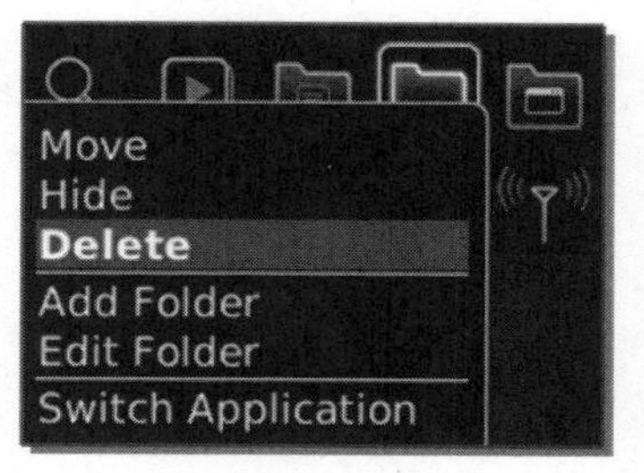

Setting the Date, Time, and Time Zone

Sometimes you'll need to adjust the date, time, and time zone. While you can set the time and time zone in the **Setup Wizard**, you can also adjust it directly using the **Options** icon. To do so, follow these steps.

1. Click your **Options** icon or press the letter **O** from your **Home** screen to start **Options**. (See "Home Screen Hotkeys" in Part 4.)
2. Press the letter **D** to jump down to **Date/Time**, or scroll down and click on it.
3. Click the **trackpad** to see all the time zones, and glide up or down to select the one that's appropriate.

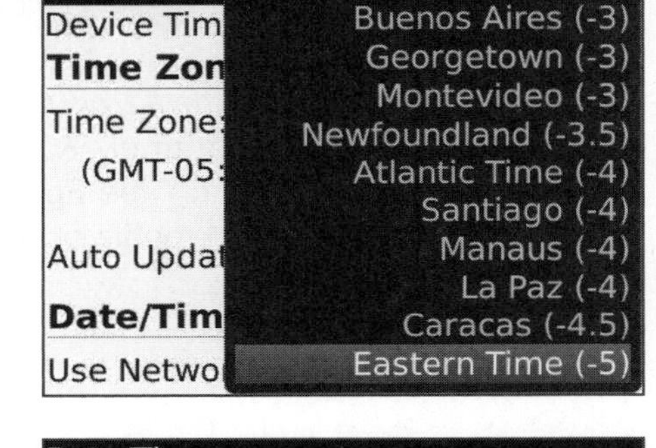

4. Scroll down to set the **Auto Update Time Zone**. You can choose the default **Prompt** (which asks you if you want to change the time zone whenever it detects you're in a new one), **Off** (which never adjusts), or **On** (which adjusts the time zone without asking you). We recommend leaving it set to **Prompt**.

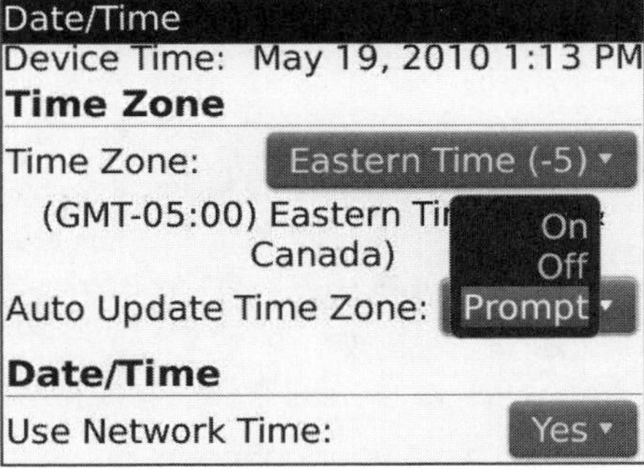

5. Roll down to **Use Network Time** and select **Yes** or **No**. The default is **Yes**, which uses the Network Time shown at the bottom of the screen. We recommend leaving this set to **Yes** unless you want to manually override the time. To manually set the time, select **No**.

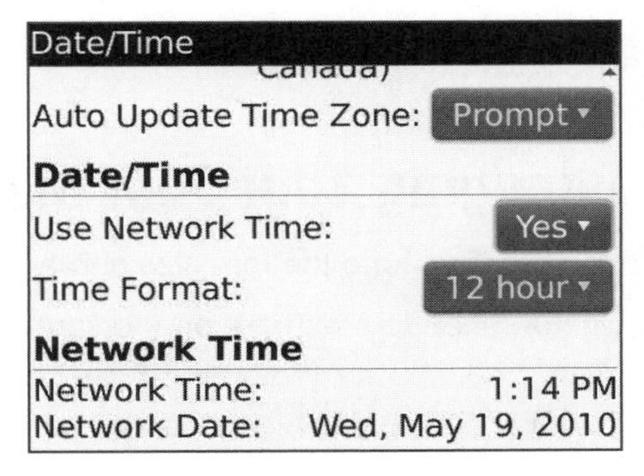

6. If you selected **No** for **Use Network Time**, you can manually adjust the **Time** and **Date** by clicking and rolling up or down.

> **TIP:** You can also use the Number keys to change the years, date, hours, or minutes. For example, simply type 26 in the minute field to jump right to 26. Another trick is using the space key—tap it to advance one number.

7. Whether you prefer the **12 hour** (7:30 AM/4:30 PM) or **24 hour** (07:30/16:30) format, you set that in **Time Format**. Tap the space key to toggle between the two options.

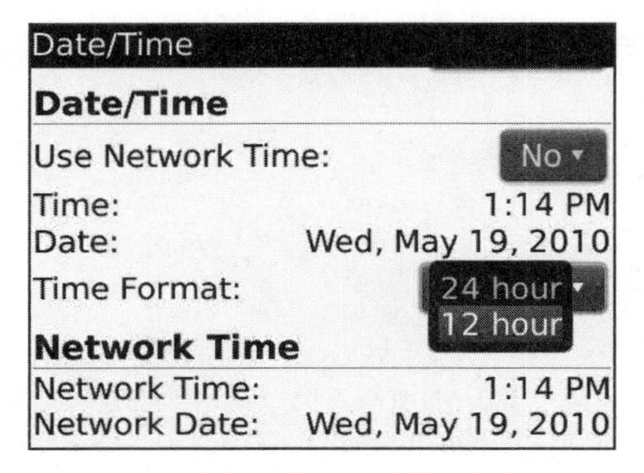

> **TIP:** You can force the Network time to update Time Zones and the Network Time if you press the Menu key and select Update Time or Get Time Zone Updates.

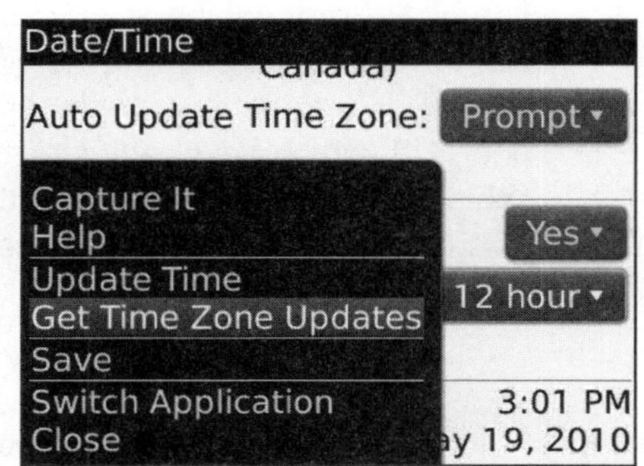

8. To save your changes, press the **Menu** key and select **Save**.

Changing Your Font Size and Type

You can fine-tune the font size and type on your BlackBerry to fit your individual needs.

Do you need to see more on the screen and don't mind small fonts? Then go all the way down to a micro-size **7-point font**.

Do you need to see bigger fonts for easy readability? Adjust the fonts to a large **14-point font** and make it bold.

Here's how to adjust your font size and type:

1. Click on the Options icon. or press the letter **O**.
2. Click on **Screen/Keyboard** (see Figure 8-1).

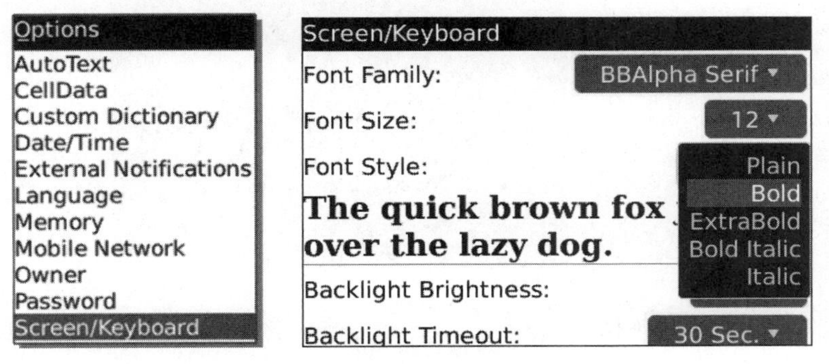

Figure 8-1. *You can change the font size.*

3. Click the **trackpad** to select a different font family, size, style or type as shown.
4. You can see a preview of your choices so you can make sure they fit your needs.

Changing Your Currency Key

You'll notice that the key next to your **Enter** key in the lower right corner of your keyboard has a currency symbol on it. It may be a dollar, pound, Euro, or Yen sign. You can change the currency character this key types.

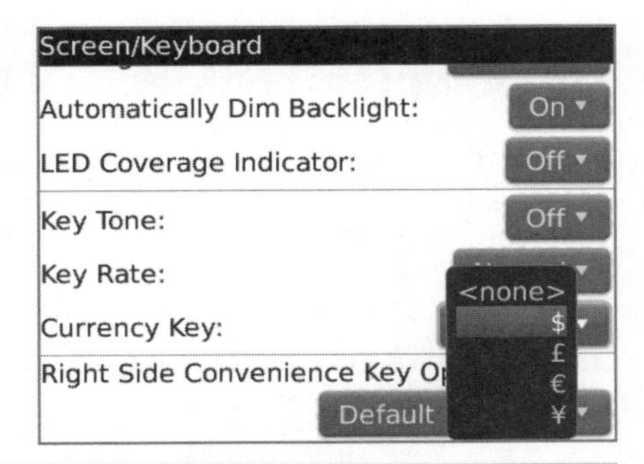

1. Click on your **Options** icon.
2. Click on **Screen/Keyboard**.
3. Roll down and change the **Currency Key** setting as shown.

TIP: This key also doubles as the **Speakerphone** hotkey. Just tap it to turn on or off your **Speakerphone**.

Home Screen Preferences and Options

On your BlackBerry, you can customize many things about your **Home** screen by pressing the **Menu** key from the **Home** screen and selecting **Options**.

This brings up the **Home Screen Preferences** screen. Glide the **trackpad** down to see all the settings options, as shown in Figure 8-2.

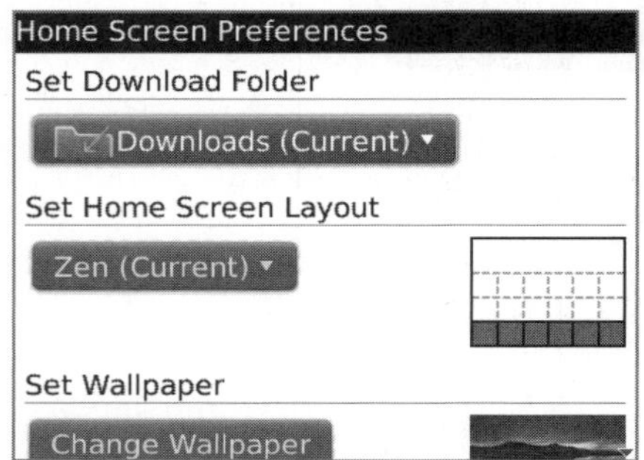

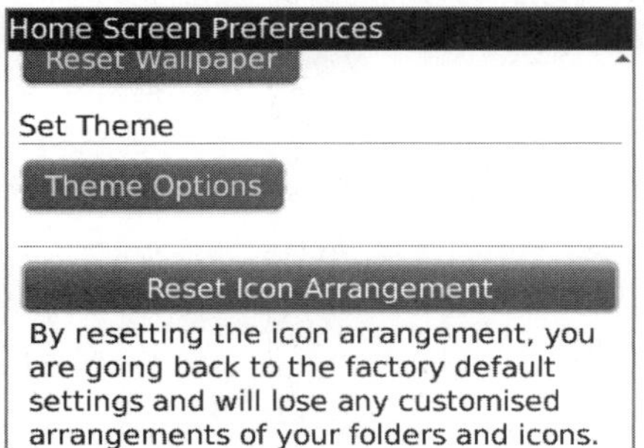

Figure 8-2. *Setting Home Screen Preferences*

Set Download Folder

Use **Downloads** (the default) or change to any other folder—even the **Home** folder. This can be convenient if you download many apps and you want easy access to those icons from your **Home** screen.

Set Home Screen Layout

The **Zen** layout is the default. With just a few icons on the bottom of the **Home Screen**, you get a nice view of the background wallpaper. The **Today** layout gives you a preview of recent Messages, Calls, and upcoming Calendar events, as Figure 8-3 shows.

Figure 8-3. *Home screen layout choices*

Set or Reset Wallpaper

Change Wallpaper lets you select any picture on your BlackBerry to set as your background, or you can take a picture with the camera and use that. Click the Change Wallpaper button and navigate to a picture you would like to select.

You can also change your wallpaper from the **Pictures** icon and from the **Camera** icon, see page 184.

Reset Wallpaper changes your wallpaper image back to the default picture.

Set Theme

Theme Options let you change the entire look and feel of your BlackBerry. Read the section on Themes below to learn more.

Reset Icon Arrangement

Click the **Reset Icon Arrangement** button with the **trackpad** to reset all your icons and folders. This can be useful if you have lost something or have just gotten tired of all the changes you made.

Selecting New Themes

You can customize your BlackBerry and make it look truly unique. One way to do this is to change your BlackBerry's **Theme**. Changing the Theme generally changes the layout and appearance of your icons and the font type and size you see when you click on an icon. Many different **Themes** are already included on your BlackBerry, and dozens more available for download at various web sites or from App World (learn how on page 431).

Carrier-Specific Themes: Depending on your BlackBerry wireless carrier (phone company), you may see various customized **Themes** not shown in this book.

More Standard BlackBerry Themes: Most of the standard Themes shown below are preinstalled on every BlackBerry.

Glide and click on the **Options icon** on your BlackBerry. You may have to press the **Menu** key to see all your icons and then locate the **Options** icon. Or, if you have **Home** screen hotkeys enabled, just tap the letter **O** (see page 541).

1. In **Options**, scroll down to **Theme** and click.

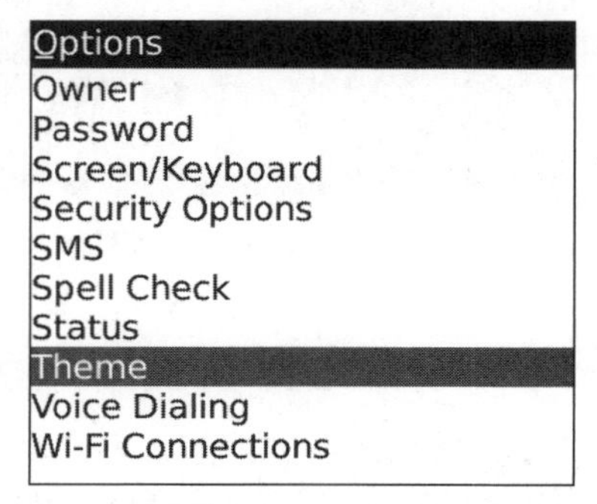

> **TIP:** Press the letter **T** to jump to the first entry starting with T, which should be **Theme**.

2. On the **Theme** screen, just glide and click on the Theme you want to make active. Your currently selected **Theme** is shown with the word (**Active**) next to it.

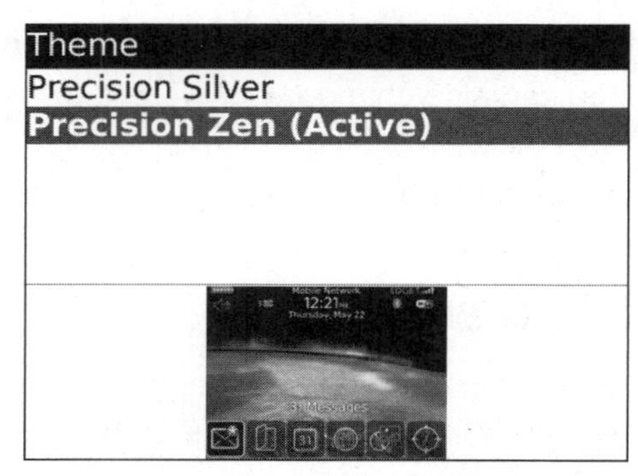

3. Press the **Escape** key to get back to the **Home** screen to check out your new theme.

Precision Silver Theme

This is like the Zen theme with just six icons on the main **Home** screen; however, all icons are a gray monotone. Unless your BlackBerry provider has specifically removed these Themes, they should be available on your BlackBerry.

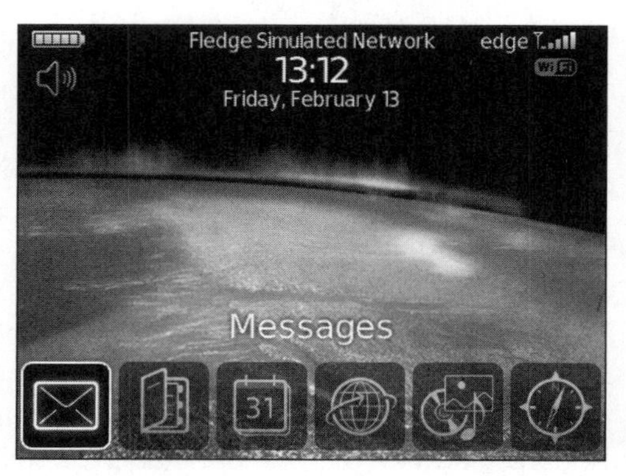

Precision Zen Theme

This looks very similar to the Precision Silver theme, but it has full-color icons; only six are shown on the main **Home** screen. These two Themes look very similar, but the **Zen Theme** has some slight color to the icons.

Your screen should look similar to one of these, with some possible slight deviations.

Downloading New Themes from BlackBerry App World

Figure 8-4 shows two examples of the many new **Themes** you can download from App World on your BlackBerry.

Figure 8-4. *Customize your Home screen with new themes.*

The one on the left is called "Sky2 PrimeTheme" from DreamTheme and the one on the right is "Animated Winter Wonderland" from Motek Americas Inc. It's animated so you can watch the snowflakes fall on the home screen. A number of the themes are animated so you see some movement, such as the wind blowing the leaves of a palm tree and the waves of the ocean. Themes range from free to about US $9.99, with many in the $4.99 to $6.99 range.

> **NOTE:** If you need help getting started with BlackBerry App World, check out our chapter starting on page 431.

1. Start **App World** by clicking on the icon.

2. Click on the **Categories** icon in the lower left corner. (It looks like folders.)
3. Notice that when this image was taken, the **Sky2** theme was one of the Featured Items.

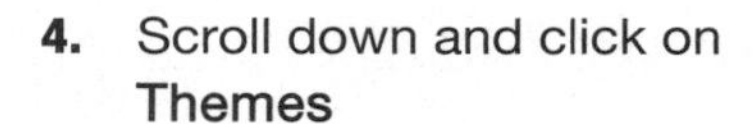

4. Scroll down and click on **Themes**

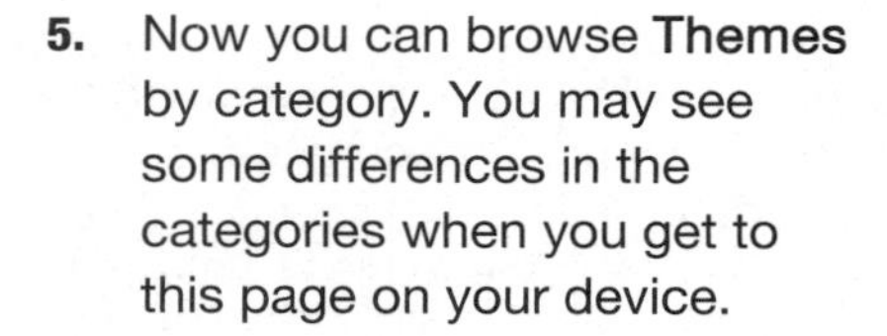

5. Now you can browse **Themes** by category. You may see some differences in the categories when you get to this page on your device.

6. Within each **Category**, if you know the name of a theme, you can **Find** it by typing a few letters of its name.
7. Click on the **Theme** you want to check out.

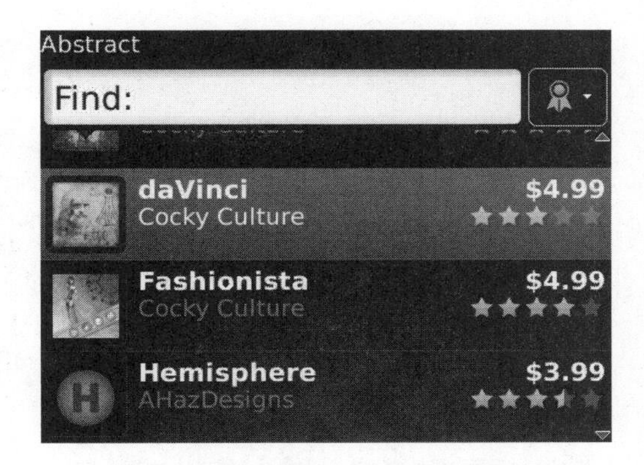

8. Each **Theme** will have a description, rating (1-5 stars), reviews, and screenshots.

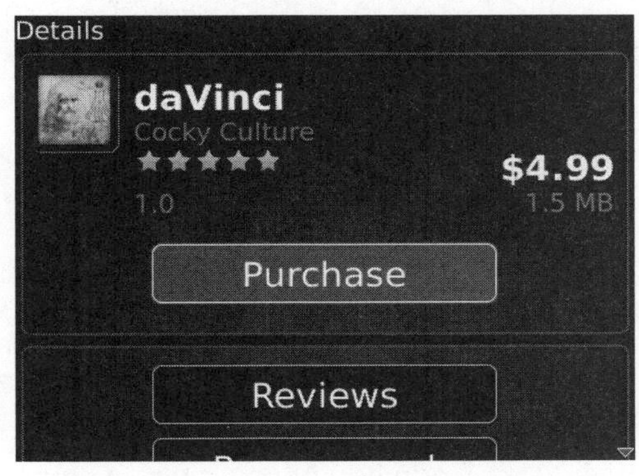

9. We recommend checking out the screenshots before you purchase.
10. Just roll the **trackpad** back and forth to see the various screenshots. Press the **Escape** key to return to the main page for the **Theme** to buy or download it.

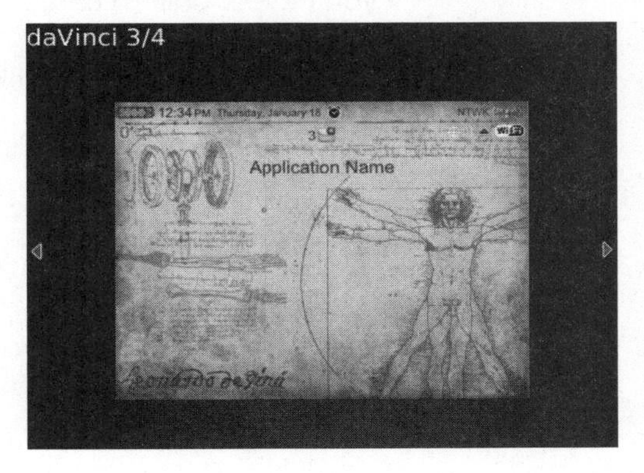

Most **Themes**, after you download and install them using App World, will ask if you want to **Activate the Theme**. If you want to start using the **Theme** right away, say **Yes**. If you want to change **Themes**, see page 179.

CAUTION: The authors have downloaded many **Themes** on their BlackBerry smartphones. Some can cause problems with your BlackBerry.

TIP: Besides App World, you can also download **Themes** from some vendors and BlackBerry community sites, such as:

bPlay: `http://www.bplay.com`

CrackBerry.com: `www.crackberry.com`

BlackBerry Forums: `www.blackberryforums.com`

Pinstack: `www.pinstack.com`

BlackBerry Cool: `www.blackberrycool.com`

Also, try a web search for "BlackBerry Themes." There are probably new sites all the time!

Changing the Home Screen Background Image or Wallpaper

Now that you have a Theme you like and have adjusted the font size and type to your preferences, you may also want to change the background image on your **Home** screen, which is also called the **Home screen image** or **wallpaper**. You saw above how to download new Themes; use the same steps to download new wallpaper.

If you like, since you have a built-in camera, you can simply snap a picture and immediately use it as wallpaper. In fact, you can use any image that is stored on your BlackBerry—either in the BlackBerry's main memory or on the memory card—as wallpaper. Grab a picture of your favorite person, a beautiful sunset, or any landscape for your own personalized BlackBerry wallpaper.

Changing Your Wallpaper or Home Screen Image Using a Stored Picture

1. Click on the **Media** folder.
2. Highlight the **Pictures** icon and click on it.

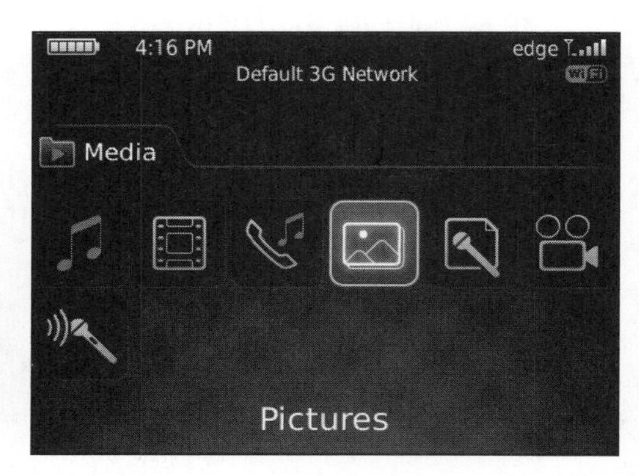

TIP: You can also change the wallpaper by pressing the **Menu** key from the **Home** screen and selecting **Options,** as shown on page 178.

3. Once in **Pictures**, use the **trackpad** to navigate to the location of the picture you wish to use—either in the **All Pictures**, **Picture Folders,** or **Sample Pictures**. You may also see an option to select the **Camera**, which you can use to take a new picture.

4. Highlight the thumbnail of the picture you want to use as your wallpaper, and then press the **Menu** key to select **Set As Wallpaper.**

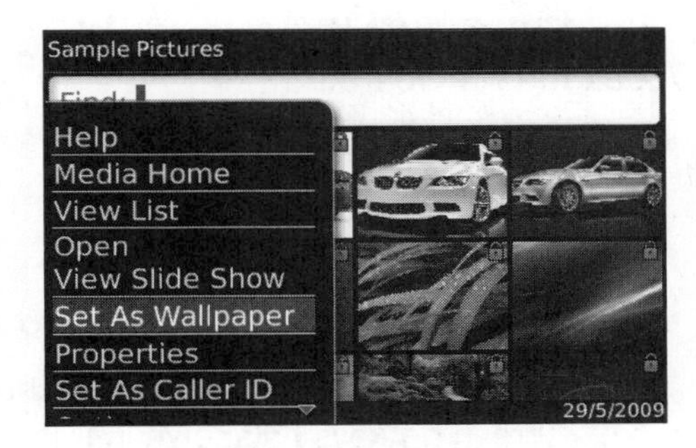

5. Now press the **Escape** key a few times to check out your new wallpaper on your **Home** screen.

Using a Picture or Image Directly from Your Camera

Take the picture. (Learn all the details about the camera on page 381.)

Click on the **Crop** icon as shown and select **Set as Home Screen Image**.

Keyboard Lock and Standby Mode

Do you ever need to put your BlackBerry into your pocket, purse, or bag but want to make sure no keys are accidentally pressed? If so, you can use either the **Keyboard Lock** key or the **Mute** key to lock the keyboard.

Enable or Disable Standby Mode:

1. Press and hold the **Mute** Key on the top right edge of your BlackBerry until you see the **Entering standby mode** message. (Figure 8-5)
2. To exit **standby mode**, just tap the **Mute** key again.

Lock the Keyboard:

You can do this in several ways:

1. Tap the **Keyboard Lock** key on the top edge of your Bold. (Figure 8-5)
2. Press and hold the letter **A** on your keyboard.
3. Tap the letter **K** key on your keyboard if you've turned on **Home** screen hotkeys as shown in Part 4 of this book.

Unlock the Keyboard:

If you have password security enabled (see page 516), simply type your password to unlock the keyboard.

If you don't have password security enabled, press the **asterisk** key **(*)**, then the **Green Phone** key to unlock.

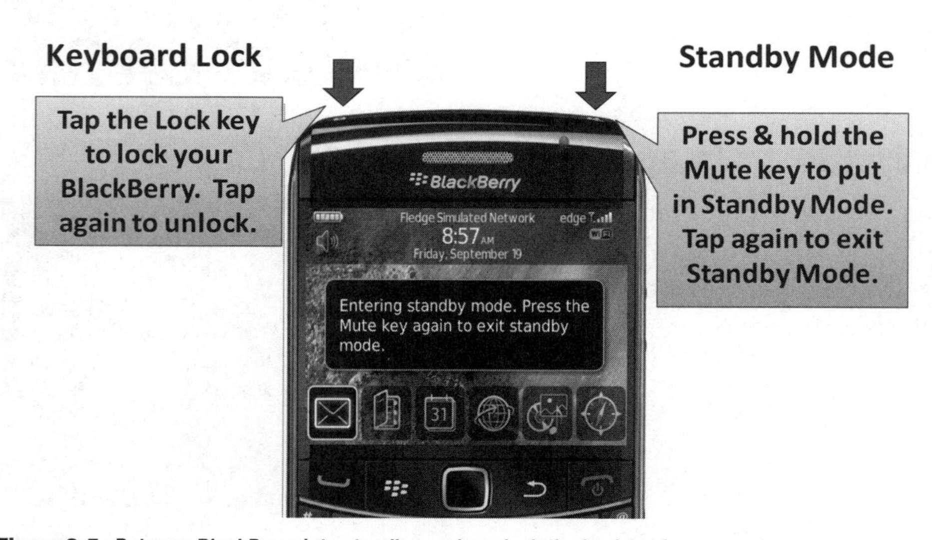

Figure 8-5. *Put your BlackBerry into standby mode or lock the keyboard.*

What If I Receive a Phone Call When in Standby Mode?

Don't worry; your BlackBerry will still ring or vibrate to notify you of an incoming call when it's in standby mode.

What's the difference between the two options? If you have password security enabled (page 516), pressing the **Lock** key will force you to enter your password to unlock it. When you use the **Mute** key to enter and then exit **standby mode**, you won't have to enter your password (unless the security timeout has elapsed). We have also read on some of the blogs that **standby mode** is supposed to save your battery more than the **Keyboard Lock mode**.

Changing Your Convenience Keys

Figure 8-6 points to two keys on the middle of the sides of your BlackBerry. These are actually programmable keys called **convenience** keys because each one can be set to conveniently open any app on your BlackBerry, even new third-party apps you add yourself.

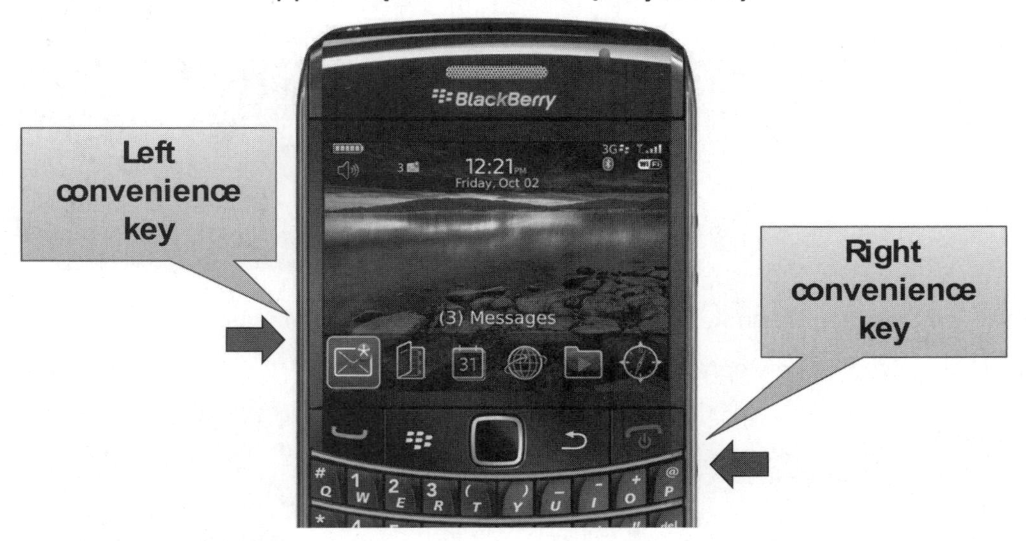

Figure 8-6. *Convenience keys*

To Change Your Convenience Keys

1. Click on the **Options** icon (press the **Menu** key if you don't see it.)

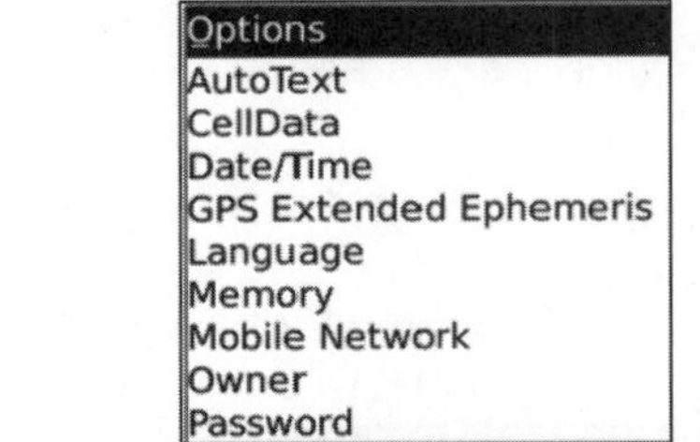

2. Press the letter **S** a few times to jump down to the **Screen/ Keyboard** item and click on it.
3. Scroll down until you see **Right Side Convenience Key Opens:** and **Left Side Convenience Key Opens:**.

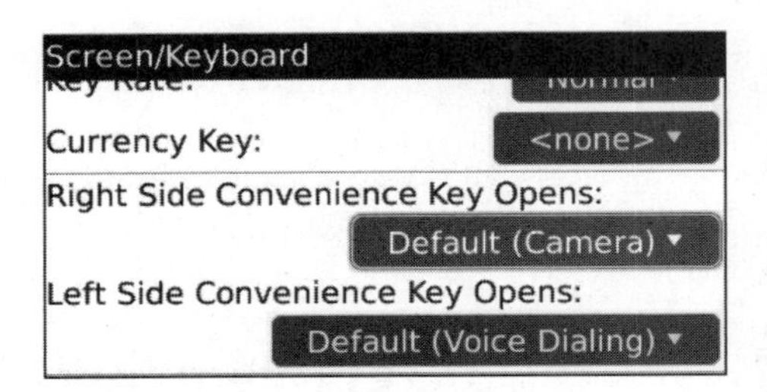

4. To change the applications these keys open, just click on the item to see the entire list. Then glide and click on the app you want.

5. Then press the **Menu** key and select **Save** to save your changes.
6. Now, give your new convenience keys a try.

TIP: The convenience keys work from anywhere, not just the **Home** screen. Also, you can set your convenience keys to open any app, even newly installed ones!

After you install new apps, you'll notice that they show up in the list of selectable apps in the Screen/Keyboard options screen. So if your newly installed stock quote, news reader, or game is important, just set it as a convenience key.

Understanding That Blinking Light

One of the features that BlackBerry users love is the little LED that blinks in the upper right corner to repeatedly notify you of something. This light may blink in different colors:

- Red when you receive an incoming message (MMS, SMS, or email) or the calendar alarm rings.
- Blue when connected to a Bluetooth Device.
- Green when you have wireless coverage.
- Amber if you need to charge your BlackBerry or it is charging.

Red Message LED

1. On the **Home** screen, click the **Sounds** icon.
2. Select **Set Ring Tones/Alerts** at the bottom of the list of profiles. (See page 191 for detailed help on profiles.)
3. Now click on **Messages** and then **Email** to see the screen shown here. Notice that one of the options is **LED** and you can choose to disable the LED by selecting **Off.**

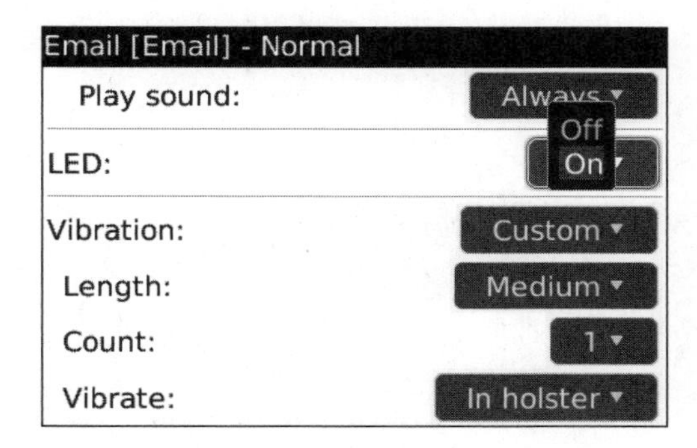

Blue Bluetooth LED

To enable or disable the blinking blue light (which can be a little annoying), follow these steps.

1. From your **Home** screen, click on the **Options** icon.
2. Click on **Bluetooth**.
3. Press the **Menu** key and select **Options.**
4. Go down to **LED Connection Indicator** and set to **On** or **Off**.
5. Press the **Menu** key and **Save** your settings.

Green Coverage Indicator LED

Click on the **Options** icon and scroll to **Screen/Keyboard** (or press the letter **S** to jump there) and click. Scroll down to **LED Coverage Indicator** and select either **On** or **Off**

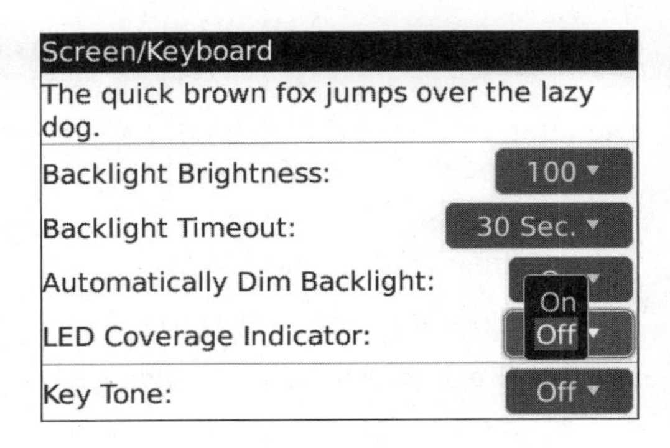

Chapter 9

Sounds: Ring and Vibrate

Your BlackBerry can be customized in many ways, notably in the area of sounds and ringtones. You can set up different sound profiles according to your needs. In this chapter, we will explain how you can create and adjust the various profiles.

You will learn how to make your BlackBerry ring, vibrate, ring and vibrate, or made no sound at all.

You'll see that you can customize your BlackBerry for individual situations and even individual contacts by setting unique ringtones and vibrations so you can immediately tell who is calling, texting, or even emailing you.

Understanding Sound Profiles

You can set up your BlackBerry to meet your individual needs—everything from ring tones to vibrations to LED notifications can be adjusted. Traveling on an airplane but still want to use your calendar or play a game without disturbing others? (See page 16 to turn off your radio.) No problem. In a meeting or at the movies and don't want the phone to ring, but still need to be notified when an email comes in? Again, no problem.

Virtually any scenario you can imagine can be dealt with preemptively by adjusting the profile settings.

Changing Your Profile

To set or change the profile settings:

Depending on your selected Theme and BlackBerry carrier (phone company), how you get to your **Sounds icon** may be slightly different.

On our BlackBerry, this speaker icon is the **Sounds icon**. If you don't see it on your screen, press the **Menu** key to see the entire list of icons, then scroll to the **Sounds icon** and click on it.

TIP: Pressing the letter F on your keyboard should start Profiles if you turned on hotkeys (see page 541).

Seven basic preloaded settings are available: Normal, Loud, Medium, Vibrate Only, Silent, Phone Calls Only, and All Alerts Off. You'll see the word "**Active**" next to one of them.

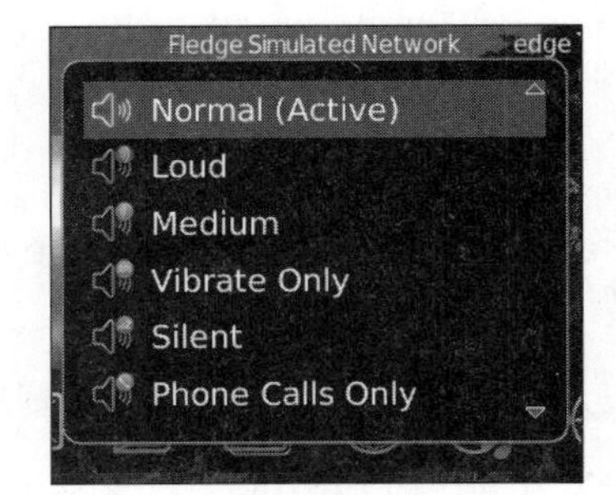

For most users, **Normal** will be the active profile that rings to announce a phone call and either vibrates or plays a tone when a message arrives.

Loud increases the volume for all notifications.

Medium is between Loud and Quiet.

Vibrate Only enables a short vibration, good for meetings, movies, or other situations where cell phone rings are discouraged.

Silent displays notifications on the display and via the LED.

Phone Calls Only turns off all email and SMS (text) notifications.

All Alerts Off turns off all notifications.

Fine-Tuning Your Sound Profiles

There may be some situations where you want a combination of options that one profile alone can't satisfy. The BlackBerry is highly customizable, so you can adjust your profile options for virtually any situation. The easiest way to accomplish this is to choose a profile that is closest to what you need and edit it as shown here.

To find the advanced **Profiles** menu:

1. Click on the **Sounds icon** as you did above.
2. Scroll down to **Set Ring Tones/Alerts** and click.

3. Now you can adjust the tones and alerts for the **Phone**, **Messages**, **Instant Messages**, **Reminders**, or **Browser** by choosing the field to edit.

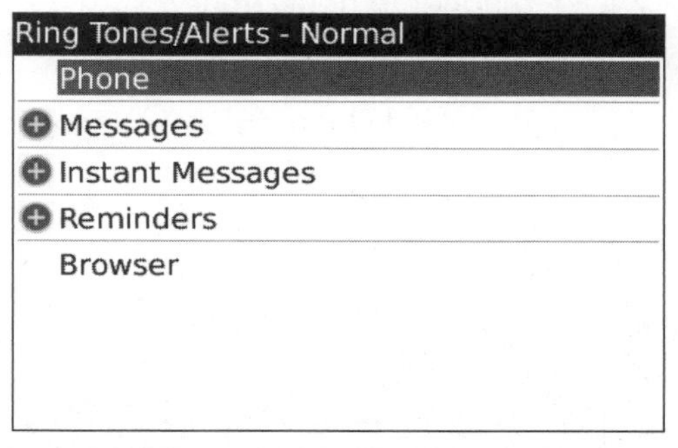

4. There are profile settings for almost every alert you can have on your BlackBerry:

   BlackBerry Messenger (Alert and New Message), Browser, Calendar, Level 1 Messages (High Priority), Email, Reminders, Phone, SMS and Tasks.

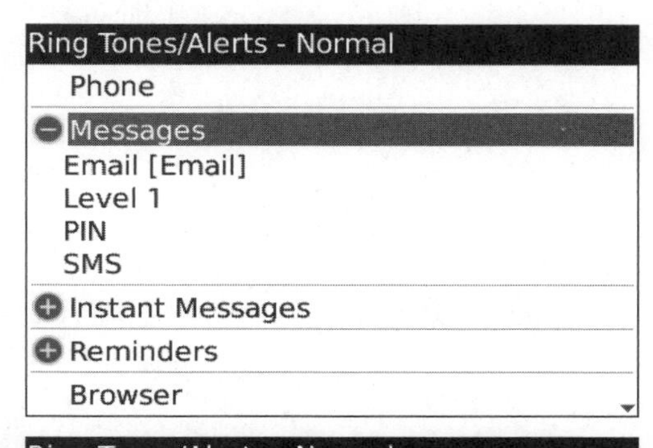

> **TIP:** Some new applications you install—some news readers, for example—can also have profile categories in this screen.

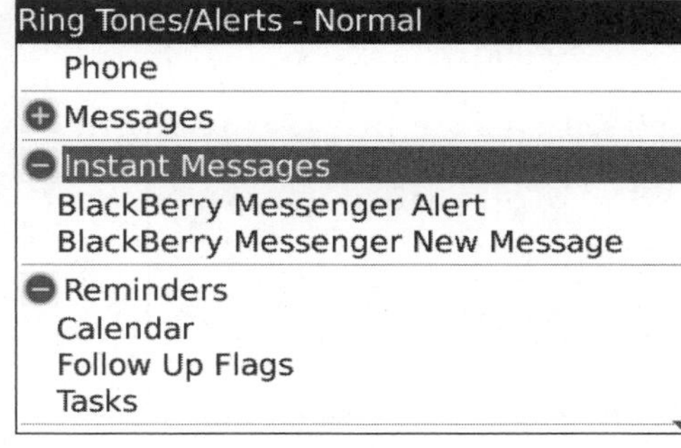

5. For example, choose **Email** and notice that you can make adjustments for your BlackBerry both **Out of holster** and **In holster**.

   A holster may be supplied with your device or sold separately. This is a leather or plastic carrying case that clips to your belt. It uses a magnet to notify your BlackBerry it is In Holster and should turn off the screen immediately (among other things).

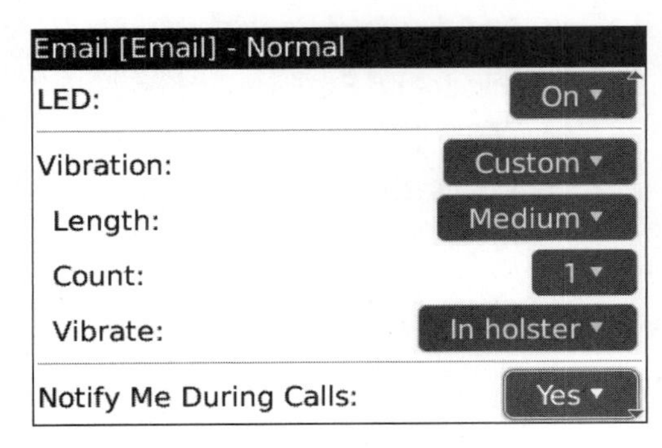

Vibrate + Tone = Vibrate first, then ring.

Count = Number of times to repeat the ring tone or LED.

LED = Red LED light flashes.

TIP: We really like to configure the **Vibrate** setting to **Always** for almost everything because it allows us to grab the BlackBerry most times before it starts ringing!

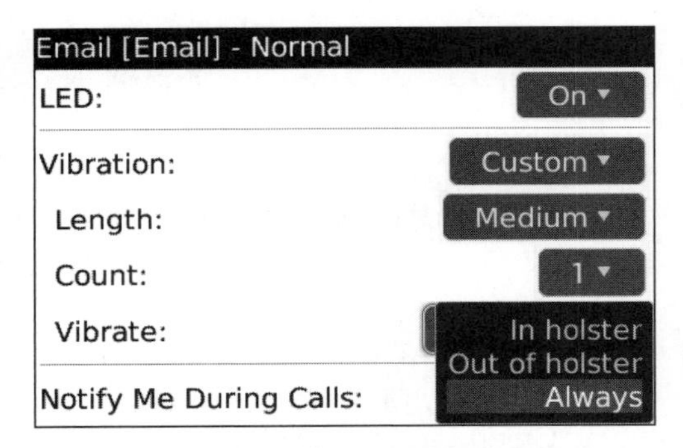

6. If you want to be notified during phone calls of an incoming message or alert, select **Yes** for the **Notify Me During Calls** setting.

CAUTION: If you set a **Tone,** change the volume to something other than **Mute** to hear it.

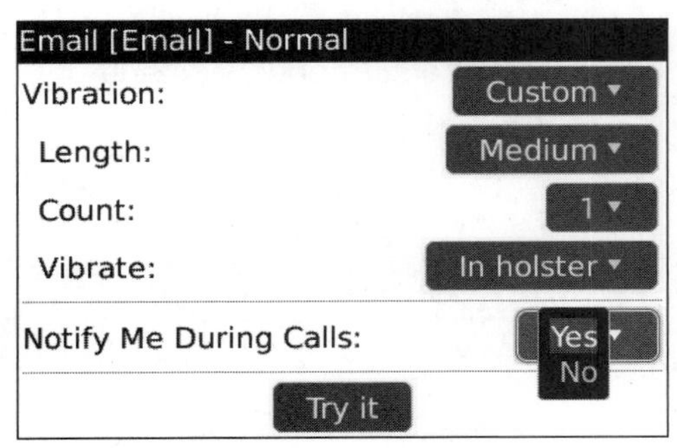

Now every time you receive a new email, you will experience this new profile setting.

TIP: If you have several email accounts integrated with your BlackBerry, you can customize each one with a separate profile (ring, vibrate, or mute).

Normal
Calendar
Level 1
Messages [info@blackberrymadesimple.c...
Messages [martin.trautschold@gmail.com]
Messages [martin@blackberrymadesimpl...
Messages [martinbb2008@att.blackberry...
Messages [orders@blackberrymadesimpl...
Messages [videocontact@blackberrymad...
Messages [Web Client]
MMS
Phone

Finding a Louder Ringtone

Sometimes, the standard ringtones just aren't loud enough for you to hear, even when you turn the Volume up to Loud. Luckily, you can download a new ringtone from mobile.blackberry.com to help with this problem.

Open your BlackBerry web browser. If you can type in a web address, type in "**mobile.blackberry.com**" and click the **trackpad**. If you can't type a web address, press the **Menu** key and select **Go To...** then type **"mobile.blackberry.com"**

1. Scroll down the home page until you see the **Personalize** topic, then click on the **Ringtones** link.
2. You will have to **Accept** the **Terms and Conditions** to continue.
3. Once you do, you'll see a list of ringtones. Click on any one to give it a try.

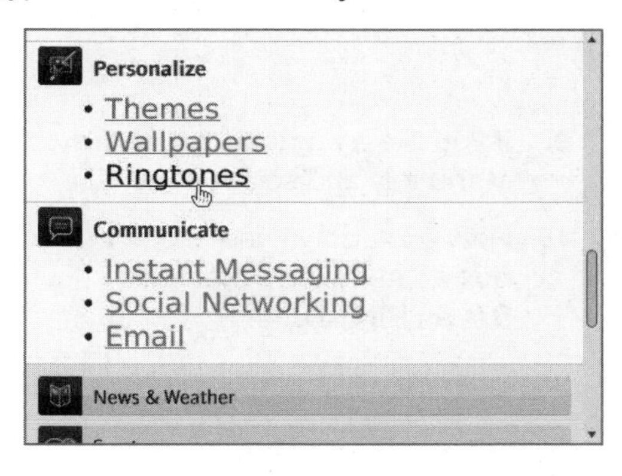

4. At the top of the page you may see the **Top 5 Ringtones**, as shown to the right.
5. Click on any **Download** link to download and try out the ringtone.

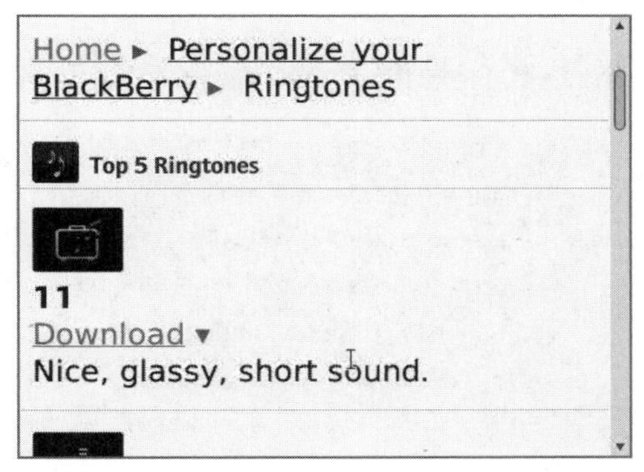

6. If you scroll further down the page, you'll come to a section labeled **Ringtone Categories**.

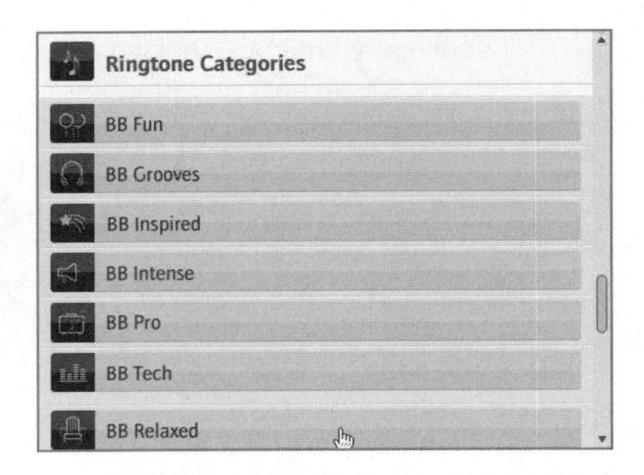

> **TIP:** The BB Intense and BB Pro categories seem to have some of the louder ringtones.

7. When you click on a ringtone, you can **Open** it to listen or **Save** it on your BlackBerry.
8. Go ahead and **Open** a few to test them out. To get back to the list and try more ringtones, press the **Escape** key.

9. If you like a ringtone, press the **Menu** key and select **Save**.
10. Now glide down and check the box at the bottom that says **Set as Ring Tone.**

That's it. Next time you receive a phone call the new louder ringtone should play.

> **TIP:** New ringtones are available on many BlackBerry user web sites like `www.crackberry.com`, where many are free. You can also purchase ringtones at web stores like `www.CrackberryAppStore.com`.

Also check out the other web stores and discussion sites listed on page 480.

Hearing a Different Tone When Someone Special Calls

You may decide you want to hear a different ringtone when someone special calls. You can do this in **Profiles** and in your **Contacts** (Address Book) by assigning a **Custom Phone Tune** to a particular contact. Below we show you the **Profiles** method, which has the added benefit of allowing you to assign as many names as you want to a single ringtone.

To set a custom notification profile:

1. Click the **Sounds icon.**
2. Scroll down and click **Custom Profiles.**

3. Press the **Menu** key and scroll to **New** and click.
4. Type in any name for this new **Profile** in the field marked **Name**, like **Boss.**
5. You can set each individual tone or alert as you did above.

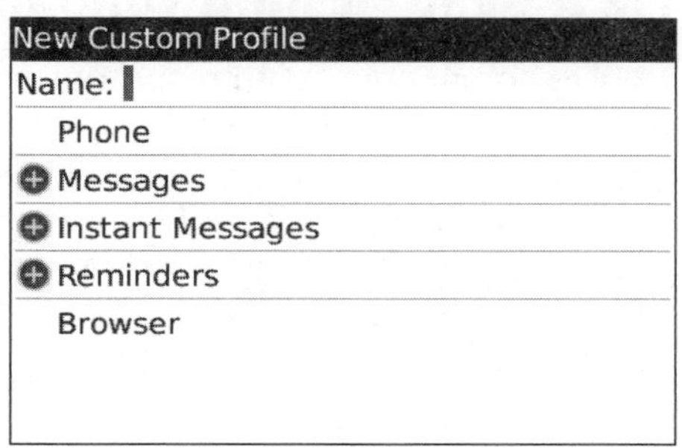

6. You can also set a **Contact Alert** from the **Profiles** menu. Just click on the **Sounds icon** and scroll down to **Contact Alerts** and click the **trackpad**. Then, click on **Add Contact Alert** and press the **trackpad** and select **Add Name.**

7. In this example, I want the phone to ring loudly when my friend Martin calls so I don't miss the important call.

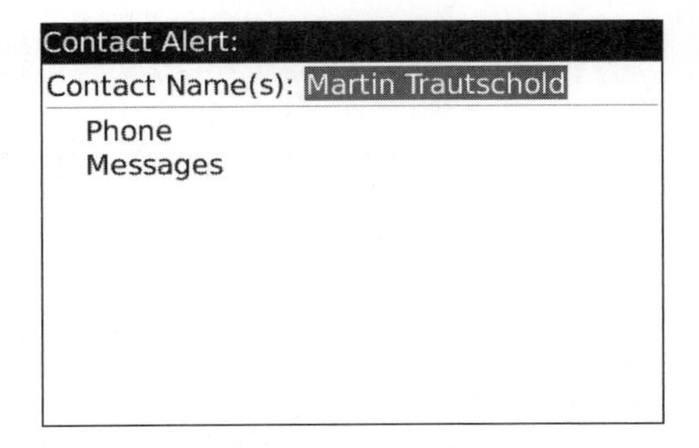

8. You now see Martin's name listed next to the **Messages** field.

9. All the various options for customizing his ring tone, volume, and so forth are now available for me to set.

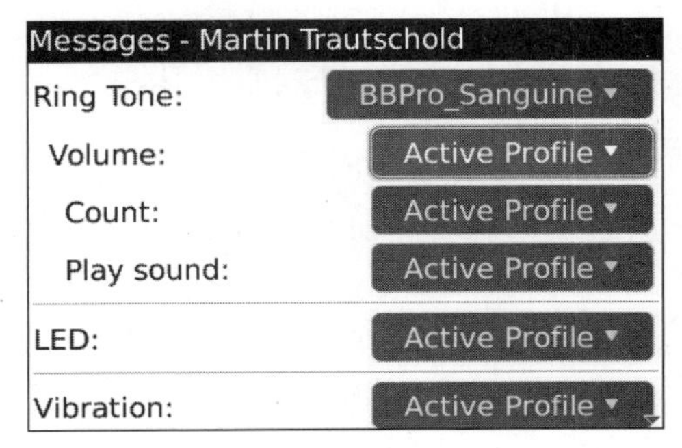

Tie a Custom Ring Tone to One Person

To give someone a custom ringtone so you know who is calling without looking at your BlackBerry screen, follow these steps:

1. Start your **Contacts** app.

2. Type a few letters to find the contact you wish to edit, for example, **Ma Tr** to find **Martin Trautschold**.

3. Click the **Menu** key and select **Edit**.

4. Now, in the **Edit Contact** screen, scroll down to **Custom Ring Tones/Alerts**.

5. Click on the **Phone** field and all the available ring tones on the BlackBerry will be displayed.

6. Just scroll up to the top. You can even **Browse** your music and turn a song into a ringtone.

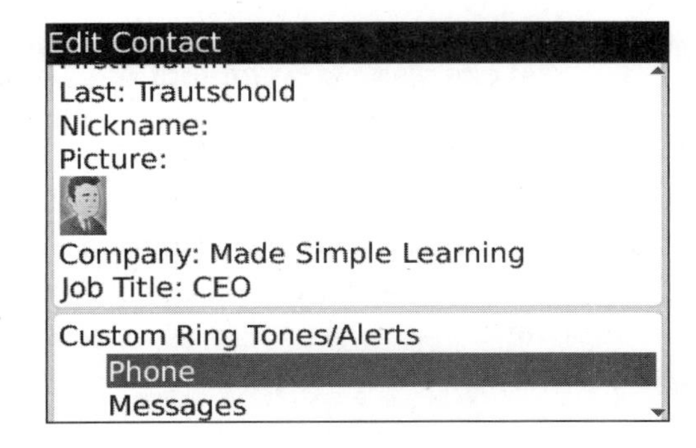

7. Now click the **Browse...** button to select the custom ringtone for this person. You can pick from any preloaded ringtones or any ringtone you have placed in the ringtones folder.

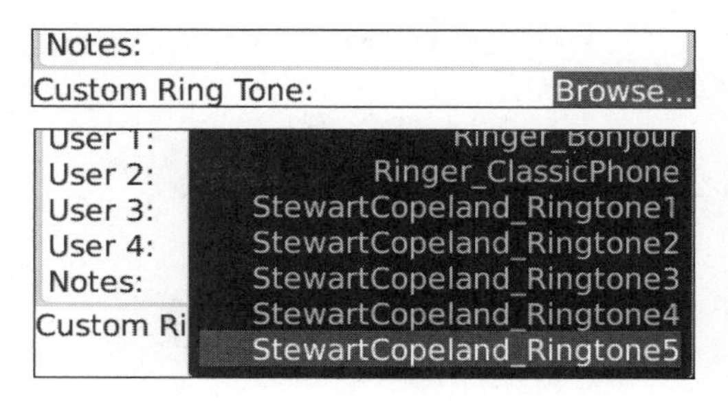

8. Once you select the ringtone, you'll see it listed at the very bottom of the contact entry.

9. Press the **Menu** key and select **Save** when you are done to save the changes.

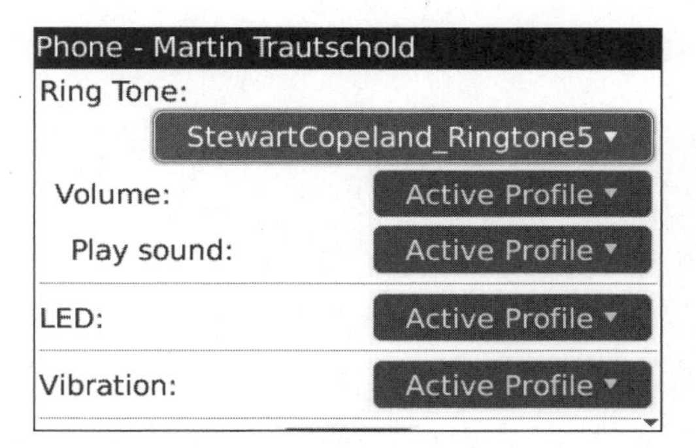

Changing a Custom Ring Tone

1. Edit the contact entry as shown above.

2. Glide to the **Custom Ring Tone** field and click the **trackpad**.

3. Select any other ring tone for the contact.

Chapter 10

Phone and Voice Dialing

Sometimes, it is easy to forget that your BlackBerry is first and foremost a phone—and a very capable phone at that. There are many extra features on your phone, as well as the ability to use voice dialing to place your calls.

In this chapter, we'll show you how to place and answer calls, mute calls, use voice dialing, and how to use the call log. We'll also show you how to set up and use speed dial on your BlackBerry.

Basic Phone Features

Let's see how to quickly start using all the basic phone features of the BlackBerry. Figure 10-1 identifies the important keys you'll need.

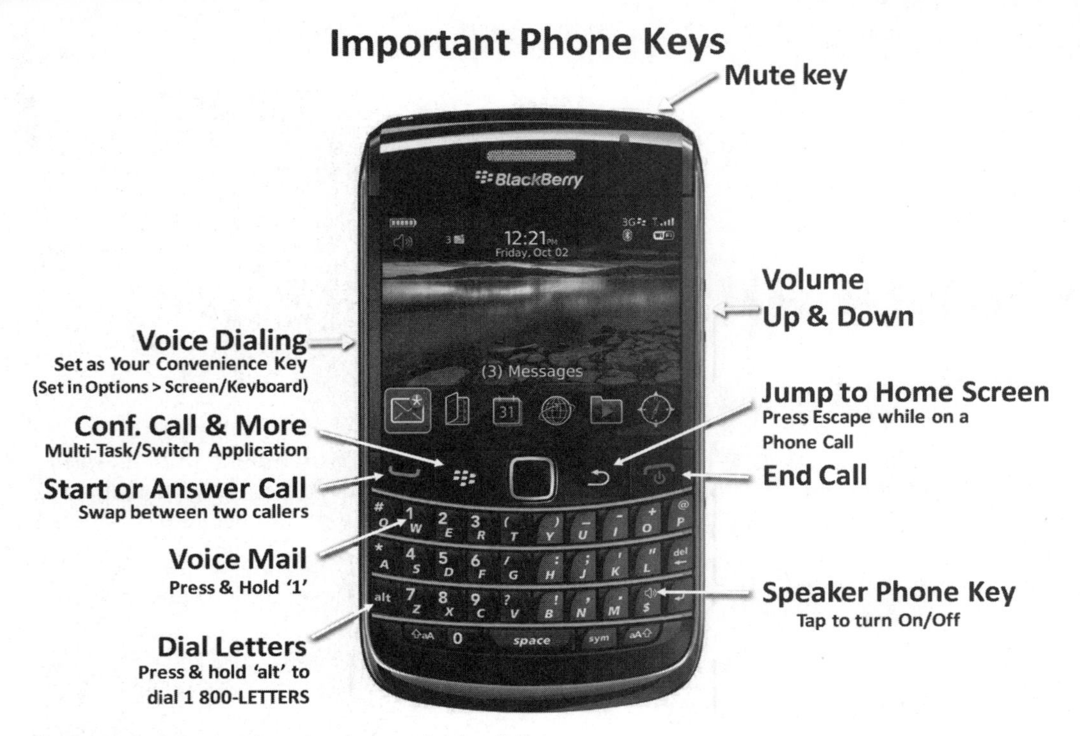

Figure 10-1. *Important keys for phone and voice dialing.*

Call Any Underlined Phone Number

You may have noticed that most phone numbers you see on your BlackBerry are underlined, as shown in Figure 10-2. This means you can place a call to that number. This feature works anywhere, in an email signature, calendar event, MemoPad item, or a web browser screen.

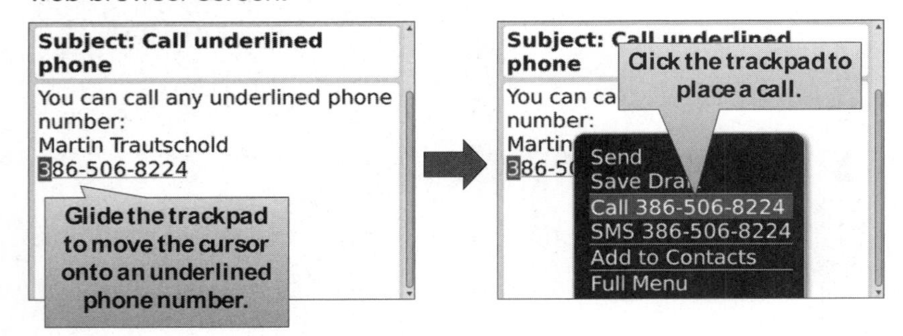

Figure 10-2. *You can call any underlined phone number*

Changing Your Ring Tone

To set a song as a ringtone, follow these steps.

1. Click on the **Music** icon, it may be in your **Media** folder. (Learn more about getting around the Music app in chapter 21 "Your Music Player").
2. Navigate to and click on the song you want to use as a ring tone to play it.
3. Press the **Menu** key and select **Set as Ring Tone**.
4. The next time you receive a call, your favorite song will play.

To use a pre-loaded ring tone as your phone ring tone.

If you want to use a pre-loaded ringtone, then instead of going to Music in your Media icon, you would go to the Ring Tones app.

1. Open your **Media** folder.
2. Click on **Ring Tones**.
3. Then select **All Ring Tones**, **My Ringtones**, or **Preloaded Ring Tones**.
4. Scroll down the list and click the trackpad to listen to the selected ring tone.
5. You can also type a few letters to find a ring tone that matches the letters you type.

TIP: Type in "ringer" to find all the phone ring tones.

6. Once you find what you want, press the **Menu** key and select **Set As Ring Tone**.

Adjusting the Volume on Calls

You may sometimes have trouble hearing a caller. The connection may be bad (*because of the caller's old-fashioned—i.e., non-BlackBerry—phone*) or you may be using a headset. Adjusting the volume is easy. While on the phone call, simply use the two volume keys on the right side of the BlackBerry to adjust the volume up or down.

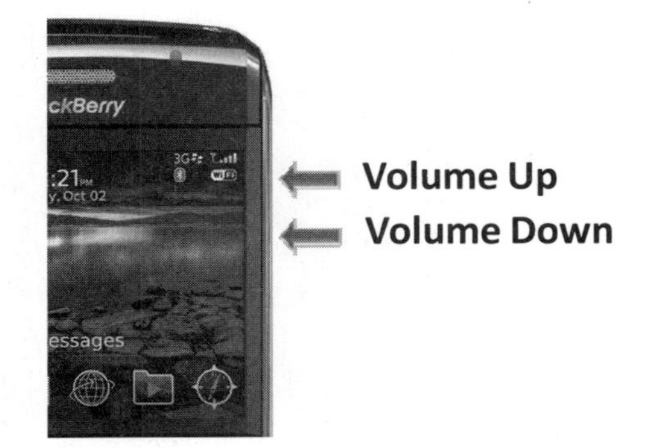

Muting Yourself on the Call

You may want to be able to mute yourself on a call. It might be so you can discuss something in private, or you may just want to be quiet as you listen to a conference call. To mute or unmute the call, just tap the **Mute** key on the top of the BlackBerry.

What's My Phone Number?

You have your phone, and you want to give your number to all your friends—you just need to know where you can get your hands on that important information. Here's an easy way to do this:

Press the **Green Phone** key and read **My Number** at the top of the screen.

This should be your phone number. In the image, the phone number of this BlackBerry is **1 519 888 7465**.

11:44 AM 3G
GSM Test Network 2
My Number: 1 519 888 7465
David Parker (M) 9:05a
+16065551923 9:04a
3865555712 9:04a
Gary Mazo (W) 9:00a
Martin Trautschold (W) 12/13

Adding Pauses and Waits in Phone Numbers

Some of the phone numbers you'll enter in your address book may require either a pause or a wait. This might be when you're dialing a conference call number, entering your password and PIN number for a voice mail access system, or want to auto-dial an extension at the end of a number, but need the extra pause. If you need more than a 3-second pause, just add a few more pauses--you can put as many pauses together as you need (Figure 10-3).

Pause = 3 second pause, then continues dialing automatically.

Wait = Waits for you to click the **trackpad**, then continues dialing.

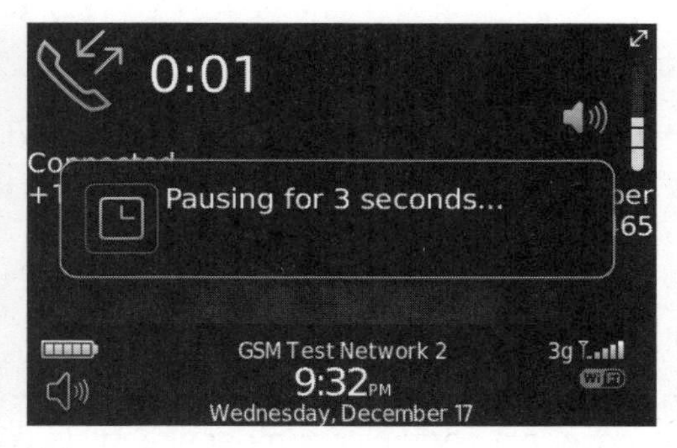

Figure 10-3. *Adding a pause when dialing.*

To add pauses and waits, press the **Menu** key (or sometimes the **trackpad** works) and select **Add Pause** or **Add Wait** from the menu.

> **TIP:** Typing a phone number, then an x and the extension is the same as adding a pause. For example: **1 800 555 1212 x 1234** is the same as adding a pause after the phone number and before the extension.

TIP: If you frequently have to dial a number that has several pieces (like when you call for your work or home voice mail messages), you can add an entry into your contact list (address book) with pauses or waits (see page 274) and assign this new entry to a speed dial (see page 216).

Changing Your Phone Ringtone

To select any of your **songs** or **preloaded ringtones** on your BlackBerry as a new phone ringtone, please check out the steps in our Advanced Phone section on page 221.

Making a Call—Just Dialing a Phone Number

Press the **Green Phone** key at any time to get into the **Phone** application.

Just start dialing numbers.

First, the BlackBerry will try to match the letters you are typing to **Address Book** entries. If it can't find any, it will just show you the digits you have typed.

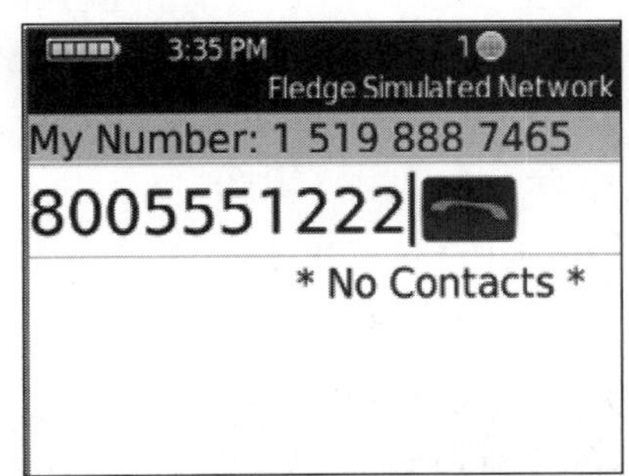

You will notice a small image of the **Green Phone** key immediately after the cursor. Once all the numbers are punched in, just press the **Green Phone** key and the call will be placed.

Answering a Call

Answering a call couldn't be easier. When a call comes in, the number will be displayed on the screen. If you have that particular number already in your **Address Book**, the name and/or picture will also be on the screen (if you've entered that information.)

When a call comes in:

Push either the **Green Phone** key or click the **trackpad** to answer the call.

If you are using a Bluetooth headset, you can usually click a button on the headset to answer the call, see page 441.

Calling Voice Mail

The easiest way to call voice mail is to press and hold the number **1** **(W)** key. This is the default key for voicemail.

To set up voice mail, just call it and follow the prompts to enter your name, greeting, password, and other information.

When Voice Mail Does Not Work

Sometimes, pressing and holding the **1** key does not dial voice mail. This happens if the voicemail access number is incorrect in your BlackBerry. You'll need to call your phone company (wireless carrier) and ask them for your local voice mail access number.

This sometimes happens if you move to a different area or change cell phones, then restore all your data onto your BlackBerry.

Once you have the new voice mail phone number from the carrier, you need to enter it into your BlackBerry.

1. Start your **Phone** by pressing the **Green Phone** key. Press the **Menu** key and then the letter **O** to jump down to the **Options** item and select it.

2. Click on Voice Mail

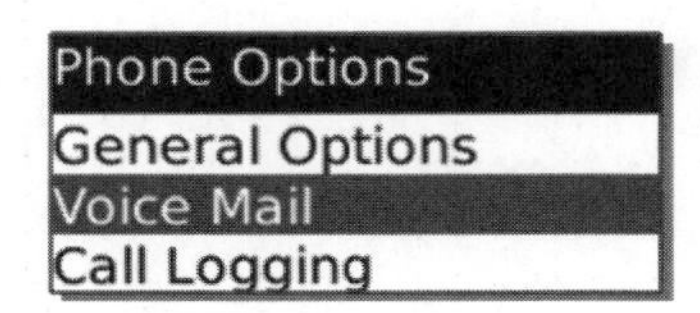

3. Enter the phone number you received into the Voice Mail **Access Number**.

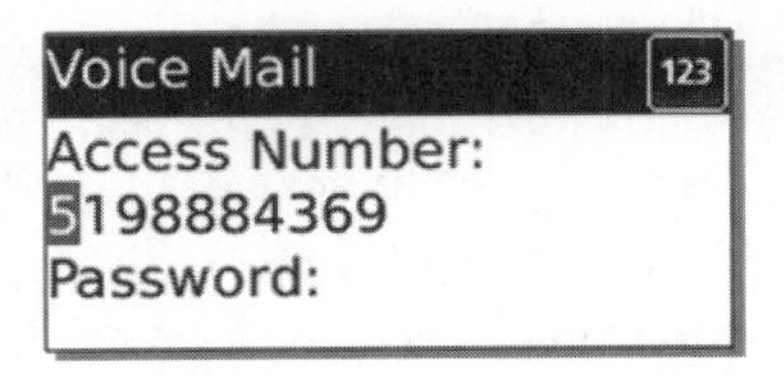

TIP: You can also enter your voice mail password if you like.

Why Do I See Names and Numbers in My Call Logs?

You'll see both phone numbers and names in your phone call logs. When you see a name instead of a phone number, you know that the person is already entered in your BlackBerry **Address Book**.

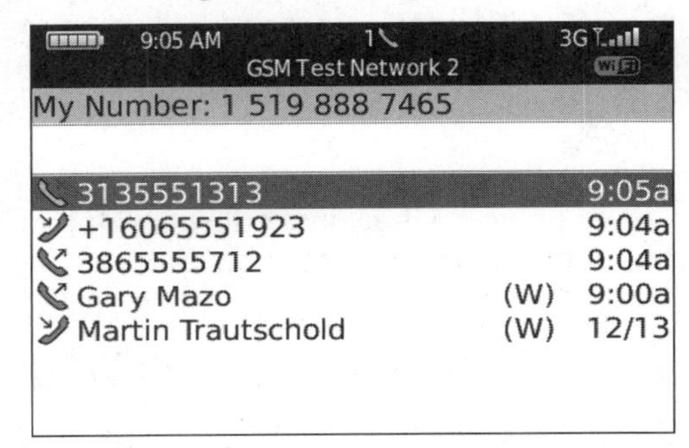

It is easy to add entries to your **Contacts** right from this phone call log screen. Below we show you how.

How Can I See Missed Calls on My Home Screen?

Many of the Themes show your missed calls with a phone icon with an X next to it, and a pop-up window as shown to the right. Here's an image with one missed call showing on the **Home** screen.

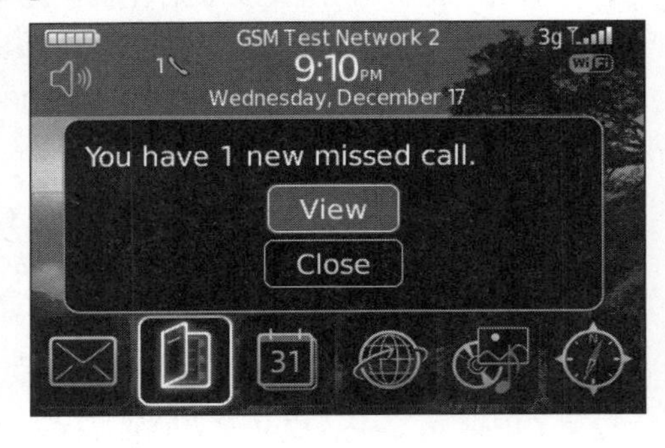

Quickly Dial from Contact List

You can quickly dial from your contact list by pressing and holding the **Green Phone** key from just about anywhere, except when you are already in the **Phone** app or have a phone number underlined. (Pressing it in those cases will start a phone call to the highlighted number or person.)

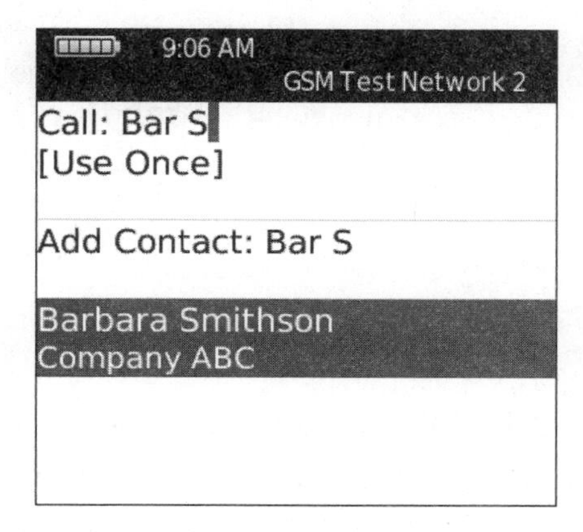

When you let go of the **Green Phone** key, you should see your contact list. Find an entry by typing a few letters of the first name, last name, nickname, or company name.

Add New Contact Entry from Phone Call Logs

If you see a phone number in your call log screen, there's a good chance you'll want to add that phone number as a new **Contact** entry.

Call log entries are generated whenever you receive, miss, ignore, or place a call from your BlackBerry.

1. Get into the call log screen by tapping the **Green Phone** key once.
2. Highlight the phone number you want to add to your **Address Book** from the **Call Log** screen.
3. Now press the **Menu** key and select **Add to Contacts**

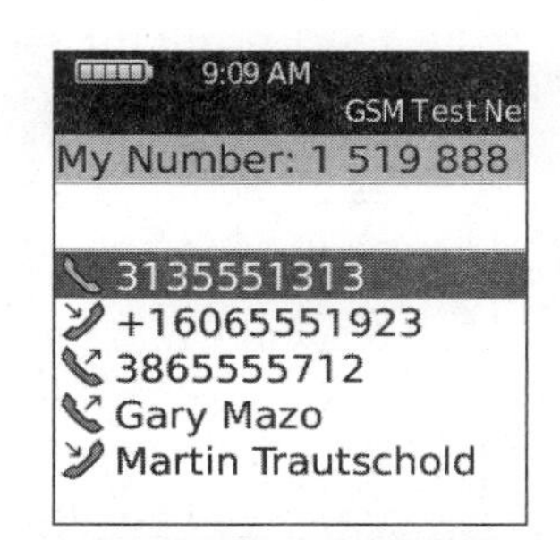

4. Type in as much information as possible. The more you add, the better your BlackBerry can help you communicate!

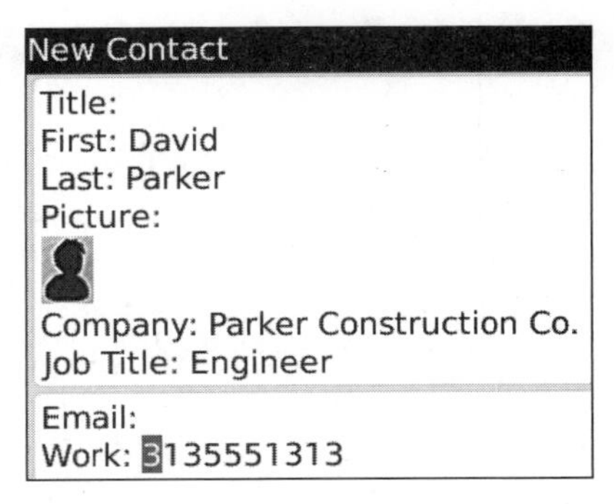

5. Notice that the BlackBerry puts the phone number into the **Work** field.

6. Is this phone number not a work number? Then just cut and paste it into another field.

Cut and Paste to Move a Phone Number

1. With the cursor at the beginning of the phone number you want to cut, press the **Shift** key to start highlighting the phone number.
2. Glide the **trackpad** down one click to highlight the entire number.
3. With the phone number highlighted as shown, press the **trackpad** and select **Cut.**
4. Now move to the correct phone field, e.g., the **Mobile** field shown to the right, and click the **trackpad** to **Paste** the number.

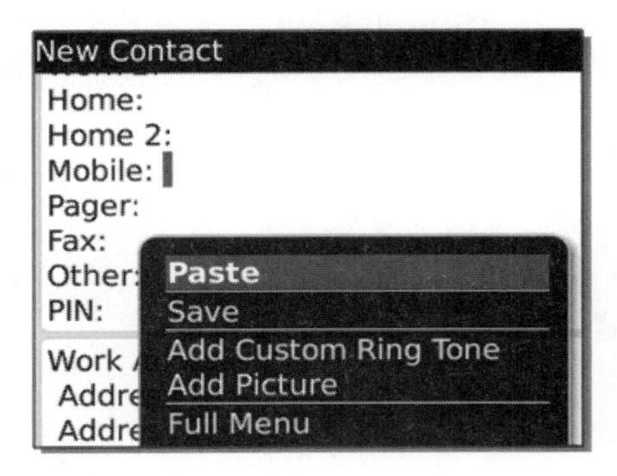

5. Enter the rest of the information for this person. When you're done, press the **Menu** key and select **Save.**

6. You'll see that **David Parker's** name replaces his phone number in the call logs since he is now in your **Contacts**.

> **TIP:** Learn more about entering new addresses on page 272.

Ignoring and Muting Phone Calls

Sometimes, you can't take a call and you need to ignore or perhaps mute the ringing of an incoming call. Both of these options can be achieved quite easily with your BlackBerry.

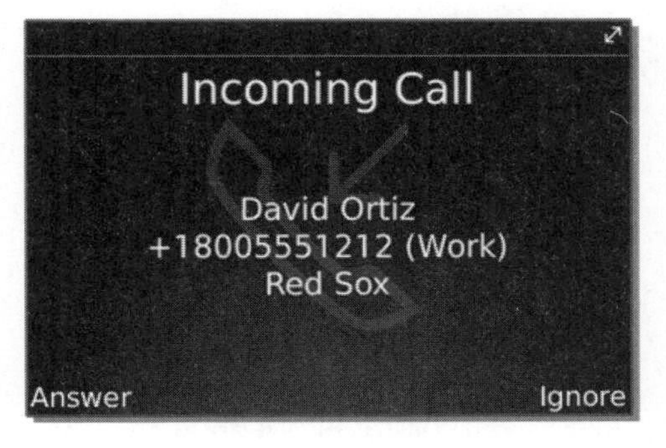

Ignore a Call and Immediately Stop the Ringing.

When the phone call comes in, simply press the **Red Phone** key to ignore the call, send to voice mail, and stop the ringer.

> **TIP:** Need to silence the ringer but still want to answer the call? Just gliding the **trackpad** up and down will give you a few more seconds to answer the call before the caller is sent to voice mail. Also, if the ringing or vibrating started while your BlackBerry was still in the holster (carrying case), simply pulling the BlackBerry out of the holster should stop the vibrating and ringing, but still give you time to answer.

Ignoring a call sends the caller to your voice mail.

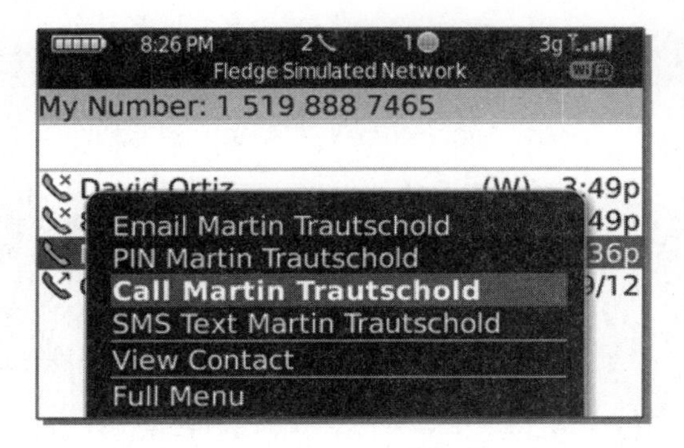

The **Missed Call** will be displayed on your **Home** screen. Click on the **Missed Call** with the **trackpad** and a small menu pops up that allows you to do various things depending on whether the phone number is already in your **Address Book**.

Benefits of Adding People to Your Contact List / Address Book

- Call any number for a person who is entered into your address book.
- Send an email if the person has an email address entered.
- Send an SMS text message.
- Send an MMS multimedia message, with pictures or songs.
- Send a PIN message.
- View the contact information.

Muting a Ringing Call

If you'd prefer not to send a call immediately to voice mail and simply let it ring a few times on the caller's end, but you don't want to hear the ring (perhaps you are in a movie theatre or a meeting), when the call comes in, press the **Mute** key on the top of your BlackBerry. The **Mute** key has the small speaker icon with the line through it. All this does is silence the ring.

You can still pick up the call or let the caller go to voice mail.

Using the Call Log

The **Call Log** is an especially useful tool if you make and receive many calls during the day. It can be hard to remember whether you added a particular individual to your **Address Book**—but chances are you'll remember that she called yesterday. Here's a perfect situation to use your Call Log to access the call, add the number into your Address Book and place a return call.

Checking Your Call Log

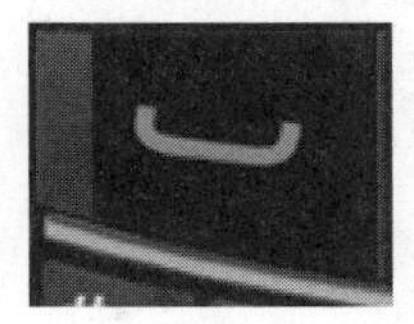

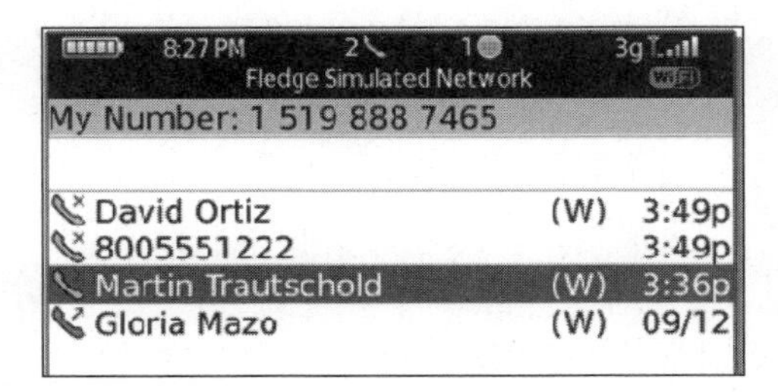

Just tap the **Green Phone** key from anywhere to see your call logs—a sequential list of calls placed, missed, or received.

> **TIP:** Change the way call logs are displayed by pressing the **Menu** Key and then selecting **Options**. Then select **General Options** and change the **Phone List View** to the option you desire—Most Recent, Most Used, or Name.

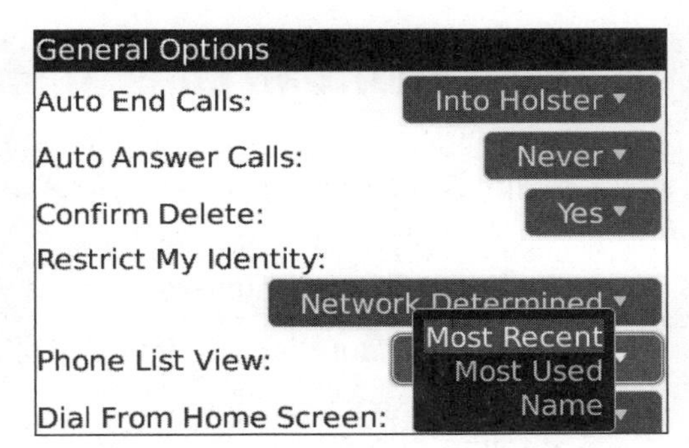

Placing a Call from the Call Log

Go to the **Call Logs** as you did above and scroll through the list.

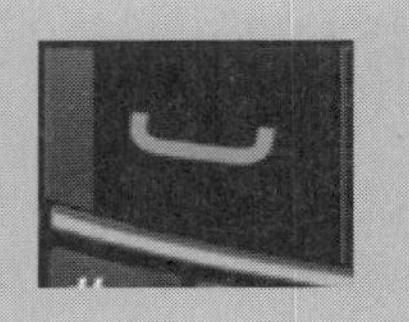

TIP: If you want to call the number listed—in this case, Susan's mobile number **(M)**—don't press the **trackpad**, press the **Green Phone** key instead to immediately start the call.

If you want to call one of the other numbers for this person, either press the **Menu** key or click the **trackpad**.

You'll be given a choice as to which number to call. **Work**, **Mobile,** or others in your contact list for this person.

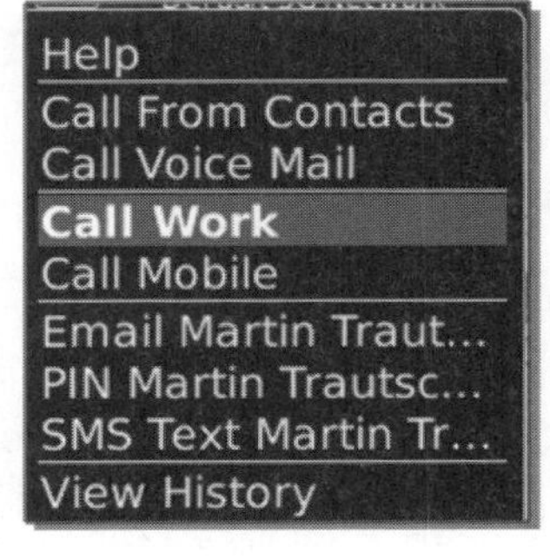

To Show Your Call Logs in Messages (Inbox)

It can be useful to show calls made, received, and missed in your message list for easy accessibility. This allows you to manage both voice and message communication in a single unified inbox.

1. Press the **Green Phone** key to see your call logs.
2. Press the **Menu** key and scroll down to **Options** and click.

TIP: If you press the letter **O**, you'll jump down to the first menu item starting with O. This should be Options.

3. Scroll to **Call Logging** and click.
4. Under Show These Call Log Types in Message List, select either:
   - **Missed Calls** (see only missed calls)
   - **All Calls** (see all placed, missed, received)
   - **None** (this is the default; you don't see any calls)
5. Press **Menu** and select **Save**.

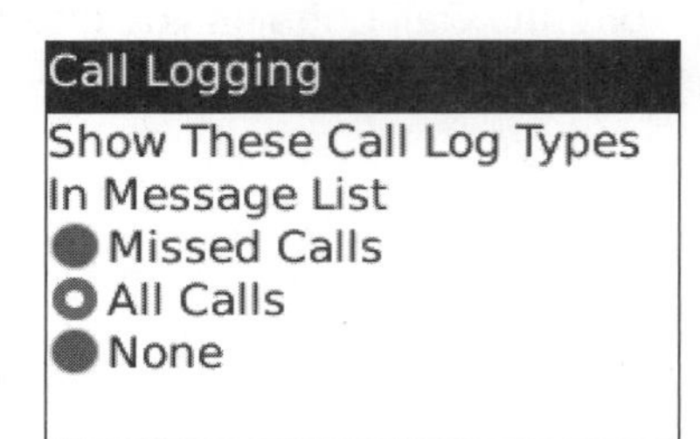

To Add a Note to a Call Log

1. Press the **Green Phone** key to get into the phone logs if you are not already there. Press the **Menu** key and select **View History.**
2. Select the item to which you want to add a note by gliding the **trackpad** up or down.
3. Press the **Menu** key again and click **Add Notes**.
4. When you are done typing, click the **trackpad** and select **Save.**

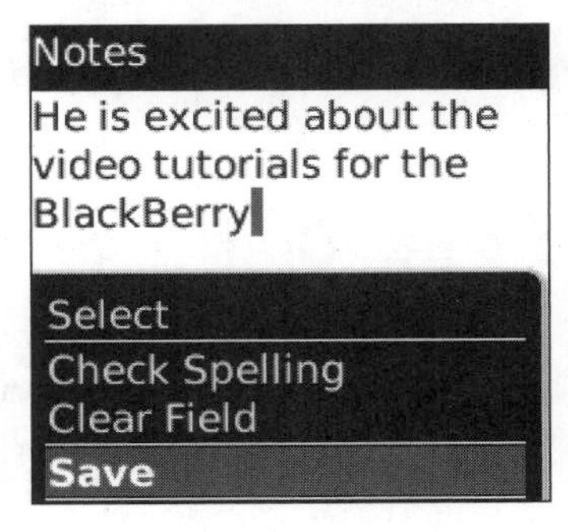

TIP: You can even **Add Notes** while you are still talking on the phone.

You may want to use the speakerphone or your headset so you can hear while typing.

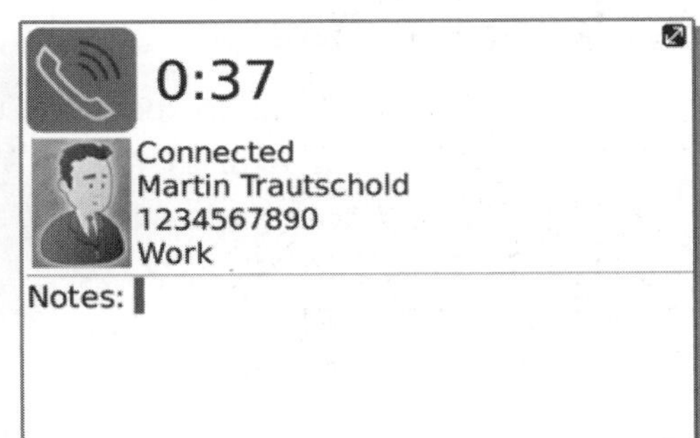

To Forward a Call Log

1. Go to your **Call Log** and highlight the log entry you wish to forward.
2. Press the **Menu** key and click **View History** just as you did above.
3. While viewing the call history entry, press the **Menu** key again and select **Forward**.

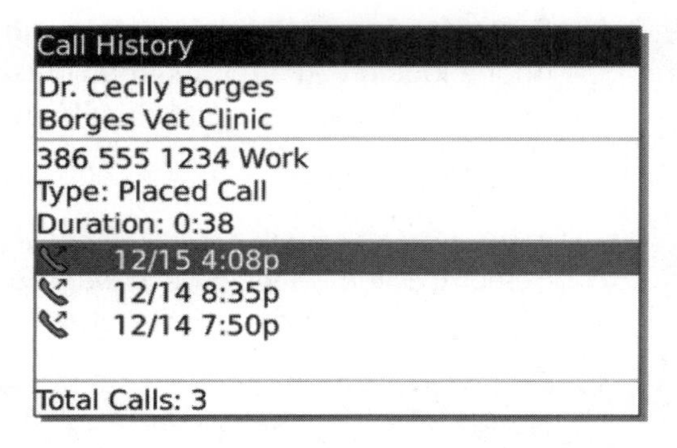

4. Type your email message or make changes to the notes, then click the **trackpad** and select **Send**

Speed Dial

Speed dialing is a great way to call your frequent contacts quickly. Just assign them a one-digit number key (or character key) that you hold, and their number is automatically dialed. You have up to 26 speed dial entries.

Set Up Speed Dial

There are a few ways to set up speed dialing on your BlackBerry.

> **TIP:** The easiest way to set a speed dial letter on your keyboard is to press and hold it from your **Home** screen. You'll be asked if you want to set it as a **Speed Dial**. Select **Yes** and choose the person from your **Contacts** list to assign.

Option #1: Use Call Logs

1. Press the **Green Phone** key to see your call logs.
2. Highlight the call log entry (either phone number or name) that you want to add to speed dial and press the **Menu** key.
3. Select Add Speed Dial.

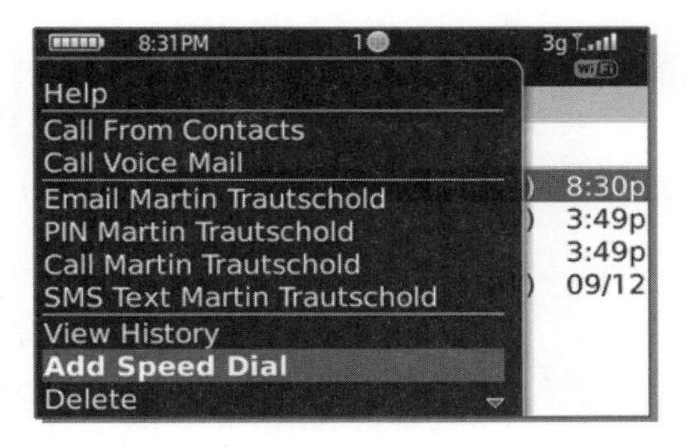

> **NOTE:** If this person is already on speed dial, you'll see a **Remove Speed Dial** option instead.

4. You may be asked to confirm that you want to add this speed dial number with a pop-up window like this.

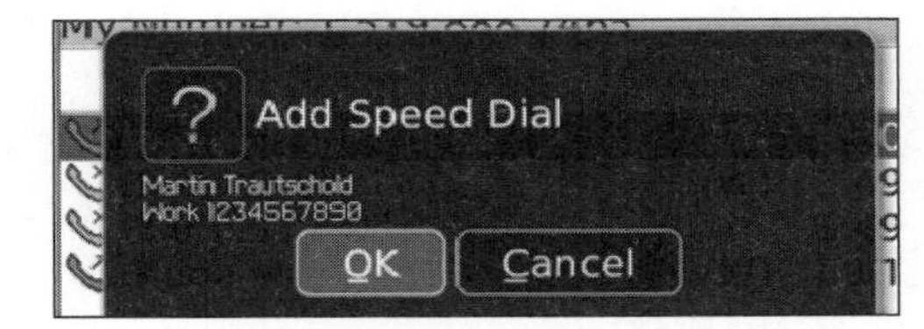

5. In the **Speed Dial Numbers** list, use the **trackpad** and move the phone number to any vacant slot.
6. Once you've chosen the correct speed dial key, just click the **trackpad**. The number or symbol you selected is now set as the speed dial key for that phone number.

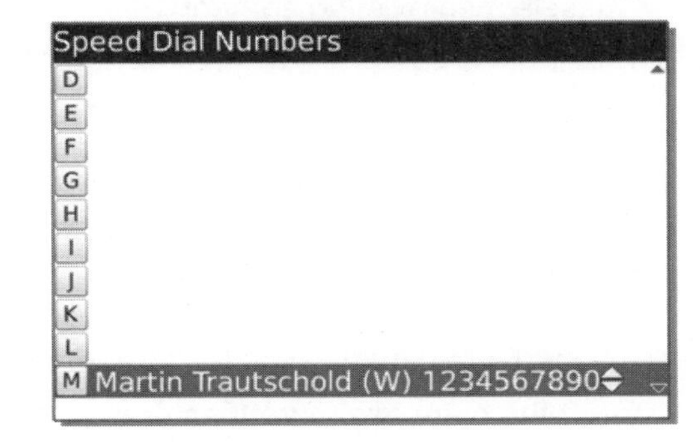

> **TIP:** You may want to reserve the **H** key to be your **Home** number.

Option #2: Press and Hold a Key

Follow these steps to easily assign a key for speed dial.

1. Press and hold any letter key from your **Home** screen that is not reserved (1, A, and Q are reserved) or already assigned.

2. You will be asked if you want to assign this key to a speed dial number. (Figure 10-4)

Setup Speed Dial

Press & Hold any of your letter keys to setup Speed Dial.

The only exceptions are:

W/1 key	Voice Mail
Q key	Turn on/off Quiet profile (Vibrate mode)
A key	Locks the keyboard

Figure 10-4. *Setting up speed dial*

3. Select **Yes** to assign it.

4. Now you'll be shown your **Contacts** list. You can either select an existing entry or click on **[Use Once]** at the top of the list to type in a new phone number that is not in your Address Book.

5. If you are using an existing contact entry, click on the selected name.

6. If the entry has more than one phone number, select the phone number, such as Home, Mobile or Work.

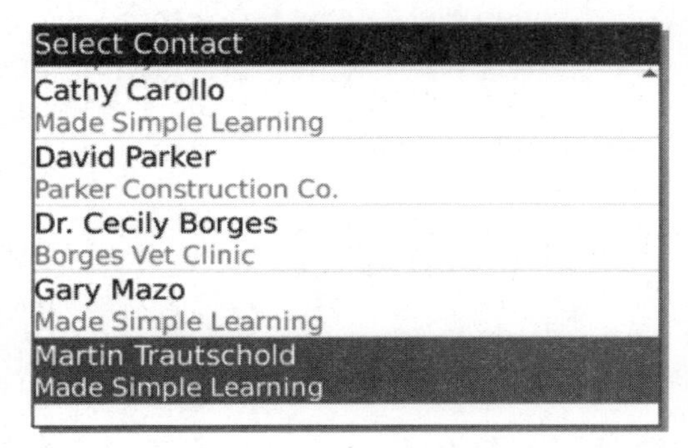

7. After selecting a number or entering a new one, you'll see the **Speed Dial Numbers** screen.

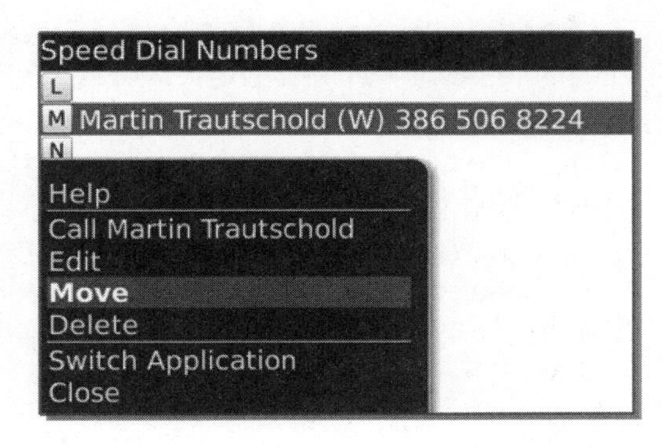

8. If you want to move this entry to a different letter in the Speed Dial list, press the **Menu** key and select **Move**.
9. Press the **Escape** key to back out and save your settings.
10. Give your new speed dial a try by pressing and holding the letter from your **Home** screen or **Messages** app.

Option #3: Use a Contact Phone Number

1. Tap the **Green Phone** key and start entering a contact name or number.
2. When you see the contact listed, scroll to it and highlight it.
3. Press the **Menu** key and select **Add Speed Dial** and follow steps to select the speed dial letter as shown above.

Voice Dialing

One of the powerful features of the BlackBerry is the **Voice Command** program for voice dialing and simple commands. Voice dialing provides a safe way to place calls without having to look at the BlackBerry and navigate through menus. **Voice Command** does not need to be trained as on other smartphones—just speak naturally.

Using Voice Dialing to Call a Contact

The left **convenience** key is (usually) set for **Voice Command**—simply press this key. What is a **convenience** key and how do you set it? See page 187.

The first time you use this feature, the BlackBerry will take a few seconds to scan your Address Book.

When you hear "**Say a command,**" just speak the name of the contact you wish to call using the syntax "**Call Martin Trautschold.**"

You will then be prompted with "**Which number**." Again, speak clearly and say "**Home,**" "**Work,**" or "**Mobile.**" Say **Yes** to confirm and the BlackBerry will dial.

Using Voice Dialing to Call a Number

1. Press the left **convenience** key as you did above.
2. When you hear "**Say a command**," say "**Call**" and the phone number. For example, "**Call 386-555-8888.**"
3. Depending on your settings, you may be asked to confirm the number you just spoke or the BlackBerry may just start dialing.

Chapter 11

Advanced Phone

Now that you have the basics down for using your BlackBerry as a phone, let's look at some more advanced phone topics.

In this chapter we will show you how to set unique caller ID's for contacts, how to use advanced voice dialing options, and how to use call forwarding and call waiting.

We'll also show you how to set up and manage conference calling on your BlackBerry.

Advanced Phone Topics

For many, the basic phone topics discussed in the previous chapter will cover most of your phone needs. Others, however, will need just about every phone feature the BlackBerry offers. Let's get started!

Using Your Music as Ringtones (Phone Tune)

The BlackBerry supports using many types of audio files as a ringtones. You can set one general ringtone (**Phone Tune**) for everyone, or set up individual tones for your important callers.

> **CAUTION:** Place your ringtones in a **ringtone** folder. In some BlackBerry handhelds. When you are attempting to set a ringtone for a specific person in the Address Book or in Sounds, you can only browse to the **ringtone** folder, not the Music folder.

To set one song (MP3) as your general **Ringtone:**

1. Navigate to your list of music
2. Find the MP3 file you want to use as the general phone tune.
3. Press the **Menu** key.
4. Click on Set As Ring Tone.

TIP: Set up unique ringtones for each of your important callers so you'll know when each of these people is calling without looking at your BlackBerry screen.

More with Voice Command

The last chapter concluded with an overview of **Voice Command.** This powerful tool not only enables basic phone calls, it's also great for using other functions of the BlackBerry without having to push buttons or input text.

Other Commands

You can use the **Voice Command** software to perform several functions on the BlackBerry. These are especially useful when you can't look at the screen (while driving) or are in an area where coverage seems to fade in and out.

The most common voice commands are:

- **Call Extension** calls a specific extension.
- **Call (person) Home** calls the contact at his home number.
- **Check Battery** checks the battery status.
- **Check Signal** lets you know the strength of your wireless signal—whether there's no signal or it's low, high , or very high.
- **Turn Off Voice Prompts** turns off the "Say a command" voice and replaces it with a simple beep.
- **Turn On Voice Prompts** turns the friendly voice back on.

Changing Your Voice Dialing Options

You can control various features of voice dialing by clicking the **Options** icon and selecting **Voice Dialing**, as shown in Figure 11-1.

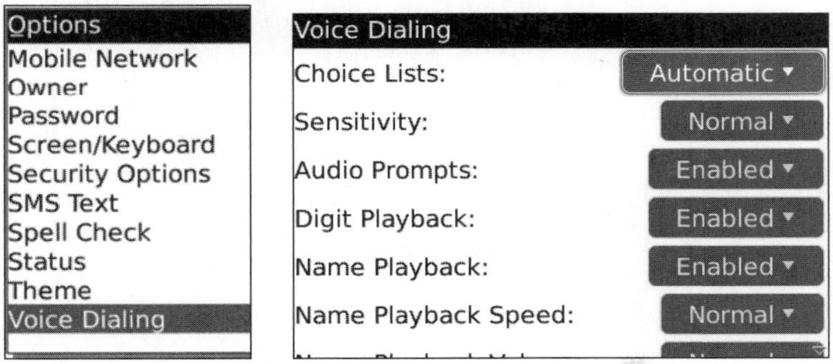

Figure 11-1. *Voice Dialing options screen*

Choice Lists—Change this if you don't want to be confronted with lots of choices after you say a command. Your options here are **Automatic** (default), **Always On,** or **Always Off**.

Sensitivity—you can adjust the acceptance/rejection ratio of voice commands by adjusting the field that initially reads **Normal**. You can go up to **3 (Reject More)** or down to **-3 (Reject Less)**

Audio Prompts—can be enabled or disabled from this screen or by saying **Turn Prompts On** (or **Off**).

Digit Playback—which repeats the numbers you say, and **Name Playback**—which repeats the name you say, can also be enabled or disabled.

Finally, you can adjust the **Playback Speed** and **Playback Volume** of the Voice Dialing program.

Voice Dialing Tips and Tricks

There are a few ways to speed up the voice command process. You can also customize the way voice dialing works on the BlackBerry.

To Make Voice Dialing Calls Quicker

When using **Voice Command**, give more information when you place the call. For example, if you say "**Call Martin Trautschold, Home**," the voice dialing program will only ask you to confirm that you are calling him at home.

The call will then be placed.

Give Your Contacts Nicknames

Make a short entry for a contact—especially one with a long name.

For example, **Gary Mazo's** complete information is in my contact list, but I have also given him the nickname **gg**.

I can simply say: "Call gg."

> **TIP:** This nickname also works when you are addressing emails, SMS text messages, and more.

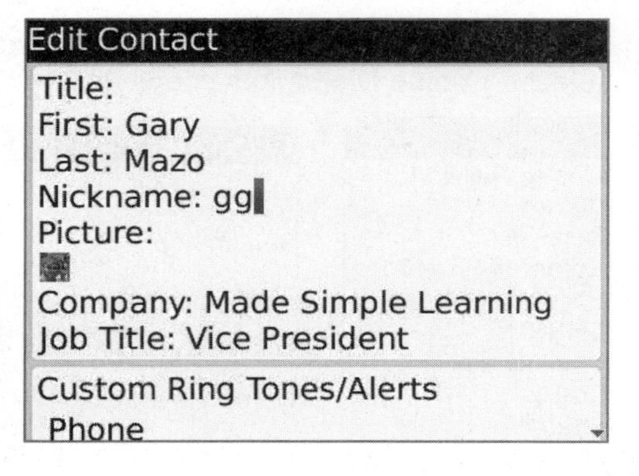

Call Waiting—Calling a Second Person

Like most phones these days, the BlackBerry supports call waiting, call forwarding, and conference calling—all useful options in the business world and in your busy life.

Enabling Call Waiting

Chances are this is already enabled, but in case it's not:

1. Press the **Green Phone** key to get into the **Phone Options** screen.
2. Press the **Menu** key.
3. Click on **Options.**
4. Click on Call Waiting
5. Make sure the **Call Waiting Enabled** field is set to **Yes**.
6. Press the **Menu** key and select **Save**. To turn off or disable **Call Waiting**, just repeat these steps and set the field to **No**.

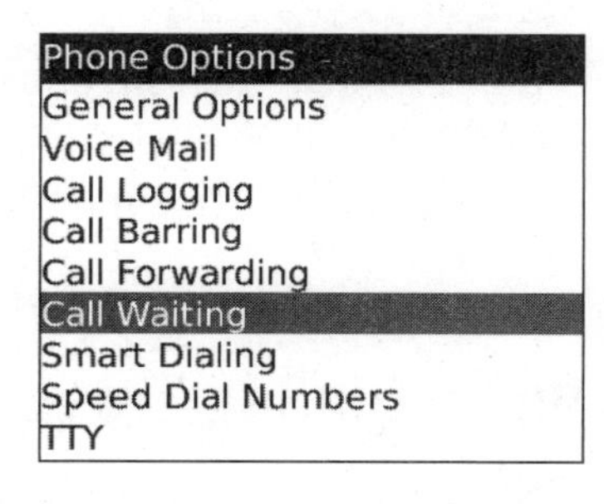

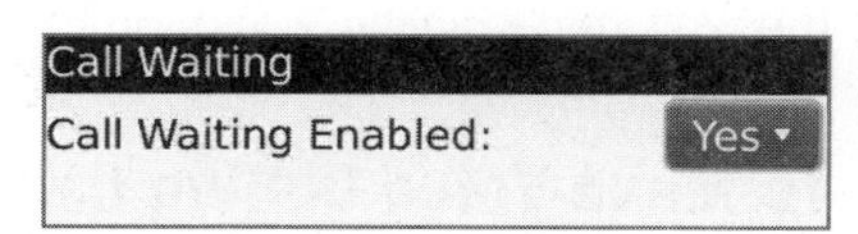

Using Call Waiting

Suppose you start a phone call with someone or receive a call.

Now you can make or receive a call from a second person.

Press the **Green Phone** key while on a call to dial a second phone number, or you can call someone else from your BlackBerry Address Book. This will put the first caller on **Hold.**

TIP: If a second person calls you while you are speaking to a first caller, just press the **Green Phone** key to answer the second caller—the first caller will still be waiting for you **On Hold**.

Press the **Green Phone** key to toggle between calls (Figure 11-2).

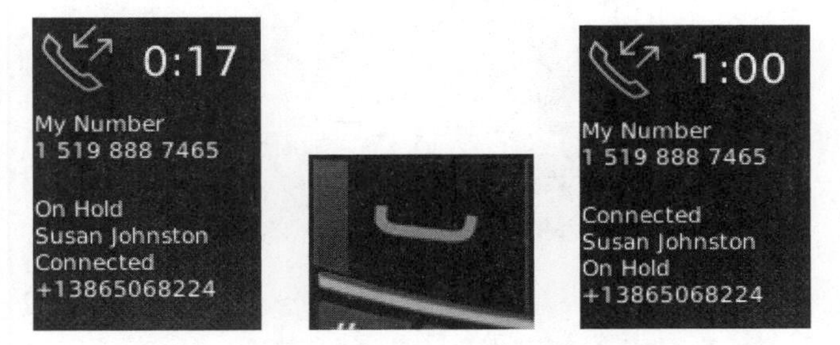

Figure 11-2. *The **Green Phone** key lets you toggle between calls.*

Working with a Second Caller

When you are speaking to a person on the phone and your phone rings again with a second caller, you can do a number of things. It just takes a little practice to do things smoothly.

Option 1: Answer and Put the First Caller on Hold

This is probably the easiest option.

1. Just press the **Green Phone** key. (This is Answer - Hold Current.)

2. To swap between the callers, just press the **Green Phone** key again.
3. With two callers on the phone, pressing the **Menu** key allows you to do a number of other things, including conference calling.

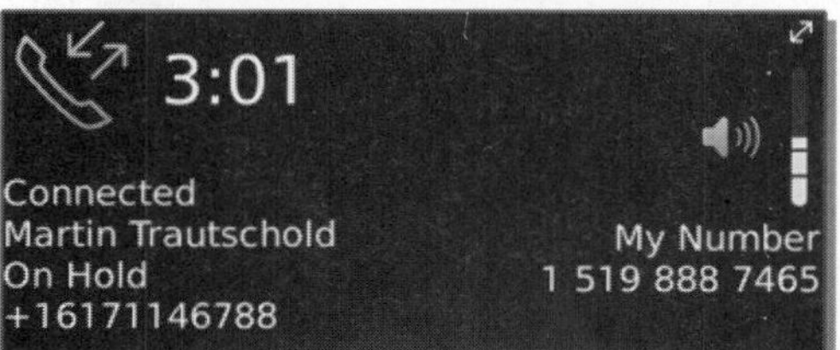

Option 2: Hang Up the First Caller and Answer the Second

Click the **trackpad** and select **Answer - Drop Current** to hang up the first caller and answer the second caller.

Option 3: Press the Red Phone to Send the Second Caller to Voice Mail

Just press the **Red Phone** key to send the second caller to voice mail.

Conference Calling

Conference calling is very helpful when you need to get people together and share ideas or to have two people talk to each other so they can transfer information directly.

Here's a good example where conferencing two parties was a faster (and safer) way to transfer needed information. One of the authors, Martin, was trying to lease a car. The car dealer left a voice mail for Martin to call the insurance company to approve the proof of insurance being faxed to the dealer. Martin called the insurance company, surprised that they did not have the dealer's fax number.

Instead of hanging up and calling the dealer to get the fax number and then calling the insurance company back, Martin did a quick **conference call** between the dealer and insurance company. The conference call allowed the dealer's fax number to be immediately relayed to the insurance company along with any special instructions and approvals.

To Set Up a Conference Call

1. Place a call as you normally would.
2. While on the call, press the **Green Phone** key.

(If this does not show you a **New Call** screen, press the **Menu** key and select **New Call**.) Now make a second call, either by choosing a contact from your **Contacts** list or by typing in a phone number and placing the call.

While on the second call, press the **Menu** key and scroll to **Join Conference** and click.

You can add more than two callers to the conference call. Just repeat the process starting with another **New Call** (pressing the **Green Phone** key), as shown in Figure 11-3.

Figure 11-3. *Adding people to a conference call*

3. **Join** the calls as you did above. Repeat as needed.
4. To speak with only one of the callers on a conference call, press the **Menu** key while on the call, then select **Split Call**. You will then be able to speak privately with that one caller. See Figure 11-4.

Figure 11-4. *Splitting calls when on a conference call*

To End or Leave a Conference Call

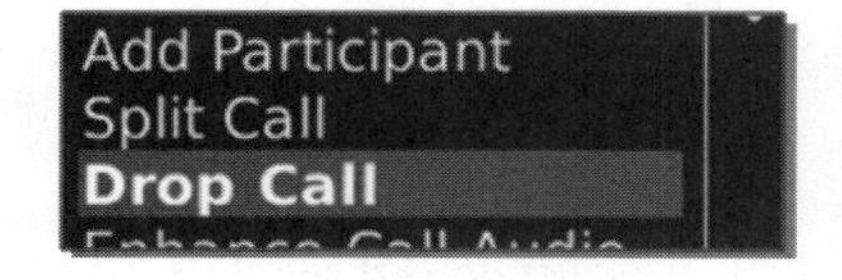

To hang up and end the conference call for all, press the **Red Phone** key or press the **Menu** key and select **Drop Call.**

Dialing Letters in Phone Numbers

You can dial letters like **"1-800-CALLABC"** when you are on a phone call or even put phone numbers with letters in your **Address Book**. To do dial while on the phone, follow these steps:

1. Press the **ALT** key (lower left-most key).
2. Type the letters on your keyboard.

Hold **ALT**, then type "**TRAUTS**":

So when you hear "In order to use this phone directory, please dial the first three letters of the person's last name to look them up," now you just press the **ALT** key and type the letters!

Typing phone numbers with letters in your Address Book or in the phone: Use the same technique. If you had to enter **1-800-CALLABC** into your address book, you would follow these steps:

1. Type in 1 800 normally.
2. Press the **Alt** key and type **CALLABC** from your keyboard.

> **TIP:** When you are not on a phone call (e.g., when you're editing a contact), pressing and holding a letter key (without holding ALT) will also produce a letter.

More Phone Tips and Tricks

As with most features on the BlackBerry, there is always more you can do. These tips and tricks will make things go even quicker for you.

- To place an active phone call on hold and answer a second incoming phone call, press the **Green Phone** key.
- To view and dial a name from your **Address Book**, press and hold the **Green Phone** key.
- To insert a **plus** sign when typing a phone number, hold the number zero **0**.
- To add an extension to a phone number, press the X key, then type the extension number. It should look like this: **8005551212x1234.**
- To check your voice mail, press and hold the number 1.
- To view the last phone number you dialed, scroll to the top of the **Phone** screen, then press the **ENTER** key. Press the **Green** phone key to dial the number.

Finding Flagged Items

After you start flagging email and other messages, you may want an easy way to locate all your flagged items. Follow these steps.

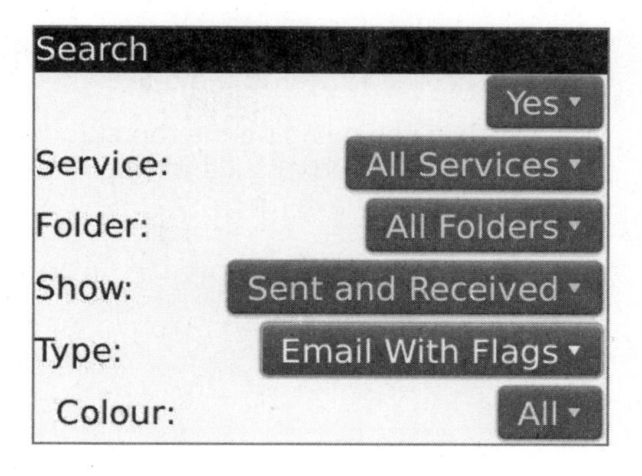

1. In your email inbox (**Messages**), press the hotkey letter **S** to bring up the **Search** window.
2. Roll down to the bottom where it says **Type:** and select **Email With Flags**.
3. Select either a specific **color** or **All** colors.

You can also **Save** your **Search** if you use it often. You can even set a **Shortcut** key for your search by following these steps.

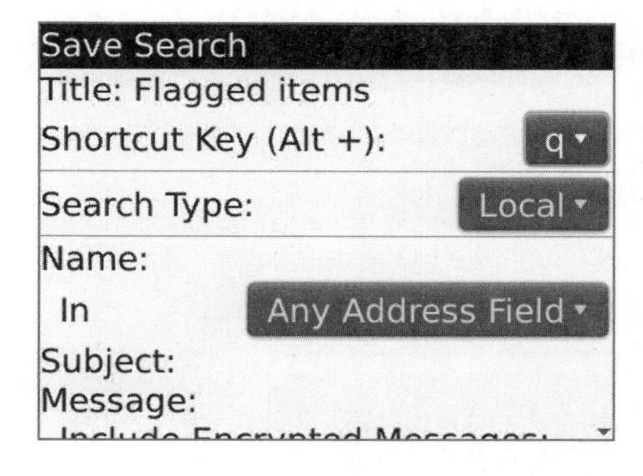

1. Before you execute your search, press the **Menu** key and select **Save**.
2. Enter a **Title** for your search.
3. Set a **Shortcut** key combination (**Alt** + some letter).
4. Try **Alt** + **Q**.
5. Now, in your email inbox, you can quickly find all flagged items by pressing **Alt** + **Q**.

Ringing Alarms for Flagged Item with Due Dates

When a flagged item alarm rings, you'll see a pop-up screen similar to the one shown.

You' will also see a flag on the top status bar with a number next to it showing how many flag alarms have rung.

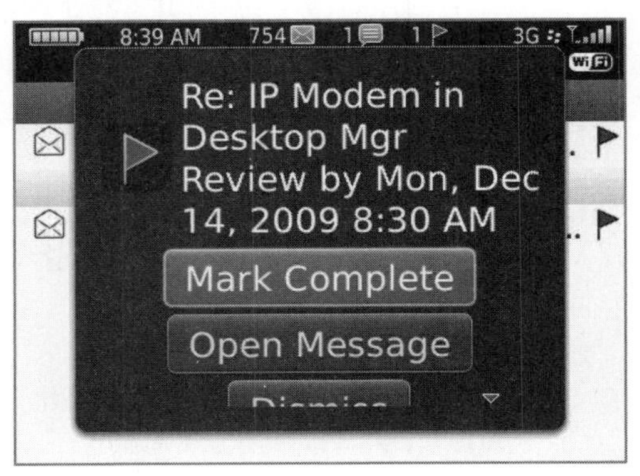

Attaching a Contact Entry (Address Book Entry) to Email

At times, you might need to send someone an address contained in your BlackBerry Contacts.

Start composing an email by pressing the **L** key or selecting **Compose Email** from the **Menu** in the **Messages** app.

Press the **Menu** key and scroll to **Attach Contact**, as shown in Figure 12-2.

Figure 12-2. *You can attach contact information to an email.*

Either type in the name of the contact or use the **trackpad** to scroll and click on a contact name. You will now see the attached contact shown as a little address book icon at the bottom the main body field of the email.

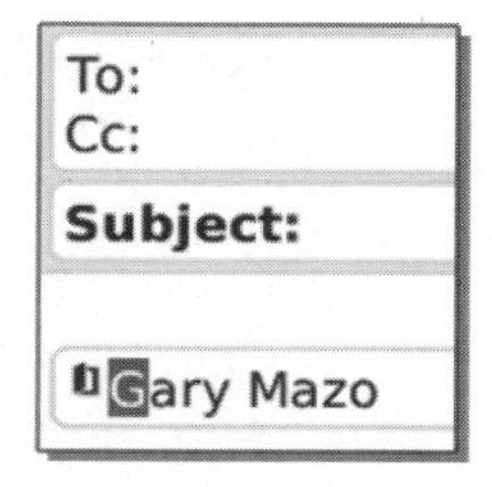

Viewing Pictures in Email Messages You Receive

On some email messages, you may see blank spaces where images should be. If you see this, press the **Menu** key and select **Get Image** to retrieve just one image, or **Get Images** to retrieve them all. You may see a warning message about exposing your email address; click **OK** or **YES** to get the image.

Attaching a File to Email

The BlackBerry is a powerful business tool. As such, there are times you might need to attach a file to an email you send from the BlackBerry, much as you would when emailing from your computer.

> **NOTE:** Depending on the version of your BlackBerry software, the **Attach File** menu option may not be available for you.

1. Start composing an email message and press the **Menu** key.
2. Select **Attach File** from the menu.
3. Next, locate the directory in which the file is stored. Your two initial options are **Device Memory** or **Media Card**.

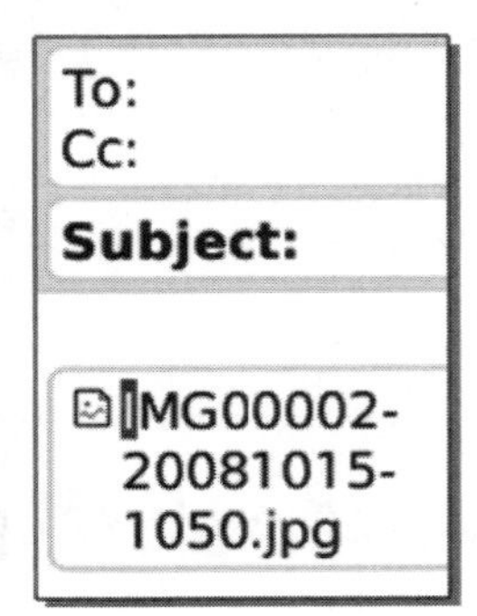

Figure 12-3. *Attach a picture to an email*

4. Use the **trackpad** to navigate to the folder where the file is stored. Once you find the file, simply click on it and it will appear in the body of the email. (See Figure 12-3.)

TIP: You can add several files to your email message; simply repeat the procedure for each.

Setting the Importance of the Email

Sometimes, you want your email to be noticed and responded to immediately. The BlackBerry lets you set that importance so that your recipient can better respond.

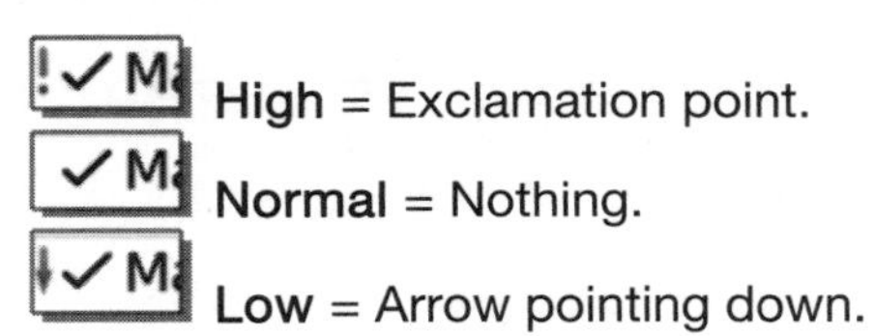

High = Exclamation point.

Normal = Nothing.

Low = Arrow pointing down.

It's easy to set the importance of a new email as you are writing it.

1. Press the **Menu** key and select **Options.**
2. Click the **Importance:** field and select **Normal, High,** or **Low.**
3. Press the **Menu** key and click **Save.**

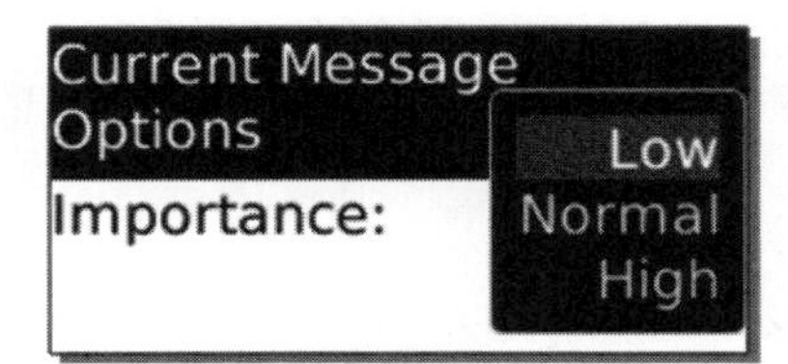

Verifying Delivery of Email Messages

On some, but not all email messages you send, you'll see a little **D** inside the checkmark. These are messages that have been confirmed delivered to the recipient. It does not mean they have opened the message, just that it was successfully delivered to their inbox. Notice in Figure 12-4 the D inside the two check marks.

Figure 12-4. *Verifying delivery of your email messages.*

Spell Checking Your Email Messages

Please see page 158 to learn how to enable spell checking on the email messages you type and send. The spell checker may not be turned on when you take your BlackBerry out of the box the first time.

Working with Email Attachments

What makes your BlackBerry more than just another pretty smartphone are its serious business capabilities. Often, emails arrive with important documents attached, such as Microsoft Word files, Excel spreadsheets, or PowerPoint presentations. Fortunately, the BlackBerry lets you open and view these attachments and other common formats (see Table 12-1) wherever you might be.

Table 12-1. *Supported Formats for Email Attachments*

■ Microsoft Word (DOC) ■ Microsoft Excel (XLS) ■ Microsoft PowerPoint (PPT) ■ Corel WordPerfect (WPD) ■ Adobe Acrobat PDF (PDF) ■ ASCII text (TXT) ■ Rich Text Format files (RTF)	■ HTML ■ Zip archive (ZIP; password-protected ZIP files are not supported) ■ Voice Mail Playback (MP3, up to 500KB file size) ■ Image Files (JPG, BMP, GIF, PNG, TIFF; multi-page TIFF files are not supported)

NOTE: Additional file types may be supported in newer versions of the system software running on your BlackBerry.

Features available for viewing attachments:

- Pan, zoom or rotate images.
- Save images to your BlackBerry.
- Show or hide tracked changes in Microsoft Word.
- Jump to another part of the file instead of paging through it
- Show images as thumbnails at the bottom of the email message.

Using Documents to Go to View and Edit Email Attachments

Your BlackBerry comes with the **Documents to Go** program from DataViz. This is an incredibly comprehensive application that allows you to not only view, but also edit Word, PowerPoint, and Excel documents, and it preserves the native formatting. This means that the documents can open on your BlackBerry and look just like they do on your computer.

How Do You Know if You Have an Email Attachment?

If your email has an attachment, you'll see an envelope with a paperclip:

= Has Attachment

= No Attachment (or it has an attachment that can't be opened by the BlackBerry)

Gary Mazo 10
files for the me
Martin Tra... 9
Book Cover Ide
Martin Tra... 9
Planning Sprea
'Kevin J. M... 9
Away from the
Martin Tra... 9
RE: URGENT: Cr

To Open an Attached File

1. Navigate to a message showing the attachment icon and click on it.
2. At the very top of the email, you'll see an indication of the number of attachments, such as **1 Attachment** or **2 Attachments.**
3. Click the **trackpad** and select **Open Attachment**

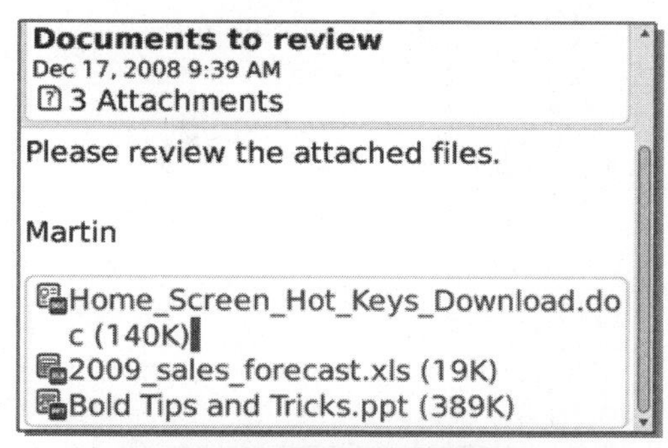

4. If the document is in one of the Microsoft Office formats, you'll be presented with the options to **View** or **Edit with Documents to Go.**
5. For a quick view, without the ability to edit the document, select **View.**
6. The **View** mode document is shown to the right.

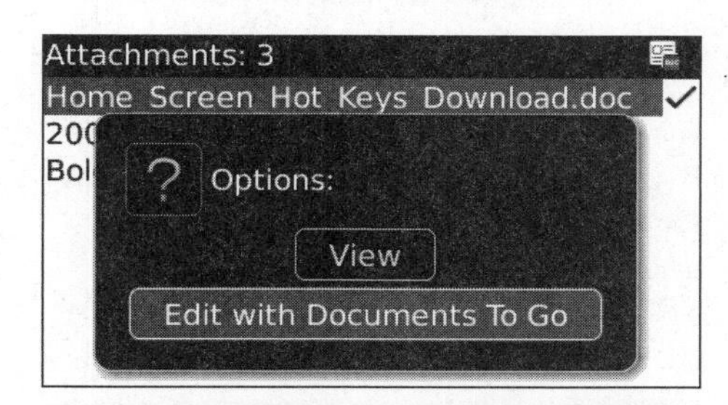

7. To really see the document the way it was meant to be seen, and be able to edit it, select **Edit with Documents to Go**.

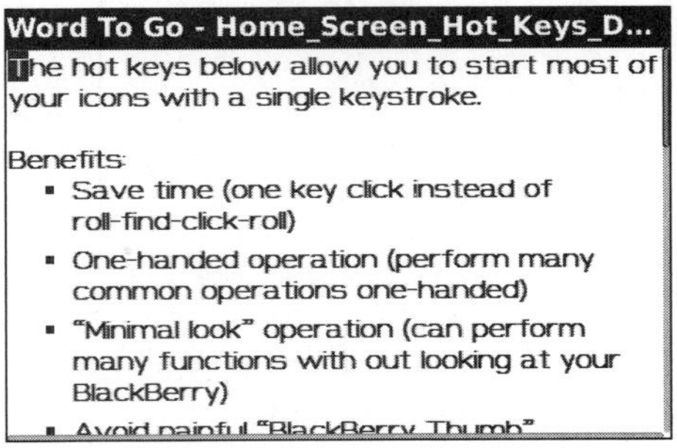

If you get an error message such as "**Document Conversion Failed,**" it's very likely that the attachment is in a format that is not viewable by the BlackBerry Attachment Viewer. Check out the list of supported attachment types in Table 12-1 e.

Email Attachment Viewer Tips and Tricks

As with all features of your BlackBerry, there are some shortcuts and tricks that might prove helpful when working with attachments using the generic BlackBerry attachment viewer:

- To search for specific text inside an attachment, press **F**.
- To switch between showing tracked changes and showing the final version, press **H**.
- To jump to the top of the attachment, press **T**.
- To jump to the bottom of the attachment, press **B**.
- To change the width of a column (spreadsheet or table only), press **W**.
- To **Go To** a specific cell (spreadsheet or table only), press **G** and then type the cell name, e.g., **C3**.
- To view the contents of a cell (spreadsheet or table only), press the space key or click on the **trackpad**.
- To view a slide show presentation (Pictures or PowerPoint presentation only), press **A**.
- To stop the slide show presentation, hold the **Escape** key.

Editing with Documents to Go

When you select **Edit with Documents to Go,** the document will open on your screen. You can scroll through it just as if you were reading a Word document on your computer.

What are all those asterisks in the menus, e.g., *Check Spelling

These are items available only in the Premium edition of Documents to Go. If you'd like, you can upgrade right from one of the menu items in the application. Press the **Menu** key and select **Try Premium Features.**

If you want to make changes to the document, press the **Menu** key and select **Edit Mode** from the menu.

To adjust the formatting of the document, press the **Menu** key and select **Format**. You will then see the available Formatting options (see Figure 12-5).

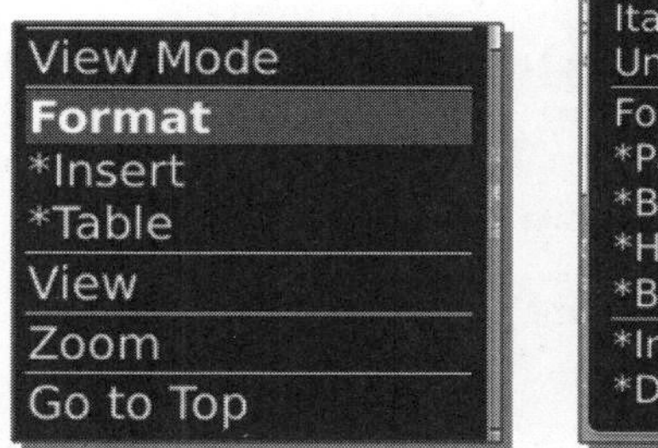

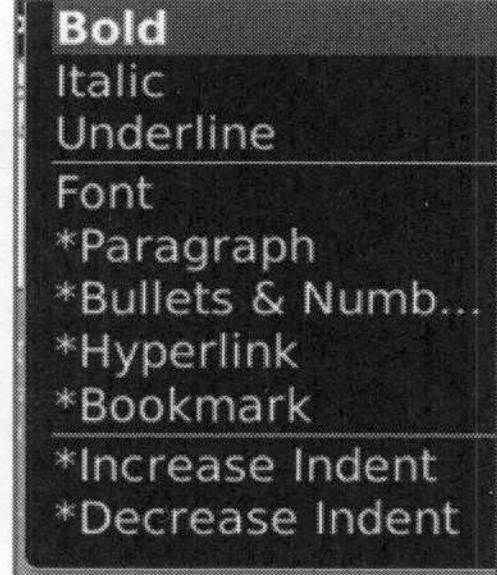

Figure 12-5. *Formatting text in Documents to Go*

> **NOTE:** Sometimes the **Documents to Go** program gets updated in **BlackBerry App World**. Every so often, check the **App World** to see if a new version is available.

Using the Standard Document Viewer

If you don't want to use the Documents to Go program, just select the **View** option when you go to open the attachment.

The document won't have the same look, but you'll be able to navigate through it quickly.

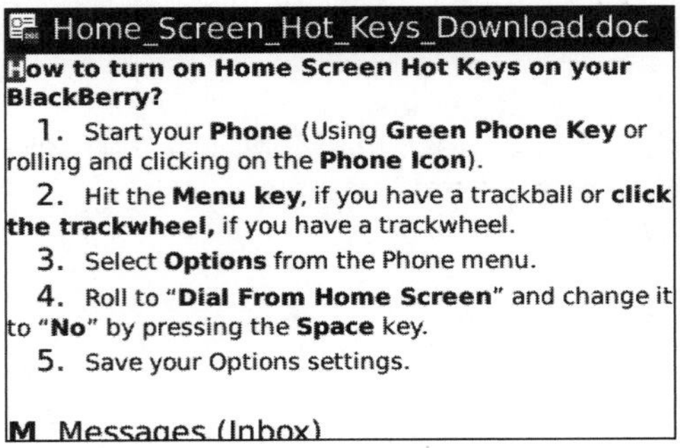

Home_Screen_Hot_Keys_Download.doc

How to turn on Home Screen Hot Keys on your BlackBerry?

1. Start your **Phone** (Using **Green Phone Key** or rolling and clicking on the **Phone Icon**).
2. Hit the **Menu key**, if you have a trackball or **click the trackwheel,** if you have a trackwheel.
3. Select **Options** from the Phone menu.
4. Roll to "**Dial From Home Screen**" and change it to "**No**" by pressing the **Space** key.
5. Save your Options settings.

M Messages (Inbox)

Using Sheet to Go or Slideshow to Go

These apps are for editing Microsoft Excel or PowerPoint documents. Follow the same steps you took earlier when you opened the word processing document.

Sheet to Go - View Mode:

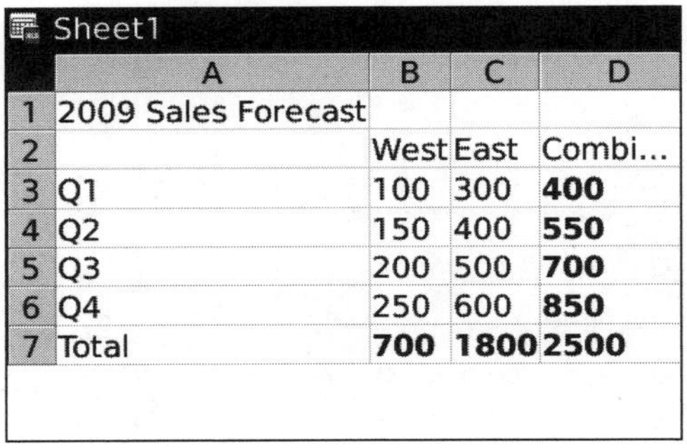

Sheet1

	A	B	C	D
1	2009 Sales Forecast			
2		West	East	Combi...
3	Q1	100	300	**400**
4	Q2	150	400	**550**
5	Q3	200	500	**700**
6	Q4	250	600	**850**
7	Total	**700**	**1800**	**2500**

Here's a spreadsheet opened with **Sheet to Go**—you can view formulas and edit!

Sheet to Go - Edit Mode:

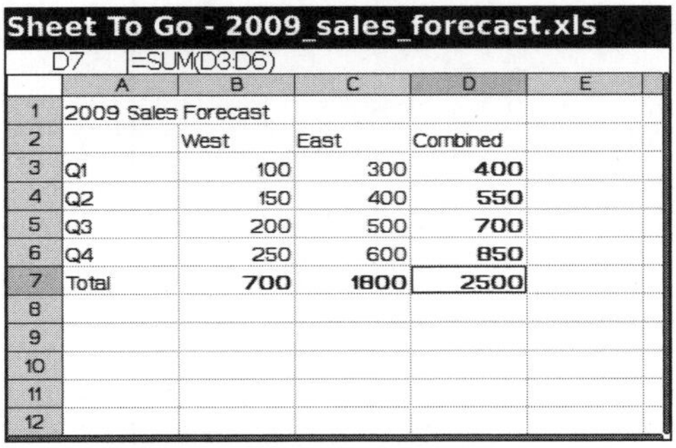

Sheet To Go - 2009_sales_forecast.xls

D7 =SUM(D3:D6)

	A	B	C	D	E
1	2009 Sales Forecast				
2		West	East	Combined	
3	Q1	100	300	**400**	
4	Q2	150	400	**550**	
5	Q3	200	500	**700**	
6	Q4	250	600	**850**	
7	Total	**700**	**1800**	**2500**	
8					
9					
10					
11					
12					

Here is a Microsoft PowerPoint document opened with **Slideshow to Go**.

Just click the **trackpad** and select **Edit Slide Text** to change any slide text items.

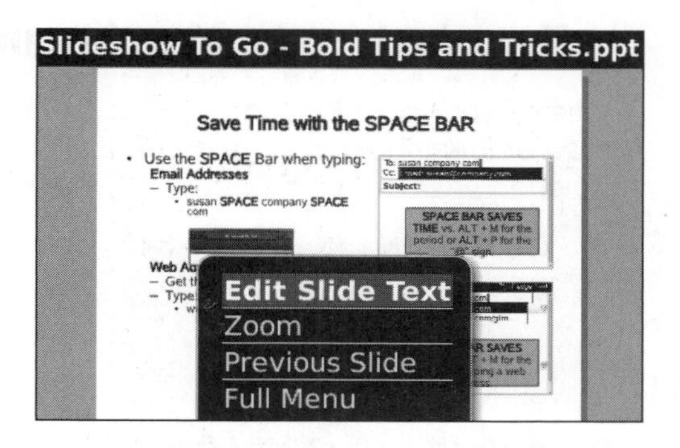

Just click and then type your changes.

You can save a copy on your BlackBerry, or send it via email.

If you **Send via Email**, you'll see a new email message screen come up with the edited file as an attachment. **You can truly get work done on the road with your BlackBerry!**

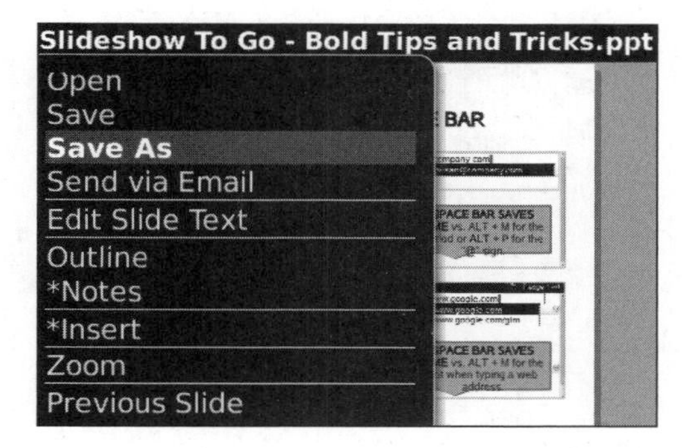

Send the edited file and you're done.

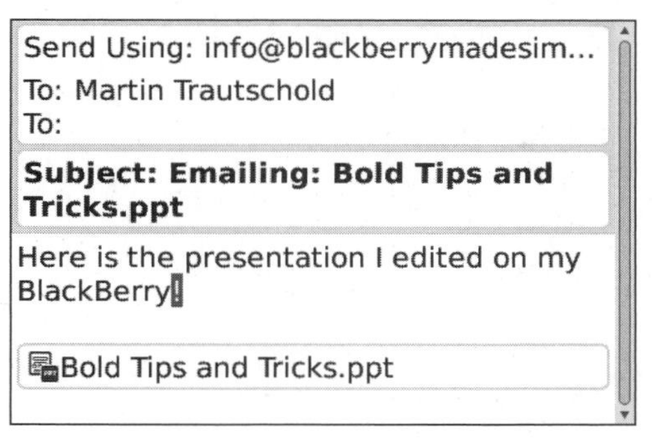

Finding Text in an Attachment

Occasionally, you might find the need to search an attachment for a particular word, phrase or idea. The "Find" features built in to attachment viewing can save you lots of time when searching for something specific.

1. Open up the attachment as described above.

2. To find text in an email attachment, press the **trackpad** and select **Find** or use the shortcut key **F**.

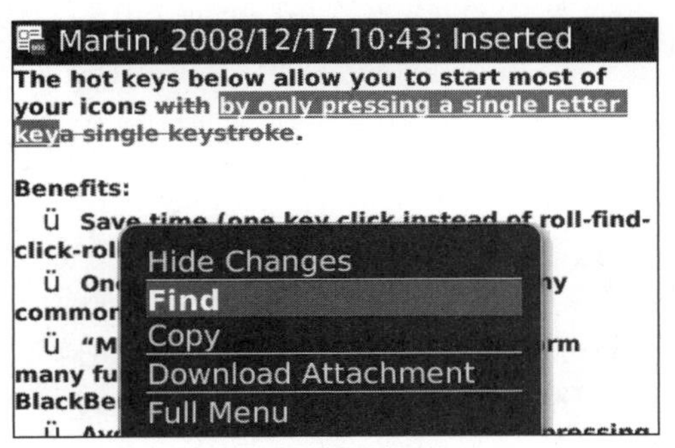

3. Type in the text to search for and select whether you want the search to match the case of your search term. Finally, click the **trackpad** to start the search.

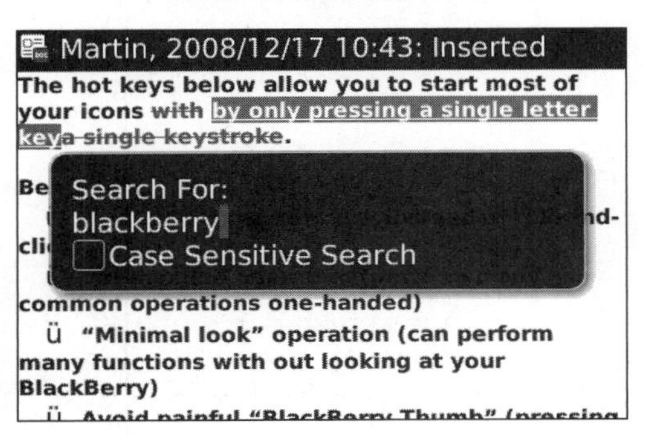

TIP: To quickly find the same text again later in the document, press the **F** key again. To search for different text, click the **trackpad** and select **Find**.

To change the way the attachment looks on the BlackBerry:

1. Open the attachment as described above.

2. Press the **Menu** key and select **Options.**

3. Choose a new **Font** from the **Font Family** to change the display font of the document.

Opening a Picture

Open a message with pictures attached.

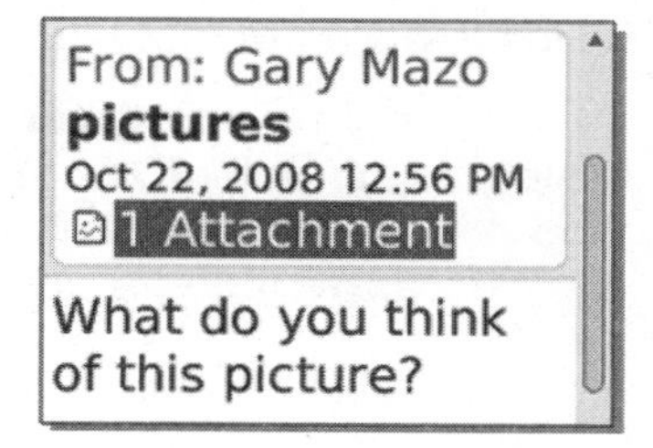

1. Click on the [1 Attachment] or the like at the top of the email message.

2. Select **Open Attachment** or **Download Attachment** (to save it on your BlackBerry). Then click on the file name of any image file to open it.

TIP: Once you've opened the pictures, the next time you view that email you'll see the thumbnails of all the pictures at the bottom of the message. You can then just glide down and click on them to open them.

3. To save the picture, press the **Menu** key or click the **trackpad** and click on **Save Image** (Figure 12-6).

4. The picture will be saved where you specify, either on your media card or the main device memory.

Figure 12-6. *Save an image you received in an email.*

Other menu options include **Zoom** (to expand the image); **Rotate,** (to rotate the image); **Send as Email** (email the picture as an attachment); or **Send as MMS** (embed the image in the email message).

To save the image as a Caller ID picture in **Contacts**, select **Set as Caller ID** from the **Menu** and then begin to type in the contact name. Navigate to the correct contact and save as prompted.

Searching for Messages (Email, SMS, MMS)

To search your **Messages**, press the hotkey letter **S** to bring up the **Search** window. There are lots of reasons why you'll need to search your Inbox. Here are some common scenarios:

- You are trying to find that funny email from Susan you received a few days ago to forward to a friend. Type **Susan** in to the **Name** field.
- You want to find all invoice-related emails. Type **invoice** in the **Message** field.
- You need to quickly find that email about your trip to San Francisco. Type **San Francisco** in the **Message** field.

You might find that you use messaging so often, since it is so easy and fun, that your messages start to really collect on your BlackBerry.

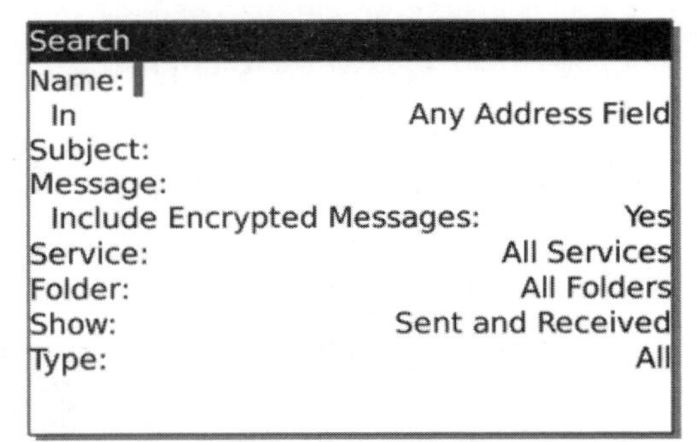

TIP: Need to search other places like your Contacts, Calendar, or Tasks? Use the **Search** app as shown on page 509.

When you need to find a message quickly, you won't want to scroll through all those messages in your in box. To find what you're looking for, you can search the entire message for a word or phrase, or you can search by the sender, the recipient, or the subject.

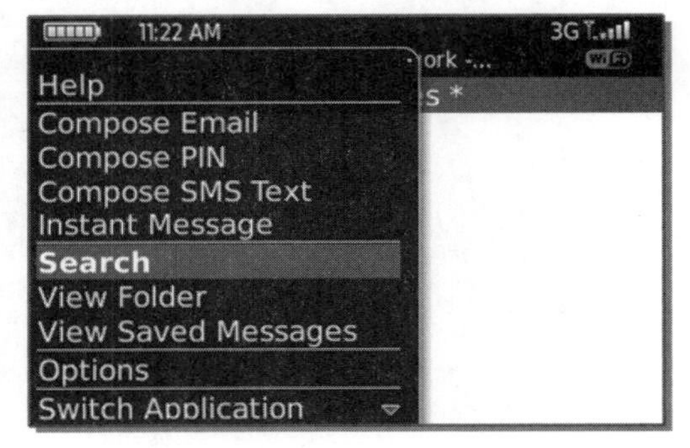

Using the Search Sender or Search Recipient Command

TIP: The Search, Search Sender, Search Recipient, and Search Subject commands work on SMS messages, Email, MMS—anything in your messages inbox!

Sometimes, you have many messages from one particular sender, and you want to see only the list of your communication with that particular individual.

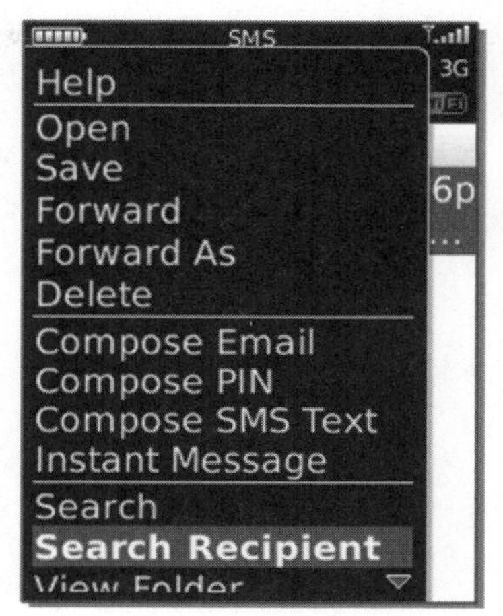

1. From the messages list, scroll to any message from or to the person you wish to search and press the **Menu** key.
2. Select **Search Recipient** or **Search Sender**.
3. Only the list of messages sent by or sent to that particular person (in this case, Martin) is now displayed.
4. Now that you only see messages to or from Martin, it is easy to scroll up or down and quickly find the message you are seeking.

Using the Search Subject Command

You might be having an SMS conversation with several people about a particular subject, and later you want to see all the messages about that subject on your device.

1. Navigate to any message that has the subject displayed that you are searching for.
2. Press the **Menu** key and select **Search Subject**.
3. All the messages with the same subject are now displayed. It is easy to navigate to the one you wish to read.

Chapter 13

Advanced Email Topics

Given the email power of the BlackBerry, there might be some things you want to do right from your handheld that you used to do only from your computer. You can write emails in other languages, select any of your integrated email accounts to send from, and easily create and select various email signatures and auto-signatures.

Switching Input Language for Email and More

Let's say you have a client in Latin America and you wish to compose your email messages in Spanish. Because of the spell checking feature and special characters and accents, you will want to change your typing or input language selection to the one in which you are composing the email message.

NOTE: Don't see the **Switch Input Language** option or it does not do anything when you select it?

During the setup wizard process, the BlackBerry will remove unused input languages based on your selections.

If you removed all languages except your display language, then the **Switch Input Language** menu item won't be visible or it will not do anything when you select it.

If you are a Windows PC user, you can use the application loader inside Desktop Manager to add back languages you have previously removed (see page 89). If you are a Mac user, please contact your service provider and ask them how to put languages back on your BlackBerry.

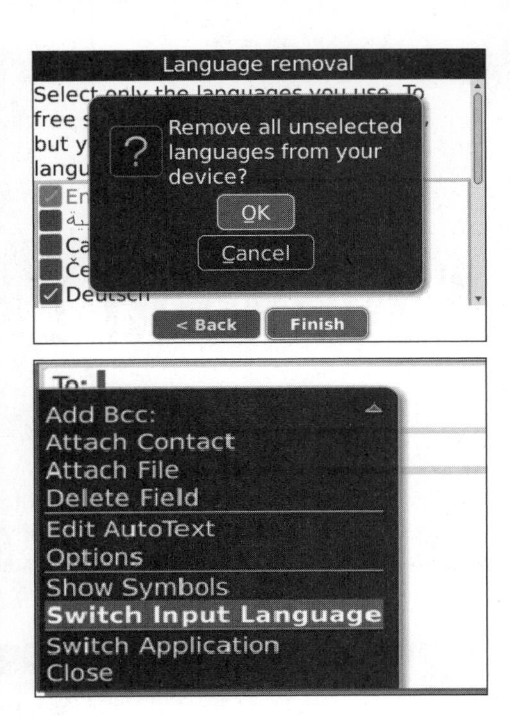

1. Start composing a new email message.
2. Press the **Menu** key and select **Switch Input Language**.
3. Scroll down to the input language you want, e.g., **Español**, and click.
4. When you begin typing, you will now have the Spanish language dictionary loaded and you can type your email message in the new language.

Inbox Housecleaning

After a while, you may have so many messages that your mailbox gets a little unwieldy. Here's how to get rid of older messages.

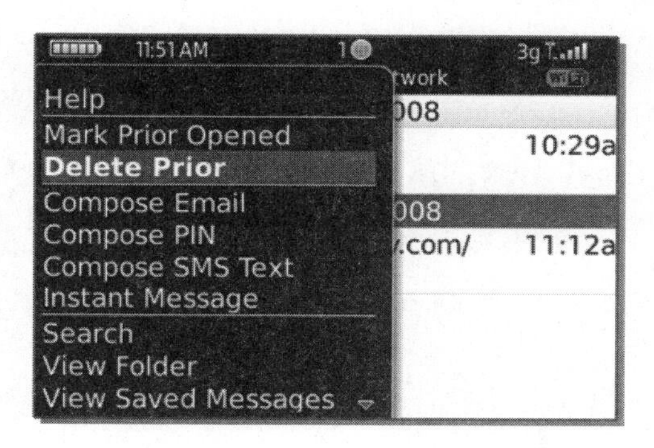

1. Start your **Messages** app.
2. Make sure you are in the Messages list (Inbox).
3. Highlight the date row separator; it will say the day of the week, followed by the date such as **Sat, Jun 19, 2010** shown here:

   under the most recent message you want to keep.
4. Press the **Menu** key and select **Delete Prior**.
5. All older messages will then be deleted

To delete an individual message, just click on the message, press the **Menu** key and choose **Delete**.

> **CAUTION:** If you have turned on Call Logs in your Messages Inbox, **Delete Prior** will also delete all your **Call Logs**.

Sending from a Different Email Account

Like many of us, you might have separate email accounts for business and personal matters, or several just for work. You can easily change which email account you use to send an email on your BlackBerry.

Start composing an email. Scroll to the top where it says "**Send Using.**"

You'll see the **Default** account, but you can click on **Default** and all of your email accounts will show up in the window.

Just select the email account to use to send this particular email.

Changing Your Default "Sent From" Email Address

You can change your default **Sent From** email address on your BlackBerry. Here's how:

1. Click on the **Options** icon from the **Home** screen. You may find the **Options** in the **Setup** folder if you can't find it from your **Home** screen.
2. Click on Advanced Options.
3. Click on **Default Services** (if you don't see this item, click on **Message Services**)
4. You'll see a screen that says **Messaging (CMIME):** or something similar.

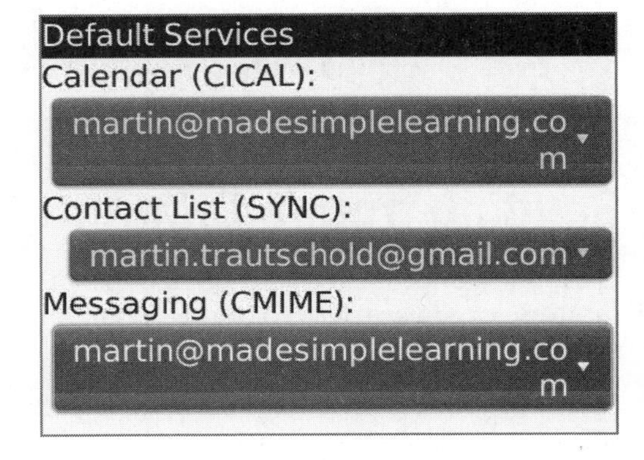

5. Click on that item to see a list of all your integrated email accounts.
6. Select your new default email account for sending new messages you compose on your BlackBerry.

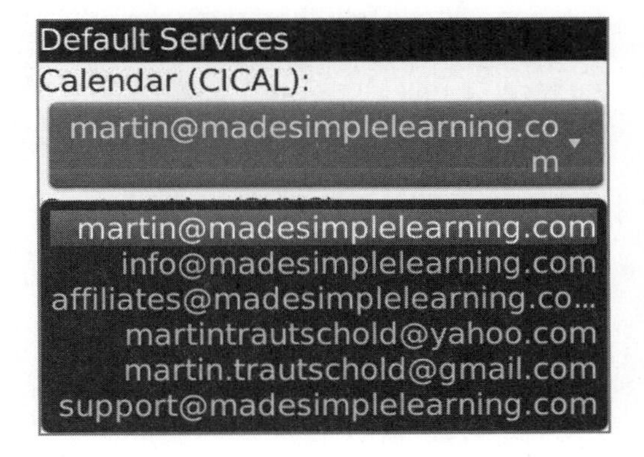

7. Press the **Menu** key and select **Save.**

Now compose a new email message. Notice that your new default email account appears in the **Send Using:** field.

Changing the Way Your Email Looks and Functions

You can change many of the more advanced options for email by doing the following:

1. Navigate to the **Messages** icon and click.
2. Press the **Menu** key, scroll down to **Options** and click.
3. Click on General Options.

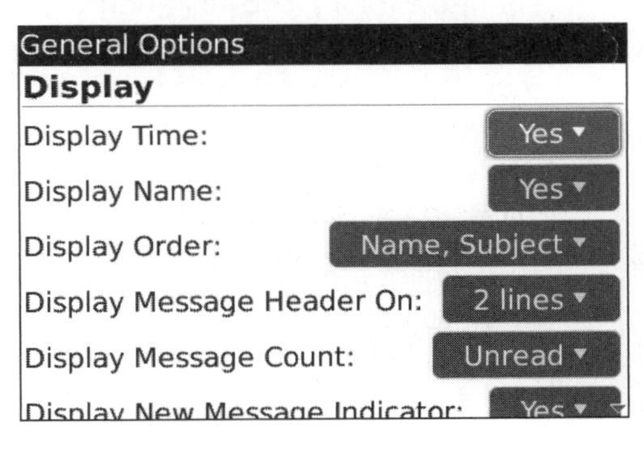

Probably the best way to see more messages on your list is to set the **Display Message Header On** to 1 line. (See below)

1 line

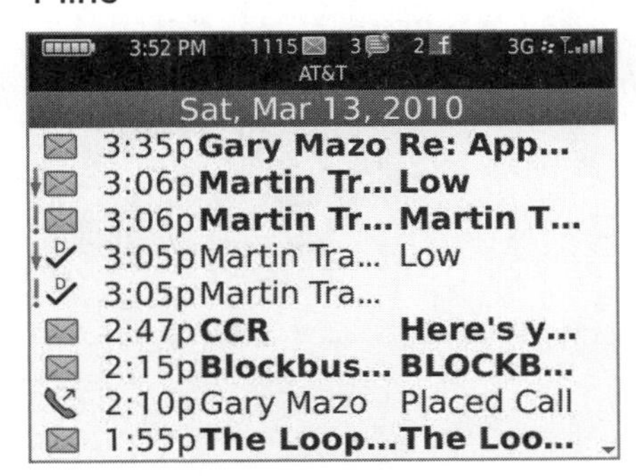

2 lines

Sat, Mar 13, 2010
Gary Mazo 3:35p
Re: App Loader in DM for Mac
Martin Trautschold 3:06p
Low
Martin Trautschold 3:06p
Martin Trautschold Made Simp...
Martin Trautschold 3:05p
Low
Martin Trautschold 3:05p

You can choose whether to display the time, name, message header, new message indicator, confirm the deletion of messages, hide file and sent messages, and change the level of PIN Messages. (Learn more about **PIN Messages** on page 341.)

Click on the desired change. When done, press the **Menu** key to save your changes and they will be reflected in your email screen.

Email Reconciliation

Depending on the type of email accounts you have set up and your messaging services, you may be able to wirelessly share your actions (**Delete** and maybe even **Open** actions) between your main email inbox and your BlackBerry.

1. Open your **Messages**.
2. Press the **Menu** key and select **Options**.
3. Select Email Reconciliation.

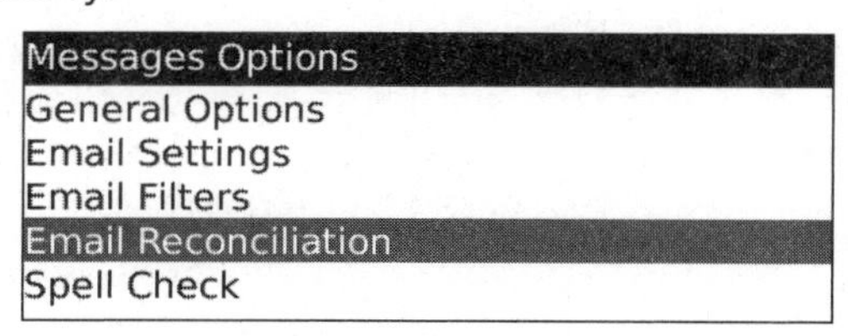

4. On **the Email Reconciliation** screen, you can select a particular email address to customize in the **Message Services** drop-down list.
5. Click on the email address and select the account you want to work with.

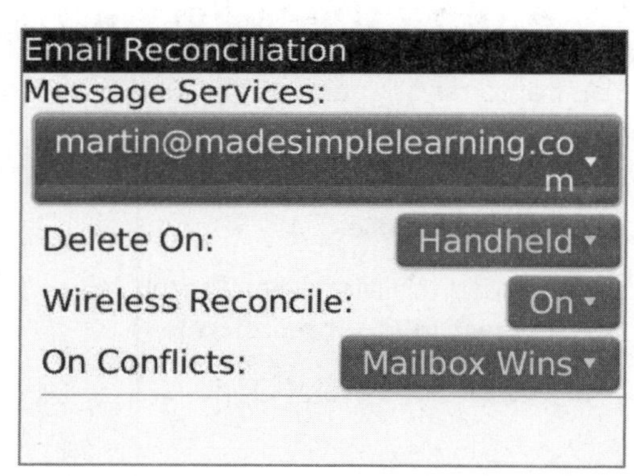

Delete On:

- **Prompt** = ask you each time.
- **Handheld** = any deletion on your BlackBerry is not sent to your main mailbox.
- **Mailbox & Handheld** = deletions are synchronized or shared between your mailbox and your BlackBerry.

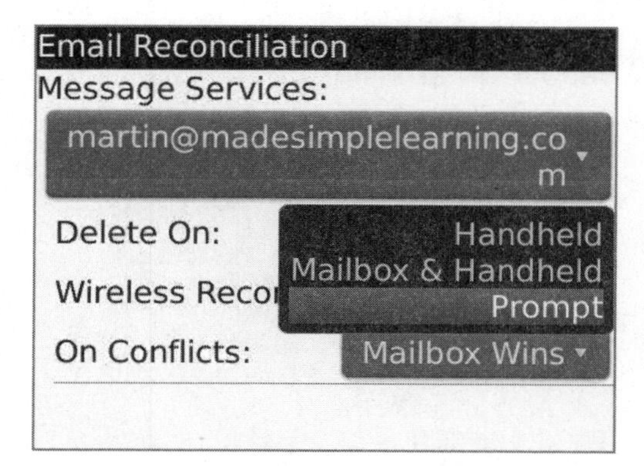

Wireless Reconcile:

- **On** = Yes, share deletions and other information between your BlackBerry and main mailbox.
- **Off** = No, don't share information.

On Conflicts

Lastly, you can choose whether your server or your handheld wins if there is a reconciliation conflict. The default setting is **Mailbox Wins,** which means that your main email box wins over your BlackBerry.

TIP: If you want to be able to delete email from your BlackBerry and not have it also delete from your main email inbox, set **Wireless Reconcile** to **Off**.

CAUTION: If you choose **Mailbox and Handheld** and set **Wireless Reconcile** to **On**, whenever you delete email from your BlackBerry, it will also be deleted from your regular email inbox, and vice versa.

Easily Adding Signatures to Your Emails

There are various ways to set up email signatures for messages you compose on your BlackBerry. Probably the easiest one is to use is the one right in the email setup program. You can also set up signatures from your carrier's BlackBerry Internet Service web site. And you can create custom **AutoText** signatures so you can select a specific signature whenever you need it.

Set Up Custom AutoText Signatures

TIP: This AutoText option works for both personal/Internet email users and corporate (BlackBerry Enterprise Server) users.

If you want to be able to select different signatures as you are composing your messages, use the **AutoText** feature to create the various signatures you need.

1. Locate and click on the **Options** icon .
2. Select **AutoText** (near **the top** of the list) and click.
3. Press the **Menu** key and select **New**.

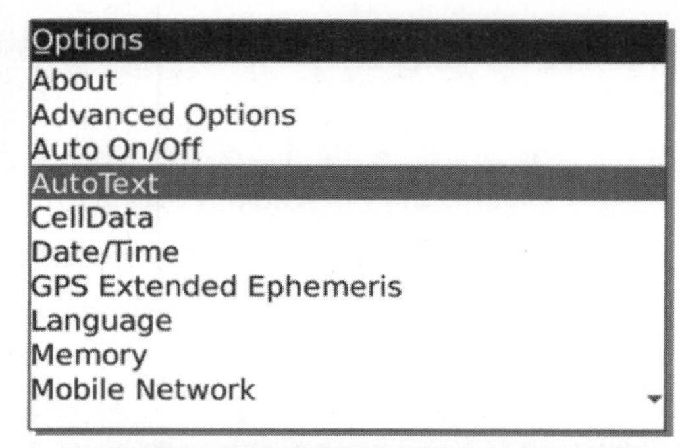

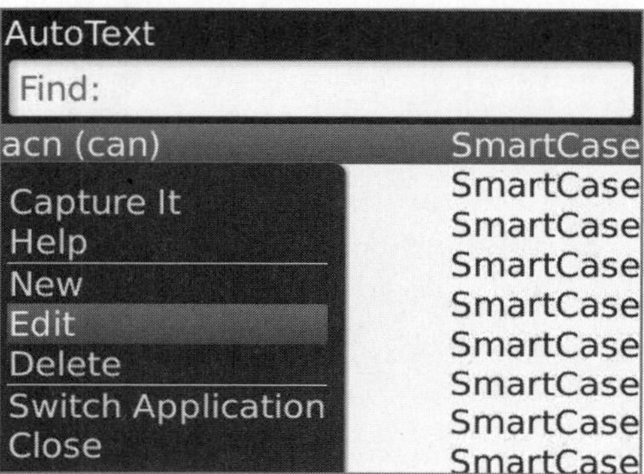

4. In the **Replace** field, type any combination of letters. We recommend using your initials. If you plan to have several different signatures, say one for work and one for personal use, you might want to use a number or extra letter after your initials, like "(initials)w" for work and "(initials)p" for personal.
5. In the **With** field, type in your full email signature exactly as you'd like it to appear in your emails.
6. Choose **SmartCase** if you want the BlackBerry to capitalize the letters according to the correct context in the sentence when they are replaced. Select **Specified Case** to replace these letters exactly as you have entered them in the **AutoText** entry. For example, if you entered "DeSoto" with **Specified Case**, it would always replace the word as "DeSoto" never "Desoto".

7. In this example, we can put in Gary's full name, company, and email address just by setting up an AutoText for **gam**.

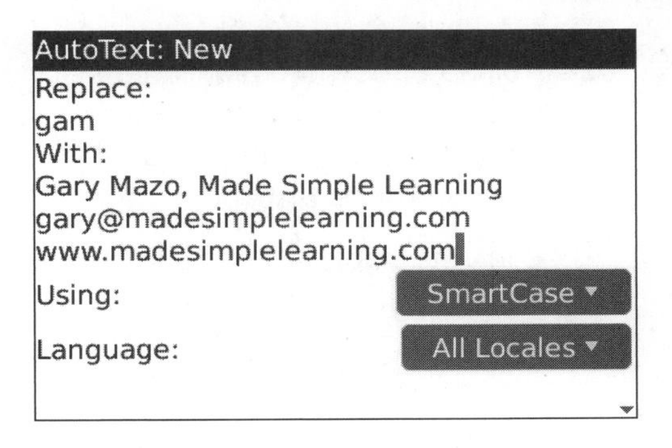

8. In the **Language:** field, select whether this AutoText signature should work in **All Locales** for every language, or in only one language. This setting can be useful if you have different signatures for different languages.
9. Press the **Menu** when you are done and select **Save**.
10. Now, each time you type in your initials and press the **Space** key, your complete signature will appear instantly.

Personal or Internet Service Email Users

If you have integrated personal or Internet email (POP3 and IMAP 4 accounts) to your BlackBerry, you have a few options for email signatures.

Setting Up Signatures from Your BlackBerry Carrier's Web Site

This option is described in detail on page 44. Using this feature, you have the ability to add a unique **Auto Signature** to each of your integrated email accounts.

Setting Up Email Auto Signatures on Your BlackBerry

From the BlackBerry **Email Settings**:

1. Go to your **Setup** folder.
2. Click on the **email settings** icon.
3. Login, if requested. Your BlackBerry will connect to your carrier.

4. Roll up or down to highlight the email account to change your email signature from the list of your accounts.

5. Click the **trackpad** and select **Edit** form the short menu.

6. Scroll down to **Signature** and type in your email signature.

7. When you're done, select **Save**.

8. Press the **Escape** key to exit out of Email Settings.

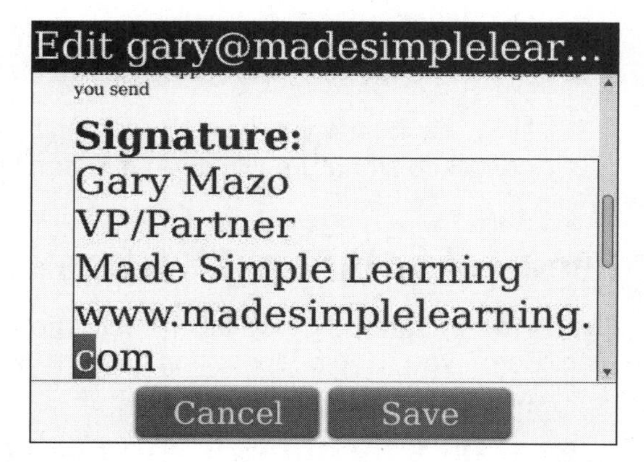

Corporate (BlackBerry Enterprise Server) Email Users

If you are a BlackBerry Enterprise Server user, you can set up your signature right in your **Messages** app.

1. From your **Messages** list, press the **Menu** key.

2. Select **Options.**

3. Click on **Email Settings**

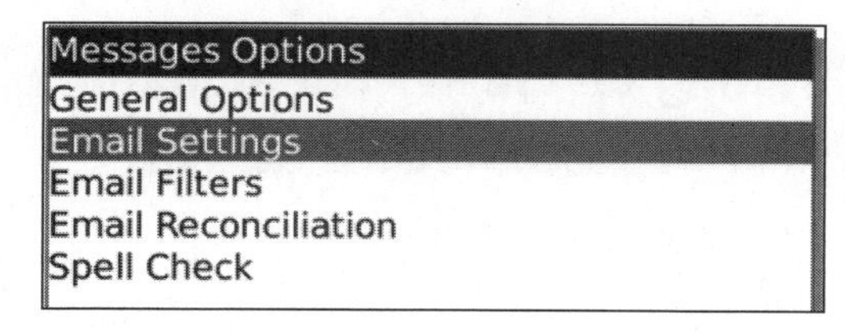

4. Scroll down to the **Use Auto Signature:** field and change **No** to **Yes**.
5. Scroll down and type your email signature in the text field.
6. Press the **Menu** key and click **Save**.

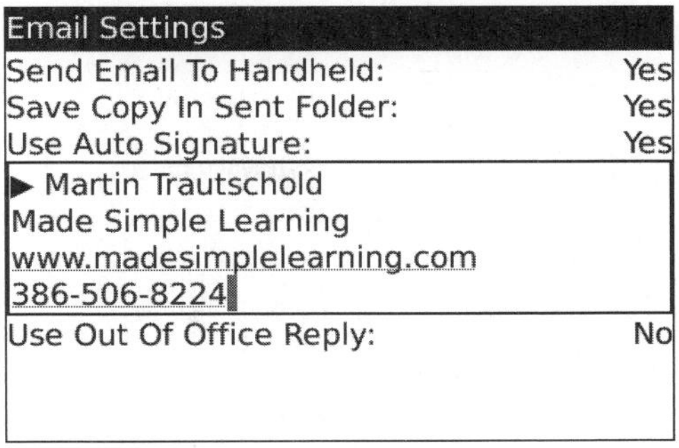

Email Settings
Send Email To Handheld: Yes
Save Copy In Sent Folder: Yes
Use Auto Signature: Yes
Martin Trautschold
Made Simple Learning
www.madesimplelearning.com
386-506-8224
Use Out Of Office Reply: No

NOTE: Please see page 158 to learn how to turn on spell Checking and use this feature.

Filtering Your Messages for SMS, Calls, and More

Your messages can really start to add up on your BlackBerry and it can get overwhelming. Fortunately, it's easy to set up filters so you see only the messages you want to see at a particular moment.

TIP: The shortcuts to filter your Messages Inbox are:

ALT + S = Show only SMS Text Messages

ALT + L = Show only MMS Multi-Media Messages

ALT + I = Show only incoming messages and phone calls.

ALT + O = Show only outgoing messages and phone calls.

ALT + P = Show only phone calls

ALT + V = Show only voice mail messages

Press the **Escape** key to "un-filter" to see your entire inbox again.

Just like on your computer, using email folders can help you be more organized and productive. Also, if you've saved many messages and are not sure which are email and which are SMS, using the folder commands can help.

1. Start your **Messages** app.
2. Press the **Menu** key.
3. Scroll down to **View Folder** and click. You will now see a list of all the message folders on your device.
4. Choose **SMS Inbox** and you'll see only the SMS messages.
5. These folders also show your missed calls, MMS messages, WAP Push messages, and Browser Messages.

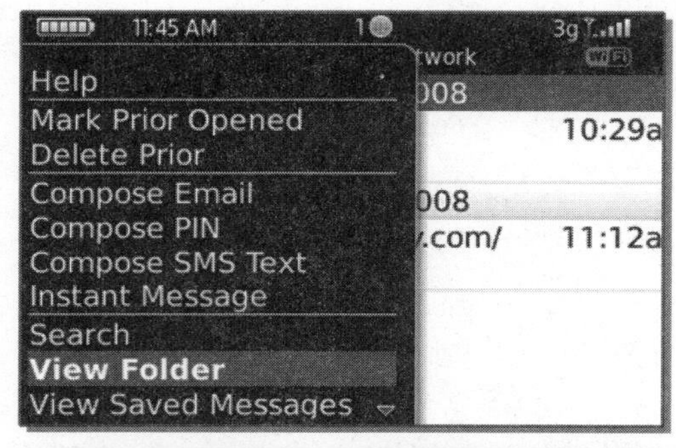

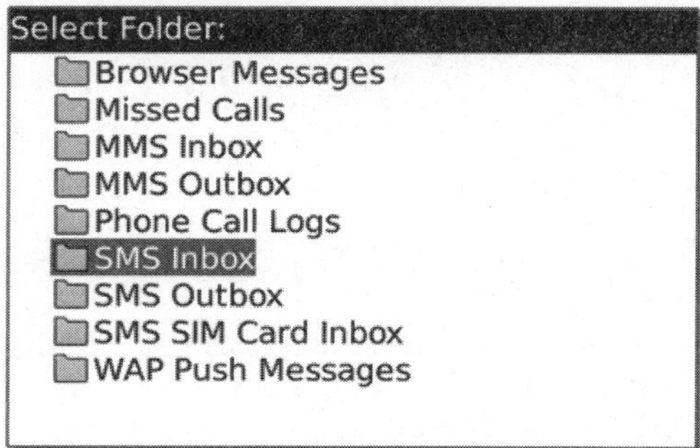

Filing a Message in a Folder

Make sure that Wireless Synchronization is turned on (as described in the "Email Reconciliation" section above). This is necessary in order to file messages.

1. Click on the **Messages** icon from your **Home** screen.
2. Highlight the message you want to file.
3. Press the **Menu** key and select **File**
4. Choose the folder you want to use to store the message.

Changing Folder Names or Adding Folders

To do this, you must be using your BlackBerry and a BlackBerry Enterprise Server with Wireless Synchronization. If you are unsure whether you are using Wireless Synchronization, most likely you are not using it.

On the desktop (or notebook) computer you use to sync your BlackBerry, simply change or add a folder to the email client you use to sync the BlackBerry.

Changes you make on the desktop or notebook will be reflected in the folders available on the BlackBerry.

Email Message Filters (Only for BlackBerry Enterprise Server Users)

If your email comes through a BlackBerry Enterprise Server, you can set up filters on your Bold by following the steps below.

> **NOTE:** If you are a BlackBerry Internet Service user with Personal or Internet email, you can set up filters using the BlackBerry Internet Service web site. See page 55 in Chapter 1 "Email Set Up" to learn how.

While receiving your email messages on your BlackBerry is a wonderful thing, there might be some email messages that, for whatever reason, you don't want sent to your BlackBerry. Fortunately, you can use an email filter to see just which messages you want sent to your BlackBerry and which ones stay on the server.

Create Email Filter

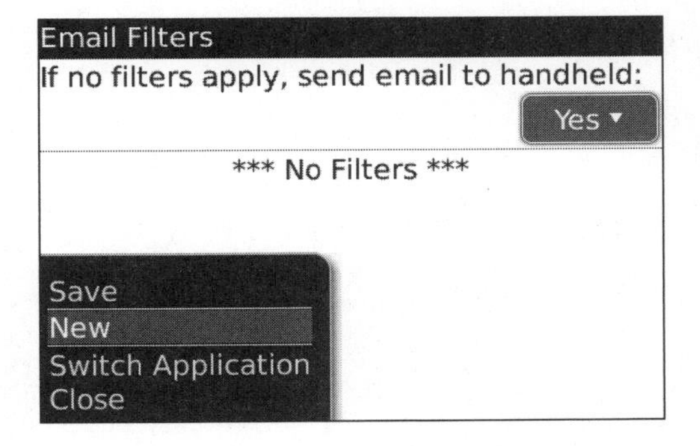

1. Click your **Messages** icon. (Or press **M** if you have enabled Home screen hotkeys.)
2. Press the **Menu** key.
3. Scroll down to **Options** and click.
4. Click **Email Filters**, press the **Menu** key, and click **New**.
5. Press the **Menu** key and choose **Save**.

Use Email Filters

1. Click your **Messages** icon.
2. Press the **Menu** key and select **Email Filters**.
3. Just use the trackpad and click the radio button next to the filter you wish you use.

Chapter 14

Your Contact List

Your BlackBerry excels as a contact manager. You will turn to your **Contacts** app, perhaps more than any other app on the device. From a contact, you can email, sent a text message, call, fax or even "poke" someone if you have the **Facebook** app installed.

One rule of thumb you will hear us say often is to add anything and everything to your contacts. Whenever someone calls, add them to your contacts. Add contacts from emails and messages—you will always go back to them.

The Heart of Your BlackBerry

Your address book is really the heart of your BlackBerry. Once you have your names and addresses in it, you can instantly call, email, send text (SMS) messages, PIN-to-PIN BlackBerry Messages, BlackBerry Messenger (BBM) or even pictures or Multimedia Messages (MMS). Since your BlackBerry came with a camera, you may even add pictures to anyone in your address book so that when they call, their pictures show up as Picture Caller ID.

Picture Caller ID
"Gary with the warm blue glow"

Getting Contacts from Your SIM Card onto Your Contact List

If you are using your SIM (Subscriber Identity Module) card from another phone in your BlackBerry and have stored names and phone numbers on that SIM card, it's easy to transfer your contacts into your **Contact** list.

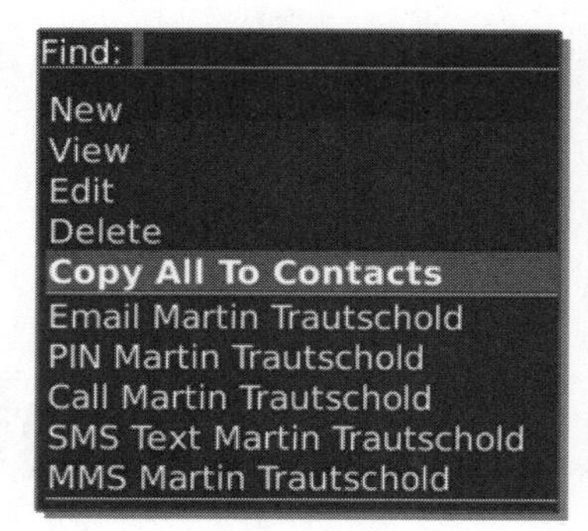

1. Click on your **Contacts** icon.
2. Press the **Menu** key and scroll to **SIM Phone book** and click.
3. Press the **Menu** key and scroll to **Copy To All Contacts**.
4. Then you will see a screen that says "Contacts Imported" or "No Contacts are Saved on your SIM Card."

After this process is complete you should see all the name and phone numbers from your SIM card in your Contacts list.

TIP: Your SIM Card only contains the bare minimum Name and Phone. Sometimes a SIM card stores names in ALL CAPS, so beware! You might need to do some additional clean up after the import. You should also review your imported contacts and add in email addresses, mobile/work phone numbers, and home/work addresses to make your BlackBerry more useful.

How Do You Get Your Contacts onto Your BlackBerry?

You can manually add contact addresses one at a time (see page 272). You can also load or sync up your computer's contacts with your BlackBerry.

If your BlackBerry is tied to a BlackBerry Enterprise Server the synchronization is wireless and automatic. Otherwise, you will use either a USB cable or Bluetooth wireless to connect your BlackBerry to your computer to keep it up to date. For Windows PC users, see page 65. Apple Mac computer users, see page 125.

If you use Gmail (Google Mail), you can use the BlackBerry Internet Service email set up for Google to wirelessly sync contacts (see page 46) or use the **Google Sync** program to wirelessly update your contacts on your BlackBerry with your address book from Gmail, see page 305.

When Is Your Contact List Most Useful?

Your Contacts program is most useful when two things are true:

- You have **many names** and addresses in it.
- You can **easily find** what you need.

Our Recommendations

We recommend keeping two rules in mind to help make your Contacts most useful.

Rule 1: Add anything and everything to your Contacts.

You never know when you might need that obscure restaurant name/number, or that plumber's number, etc.

Rule 2: As you add entries, make sure you think about ways to easily find them in the future.

We have many tips and tricks in this chapter to help you enter names so that they can be instantly located when you need them.

TIP: Finding Restaurants

Whenever you enter a restaurant into your Contact list, make sure to put the word "Restaurant" into the company name field, even if it's not part of the name. Then when you type the letters "rest" you should instantly find all your restaurants!

What Fields Are Searched to Find Contacts

At the top of the Contacts list, you will see a **Find:** field. As long as you know which fields in your contact entries are searched, you can easily find all your contacts. Just keep these fields in mind when you enter new information into your address book.

The following fields are used when you type in letters to search for contacts.

- First name
- Last name
- Nickname
- Company name

Add New Addresses Easily

On your BlackBerry, since your address book is closely tied to all the other icons (Messages/Email, Phone and Web Browser) you have many methods to easily add new addresses:

- **Choice 1:** Add a new address inside the **Contacts** app.
- **Choice 2:** Add an address from an email message in Messages.

- **Choice 3:** Add an address from a phone call log in the phone.
- **Choice 4:** Add a new address from an underlined email address or phone number from any source.

Choice 1: Add An Address into Contacts

The most obvious way of adding a contact is to use New Contact and type in the information right on your BlackBerry. To do this:

1. Use the trackpad and navigate and click on the **Contacts** icon.

2. Press the **Menu** key and select **New Contact** or just glide to the top and click on **Add Contact:** at the top of the Contact List.

OR

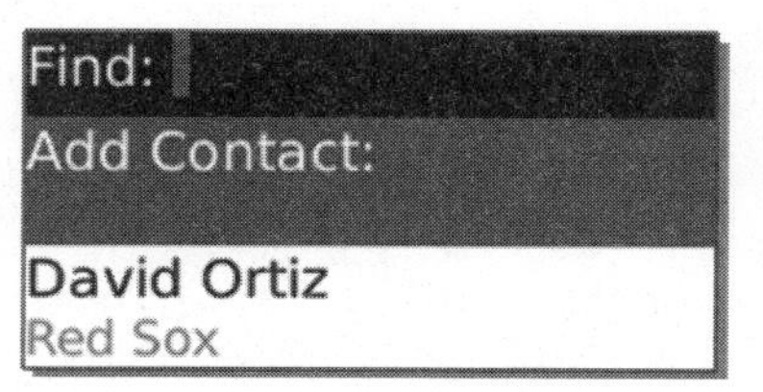

3. Add as much information as you know because the more you add, the more useful your BlackBerry will be!

Edit Contact
First: David
Last: Parker
Picture:
Company: Parker Construction Co.
Job Title: Engineer
Email: david@parkerconstruction.com
Work: 6175551234
Work 2: 1800REDSOX1

TIP: Press the **Space** key instead of typing the @ and "." in the email address.

TIP: If you add a contact's work or home address, you can easily map it to get directions right on your BlackBerry.

Need to Add More Than One Email Address for a Person?

While you are adding or editing their Contact entry, just press the **Menu** key and select **Add Email Address**

Be sure to save your changes by pressing the **Menu** key and selecting **Save.**

Enter a Phone Number with Letters

Some phone numbers have letters like **1 800-REDSOX1**. These are easier than you might think to add to your BlackBerry address book (or type while on the phone). The trick is to hold down your **Alt** key (lower left key with up/down arrows on it) and just type the letters on your keyboard. You can also just press and hold a key to see the letter appear.

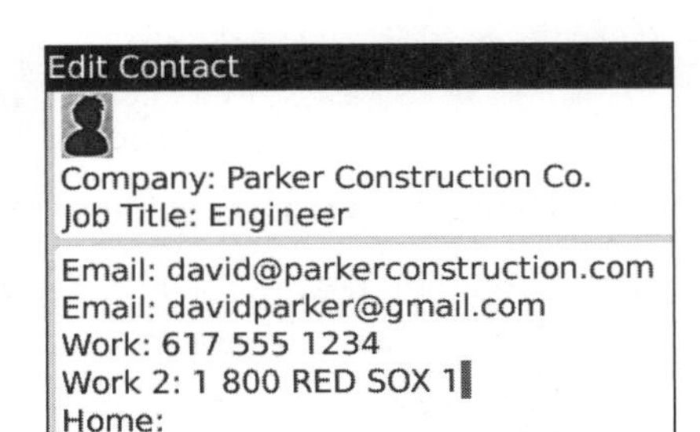

> **TIP:** Put spaces in the phone number by hitting the **Space** key — this makes it easier to read.

Enter a Phone Number with Pauses or Waits

Sometimes you need to dial a phone number that has several components, such as a dial-in number and a separate password. One example is a voice mail system for your home or office phone number. Let's say you had to dial an 800 number, wait 4 seconds, then your own number, wait 2 seconds, then your own password. You can do it all with pauses or waits. A **pause** is a 2 second pause then it continues dialing, a **wait** will wait for you to manually click a button before continuing dialing.

So to enter 1-800-555-1234, pause 4 seconds, 386-506-8224, pause 2 seconds, and enter your password 12345, you would follow these steps:

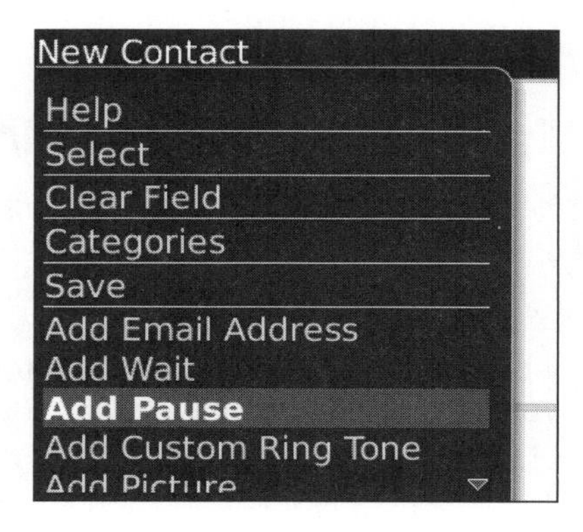

1. Type in 1-800-555-1234.
2. Press the **Menu** key.
3. Select **Add Pause** (2 seconds).
4. Press the **Menu** key and select **Add Pause** (2 more seconds) again.
5. Type 386-506-8224.
6. Press the **Menu** key and select **Add Pause.**
7. Finally type your password 12345.

When you're done, the screen should look similar to this.

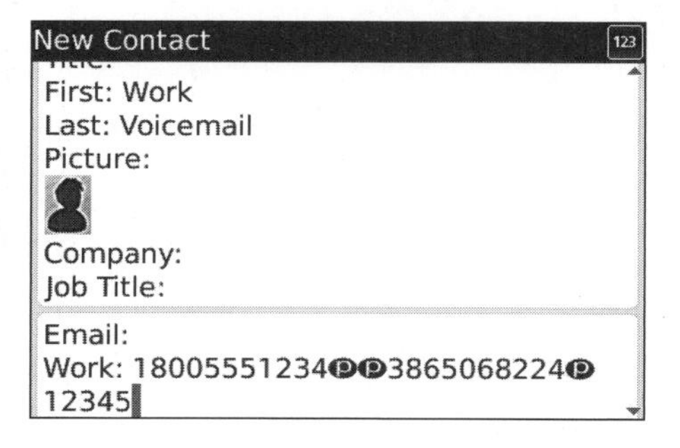

TIP: Try adding several pauses in a row or use Add wait if a single pause does not work.

Choice 2: Add an Address from an Email Message

Another easy way to update your address book is to simply add the contact information from emails that are sent to you.

1. Navigate to your message list and scroll to an email message in your inbox.
2. Click on the email message and press the **Menu** key.
3. Scroll to **Add to Contacts** and click.
4. Add the information in the appropriate fields, press the **Menu** key, and select **Save** from the menu.

Choice 3: Add an Address from a Phone Call Log

Sometimes you will remember that someone called you a while back, and you want to add their information into your address book.

1. Press the **Green Phone** key to bring up your call logs.
2. Scroll to the number you want to add to your address book.
3. Press the **Menu** key and select **Add to Contacts.**
4. Add the address information, press the **Menu** key, and select **Save.**

Choice 4: Add an Address from an Underlined Email Address or Phone Number

One of the very powerful features of the BlackBerry is that you can really add your contacts from just about anywhere. Even though the next steps are shown on the **MemoPad**, they can be applied to tasks, emails (email addresses in the To:, From:, and CC: fields and in the body of an email message), and web pages. Let's say you wrote down a contact's name and phone number in a memo, but never added it to your address book:

1. Locate and click on your **MemoPad** icon.

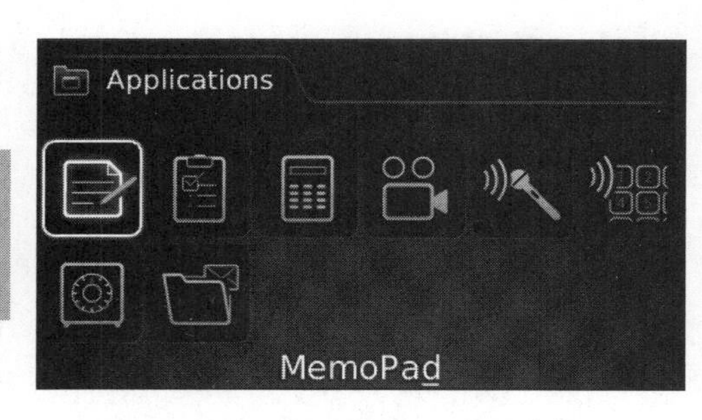

> **TIP:** Press the Hotkey D to instantly start it (see page 541).

2. Locate the memo you wrote earlier with the phone number and open it.
3. Then glide over to the underlined phone number to highlight it.
4. Press the **Menu** key or click the trackpad and select **Add to Contacts**

5. Now, type all the contact information for this person.
6. Then click the trackpad and select **Save.**

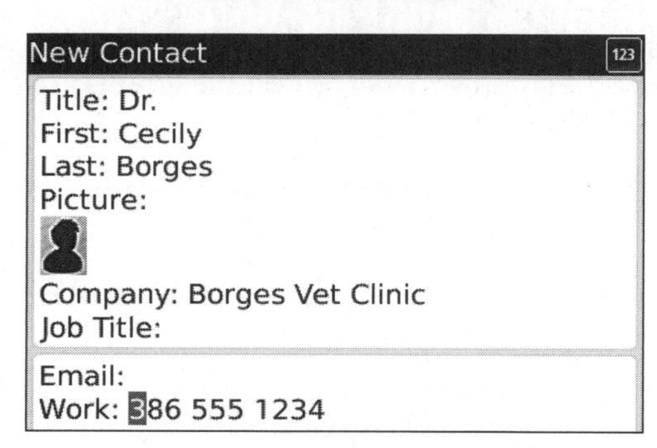

7. Now you are back in the Memo item. Try highlighting the number and clicking the trackpad. Notice that it says **Call (name)** because you have added this person to your address book.

TIP: You can also call any underlined number. See Figure 14-1 showing an underlined phone number in an email message. You will typically see phone numbers in email signatures.

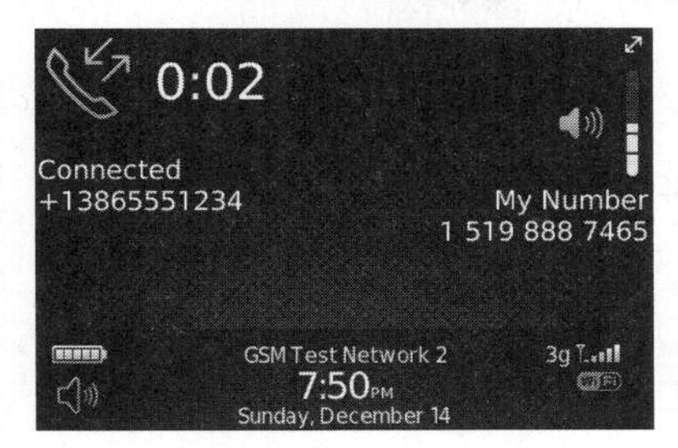

Figure 14-1. *Place a call from a phone number in an email.*

TIP: Finding that person you met at the bus stop

If you just met someone at the bus stop, enter their first and last name (if you know it), but also enter the words "bus stop" in the **Nickname** field. Then when you type the letters "bus" or "stop" you should instantly find everyone you met at the bus stop, even if you cannot remember their name!

Why Can't You See All of Your Names and Addresses (Filtered)?

If you are seeing no names, a few names, or if you just added a new name and do not see it on the list, it is very likely your contact list is **filtered**.

This means it is showing you only those names that are assigned to a particular category.

The tip-off that your list is filtered is the black bar (or other color) at the top with the category name.

In the image to the shown here, the category applied for the filter is **Business**.

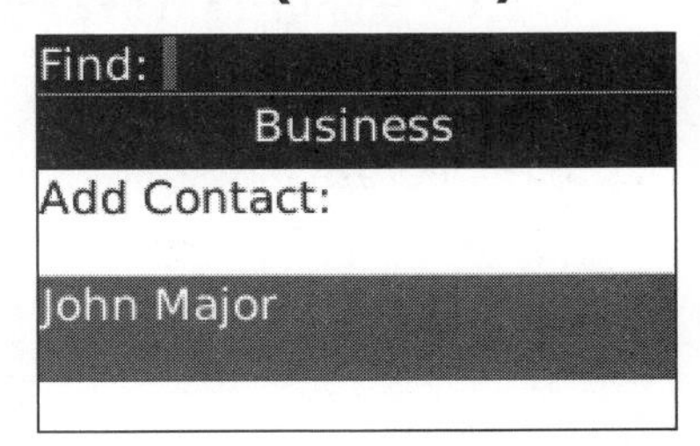

Learn how to see your entire Contact list again on page 285. Learn more about **Categories** on page 283.

Locate Names and Addresses Easily

Once you get the hang of adding contact information into the Contacts app, you will begin to see how useful it is to have all that information at your fingertips. The tricky part can be actually locating all the information you have input into the BlackBerry.

Option 1: Use the Find Feature in Contacts

Contacts has a great **Find** feature at the top that will search for entries that match the letters you type into one of these fields:

- First
- Last
- Nickname
- Company

1. Inside **Contacts**, just type a few letters of a person's first name, last name and/or company name (separated by spaces) to instantly find that person.

2. Press the letter **M** to see only entries where the **first name**, **last name**, or **company name** start with the pressed letter; doing so might bring up the following names:
   - **Martin Trautschold**: This match is on the contact's first name.
   - **Gary Mazo**: This match is on the contact's last name.
   - **Cathy Carollo**: This match is on the contact's company name, "Made Simple Learning."

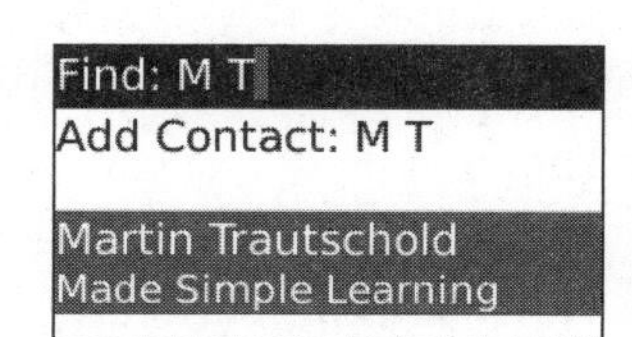

3. Then press the **Space** key and type another letter like T to further narrow the list to people with an M and a T starting their first, last, or company name.
4. In this case, there is only one match: Martin Trautschold

Option 2: Finding and Calling Someone

You can also locate people in your contact list when you are dialing in the phone.

1. Press and hold the **Green Phone** key for about 3 seconds to start the phone and go immediately into the dial from Contacts screen.

2. Now, type a few letters of someone's first name, last name, or company name.

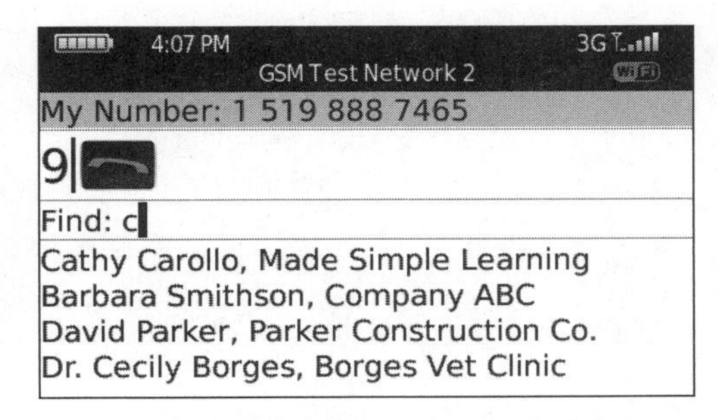

TIP: If no matching entries are found, you just see the digits being dialed.

3. Press the **Space** key and type a few more letters to further narrow the list.
4. Now, just scroll down to the correct entry and phone number (work, home, mobile, etc.).

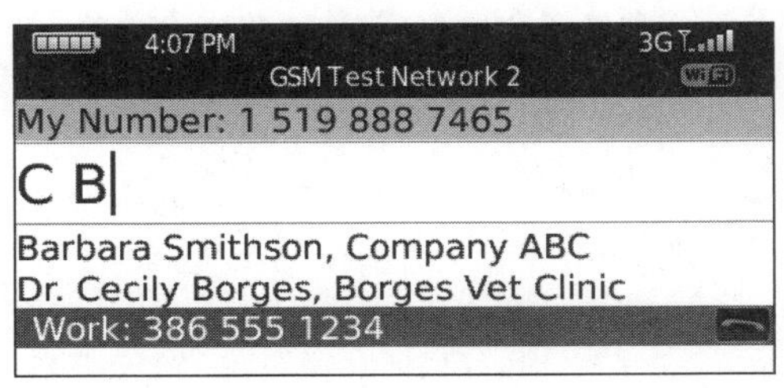

5. Press the **Green Phone** key to start the call.
6. If you decide you want to email or SMS this person instead, highlight their name and press the **Menu** key to select email or SMS. You will only see these menu options if there is an email and phone number for this contact.

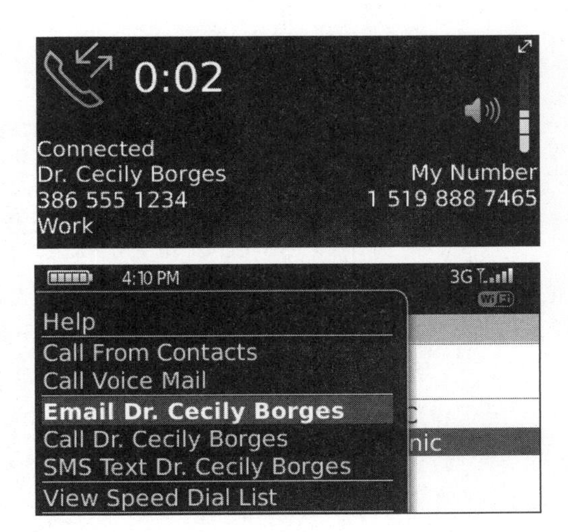

Managing Your Contacts

Sometimes, your contact information can get a little unwieldy, with multiple entries for the same individual, business contacts mixed in with personal ones, etc. There are some very powerful tools within the **Contacts** application that can easily help you get organized.

TIP: Finding Your Neighbors

If you just moved into a new neighborhood, it can be quite daunting to remember everyone's name. One tip is to add the word "neighbor" into the **Nickname** field for every neighbor you meet. Then to instantly call up all neighbors, you type the letters "neigh" to instantly find all of your new friends!

Adding More Information to Your Contact Entries (Menu Commands)

One of the first things to do is to make sure that all the correct information is included in your contacts.

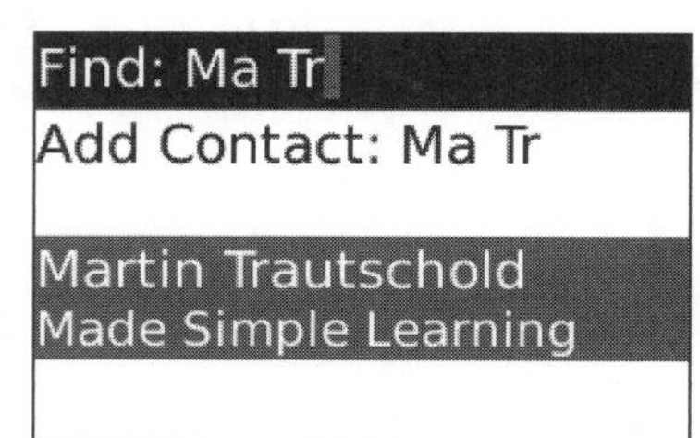

1. Select the **Contacts** app and click on it or press the **C** hotkey (see page 541). Type in a few letters of the first, last, or company name to find the contact or just scroll through the list.
2. Highlight the contact you want to manage with the trackpad.
3. Press the **Menu** key and choose **Edit** to access the detailed contact screen and add any information missing in the fields.
4. Now you can add more information to your contact to make your BlackBerry more useful.

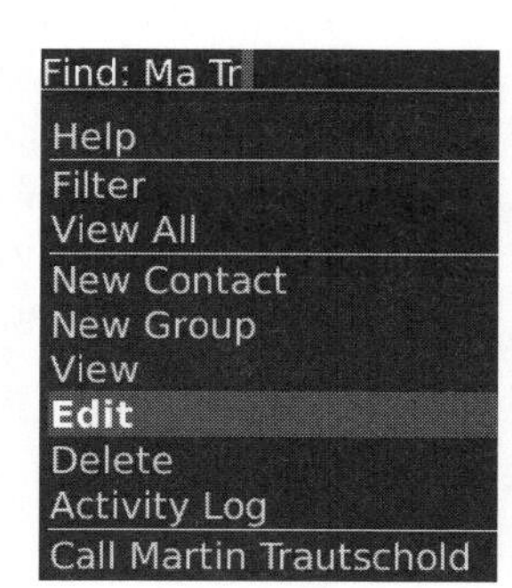

Adding a Picture to the Contact for Caller ID

It's helpful to match a face with a name. If you have loaded pictures onto your media card or have them stored in memory, you can add them to the appropriate contact in your contacts. Since you have a BlackBerry with a camera, you can simply take the picture and add it as a **Picture Caller ID** right from your camera.

1. Select the contact to edit with the trackpad as you did above.
2. Scroll down to the Picture or picture icon and click the trackpad.
3. Choose **Add** or **Replace Picture.**

4. You have the choice of finding a picture already stored on your BlackBerry or taking a new one with the camera.
5. If you want to use a stored picture, then navigate to the folder in which your pictures are stored by rolling the trackpad and clicking on the correct folder.
6. Once you have located the correct picture, click the trackpad on it and you will be prompted to **Crop and Save** the picture.

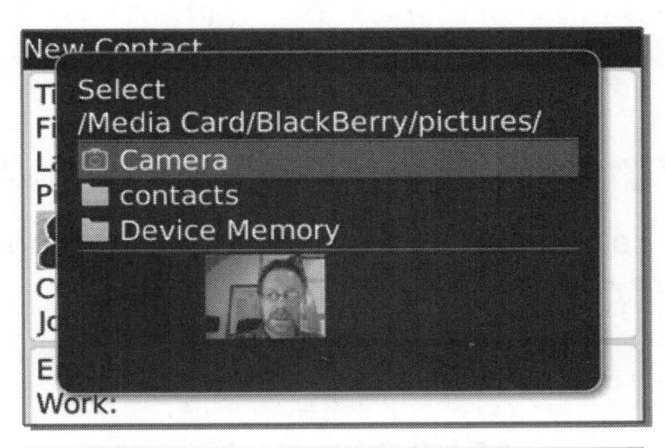

7. You can also use the camera to take a picture right now. To do this, click on the camera and take the picture. Move the viewing box to center the face, click the trackpad and select **Crop and Save.**

The picture will now appear in several places:

- On the screen when you speak with that person on the phone
- In their contact entry.
- It will also appear next to their email address.

TIP: On any menu, you can jump down to an entry by pressing the key using the key's first letter. So, to jump down to Options press the O key.

Changing the Way Contacts Are Sorted

You can sort your contacts by **First Name**, **Last Name,** or **Company Name.**

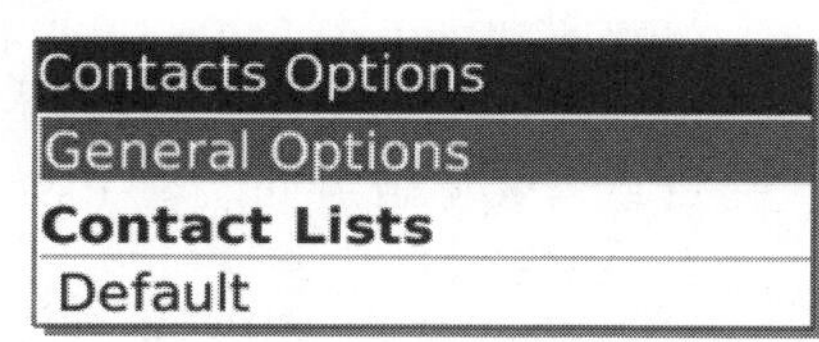

1. Click on your **Contacts** i con—but don't click on any particular contact.
2. Press the **Menu** key and select **Options.**
3. Then click on **General Options.**
4. In the **Sort By** field, click on the trackpad and choose the way you wish for your contacts to be sorted. You may also select whether to allow duplicate names and whether to confirm the deletions of contacts from this menu.

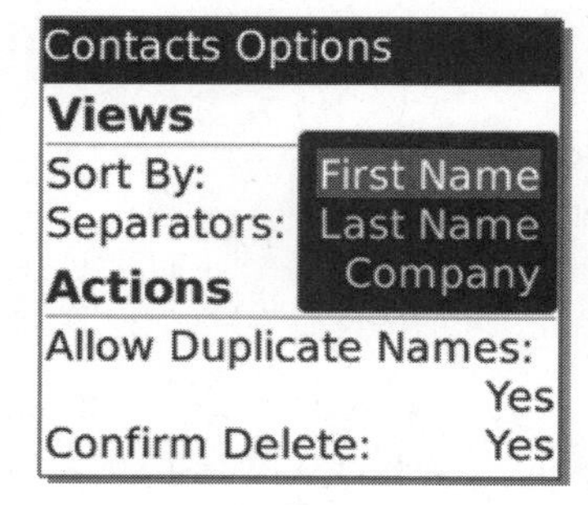

TIP: First Letter Trick for Menus/Lists

On any drop down list like "First Name, Last Name, Company" in Sort By, pressing the first letter of the entry will jump there. For example, pressing the letter L will jump to the entry starting with L.

TIP: Finding Your Child's Friend's Parents Names

Learning the names of parents of your school age children can be fairly challenging. In the First Name field, you should add in not just your son/daughter's friend's name but their parent's names as well (e.g. First: "Samantha (Mom: Susan Dad: Ron)" Then in the company name field, add in the name of your son/daughter and "school friend" (e.g. Company name: "Cece school friend") So typing your son/daughter's name in your Contact list Find field will instantly find everyone you've met at your child's school. Now you can say, "Hello Susan, great to see you again!" without missing a beat. *Try to conceal your BlackBerry when you are doing your name search.*

Organizing Contacts with Categories

Sometimes, organizing similar contacts into **Categories** can be a very useful way to quickly find people. What is even better is that the categories you add, change, or edit on your BlackBerry are kept fully in sync with those on your computer software.

1. Find the contact you want to assign to a category, click the trackpad to view the contact, click it again, and select **Edit** from the short menu.

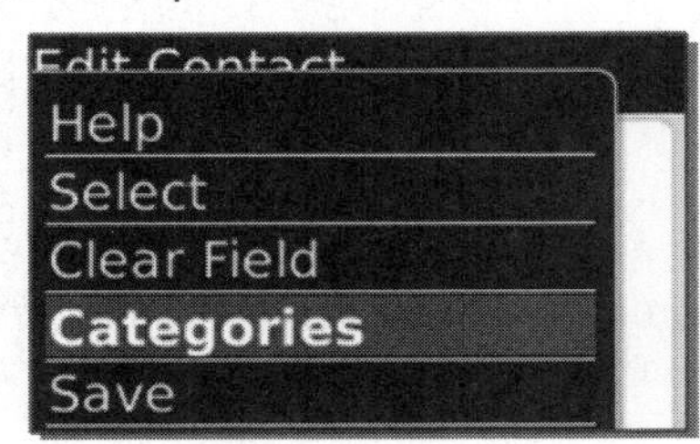

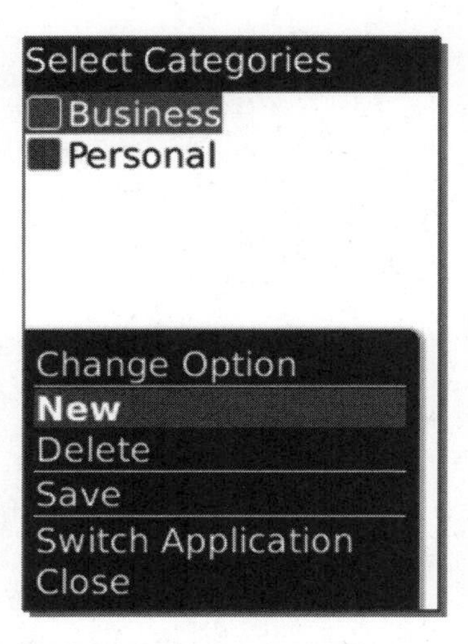

2. Click the **Menu** key and select **Categories**.
3. Now you will see the available categories. (The default categories are **Business** and **Personal**.)
4. If you need an additional category, just press the **Menu** key and choose **New**.
5. Type in the name of the new category and it will now be available for all your contacts.
6. Scroll to the category to which you wish to add this contact and click.

TIP: You can create and assign contacts as many categories as you want.

Filtering Your Contacts by Category

Now that you have your contacts assigned to categories, you can filter the names on the screen by their categories. So, let's say that you wanted to quickly find everyone who you have assigned to the business category.

1. Click on your contacts.
2. Press the **Menu** key, scroll up, and select **Filter.**

The available categories are listed. Just click the trackpad (or press the **Space** key) on the category you wish to use as your filter. Once you do, only the contacts in that category are available to scroll through.

How Do You Know When Your Contact List Is Filtered?

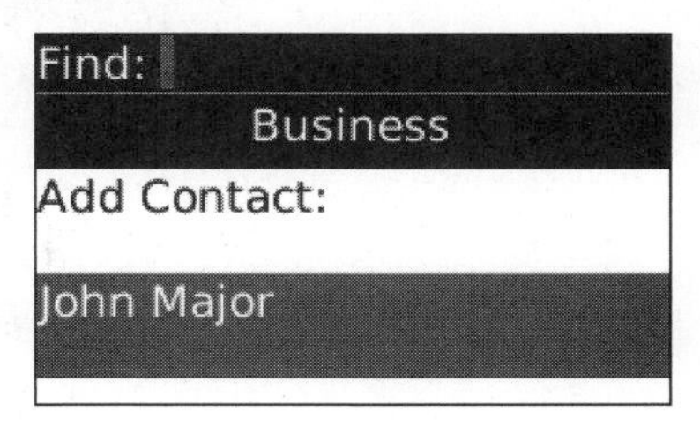

You will see a black bar at the top of your Contact list with the name of the category.

In this case, the category is **Business.**

This contact list is filtered to show you only contacts that are in the category **Business**.

Un-Filtering Your Contacts by Category

Unlike the Find feature, you cannot just press the **Escape** key to un-filter your categories. You need to reverse the Filter procedure.

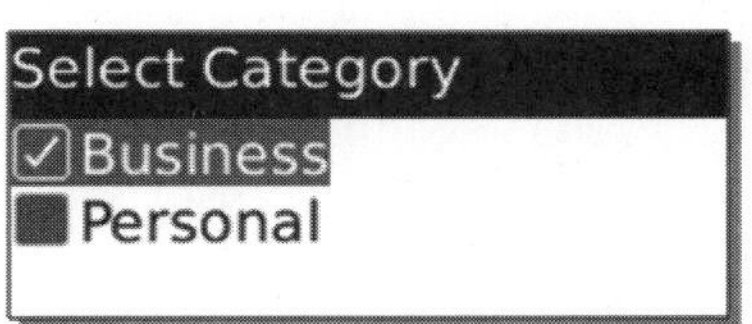

1. Inside your **Contacts** app, press the **Menu** key.
2. Select **Filter**, glide down to the checked category, and uncheck it by clicking on it or pressing the **Space** key when it is highlighted.

Using Groups as Mailing Lists

Sometimes, you need even more organizing power from your BlackBerry. Depending on your needs, grouping contacts into mailing lists might be useful so that you can send mass mailings from your BlackBerry.

Why Would You Want to Use a Mailing List?

There are many reasons to create mailing lists including:

- If you have your team members on mailing list, you can instantly notify them of project updates.
- Let's say you're about to have a baby. Put everyone in the notify list into a "New Baby" group. Then you can snap a picture with your BlackBerry and instantly send it from the hospital!

Change Your Initial View

If you prefer the **Agenda View**, **Week View**, or **Month View** instead of the default **Day View** when you open your Calendar, you can set that in the options screen. Click the drop down list next to **Initial View** to set these.

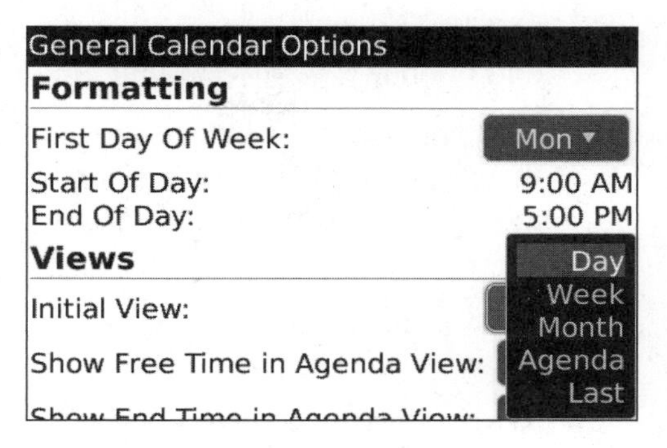

Change Your Start and End of Day Time on Day View

If you are someone that has early morning or evening appointments, the default 9am – 5pm calendar will not work well. You will need to adjust the **Start of Day** and **End of Day** fields in the Options screen. These options are up at the top of the screen, under **Formatting.**

TIP: Use the number keys on your keypad to type in the correct hours (e.g. type 7 for 7:00 and 10 for 10:00). Then glide over to AM/PM and press the **Space** key if you need to change it. This is faster than clicking the trackpad and rolling to an hour.

Copy and Paste Information into Your Calendar or Any Icon

The beauty of the Blackberry is how simple it is to use. Let's say that you wanted to copy part of a text in your email message and paste it into your calendar. A few good examples are:

- Conference Call information via email
- Driving directions via email
- Travel details (flights, rental cars, hotel) via email

1. First, glide the trackpad to select the email from which you want to copy and open it.
2. Then, highlight the text by moving the cursor to the beginning or end of the section you want to highlight.

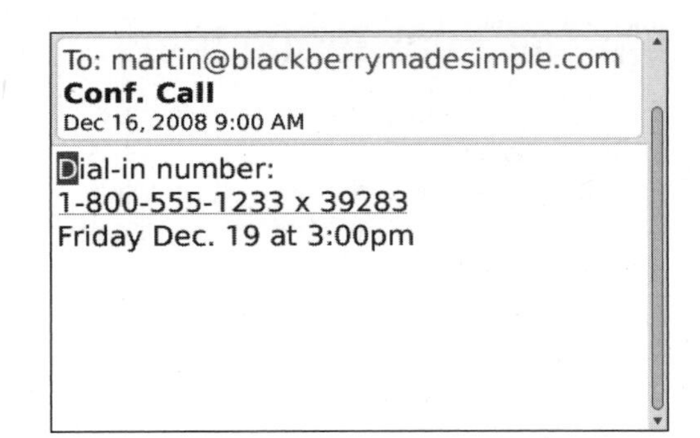

3. Next, press the **Shift** key to start selecting text to copy. You can also press the **Menu** key and press **Select.**
4. Then all you do is glide the trackpad to move the cursor to the end (or beginning) of the section you want to highlight.

   The text that you just highlighted will turn blue.
5. Press the trackpad and select **Copy**.

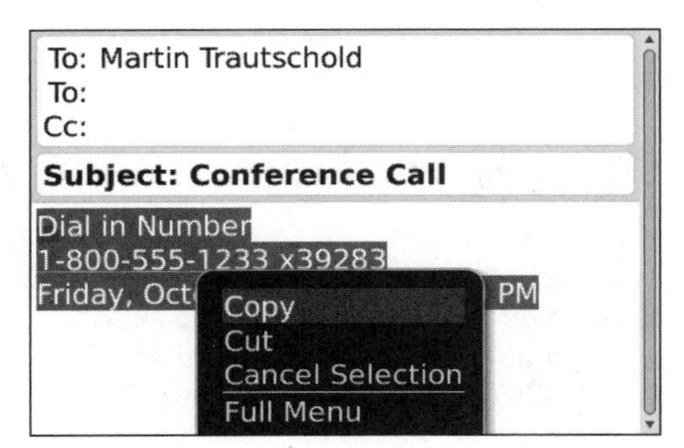

6. Press the **Red Phone** key to jump back to your **Home** screen and leave this email open in the background.

Glide to and click on the **Calendar** icon to open it.

7. Schedule a new appointment by clicking the trackpad in day view.
8. Move the cursor to the field where you want to insert the text from the clipboard. Then click the trackpad or press the **Menu** key and select **Paste.**

You will now see your information pasted into the calendar appointment.

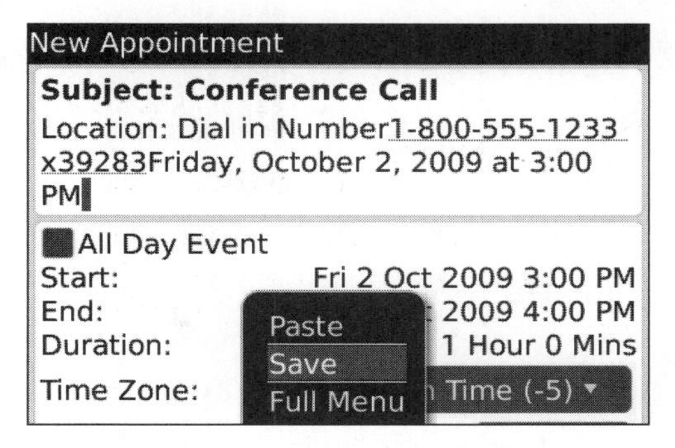

TIP: Shortcuts for Copy & Paste

Select - Press the **Shift** key (to right of the **Space** key) to start Select mode.

Copy - To copy selected text, hold the **ALT** key (lower leftmost key – up/down arrow) and click the trackpad.

Paste - To paste the copied (or cut) text, click the trackpad and select Paste.

Press the **Escape** key (to the right of the trackpad) and select **Save** when prompted. Now your text is in your calendar, available exactly when you need it. Gone are the days of hunting for the conference call numbers, driving directions, or asking yourself *"What rental car company did I book?"*

Alarms and Recurring Appointments

Some appointments occur every week, month, or year. Others are easy to forget, so setting an alarm helps remind us where to be or where to go.

To Schedule an Alarm

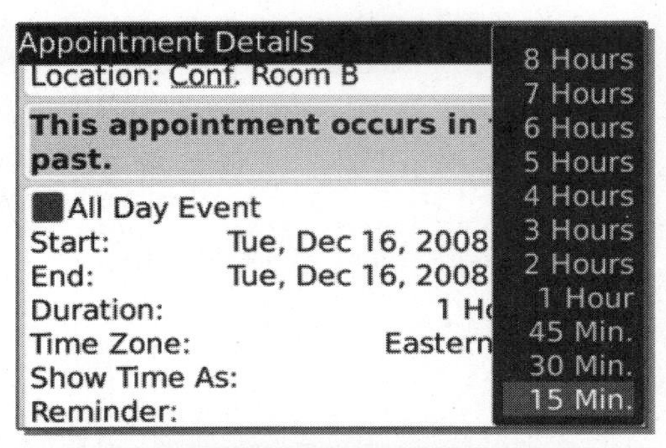

1. Navigate to the **Calendar** icon and click.
2. Begin the process of scheduling an appointment as detailed above.
3. In the **New Appoin**tment screen, scroll down to **Reminder**.
4. The default reminder is 15 minutes. Click on the highlighted field and change the reminder to any of the options listed. Press the **Escape** key or the **Menu** key and select **Save**.
5. When the calendar alarm rings, you can glide down to choose **Open**, **Dismiss,** or **Snooze** from the alarm popup window.

To Change the Default Reminder (Alarm) and Snooze Times

If you need a little more advanced warning than the default 15 minutes, or a little more snooze time than the default 5 minutes, you can change those also in the options screen.

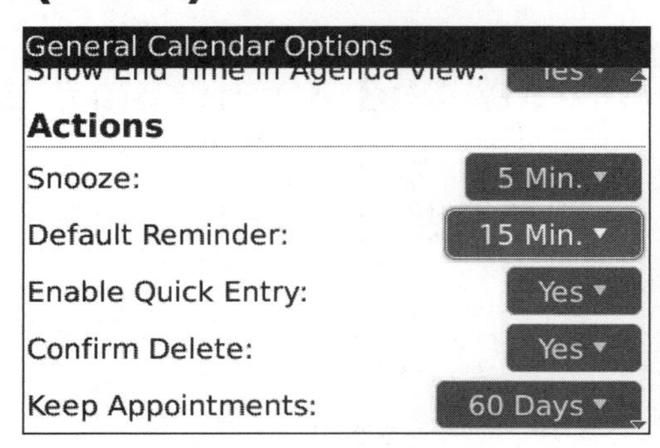

To Dial the Phone from a Ringing Calendar Alarm

What is really great is that if you put a phone number into a calendar item (like shown previously in copy/paste), you can actually dial the phone right from the ringing calendar alarm!

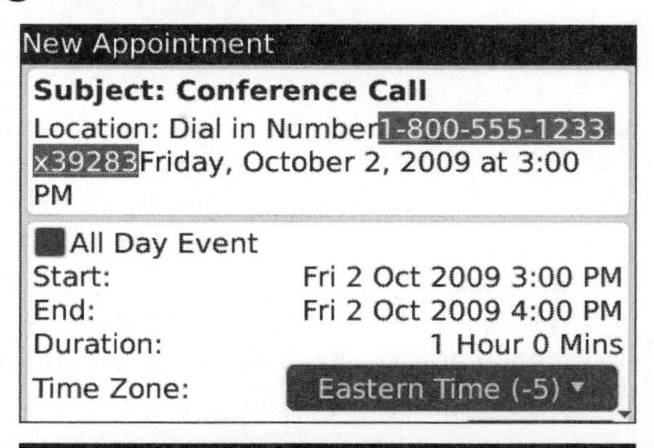

All you need to do is open the event and then click the underlined phone number.

> **TIP:** This is a great way to instantly call someone at a specified time without ever having to hunt around for their phone number!

Even better: since you formatted the number with an x between the main phone number and the access code (or extension), the BlackBerry will pause for 3 seconds and automatically dial the access code!

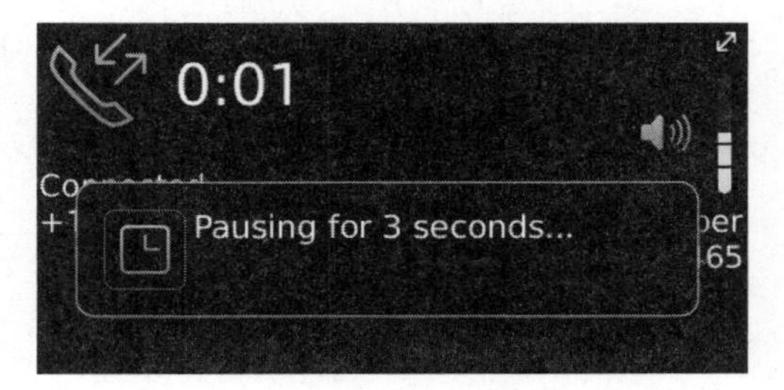

To Adjust Your Calendar (and Task) Alarms in Sound Profiles

Most BlackBerry smartphones come with the calendar alarm profile setting to be silent and not vibrate when your calendar alarm rings and your BlackBerry is out of holster (e.g. sitting on your desk)! If you are an avid user of the calendar, you will definitely want to adjust this profile setting.

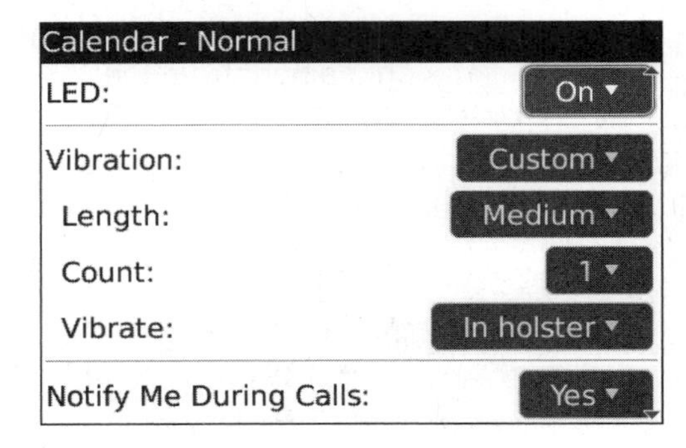

Here is how to change your Calendar profile so that it will ring and/or vibrate when the alarm rings if your BlackBerry is sitting on a desk.

1. Scroll to the upper left hand corner of your home screen and click on the **Sounds** icon (looks like a speaker).
2. Scroll down to the bottom of the list and click on **Set Ring Tones/Alerts.**

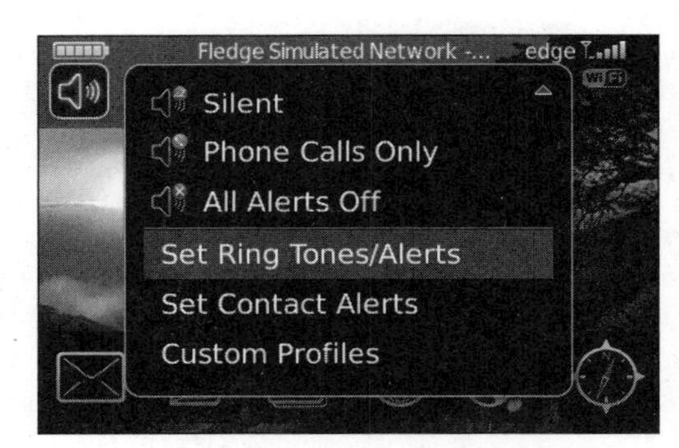

3. Scroll down to **Reminders** and click the trackpad.
4. Then, click on **Calendar** underneath **Reminders**.

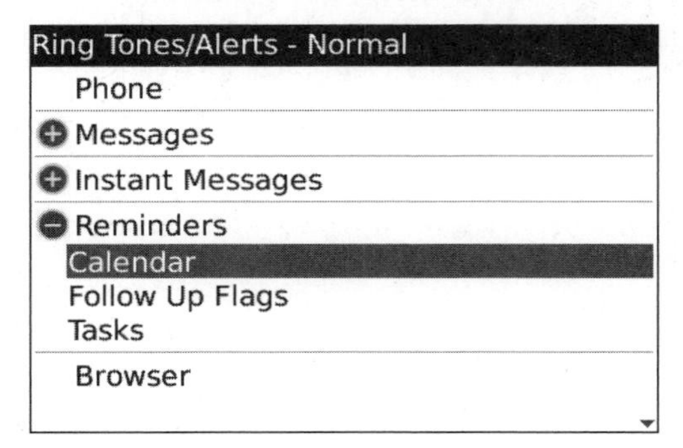

5. Click next to **Ring Tone** and select a new tone for your calendar alarms.
6. Glide down and click on any of the fields to change those settings.

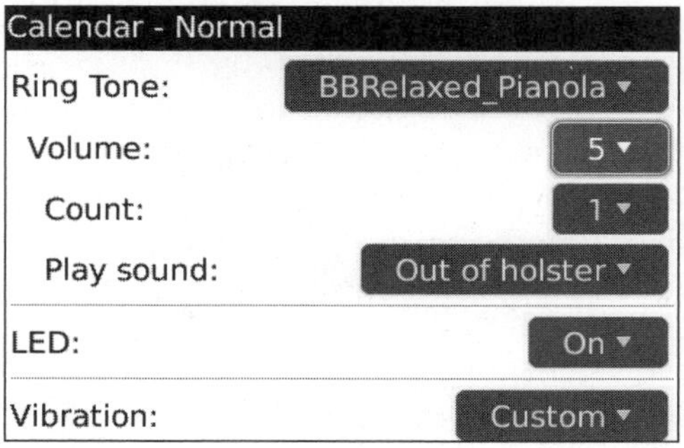

> **NOTE:** The default **Task** alarms work the same way as Mute and No Vibrate do when Out of Holster; so if you use **Task** alarms, you will want to change Tasks as well.

7. Scroll down and set **Vibration** field to **On.**
8. Set the **Vibrate with Ring Tone** field is set to **Off** so the BlackBerry vibrate before it starts to ring. This will give you the chance to catch it before it makes noise.
9. If you want to test out your settings, roll down and click on the **Try It** button.

   The combined vibration and sounds for ringing calendar alarms will help ensure you do not miss any important appointments.

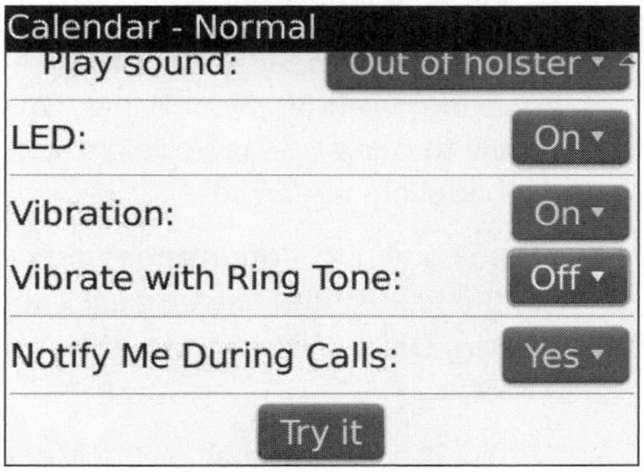

> **NOTE: Here are a few definitions to remember.**
>
> **Out of Holster** - The BlackBerry is sitting on your desk or in your hand (out of the approved holster)
>
> **In Holster** - The BlackBerry is inside the approved holster (plastic or leather).
>
> **Count** – The number of times the selected ring tone is repeated.

10. Press the **Menu** key and save your settings.
11. Then press the **Escape** key a few times to get back to your **Home** screen.

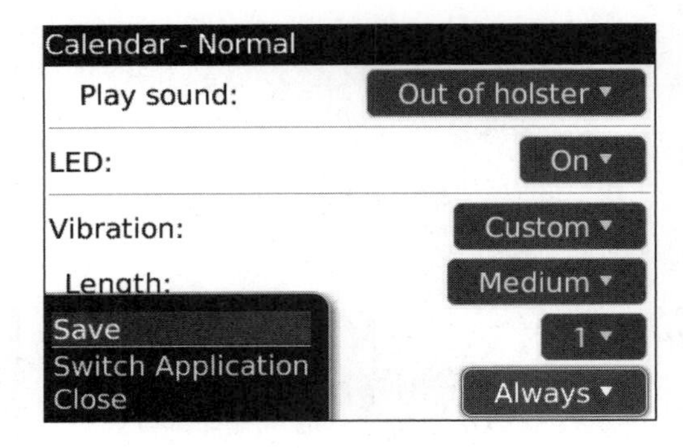

To Set a Recurring (Daily, Weekly, Monthly, or Yearly) Appointment

1. Click on your **Calendar** icon.
2. Click on an empty slot or on an existing calendar event in the **Day View** to bring up the appointment scheduling screen.
3. Scroll down to **Recurrence** and click in the highlighted field. Select either **None**, **Daily**, **Weekly**, **Monthly**, or **Yearly**.
4. Press the **Escape** key or the **Menu** key and select **Save**.

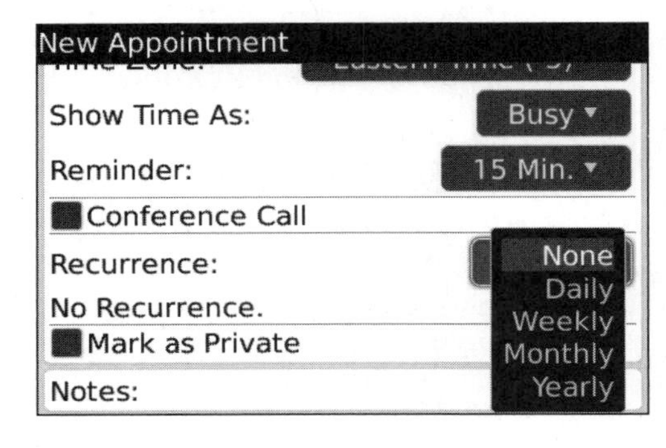

TIP: If your meeting is every 2 weeks until 10/01/2009, your settings should look like this.

TIP: Set yearly recurring reminders for birthdays or anniversaries

Set a Yearly recurring appointment with a reminder of one week (so you have time to buy the present or send a card).

New Appointment
Start: Thu 17 Dec 2009 11:00 AM
End: Thu 17 Dec 2009 12:00 PM
Duration: 1 Hour 0 Mins
Time Zone: Eastern Time (-5)
Show Time As: Busy
Reminder: 1 Week
Conference Call
Recurrence: Yearly

To Snooze a Ringing Calendar or Task Alarm

When a calendar or task alarm rings, you can **Open** it, **Dismiss** it, or **Snooze** it.

If you don't see a Snooze option, then you need to change your Options in your Calendar from **None** to some other value.

Click on **Snooze** to have the alarm ring in the specified number of minutes.

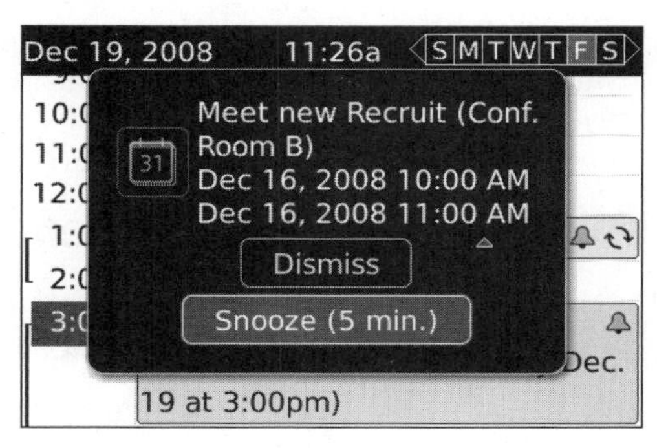

To make sure you have the Snooze option active:

1. Open the Calendar Application.
2. Push the **Menu** key and glide down to **Options**.
3. Click on **General Options**.
4. Look under the **Actions** sub heading and set the field next to **Snooze** to something other than **None.**

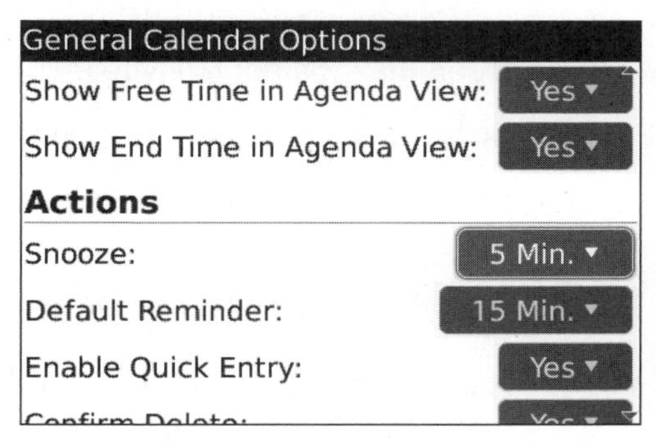

What If You Need More Snooze Time?

In this case, you should select **Open** and scroll down to the scheduled time and change that. Use the number keys on your BlackBerry to change the date or time. For example, typing "25" in the minutes field would change the time to: 25. Or you can change the hours or days by highlighting them and pressing the **Space** key.

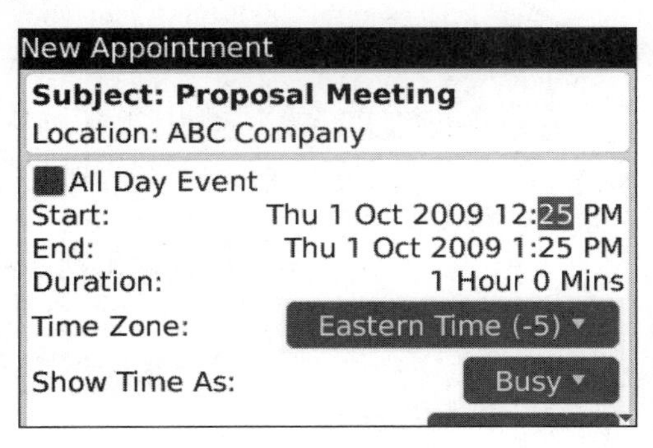

Inviting Attendees and Working with Meeting Invitations

Now with your BlackBerry you can invite people to meetings and reply to meeting invitations. If your BlackBerry is connected to a BlackBerry Enterprise Server (BES), you may also be able to check your invitees' availability for specific times as well.

To Invite Someone to Attend a Meeting

1. Create a new appointment in your calendar or **Open** an existing meeting by clicking on it with your trackpad.
2. Click the trackpad to see the short menu.
3. Click on **Invite Attendee** and follow the prompts to find a contact and click on that contact to invite them.
4. Follow the same procedure to invite more people to your meeting.
5. Click on the trackpad and select **Save**.

To Respond to a Meeting Invitation

You will see the meeting invitation in your Messages (Email Inbox).

Open up the invitation by clicking on it with the trackpad, and then press the **Menu** key.

Several options are now available to you. Click either **Accept** or **Accept with Comments**, **Tentative** or **Tentative with Comments,** or **Decline** or **Decline with Comments**.

To Change Your List of Participants for a Meeting

1. Click on the meeting in your calendar application.
2. Navigate to the **Accepted** or **Declined** field and click the Contact you wish to change.
3. The options **Invite Attendee, Change Attendee**, or **Remove Attendee** are available. Just click on the correct option.

To Contact Your Meeting Participants

1. Open up the meeting or your meeting invitation or even one of the responses from the participants.
2. Simply highlight the contact and press the **Menu** key
3. Scroll through the various ways you can contact that contact—email, PIN message, SMS, etc.—or call the contact directly.

To Send an Email to Everyone Attending a Meeting

1. Navigate to the meeting in your calendar and click on the trackpad.
2. Click on **Email all Attendees** and compose your email.
3. Click on the trackpad and select **Send**.

Using Google Sync to Sync Calendar and Contacts

> **NOTE:** You can sync your Google contacts using the BlackBerry Internet Service when you setup your email (see page 46). We would strongly suggest not trying to sync Google Contacts with Google Sync as well as BlackBerry Internet Service—you are just asking for trouble. Instead, just use Google Sync for your Google Calendar.

The great thing about using Google Sync for BlackBerry is that it provides you a full two-way wireless synchronization of your BlackBerry Calendar and Contacts with and your Google Calendar and Address Book. What this means is anything you type in on your Google Calendar/Addresses "magically" (wirelessly and automatically) appears on your BlackBerry Calendar/Contacts in minutes! The same thing goes for contacts or calendar events you add or change on your BlackBerry—they are transmitted wirelessly and automatically to show up on your Google Calendar and Address Book.

> **NOTE:** The only exceptions are those calendar events you have added on your BlackBerry prior to installing the Google Sync application—those old events don't get synced. Contacts on your BlackBerry prior to the install are synced.

This wireless calendar update function has previously only been available only with a BlackBerry Enterprise Server.

Getting Started with Gmail and Google Calendar

First, if you don't already have one, you must sign up for a free Google mail (Gmail) account at `www.gmail.com`.

Then, follow the great help and instructions on Google to start adding address book entries and creating calendar events on Gmail and Google Calendar on your computer.

Installing the Google Sync Program

1. Click on your **Browser** icon to start it.
2. Type in this address in the address bar on the top: `http://m.google.com/sync`.

> **TIP: Shift + Space** key will type the forward slash "/" in the web address.

3. Then click on the **Install Now** link on this page.

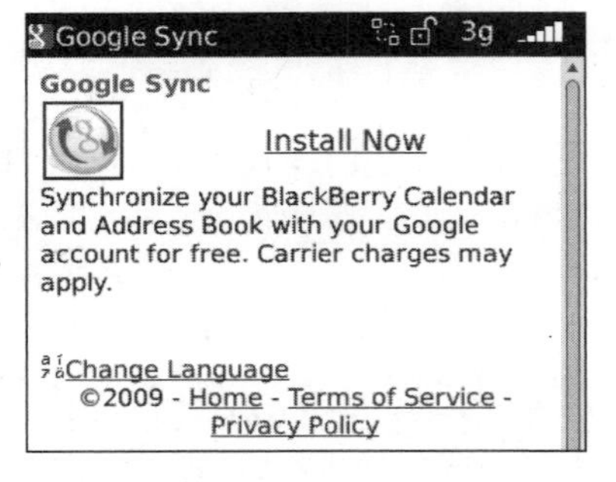

> **NOTE:** This page may look slightly different when you see it.

4. Click on the **Download** button.

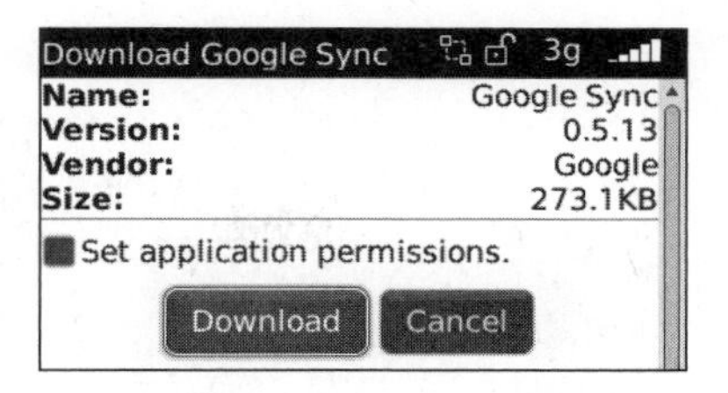

NOTE: The version numbers and size will probably be different when you see this page.

5. After you download and install it, you can **Run** it or just click **OK**. If you clicked OK and exited the browser, you may need to check in your **Downloads** folder for the new **Google Sync** icon.

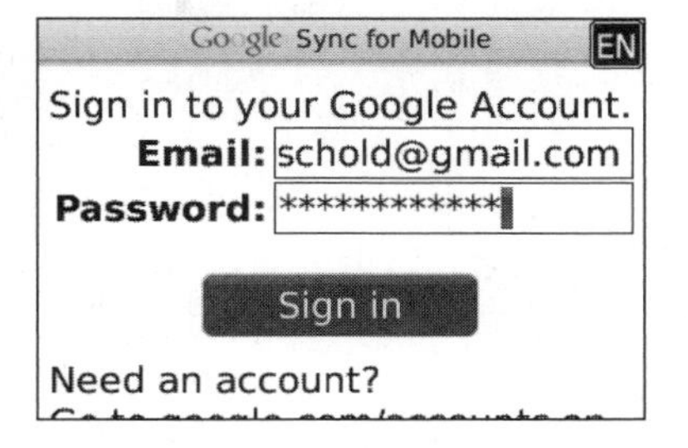

6. After you successfully login, you will see a page similar to this one describing what Google Sync will sync to and from your BlackBerry.
7. Notice at the bottom of this screen, you will see details on which fields (or pieces of information) from your BlackBerry will be shared or synchronized with your Google Address book.
8. Finally, click **Sync Now** at the bottom of this page.

Google Sync for Mobile

Welcome to Google Sync for mobile.

To get started, click the **Sync Now** button below and Google Sync will start synchronizing your handheld's calendar and contacts with Google automatically.

To access settings or initiate a manual synchronization, click on the **Google Sync** entry in the menu of your handheld's Calendar.

The first time Google Sync runs, it may take quite a while to complete the synchronization. If you have many Contacts, your Blackberry may also feel a little unresponsive for a short period, while Google Sync adds them to your Address Book.

Google Sync can synchronize the following fields with your handheld: Title, First, Last, Job Title, Company, Email, Work, Work 2, Home, Home 2, Mobile, Pager, Fax, Other, Work Address, Home Address, and Notes. Other fields will not be synchronized with your Google Contacts.

Sync Now

You will see some sync status screens…

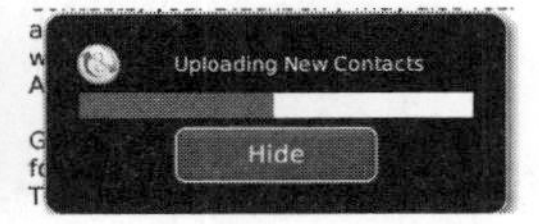

9. Then, finally, you will see a screen like this. Click **Summary** to see the details of what was synced.

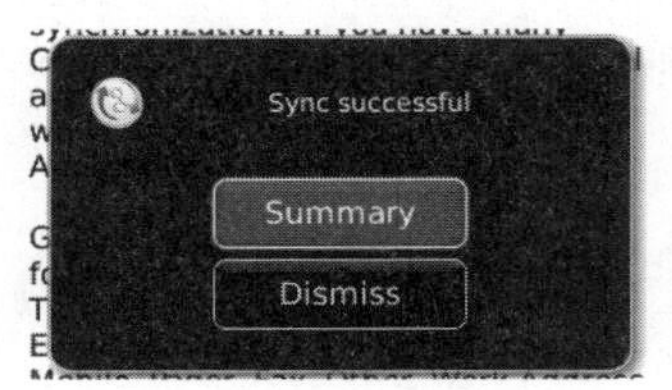

From this point forward, the Google Sync program should run automatically in the background. It will sync every time you make changes on your BlackBerry or at a minimum every two hours. From both the BlackBerry **Calendar** and **Contacts** icons, you will see a new **Google Sync** menu item. Select that to see the status of the most recent sync. You can also press the **Menu** key from the Sync status screen to **Sync Now** or go into **Options** for the sync. If you are having trouble with the sync, check out the "Fixing Problems" section on page 521 or view Google's extensive online help.

The Results of a Successful Google Sync

Here are views of the same calendar items in Google and BlackBerry.

Google Calendar on the Computer

BlackBerry Calendar

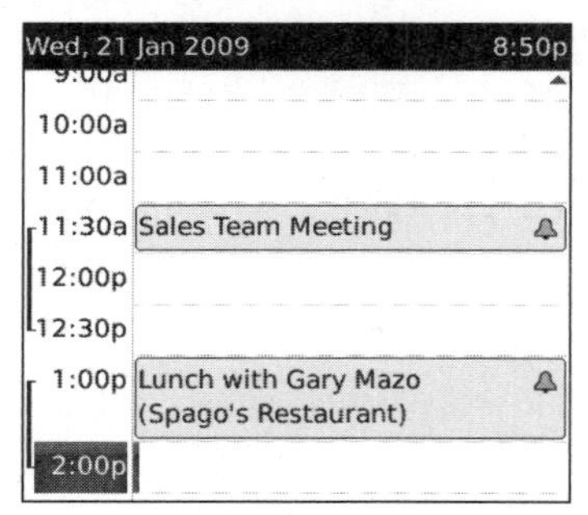

Notice the calendar events from Google Calendar are now on your BlackBerry. Anything you add or change on your BlackBerry or Google calendar will be shared both ways going forward. Automatically!

Google Address Book on your Computer

BlackBerry Contact List

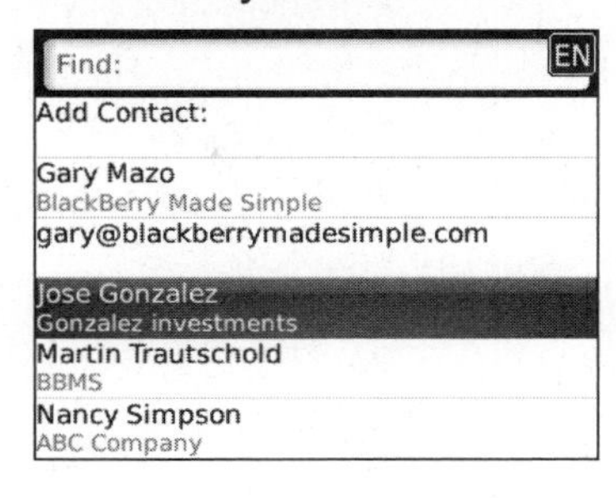

Chapter 16

Get Tasks Done

Our lives are busy. One of the reasons we use a BlackBerry in the first place is to simply keep track of what we need to do and when we need to do it. The BlackBerry helps us multitask and the **Tasks** app helps us keep track of specific tasks.

Your BlackBerry provides robust task-management functionality in the form of a **Tasks** list program. In this chapter, you'll learn how to use this application to manage **Tasks** in your life, as well as to synchronize the **Tasks** on your BlackBerry with your computer.

We will also show you how to view your tasks, categorize them, filter them and then check them off when they are completed.

The Task Icon

Like your contacts, calendar, and Memo Pad, your task list becomes more powerful when you share or synchronize it with your computer. Since the BlackBerry is so easy to carry around, you can update, check-off, and even create new tasks anytime, anywhere they come to mind. Gone are the days of writing down a task on a sticky note and hoping to find it later when you need it.

Syncing Tasks from Computer to BlackBerry

You can also mass load or sync up your computer's task list with your BlackBerry **Task** icon.

If your BlackBerry is tied to a BlackBerry Enterprise Server, the synchronization is wireless and automatic. Otherwise, you will use either a USB cable or Bluetooth wireless to connect your BlackBerry to your computer to keep it up to date. For Windows PC users, see page 65. Apple Mac computer users, see page 125.

Viewing Tasks on Your BlackBerry (Hotkey: T)

Press the hotkey letter **T** to start Tasks (see page 541 for hotkeys help). Or, locate and click on the **Tasks** icon.

You may need to press the **Menu** key to bring up all your applications on the Home Screen and you may need to click on the **Applications** folder.

Then, find the icon that says **Tasks** when you glide over it. You may need to press the **Menu** key to the left of the trackpad to see all your icons.

The first time you start tasks on your BlackBerry, you may see an empty task list if you have not yet synchronized with your computer.

Adding a New Task

Press the **Menu** key and select **New**. Then you can enter information for your new task in the screen in Figure 16-1.

> **TIP:** Keep in mind the way the **Find** feature works as you name your task. For example, all tasks for a particular Project Red should have "Red" in the name for easy retrieval.

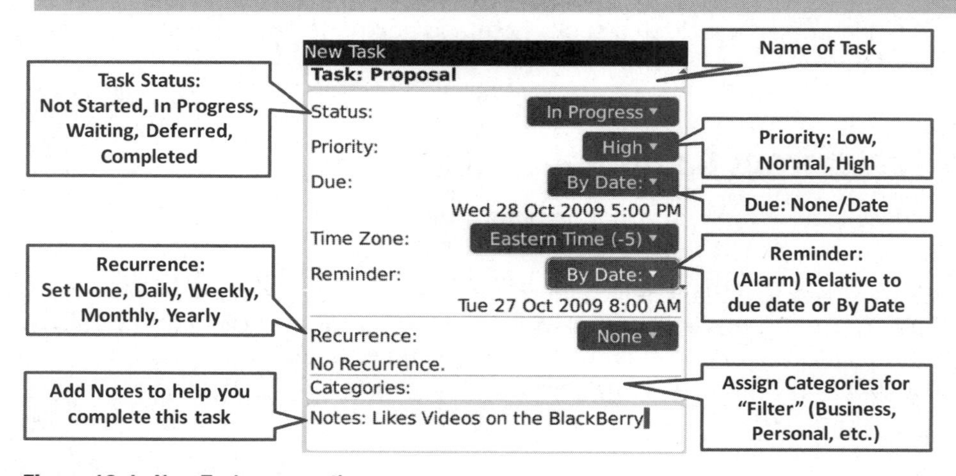

Figure 16-1. *New Task menu options.*

Categorizing Your Tasks

Like Address Book entries, you can group your tasks into Categories. And you can also share or synchronize these categories with your computer.

To Assign a Task to a Category

1. Highlight the Task, click the trackpad and **Open** it.
2. Press the **Menu** key and select **Categories** (Figure 16-2).

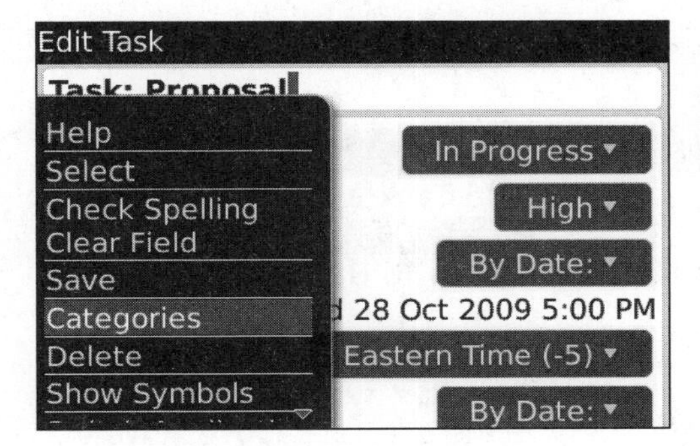

Figure 16-2. *Assign task to categories.*

3. Select as many categories as you would like by checking them with the trackpad or pressing the **Space** key.
4. You may even add new categories by pressing the **Menu** key and selecting **New.**
5. Once you're done, press the **Menu** key and select **Save** to save your Category settings.
6. Press the **Menu** key and select **Save** again to save your task.

Filtering Tasks

Once you assign tasks to categories, you can filter or show only those tasks assigned to a particular category.

1. To do this, view your **Task** list and press the **Menu** key.
2. Select **Filter.**
3. Check off a particular category for the filter by tapping the **Space** key on the appropriate check box.

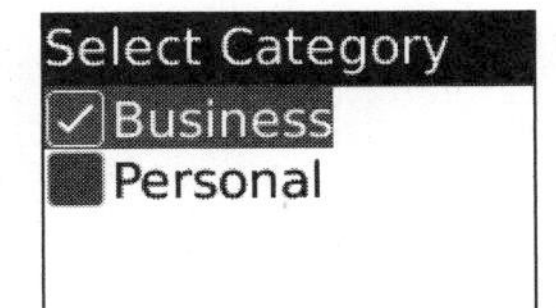

4. Immediately, you will see a black line at the top with the Category name showing the task list is filtered.

Here is the filtered task list:

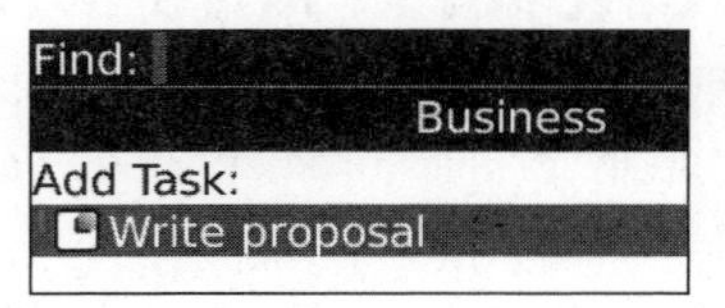

To un-filter your tasks, you need to repeat the procedure above, but un-check the selected task. Pressing the **Escape** key to exit Tasks and re-enter will not un-filter your tasks.

Finding Tasks

Once you have a few tasks in your task list, you will want to know how to quickly locate them. One of the fastest ways is with the **Find** feature.

The same **Find** feature from the address book works in Tasks.

Just start typing a few letters to view only those that contain those letters.

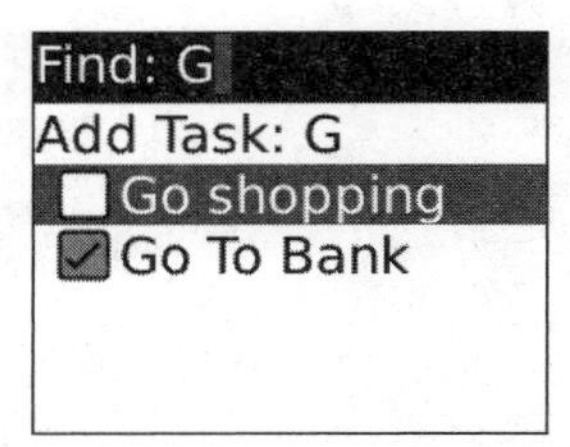

In the example above, if we wanted to quickly find all tasks that have any word starting with the letter G, then we would press the letter G.

Checking Off Completed Tasks

You can complete tasks by pressing the **Menu** key and selecting **Mark Completed**. The easier way is to highlight the task in the full list and press the **Space** key.

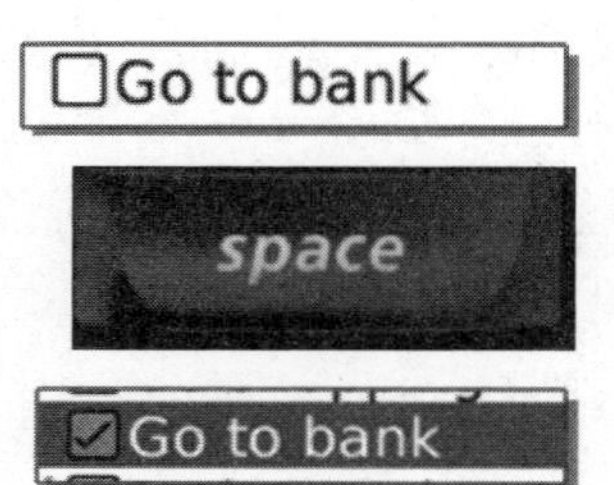

TIP: Press **Space** key again to un-check it.

Handling a Task Alarm

When a task alarm rings, you will see a pop-up window similar to the one shown. You can:

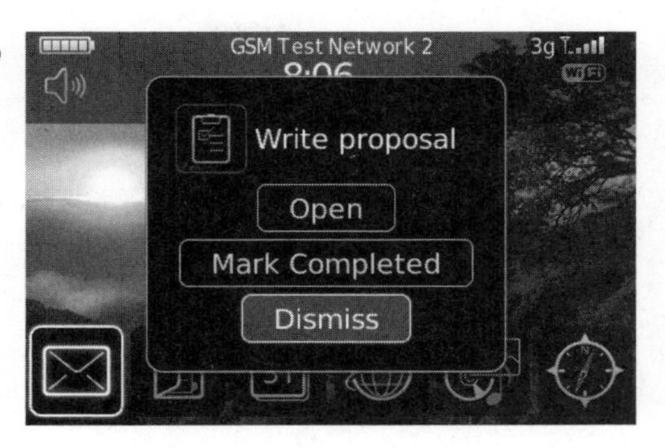

1. Open the task to change the due date, add notes, or edit anything else.
2. Click **Mark Completed** if you already got this task done.
3. Click **Dismiss** to ignore this ringing task alarm.

Adjusting the Way Task Alarms Notify You

Please see the "Calendar and Task Alarms" section on page 300 to learn how to make sure your task alarms will notify you with at least a vibrate or a vibrate and tone when your BlackBerry is sitting on your desk (**Out of Holster**).

Sorting Your Tasks and Task Options

You may sort your tasks by the following methods in your Task Options screen: **Subject (default)**, **Priority**, **Due Date**, or **Status**.

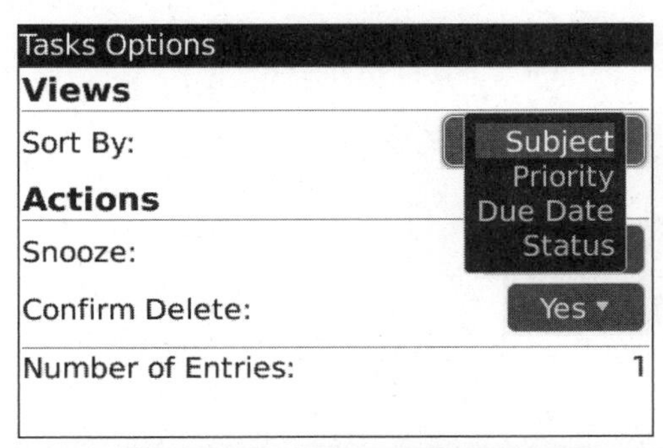

You may also change Snooze from **None** to **30 Min**.

You may also change the **Confirm Delete field** to **No** (default is **Yes**).

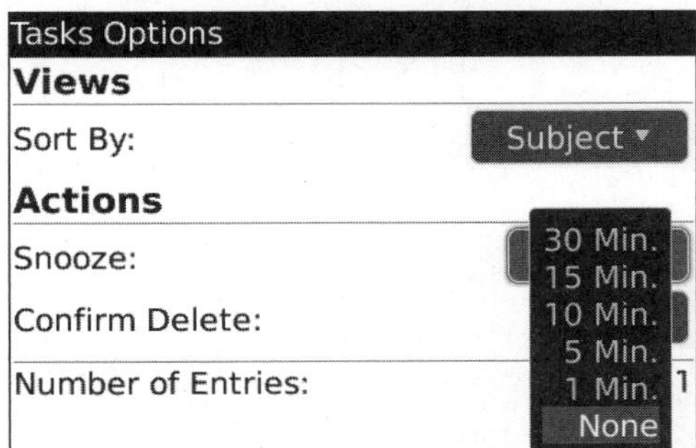

Chapter 17

MemoPad: Sticky Notes

One of the simplest and most useful programs on your BlackBerry is the **MemoPad**. Its uses are truly limitless. There is nothing flashy about this program—just type your memo or your notes and keep them with you at all times.

Using the **MemoPad** is easy and intuitive. The following steps guide you through the basic process of inputting a memo and saving it on your BlackBerry. There are two basic ways of setting up memos on the BlackBerry: either compose the note on your computer organizer application and then synchronize (or transfer) the note to the BlackBerry or compose the Memo on the BlackBerry itself.

Syncing MemoPad items (Notes) from Computer to BlackBerry

You can synchronize or share your computer's **MemoPad** notes list with your BlackBerry **MemoPad** icon.

If your BlackBerry is tied to a BlackBerry Enterprise Server, the synchronization is wireless and automatic. Otherwise, you will use either a USB cable or Bluetooth wireless to connect your BlackBerry to your computer to keep it up to date. For Windows PC users, see page 65 or an Apple Mac computer user, see page 125.

The sync works both ways, which extends the power of your desktop computer to your BlackBerry—add or edit notes anywhere and anytime on your BlackBerry and rest assured they will be back on your computer (and backed up) after the next sync.

1,001 Uses for the MemoPad (Notes) Feature

OK, maybe we won't list 1,001 uses here, but we could. Anything that occupies space on a sticky note on your desk, in your calendar, or on your refrigerator could be written neatly and organized simply using the **MemoPad** app.

Common Uses for the MemoPad

- Grocery list
- Hardware store list
- Shopping list for any store
- Meeting Agenda
- Triathlon training log (run, bike, swim)
- Packing list for your next ski or sun vacation
- Made Simple Learning Videos you want to watch
- Movies you want to rent
- Your parking space at the airport, mall, or theme park.

Adding or Editing Memos on the BlackBerry

This is the **MemoPad** icon

(your icon may look different, but look for MemoPad to be shown when you highlight it).

NOTE: You may need to first click on the Applications folder to find it. 

If you have no memos in the list, then to add a new memo, simply press in the trackpad and start typing.

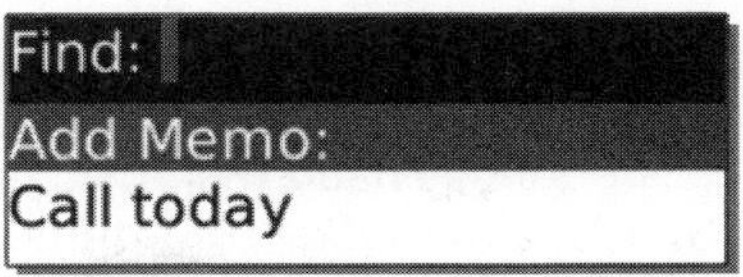

If you have memos in the list and want to add a new one, you just click on **Add Memo**.

To open an existing memo, just glide to it and click on it. You will probably be in **Edit** mode where you can change the memo. If not, click the trackpad again and select **Edit.**

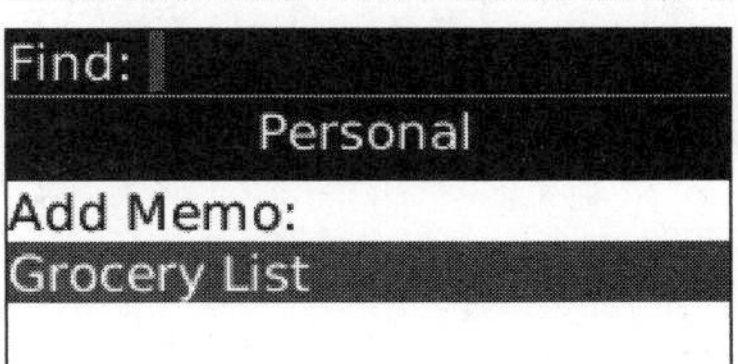

When typing a new memo, you will want to enter a title that will be easy to find later by typing a few matching letters.

1. Type your memo in the body section.
2. Press the **Enter** key to go down to the next line.
3. When you're done, click the trackpad to select **Save.**

Note that if you have copied something from another icon, like Email, you could **Paste** it into the memo. (See page 295 for details on copy & paste.)

TIP: When you pick up an item in the store, "check it off" by putting a Space in front of the item—this helps make sure you get everything, even if you have a long list.

Quickly Locating Memos

Like the Contact List and Tasks, the **MemoPad** has a **Find** feature to locate memos quickly by typing the first few letters of words that match the title of your memos. Example, typing "**gr**" would immediately show you only memos matching those three letters in the first part of any word like "**grocery**."

Ordering Frequently Used Memos

For frequently used memos, type numbers (01, 02, 03, etc.) at the beginning of the title to force those memos to be listed in order at the very top of the list. (If you use 01-09, as opposed to 1-9, the memos will stay in order after you get to 10).

Viewing Your Memos

Glide the trackpad down to a memo or type a few letters to find the memo you want to view.

Click the trackpad to instantly view the memo.

> **TIP:** Time Stamping Your Memos
>
> Typing "**ld**" (stands for Long Date) and pressing the **Space** key will insert the date "Tue, 28 Aug 2007." Likewise, typing "**lt**" (Long Time) will enter the time "8:51:40 PM" (which will be in the local date/time format you have set on your BlackBerry).

See page 166 to learn how to create a customized Time Stamp.

Organizing Your Memos with Categories

Similar to your Address Book and Task list, the **MemoPad** application allows you to organize and filter memos using Categories.

First, you must assign your memos to categories before they can be **filtered**.

One way to be extra organized with your **MemoPad** application is to utilize Categories so all your Memos are filed neatly away.

The two default categories are **Personal** and **Business** but you can easily change or add to these.

> **TIP: Categories** are shared between your address book, task list, and **MemoPad**. They are even synchronized or shared with your computer.

To File a Memo in a New or Existing Category

Just like with the Contact List (Address Book) and Task List, you can organize your MemoPad items with categories.

1. Start the **MemoPad** icon by clicking on it.
2. Locate the memo you want to file to one or more categories by rolling and clicking on it or by typing a few letters and using the **Find** feature at the top.

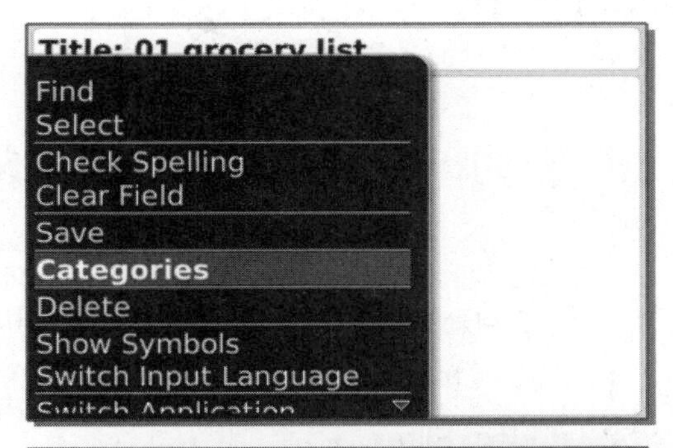

3. Press the **Menu** key again and select **Categories.**
4. Now you will see a screen similar to the right. Glide the trackpad to a category and click to check/uncheck it. You can add a new category by pressing the **Menu** key and selecting **New**.

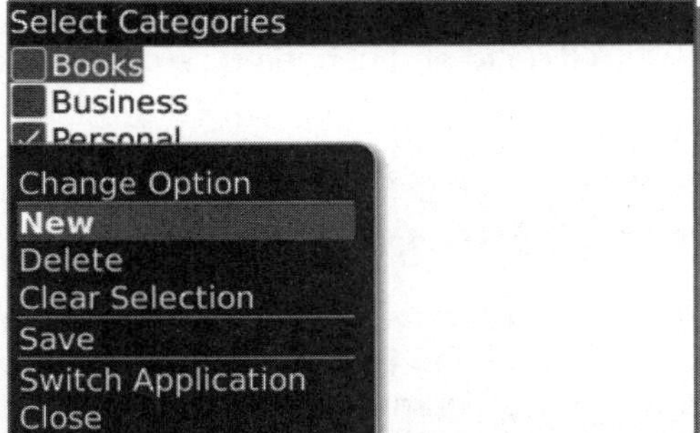

5. Then save your category settings, and save the memo.

To Filter Memos Using Categories to Only See Memos in a Specific Category

1. Start the **MemoPad** by clicking on it.
2. Press the **Menu** key and select **Filter**
3. Now glide to and click (or press **Space**) on the Category you would like to use to filter the list of memos.

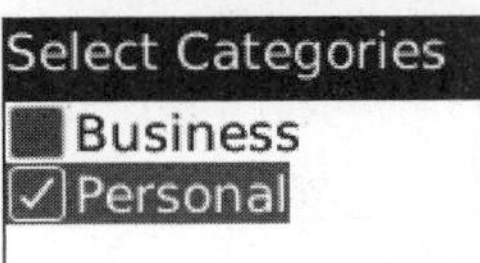

TIP: You know when your list is filtered if you see an extra black bar below the Find: field at the top. See the image to the right which is filtered by the category called **Personal**.

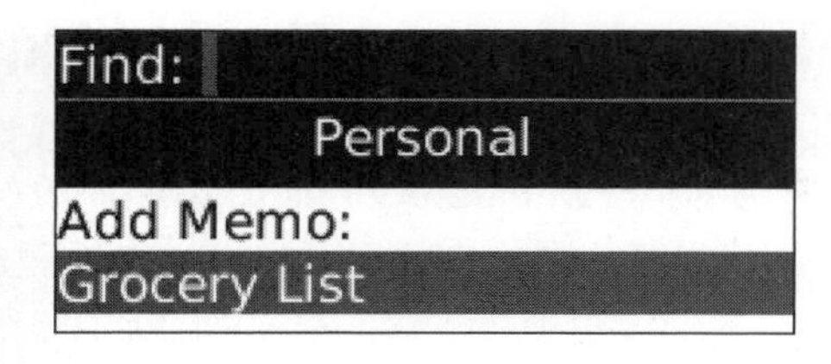

To Un-Filter (Turn Off the Filter) on Your MemoPad

You probably noticed that pressing the **Escape** key (which clears out the **Find** characters already typed) will do nothing for the Filter; it just exits the MemoPad. When you re-enter, you still see the filtered list. To un-filter or turn off the filter, you need to:

1. Press the **Menu** key.
2. Select **Filter.**
3. Glide to and uncheck the checked category by clicking the trackpad on it or pressing the **Space** key.

Multitask or Switch Applications

From almost every icon on your BlackBerry, the MemoPad included, pressing the **Menu** key and selecting **Switch Application** allows you to multitask by leaving your current icon open and jumping to any other icon on your BlackBerry. This is especially useful when you want to copy and paste information between apps.

TIP: There are two shortcuts for doing this

(1) Press and hold the **Menu** key or

(2) Press **ALT** + **Escape** together. Give it a try!

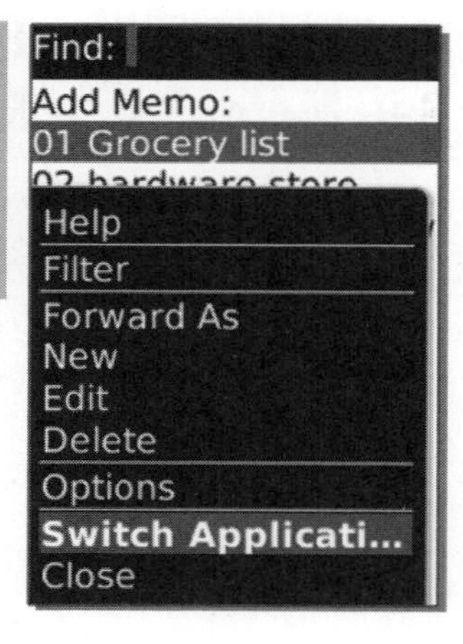

Here's how you jump or switch applications:

1. Press the **Menu** key and select **Switch Application**.

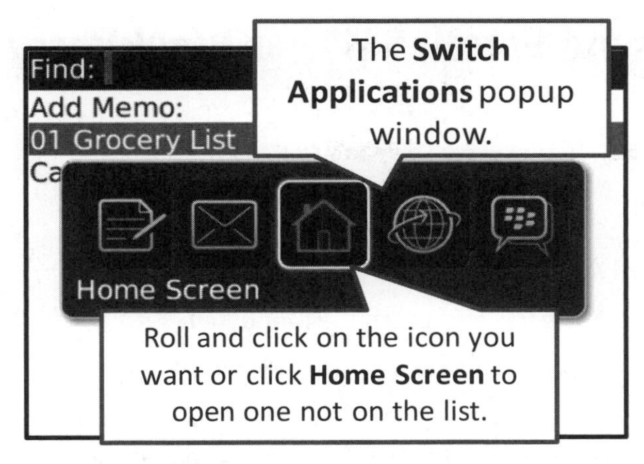

Figure 17-1. *Multitasking on the BlackBerry.*

2. You will now see the **Switch Applications** pop-up window (Figure 17-1) which shows every icon currently running.
3. If you see the icon you want to switch to, just glide and click on it.
4. If you don't see the icon you want, then click on the **Home Screen** icon. Then you can locate and click on the right icon.
5. You can then jump back to the **MemoPad** or application you just left by selecting the **Switch Application** menu item from the icon you jumped to (Figure 17-2).

Figure 17-2. *Switch applications*

TIP: Press & Hold **Menu** key or Press **ALT+ Escape** to Multitask

Press and hold the **Menu** key, or press and hold the **ALT** key and then tap the **Escape** key, are shortcuts to bring up the Switch Application pop-up window.

Forwarding Memos via Email, SMS, or BlackBerry Messenger

Say you just took some great notes at a meeting and you want to send them to your colleague.

You can send a memo item via email, BlackBerry PIN message, or SMS text message to others. To do so, use the **Forward As** command from the menu.

1. Highlight the memo you want to send and press the **Menu** key.
2. Select **Forward As**, and then select whether you want **Email**, **PIN** or **SMS Text** (Figure 17-3).

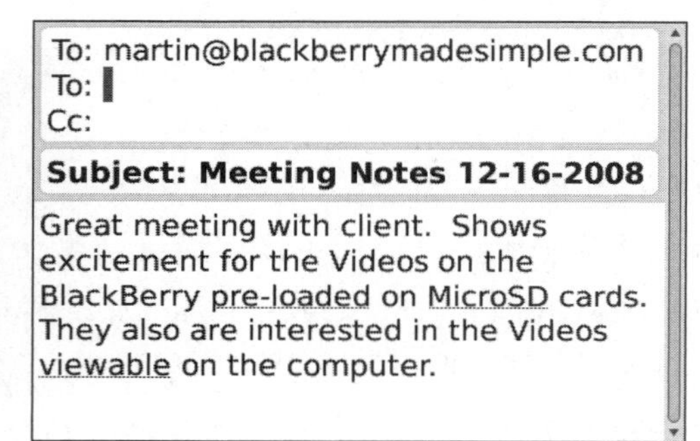

Figure 17-3. *Forward a memo as email, PIN, or SMS text*

3. Finish composing your message, click the trackpad, and select **Send.**

Other Memo Menu Commands

There may be a few other things you want to do with your memos. These can be found in the more advanced menu commands.

Start the **MemoPad** app.

Some of these advanced menu items can only be seen when you are either writing a **new** memo or **editing** an existing one.

So select **New** to begin working on a new memo or select **Edit** to edit an existing memo.

From the Editing screen, press the **Menu** key and the following options become available to you:

Find	If you are in a memo item, this will allow you to find any text inside the memo.
Paste	Suppose you have copied text from another program and want to paste it into a memo. Select and copy the text (from the calendar, address book, or another application) and select Paste from this menu. The text is now in your memo.
Select	This allows you to do the reverse— click here and select text from the memo, then press the **Menu** key again and select **Copy**. Now, use the **Switch Applications** menu item to navigate to another application and press the **Menu** key and select **Paste** to put the text in that application.
Check Spelling	Will run the BlackBerry spell checker on the currently open memo item.
Clear field	Clears all contents of the entire memo item. USE WITH CAUTION!
Save	This saves the changes in the memo.
Categories	This allows you to file this memo into either the Business or Personal categories. After selecting **Categories**, press the **Menu** key again and select **New** to create yet another category for this memo.
Delete	This deletes the current memo.
Show Symbols	This brings up the list of symbols. **TIP:** You can also just press the **SYM** key to see the symbol list.
Switch Application	This brings up the multitasking window so you can jump to another application icon. See page 322.
Close	Exit the application icon—same as pressing the **Escape** key.

Memo Tips and Tricks

There are a couple of tricks you can use to make your filing and locating of memos even easier.

Add Items for Different Stores

This can help eliminate the forgetting of one particular item you were supposed to get at the hardware or grocery store (and save you time and gas money!).

Put Numbers at the Beginning of Memo Names

This will then order them numerically on your BlackBerry. This is a great way to prioritize your memos and keep the most important ones always at the top of the list.

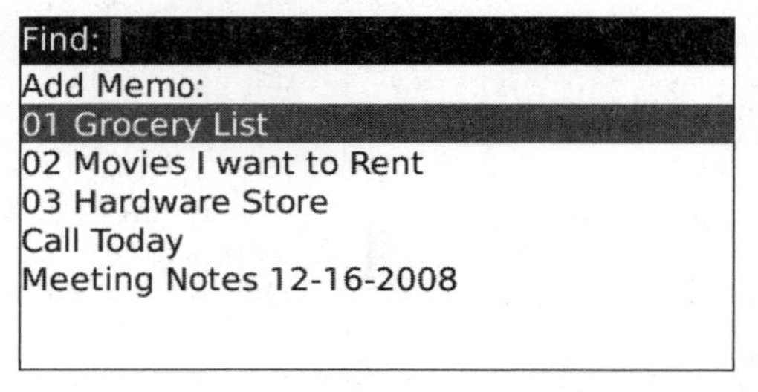

Putting a **Time Stamp** in your Memos—see page 166.

Space Key to Check Off Items Purchased

When you are checking off items you have purchased in the grocery or other store from your memo list, roll the cursor up to the beginning of the item and press the **Space** key to move the item over one space.

Each item moved over one space has been placed in your shopping cart. Now you know what you have left to buy.

In the image to the right, we have already placed the swirl bread, Miniwheats, and Oat squares into the shopping basket. Each indented item is a check mark.

This is especially helpful when you have a really long list or your list is not in the order of the aisles in the store. It ensures that you do not forget anything.

Title: 01 Groceries

Peanut butter
 Swirl bread
Apples
 Miniwheats
 Oat squares
Milk
Dog food

Chapter 18

SMS Text and MMS Messaging

As you may be aware, a key strength of all BlackBerry devices is their messaging abilities. We have covered email extensively and now turn to SMS and MMS messaging. In this chapter, we will show you how to send and view both text messages (SMS) and multimedia messages (MMS). We will show the various apps from which you can send a message and offer some advice on how to keep your messages organized.

Text and Multimedia Messaging

SMS stands for **Short Messaging Service** (text messaging) and MMS stands for **Multimedia Messaging Service.** MMS is a short way to say that you have included pictures, sounds, video, or some other form of media right inside your message (not to be confused with regular email where media is an attachment to an email message.) The BlackBerry is beautifully equipped to use both of these services—learning them will make you more productive and make your BlackBerry that much more fun to use.

> **TIP:** SMS/MMS sometimes cost extra!
>
> Watch out! Many phone companies charge extra for SMS Text Messaging and MMS multimedia messaging, even if you have an "unlimited" BlackBerry data plan. Typical charges can be $0.10 to $0.25 per message. This adds up quickly!
>
> The Solution is to check with your carrier about bundled SMS/MMS plans. For just $5 / month or $10 / month you might receive several hundred or thousand or even unlimited monthly SMS/MMS text messages.

SMS Text Messaging on Your BlackBerry

Text messaging has become one of the most popular services on cell phones today. While it is still used more extensively in Europe and Asia, it is growing in popularity in North America.

The concept is very simple; instead of placing a phone call, you send a short message to someone's handset. It is less disruptive than a phone call. It's also helpful if you have friends, colleagues, or co-workers who do not own a BlackBerry so email is not an option.

One of the authors uses text messaging with his children all the time; this is how their generation communicates. ''R u coming home 4 dinner?'' ''Yup.'' There you have it—meaningful dialogue with a seventeen year old—short, instant, and easy.

Composing SMS Text Messages

Composing an SMS message is much like sending an email. The beauty of an SMS message is that it arrives on virtually any handset and is so easy to respond to.

Option 1: Sending an SMS Message from the Message List

1. Use the trackpad and navigate to your Messages list and press the **Menu** key. (Shortcut: Glide to a date row separator—e.g. "**Mon, July 6, 2009**"—and click the trackpad).
2. Select Compose SMS Text.

3. Begin typing in a contact name (as you did when selecting an email recipient in chapter 7.)
4. When you find the contact, click on the trackpad.

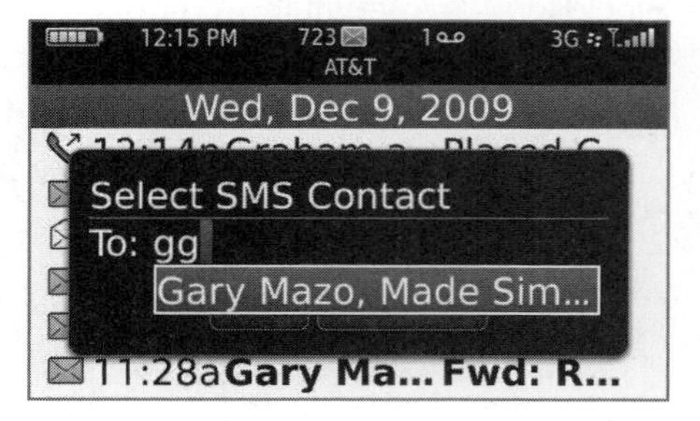

5. If the contact has multiple phone numbers, the BlackBerry will ask you to choose which number. Select the Mobile number you desire and click.

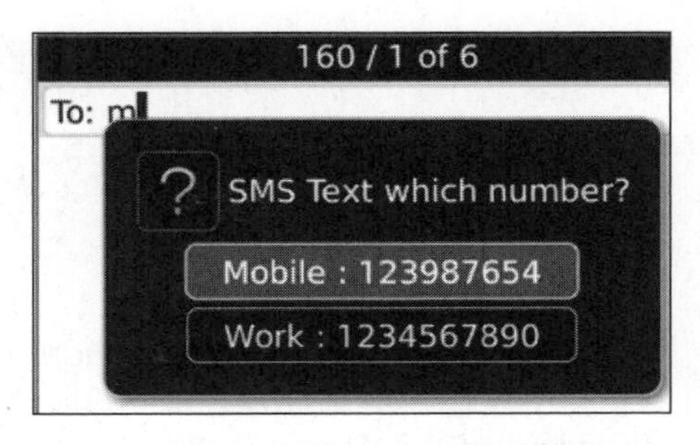

6. After you select a recipient, you can easily add more by typing a phone number or by typing a name to find matching contact entries.

7. In the main body (where the cursor is), type your message as if you were sending an email.

CAUTION: SMS messages are limited to 160 characters by most carriers. If you go over that in the BlackBerry, two separate text messages will be sent.

8. When you are done typing, click the trackpad and choose **Send**. That's all there is to it!

TIP: Roll over to the smiley face icon and click on it to see a list of icons you can add to enhance your message!

You can also use the keyboard shortcut. In the image shown, typing :D would give you a **Big Smile** icon.

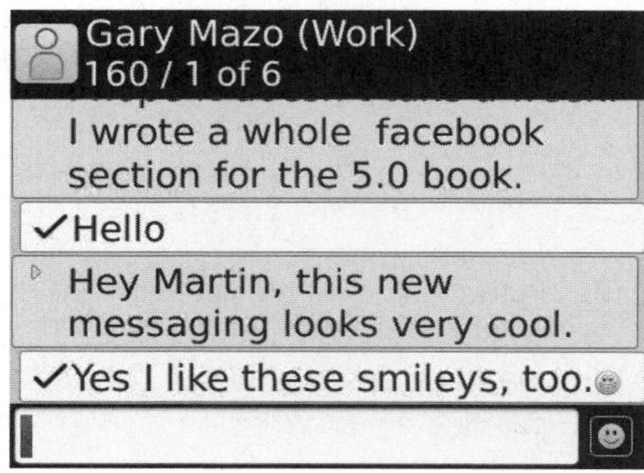

As you continue the discussion, you will see a threaded discussion with different colors for your replies and those of the person you are communicating with.

Option 2: Sending SMS Message from the Contact List

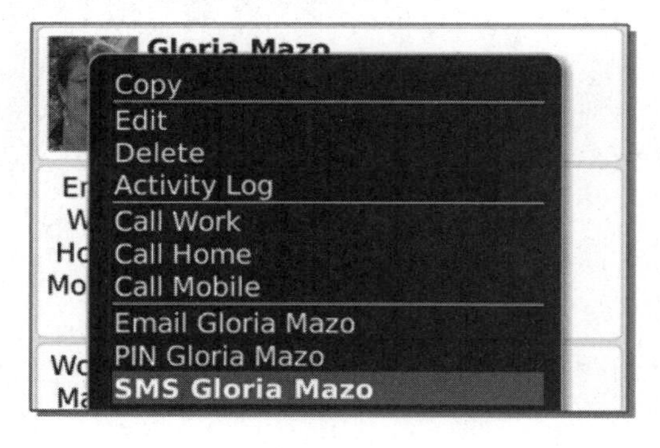

1. Click on your **Address Book** icon (it may say **Contacts** instead).
2. Type a few letters to find the person to whom you want to send your SMS message.
3. With the contact highlighted from the list, press the **Menu** key. You will see that one of your menu options is **SMS Text** followed by the contact name.
4. Select this option and follow the steps from above to send your message.

Basic SMS Menu Commands

The following table list many of the options that are available via the menu commands in SMS messaging.

Help	Gives you contextual help with SMS messaging (See page 159 for more tips on this Help feature).
Paste	Suppose you have copied text from another program and want to paste it into your SMS message. Select and copy the text (from the calendar, address book, or another application) and select **Paste** from this menu. The text is now in your SMS message.
Select	Click here (or just use the shortcut: Shift key) to begin selecting text from the SMS message, then press the **Menu** key or trackpad again to choose **Copy** or **Cut**. Now, use the Switch Applications menu item to navigate to another application and press the **Menu** key and select **Paste** to put the text in that application.
Add Smiley	Insert a smiley or emoticon (happy face, sad face, etc.)
Mark Unopened	Mark the selected message as unopened.
Check Spelling	Will run the BlackBerry spell checker on the message.
Clear Chat	Clears all messages of the entire chat session.
Delete Chat	Deletes all messages of the entire chat session.
Show Recipients	Show every recipient of SMS message you are about to send.
Send	Send this SMS message.
Save Draft	Saves the current message as a Draft, which you can later edit and send.
Add To:	Adds a second line for putting in another recipient.
Edit AutoText	Allows you to add or edit AutoText entries.
Show Symbols	Bring up the list of Symbols. **TIP:** You can also just press the **SYM** key to see the symbol list.
Switch Input Language	Change the language used for typing and spell checker.
Switch Application	Bring up the multitasking window so you can jump to another application icon. See page 322.
Close	Exit the application icon—same as pressing the **Escape** key.

Advanced SMS Menu Commands and Options

There are ways to personalize and customize your SMS messaging. These settings are found in your **Options** icon.

1. Click on your **Options** icon (it may be inside the **Applications** or **Settings** folder, if you don't see it from your Home Screen).
2. Scroll down to **SMS Text** or **SMS** and click with the trackpad.

TIP: Press the letter S on your keyboard a few times to jump down to this entry—it's usually faster than scrolling.

Usually, the options shown are adjustable; however, some BlackBerry phones are more limited. (This depends on your BlackBerry phone software version and your wireless carrier.)

Screen Options:

- **Conversation Style - Bubbles** or **Lines**—the look and feel of the threaded conversation.
- **Show Names -** Show or hide the names of the people in the conversation.

Message Options:

- **Data Coding -** Generally kept at **7 bit**
- **Service Center -** The phone number of the SMS text service center for your wireless carrier (will be different from what is shown).
- **Disable AutoText -** Generally set to **No** (this means you want to use the AutoText when typing SMS messages)
- **Leave messages on SIM card -** Change to **Yes** if you want to keep copies of your messages on the SIM card and not have them deleted. **Watch out, your SIM card will get filled pretty quickly if you use SMS a lot!**

- **Delivery reports** - Change to **On** if you want delivery confirmation for your SMS messages.
- **Network to Send Over** - Choose between **Circuit Switched, Packet Switched, Circuit Preferred** or **Packet Preferred**.

  Packet Switching is a more modern way of delivering data, but it may not work in all coverage areas. Your BlackBerry usually does a good job of choosing the correct method, but if you have problems, try switching the settings.
- **Retries – The number** 3 is the default, but adjust up or down (0 to 5) to meet your needs. Some networks may need more retries.

Opening and Replying to SMS Messages

Opening y our S MS m essages c ouldn't be e asier—the BlackBerry makes it simple to quickly keep in touch and respond to your messages.

1. Navigate to your waiting messages from either the **Message** icon on the Home Screen or your **SMS messages** icon and click on the new SMS message.
2. If you are in the midst of a dialogue with someone, your messages will appear in a threaded message format which looks like a running discussion, see Figure 18-1.
3. Just press the **Menu** key or click the trackpad and select **Reply**.
4. The cursor appears in a blank field. Type in your reply, click the trackpad, and select **Send**.

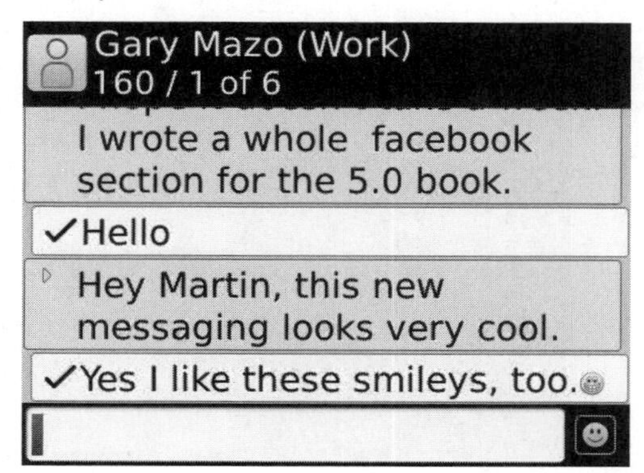

Figure 18-1. *Threaded text messaging on the BlackBerry*

Need to Find an SMS or MMS Message?

TIP: ALT + S will show only your SMS text messages in your Inbox.

Go to page 252 to learn how to search for messages. If you need to search for other information, like in your BlackBerry Messenger, Calendar, Contacts, or Tasks, go to page 509.

SMS Mailbox Housecleaning

It is possible for your SMS mailbox to get a little unwieldy. Follow these suggestions to manage and clean your mailbox.

To view and clean a particular SMS folder:

1. Start your **Messages** icon and press the **Menu** key.
2. Scroll down to **View Folder** and click. You will see a list of your available message folders. See left image of Figure 18-2.
3. To clean up your inbox, just click on **SMS Inbox** and only your SMS inbox messages will be displayed.
4. Highlight the date row separator (e.g. "Mon, Sep 4, 2007") under the most recent message you desire to keep, press the **Menu** key, and select **Delete Prior**. See right image of Figure 18-2.
5. All older messages will then be deleted.
6. To delete an individual message, just click on the message, press the **Menu** key, and choose **Delete**.
7. Choose the **SMS Outbox** folder to do the same with sent messages.

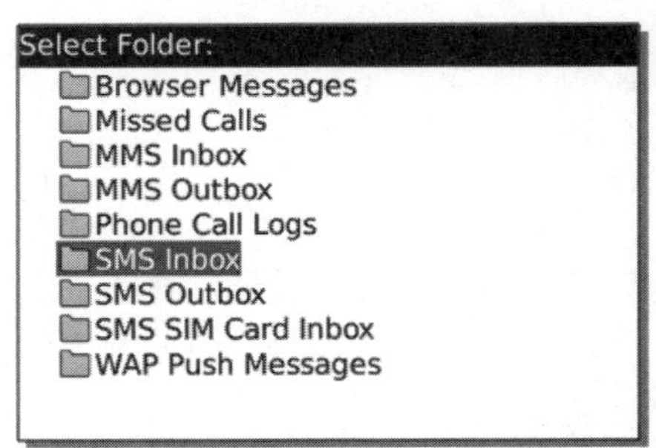

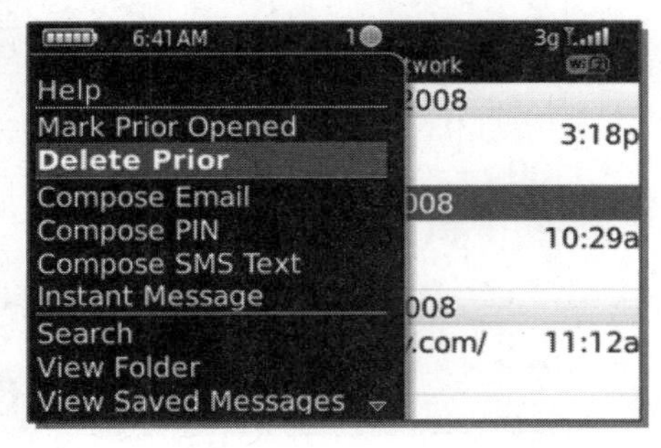

Figure 18-2. *Deleting messages from SMS inbox*

SMS Tips and Tricks

There are some quick keystrokes you can use to navigate quickly through your messages.

These shortcuts are entered when in the Message List

- Press the **C** key to compose a message.
- Press the **R** key to reply to a message.
- Press the **L** key to reply to all senders.
- Press the **F** key to forward a message.
- Push and hold the **Alt** key and press **I** to view Incoming (received) messages.
- Push and hold the **Alt** key and press the **S** key to view all SMS text messages.
- Press the **Escape** key to view your entire message list once more (un-filter it).

TIP: Most of these same shortcuts work for email messages as well.

MMS Messaging on Your BlackBerry

MMS stands for Multimedia Messaging, which includes pictures, video, and audio in the body of the message. The fact that the media is included in the body of the message is different than regular email, which can include media as attachments. The other key difference with MMS compared to email is that you can send MMS to phones that do not have email capabilities. This might be useful if not all your colleagues, friends, and family use BlackBerry smartphones!

NOTE: *Not all BlackBerry devices or carriers support MMS messaging*, so it is a good idea to make sure that your recipient can receive these messages before you send them.

Sending MMS from the Message List

Perhaps the easiest way to send an MMS message is to start the process just like you started the SMS process earlier:

1. Click on the **Messages** icon and press the **Menu** key.
2. Scroll down to **Compose MMS** and click the trackpad.
3. Press the **Menu** key and select **Attach Picture.**
4. Locate and click on a picture to attach. Depending on the phone company that supplied your BlackBerry, you may be prompted to locate the MMS file, which is often stored in your **mms/pictures** folder.

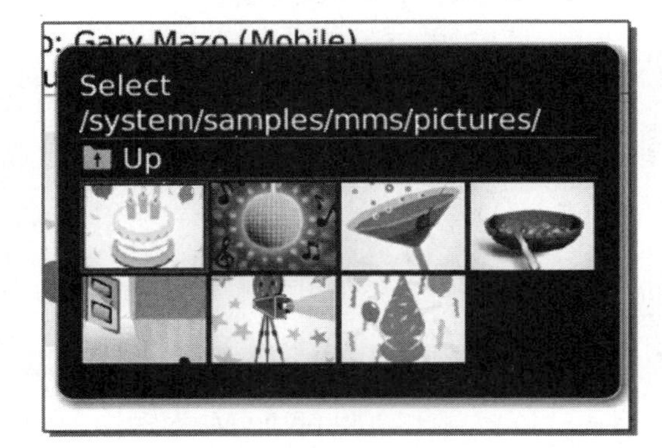

Some BlackBerry devices have a pre-loaded Birthday.mms in this folder. Click on the Birthday MMS, if you have it.

1. Then type in the recipient in the **To:** field, scroll to the appropriate contact, and click.
2. You can add a subject and text in the body of the MMS.
3. When finished, just click the trackpad and select **Send**.

There are lots of template MMS files you can download from the web and put into this folder to be selected in the future.

Sending a Media File as an MMS from the Media Icon

This is an easier way for you to send media files as MMS messages.

1. Use the trackpad and navigate to your **Media** icon and click. You will be brought to the Media screen.
2. Click on **Pictures** and find the picture that you wish to send either on your device or Media Card.
3. Highlight the picture (no need to click on it) and press the **Menu** key.
4. Scroll down to **Send as MMS** and click. You will then be directed to choose the recipient from your contacts. Find the contact you desire and click.

TIP and CAUTION: If you do not have an MMS or SMS Text Messaging service plan from your phone company, you can usually **Send as Email** for no additional cost. The only other thing to be aware of is whether or not you have an "unlimited BlackBerry data plan." If you don't have this unlimited data plan, you will want to send pictures only very rarely because they can eat up your data much faster than a plain-text email.

5. Type in a subject and any text in the message, click the trackpad and send.

Basic MMS Menu Commands

You can personalize your MMS message even more through the MMS menu.

1. When you are composing the MMS message, press the **Menu** key.
2. Scroll through the menu to see your options; you can easily add more recipients via the **To:, Cc:,** and **Bcc:** fields.

   You can also attach addresses from your Address Book. To add an address, click **Attach Address** and find the appropriate address on the next screen.
3. If you had scheduled a birthday dinner together, you might want add the appointment from your BlackBerry Calendar ("Dinner at the Fancy French Restaurant for Two"), so click the **Add Appointment** option.
4. You could even attach one of your recorded Voice Notes by selecting **Attach Voice Note** to personalize your message.

To add an audio file to accompany the picture, choose **Attach Audio** as shown in Figure 18-3 and then navigate to the folder that contains the audio file.

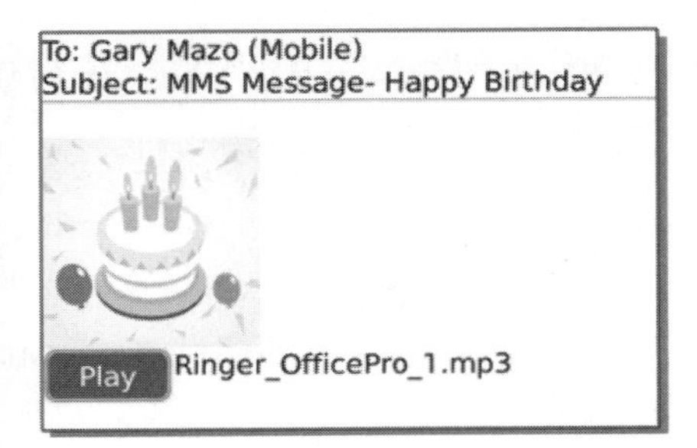

Figure 18-3 *Attach an audio file to an MMS message*

5. You can also attach other items by pressing the **Menu** key and selecting them.

Advanced MMS Commands

While you are composing your MMS message, press the **Menu** key and scroll down to **Options**.

In the **Current Message Options** screen you will see the estimated size of the MMS, which is important. If the file is too big, your recipient may not be able to download it onto their device.

You can set the importance of this MMS and the delivery confirmation options, too.

Current Message Options	
Estimated Size:	0.5 KB
Importance:	Normal
Confirm Delivery:	No
Confirm Read:	No

For more Advanced MMS commands, navigate from the Home Screen to **Options** and click. Scroll to **MMS** and click.

From this screen you can set your phone to always receive multimedia files by setting the first line to say **Always**.

You can also set your automatic retrieval to occur **Always** or **Never**.

You can select each checkbox to set your notification and message filtering options.

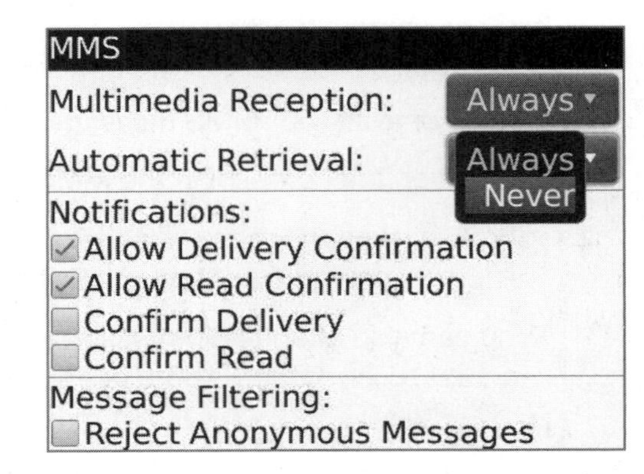

The Notifications options are as follows:

- **Allow Delivery Confirmation** allows delivery confirmation messages to be sent when you receive MMS messages from others.
- **Allow Read Confirmation** allows confirmation messages to be sent when you have opened an MMS message.

- **Confirm Delivery** means that you are requesting a delivery confirmation from people to whom you send MMS messages.
- **Confirm Read** means that you are requesting a read receipt message when your MMS recipient opens the message you sent them.

We recommend leaving the filtering options checked as they are by default.

MMS and SMS Text Troubleshooting

These troubleshooting steps will work for MMS, SMS, email, web browsing—anything that requires a wireless radio connection.

Host R outing T able—Register Now

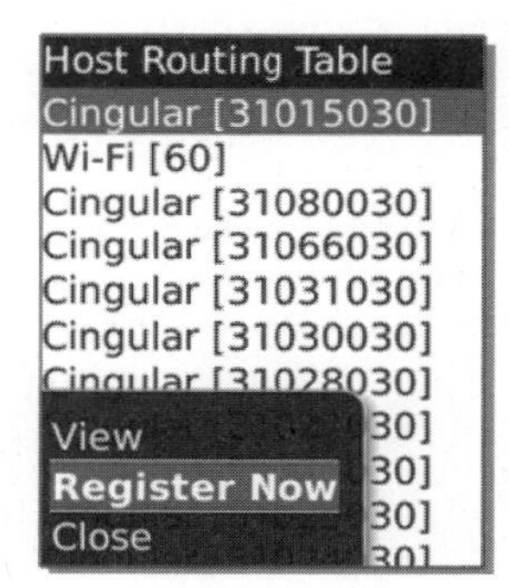

1. From your Home screen, click on the **Options** icon.
2. Click on **Advanced Options** and then scroll to **Host Routing Table** to see the screen similar to the one shown to the right. (You will see different carrier names if you are not on the carrier shown)
3. Press the **Menu** key and then click on **Register Now**.
4. While the BlackBerry is still on, do a battery pull. Take off the back of the casing, remove the battery, wait 30 seconds, and then re-install it. Once the BlackBerry reboots, you should be all set for SMS Text and MMS messaging.

Chapter 19

Even More Messaging

You already know that your BlackBerry is an amazing messaging device. What you may not realize is that there are so many more ways to use "messaging" on the BlackBerry. In this chapter, we will look at PIN messaging (using the unique identifying PIN for each BlackBerry), the famous BlackBerry Messaging (BBM), voice notes, and other instant messaging.

PIN Messaging

BlackBerry handhelds have a unique feature called PIN-to-PIN, also known as **PIN Messaging** or **Peer-to-Peer Messaging**. This allows one BlackBerry user to communicate directly with another BlackBerry user as long as you know that user's BlackBerry PIN. We'll show you an easy way to find your PIN and send it in an email to your colleague.

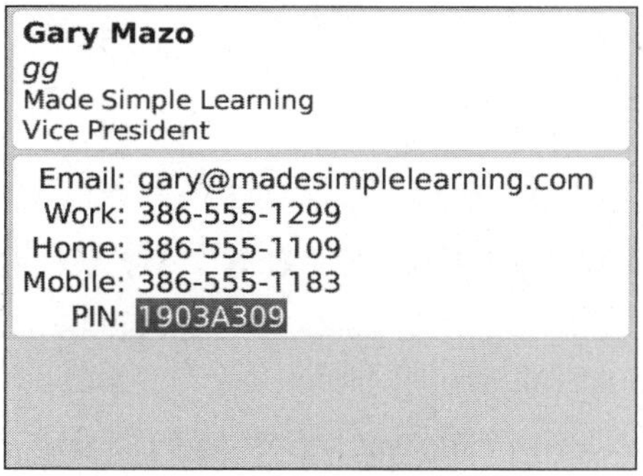

Sending Your PIN Number via Email with the Mypin Shortcut

1. Compose an email to your colleague. In the body of the email, type something like This is my BlackBerry PIN number
2. Type the code letters **mypin.**

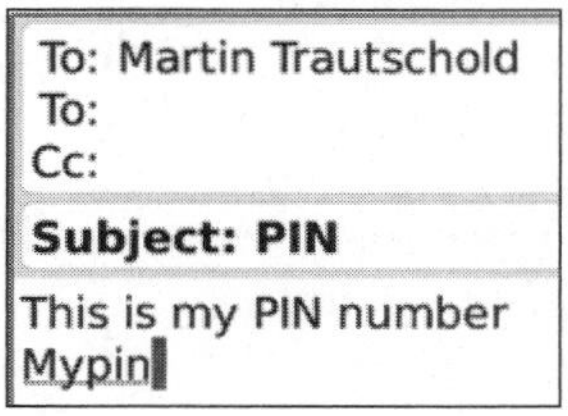

3. Press the **Space** key to see your own BlackBerry PIN number inserted right in the text replacing the **mypin** text.

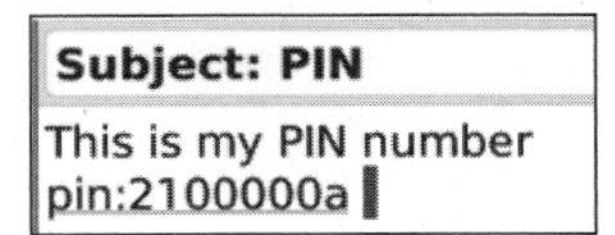

4. Click the trackpad and select **Send**.

TIP: Store your friends' PINs in your Contact List for easy access.

Replying to a PIN Message

Once you receive your PIN message, you will see that it is highlighted in red text in your Inbox.

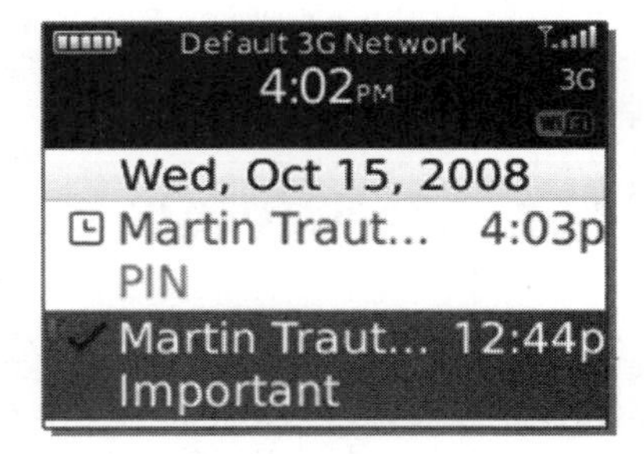

To reply to a PIN message, simply click the trackpad to open it, then click the trackpad again and select **Reply** (just like with email and other messaging).

Adding Someone's PIN to Your Contacts

Once you receive an email containing a PIN from your colleague or family member, you should put this PIN into your Contact list.

If you don't already have this person in your contact list, then highlight the PIN and press the **Menu** key to select **Add to Contacts**. Enter their name, phone number, email, and other information. See Figure 19-1.

Figure 19-1. *Add a PIN to contact information*

If you already have this person in your Contact List, then you should Copy/Paste the PIN into their contact record.

1. Highlight the PIN.
2. Click the trackpad and select **Copy**.
3. Then press the **Menu** key and select **Switch Applications** or press the **ALT + Escape** multitasking hotkey combination.

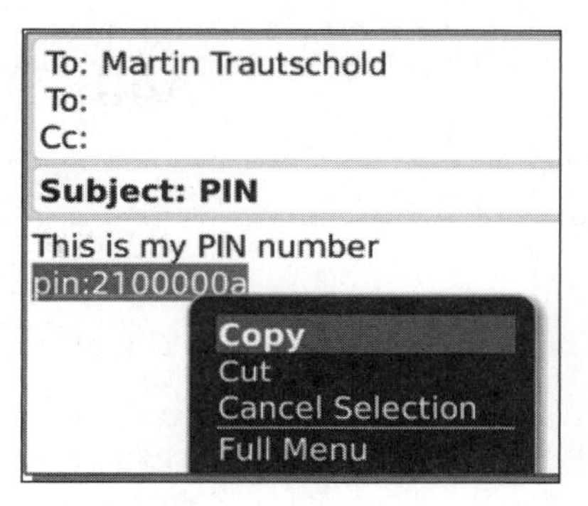

4. Locate and click on your **Contact List**, if visible in the pop-up window.
5. Otherwise, click on the **Home Screen** icon and start up your Contact list by clicking on the **Contact List** icon.

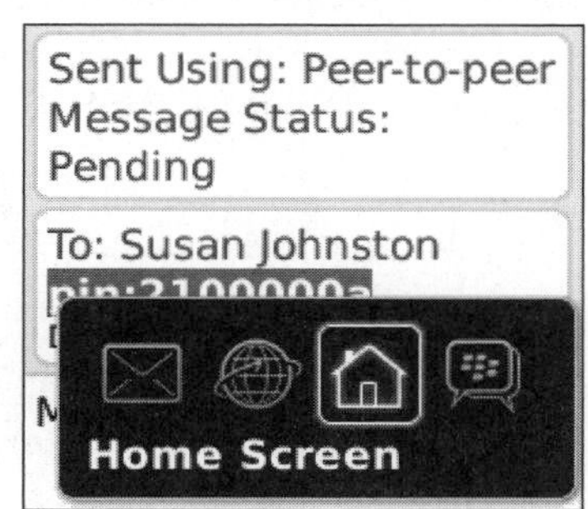

6. Type a few letters of the person's first, last, or company name to find them and then press the **Menu** key to select **Edit**.
7. Glide down to their **PIN**, click the trackpad and select **Paste**.
8. Press the **Menu** key and select **Save**.

Now, next time you search through your contacts you will have the option of sending a PIN message in addition to the other email, SMS, MMS, and phone options.

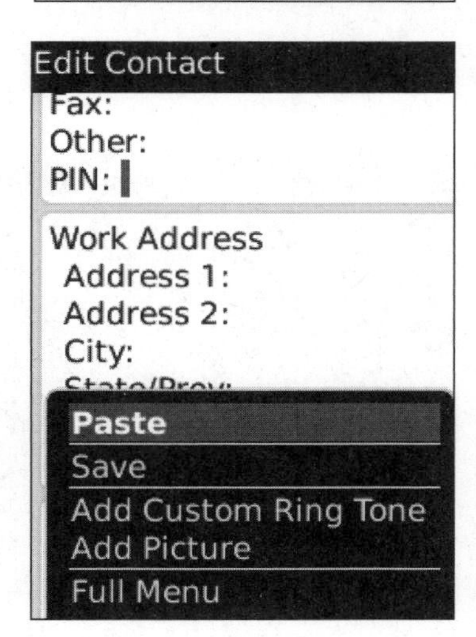

TIP: PIN Messages are free

As of publishing time, PIN Messages are still free, whereas other SMS text, BlackBerry Messenger, and MMS (multimedia messages) may be charged as extra services by your phone company.

BlackBerry Messenger

So far, we have covered email, SMS text, MMS, and BlackBerry PIN-to-PIN messaging. If you still need other ways of communicating with friends, family, and colleagues you can try BlackBerry Messenger or any one of the popular instant messaging programs like AIM (AOL Instant Messenger), Yahoo, or GoogleTalk instant messengers. The BlackBerry is really the ultimate communication tool.

TIP: Don't have the BlackBerry Messenger Icon (Figure 19-2)?

First, look in your Instant Messaging, Messaging, or Applications folders. If you don't see it, then download and install the BlackBerry Messenger icon. Go to `mobile.blackberry.com` from your BlackBerry web browser or go to your BlackBerry App World icon and follow the directions.

Figure 19-2. *BlackBerry Messenger icon*

Many users have IM programs on their PC, but some have it on their mobile phones as well. BlackBerry includes a messaging program just for fellow BlackBerry users called BlackBerry Messenger. You will find the BlackBerry Messenger icon in your **Applications** menu or your **Instant Messaging** folder.

Setting Up BlackBerry Messenger

BlackBerry Messenger offers you a more secure way of keeping in touch quickly with fellow BlackBerry users. Setup is very easy.

If you don't see the **Blackberry Messenger** icon, then press the **Menu** key (to the left of the trackpad) and glide up or down to find it.

1. Click on the **BlackBerry Messenger** icon. You will be prompted to set your **User Name**.
2. Type in your **User Name** and click **OK**. You will then be asked to set a BlackBerry Messenger password. Type in your password and confirm, then click **OK**.

3. To change your Display Name or Picture, just press the **Menu** button and select **My Profile**. Then just move the cursor next to your display name to make changes.
4. When you are done, press the **Escape** key and save your changes.

Add Contacts to Your BlackBerry Messenger Group

Once your User Name is setup, you need to add contacts to your BlackBerry Messenger Group. In BlackBerry Messenger, your contacts are fellow BlackBerry users who have the BlackBerry Messenger program installed on their handhelds.

1. Navigate to the main BlackBerry Messenger screen and press the **Menu** key. Scroll to **Invite Contact** and click.
2. Begin typing the name of the desired contact. When the desired contact appears, click on the trackpad. Choose whether to invite the contact by PIN or email.

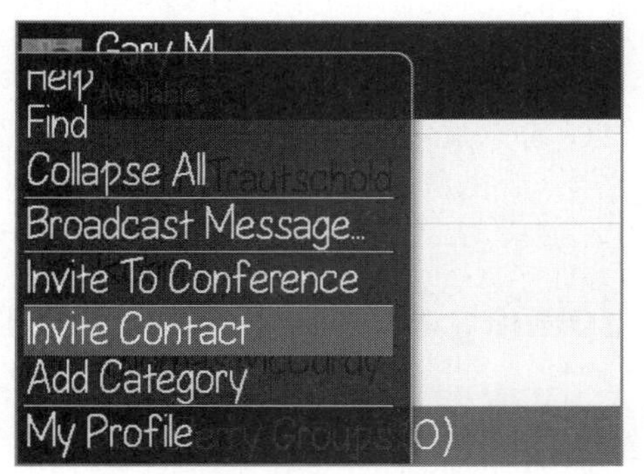

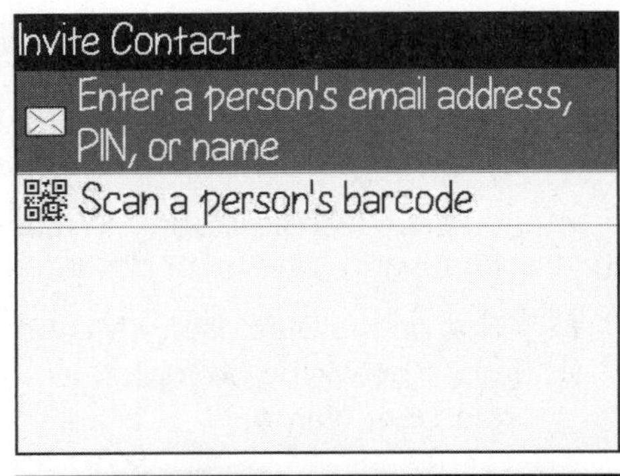

3. The BlackBerry generates a message stating: “(Name) would like to add you to his/her BlackBerry Messenger.” Click **Send** and the message is sent to this person.

4. The Message request shows up in your **Pending** group, under Contacts.
5. The contact will be listed under **Pending** until the recipient responds to your invitation.

If you do not get a response, then click on their name and use another communications method (e.g. Phone, Email, SMS Text) to ask your colleague to hurry up.

Joining a Fellow User's BlackBerry Messenger Group

You may be invited to join another BlackBerry user's Messaging group. You can either **Accept** or **Decline** this invitation.

You will receive your invitations via email or you can see them directly in BlackBerry Messenger.

1. Click on your **BlackBerry Messenger** icon.
2. Scroll to your **Requests** group and highlight the invitation and click.

3. A menu pops up with three options: **Accept, Decline, Remove.**
4. Click on **Accept** and you will now be part of the messaging group. Click **Decline** to deny the invitation or **Remove** to no longer show the invitation on your BlackBerry.

BlackBerry Messenger Menu Commands

You can do more with BlackBerry Messenger when you bring up the menu.

From your main BlackBerry Messenger ("BBM") screen press the **Menu** key. You will see the following menu commands.

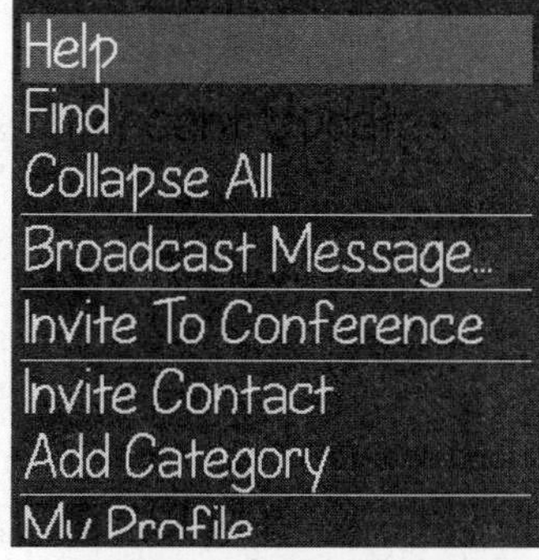

- **Help** – Use this launch the built-in Help menu.
- **Find** – Allows you to type in a contact name to search for a Messenger contact.
- **Collapse All** – Hides all group members.
- **Broadcast Message** – Allows you to send one message to as many of your Messenger contacts as you wish.
- **Invite to Conference** – Allows you to invite Messenger buddies to a Messenger conference.
- **Add Category** – Click to add a new messaging category such as Work, Family, or Friends.
- **My Profile** – Click to make yourself **Available** or **Unavailable** to your messaging buddies or change your picture or display name.
- **Options** – Click to bring up your **Options** screen (see below for details)
- **Backup/Restore Contact List** - With BlackBerry Messenger 5.0 and higher you can now create a wireless backup of your Messenger contacts and restore them if you lose them or need to change your BlackBerry.
- **Create New Group** – Allows you to create new Messenger groups to quickly communicate with more people.

- **Scan a Group Barcode** – BlackBerry Messenger now allows you to use your camera and scan a unique barcode that is part of the display profile of all BlackBerry Messenger users. Scanning the barcode essentially invites them to your Messenger group.
- **Switch Application** – Press this to multitask or jump to another application while leaving the messenger application running.
- **Close** – Exit the Messenger application.

Customizing Your BlackBerry Messenger

Below, we show you the entire **Options** screen to give you a feel for those items you can customize in your BlackBerry Messenger. We do not describe every option in detail as some are self-explanatory.

To see your BBM options, press the **Menu** key and select **Options** to see the screens shown in Figure 19-3.

- The default for **Press Enter** key **to Send** is **checked**. If you want to turn this off, **uncheck** the box. If this is **unchecked**, you will have to click the trackpad and select **Send** to send your BBM message.
- The default is a **checkmark** in the **Show Chats in Messages Application** box. This will show your BBM messages in your **Messages** (email Inbox).
- You can also choose to have the BlackBerry vibrate when someone **pings** you (the default is Yes — a **checkmark** in the box).
- To save a copy of your contact list—just click on Backup.

Figure 19-3. *BlackBerry Messenger options*

NOTE: There are far more options to adjust in version 5 than in previous versions of BlackBerry Messenger. On the Messenger Options screen, you can set the following (and much more):

Starting or Continuing Conversations

While messaging is a lot like text messaging, you actually have more options for personal expression and the ability to see a complete conversation with the Messaging program.

Your conversation list is in your main screen. Just highlight the individual with whom you are conversing and click the trackpad. The conversation screen opens.

Type in the new message and click the trackpad to send.

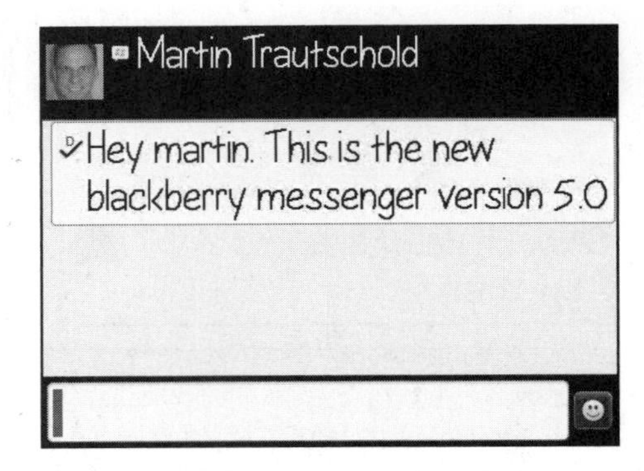

Using the Emoticons (Smiley Faces and More)

To add an emoticon to your message, click on the emoticon button next to the text input window or press the **SYM** key (next to your **Space** key) several times and navigate to the emoticon you wish to use.

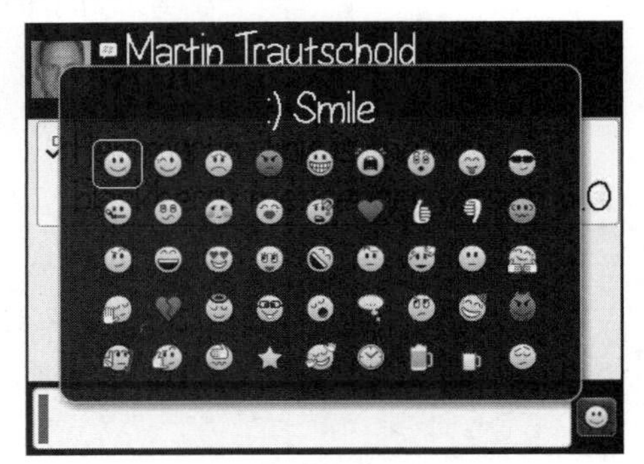

Sending Files or Voice Notes to a Message Buddy

Unless you are in a Messenger conference, you can send a file very easily (At the time of publishing of this book, you were limited to sending only files that are images, photos, or sound files (ring tones and music). The sound files are limited to very small file sizes of 15kb—about the size of a ring tone). Larger image files, like photos that are 400KB or larger, can be sent because they are compressed down to less than 15KB.

1. Click on the contact in your conversation screen and open the dialogue with that individual.
2. Press the **Menu** key and scroll and click on **Send a Picture**.

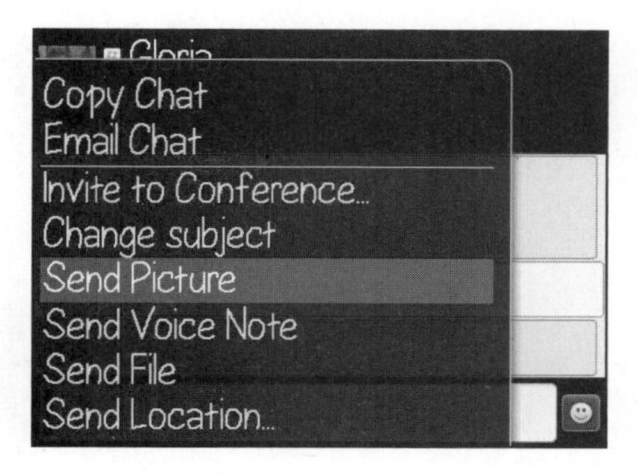

NOTE: You can also select **Send a Voice Note**, **Send a File**, or **Send Location** to send those things to a Messenger contact. With version 5 you can send up to a 6GB file or picture.

3. Using the trackpad, navigate to where the image or audio file is stored on your BlackBerry and click.
4. Selecting the file will automatically send it to your message buddy.
5. Click on the folder that contains your music and pictures—usually the BlackBerry folder.
6. Then select music, pictures, or ringtones.

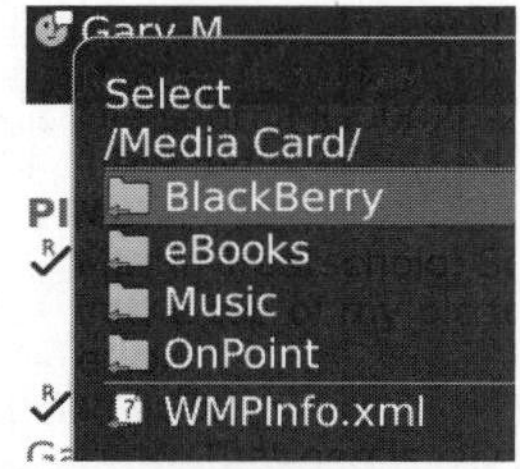

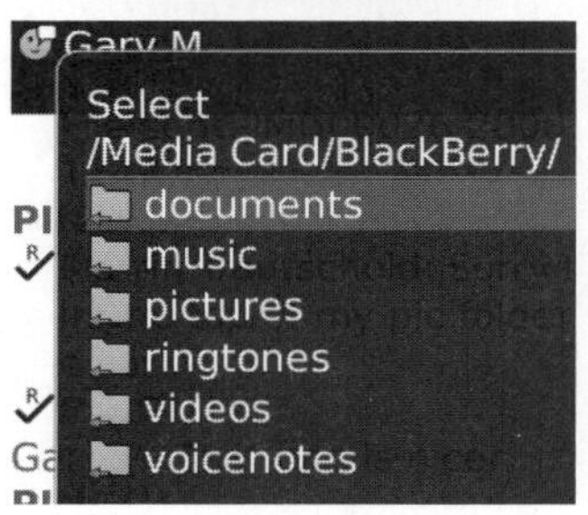

7. If you are viewing pictures, you can see either **List View** or **Thumbnail View**. List View shows the file names and very small pictures.

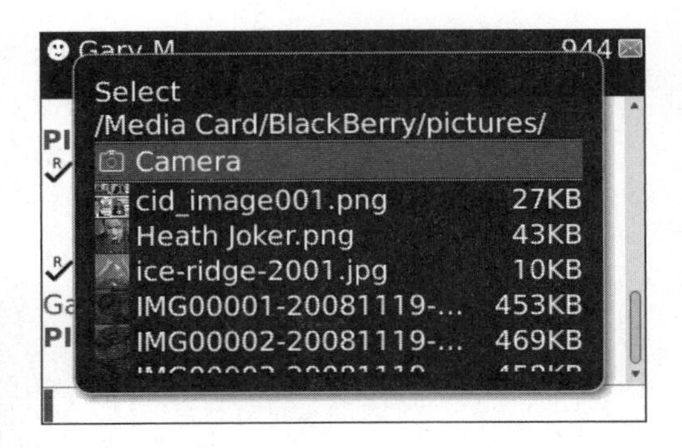

TIP: To snap a picture and send it immediately, click on the Camera at the top of the list!

8. To see the Thumbnail view, press the **Menu** key and select **View Thumbnails.**

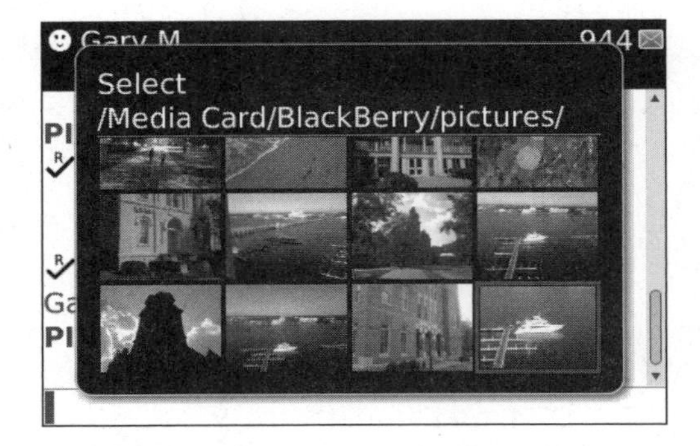

The cool thing about sending voice notes is that you can personalize a message with your own voice. Here's what a voice note looks for the person who receives it.

The receivers can either play or save the voice note on their own BlackBerry. When pressing **Play**, the Media player pops up to play the voice note.

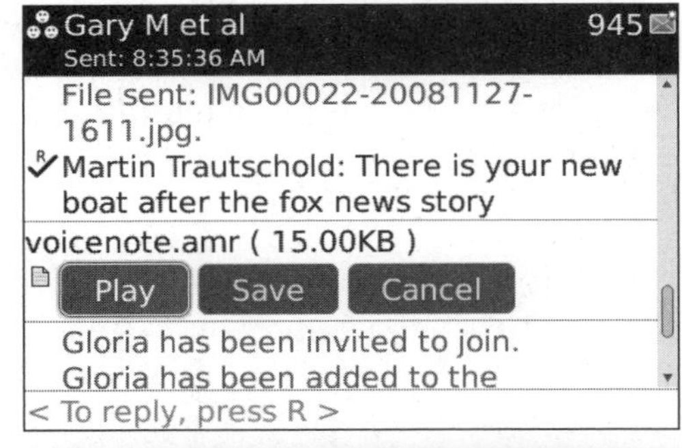

Notification of New Messenger Item or Ping

Just like email, you will see new BlackBerry Messenger items on your home screen with a red asterisk at the very top next to the time and also on the Messenger icon itself.

> **NOTE:** That is **Pixie,** Martin's family Labradoodle, pulling on her leash.

Ping a Contact

Let's say that you wanted to reach a BlackBerry Messenger contact quickly. One option available to you is to **PING!!!** that contact. When you ping a BlackBerry user, their device will vibrate once to let them know that they are wanted/needed immediately.

> **TIP:** You can set your BlackBerry to vibrate or not vibrate when you receive a ping in your BlackBerry Messenger Options screen.

And you will also see the ping in your Messages inbox.

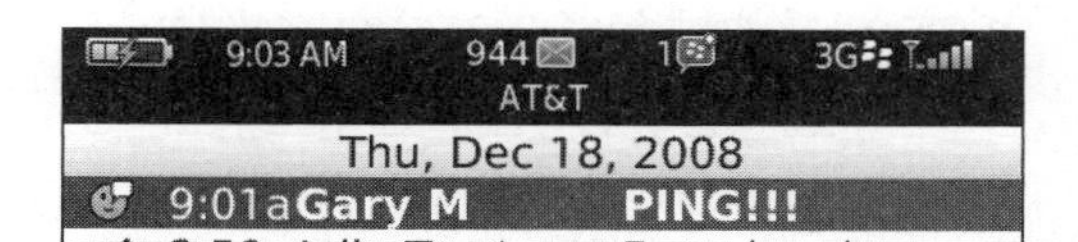

Using the My Status BlackBerry Messenger Options (Your Availability to Others)

Sometimes you might not want to be disrupted with instant messages. You can change your status to Unavailable and you won't be disturbed. Conversely, one of your contacts might be offline, so to speak, and you want to know when they become available. You can even set an alert to notify you of their availability.

1. Navigate to your main Messaging screen and press the **Menu** key
2. Scroll to **My Profile** and click.
3. Highlight the bar next to **Status** and click the trackpad.

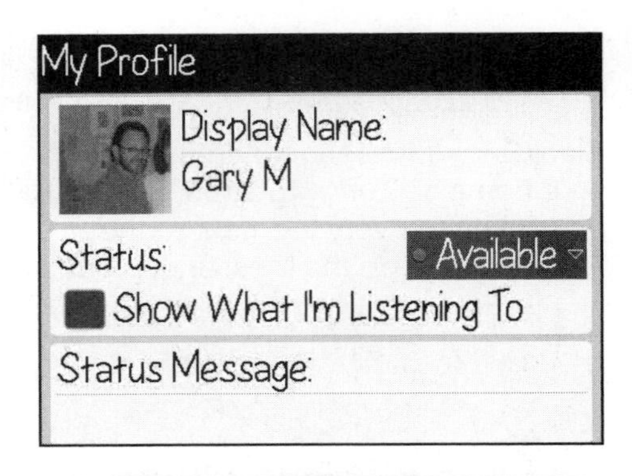

4. Choose either **Available** or **Busy**.

You can also personalize your status by selecting **Personalize Status**, then type in a description and even select a new status icon.

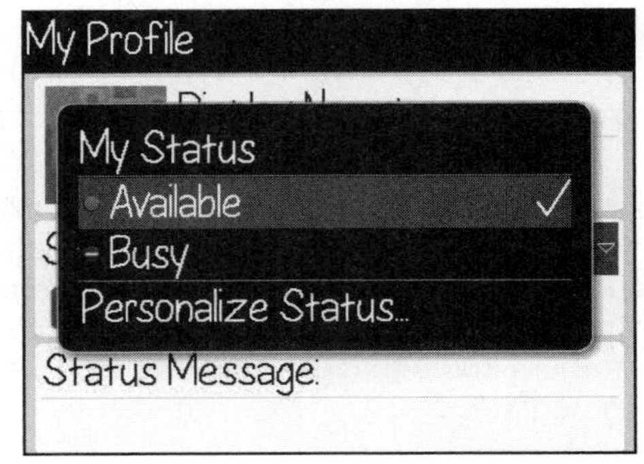

Conferencing with BlackBerry Messenger

One great new feature on BlackBerry Messenger is the ability to have a **Conference** chat with two or more of your Messenger contacts.

1. Just start up a conversation as you did before. In the image to the right, you can see that I am in a conversation with my friend Martin.

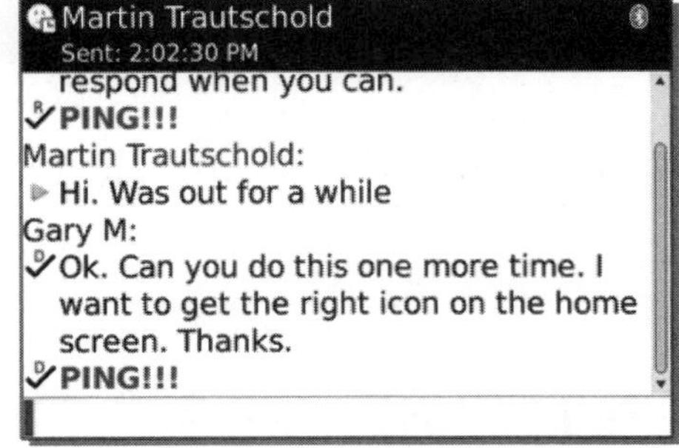

2. Now, let's say I wanted to invite Gloria to join a conversation; I press the **Menu** button and scroll to **Invite to Conference** and click the trackpad.

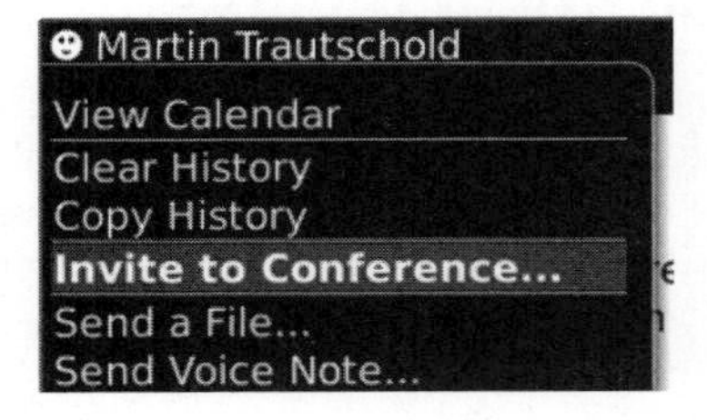

3. Then I find the contact from my Messenger list to invite Gloria.

4. An invitation is sent for her to join the conference. When she accepts, it will be noted on the Messenger screen and all three of us can have our conversation.

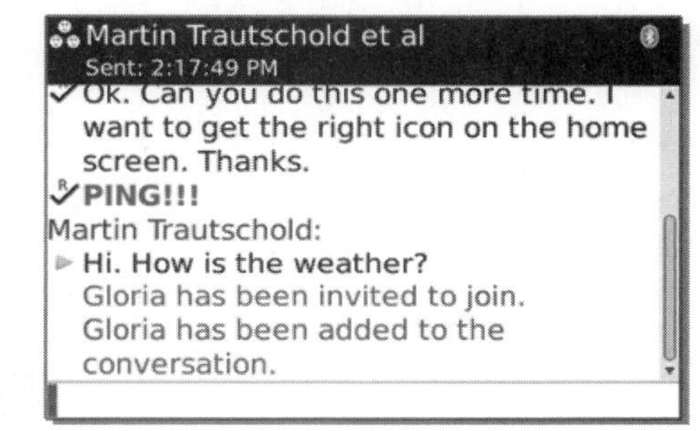

5. Notice the 3 face icons in the top left corner, showing we are in a group conversation.

Using Groups in BlackBerry Messenger

Another new feature of BlackBerry Messenger is the ability to create and use Groups. This is very useful when you don't need a Broadcast Message to everyone but just want a small group of colleagues.

1. From your main Messenger screen, click on the **BlackBerry Groups** tab.

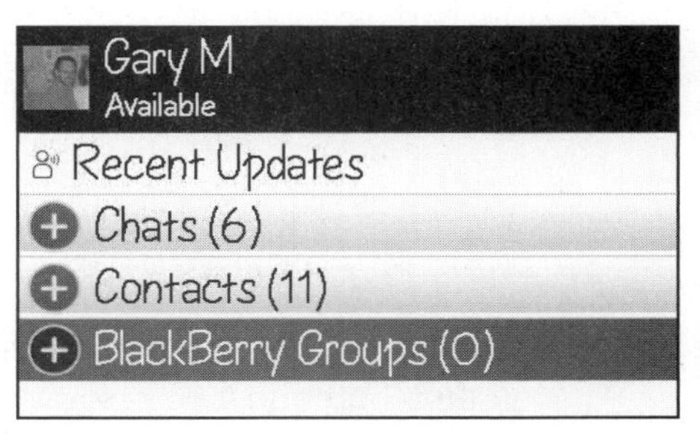

2. If you don't have any groups, you will see only the **Create a New Group** tab.

3. Give your group a Name, a Description and even choose a new Group icon if you wish.

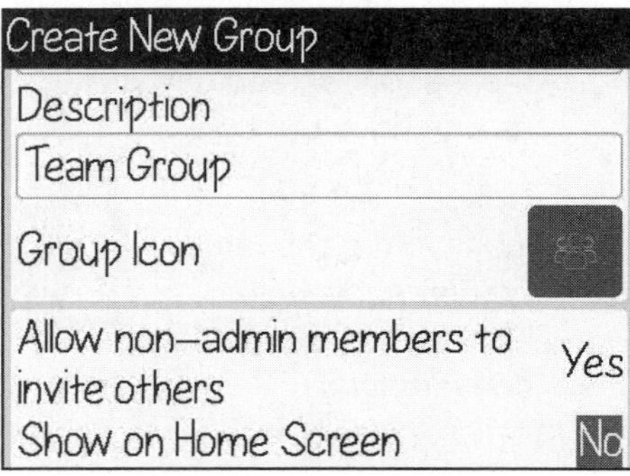

4. Click on **Create Group** and you will see your new group displayed. Click on **Members** and choose members for your group.

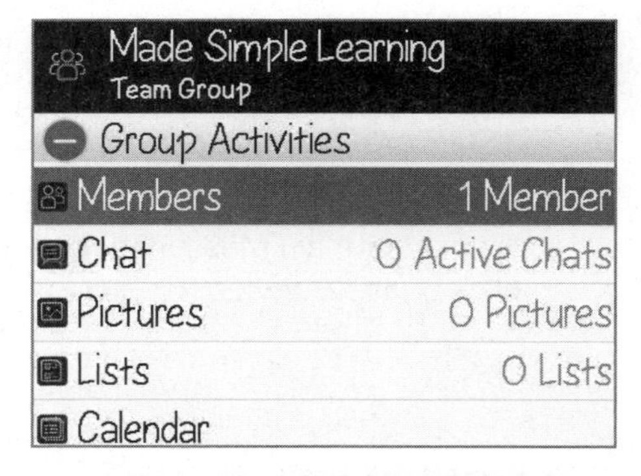

5. You can add a member's email address, scan their barcode (see below), or choose an existing BlackBerry Messenger contact to be a part of the new group.

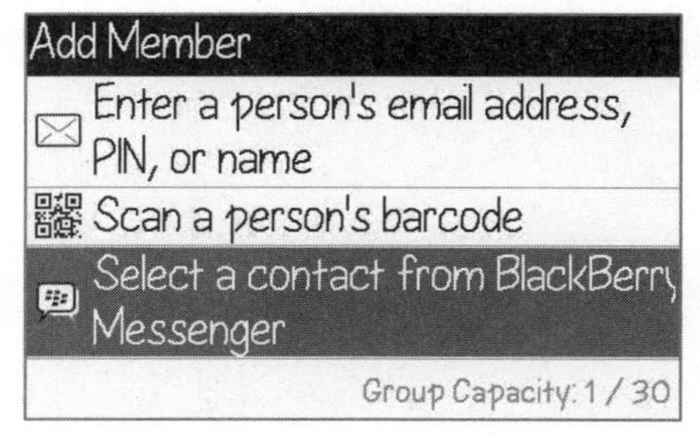

6. In this example, you want to invite Martin and Evan to be a part of the Made Simple Learning team group. They are existing Messenger Contacts, so choose **Select a contact from BlackBerry Messenger** and then place check marks next to their names.

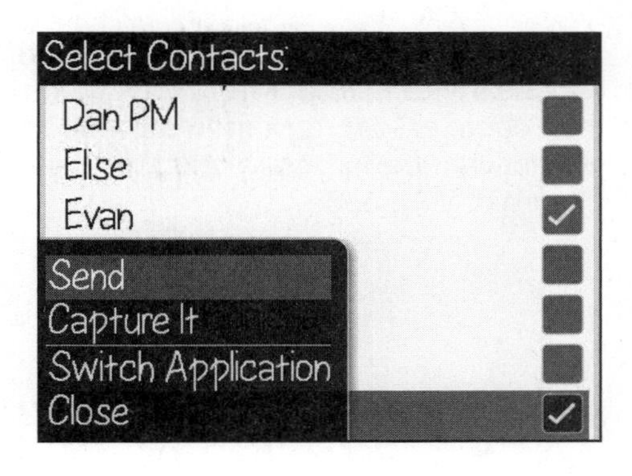

7. When you're done, press the **Menu** key and select **Send**.

Using BBM Barcodes

Perhaps the most innovative feature of the new BlackBerry Messenger is the use of a unique barcode as a means of connecting to someone as a Messenger contact.

Each BlackBerry can generate its own unique barcode. To find your barcode, just start up the Messenger application and press the **Menu** key and select **My Profile**.

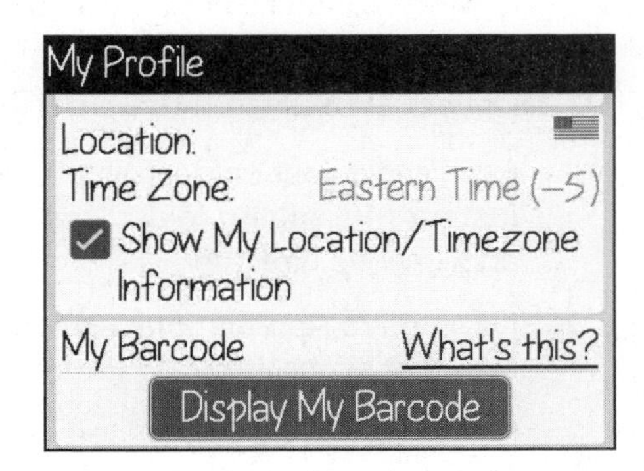

Scroll down to **Display My Barcode** and click the trackpad. The screen will now show your unique barcode. Any other BlackBerry user (with BlackBerry Messenger 5 installed AND a camera on their BlackBerry) can snap a picture of the barcode.

The other BlackBerry Messenger user needs to choose **Invite Contact** from their menu and then choose **Scan a person's barcode**. This will activate their camera and they can take a picture of your code.

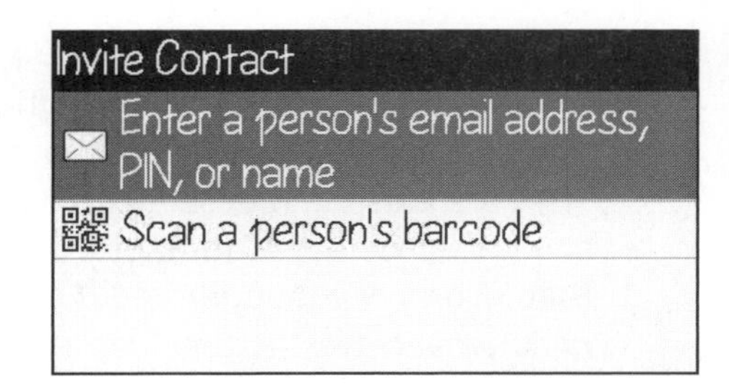

Using AIM, Yahoo, and Google Talk Messaging

After you get used to BlackBerry Messenger you will begin to see that it is a powerful way of quickly keeping in touch with friends, family, and colleagues. Realizing that many are still not in the BlackBerry world, you can also access and use popular IM programs like AOL Instant Messenger (AIM), Yahoo Messenger, and Google Talk Messenger right out of the box on the BlackBerry. Individual carriers do have some restrictions, however, and you will need to check your carrier web sites to see which services are supported.

To Install Instant Messenger Applications

1. First, see if your carrier has placed an **IM** or **Instant Messaging** folder icon in your applications directory.
2. Navigate to **Applications** and scroll for an icon that simply is called **IM**.
3. If you have an **IM folder** icon, click and follow the on screen prompts to install each application.

4. If there is no **IM folder** icon or you cannot find IM applications, use BlackBerry App World to download any number of IM applications. See page 431 for details.

Chapter 20

Adding Memory with a Media Card

As we have shown you in this book, your BlackBerry is a very capable media device. Eventually, you will need additional space to hold all those pictures, songs and videos that you will be downloading and transferring onto your BlackBerry.

How to Boost Your Memory with a Media Card

Your BlackBerry comes with about 1GB (Gigabyte) of main memory, but all that won't all be available to you. The operating system and installed software takes up some of that room and so will all your personal information. (The image of the SanDisk MicroSD card and the SanDisk logo are copyrights owned by SanDisk Corporation.)

Since your BlackBerry is also a very capable media device, you will probably want more room to store things like music files, videos, ringtones, and pictures.

That's where the MicroSD memory card comes in. You can purchase 2GB, 4GB, and even higher capacity Micro SD memory cards. At the time of publishing, the maximum allowable capacity for a BlackBerry running Operating System Software version 5.0 was 32GB. It is likely that this will increase over time.

Installing or Removing Your Memory Card/Media Card

Your BlackBerry has the MicroSD card slot next to the edge of the battery under the Battery Door. You have to remove the battery door to insert the Media Card.

Remove the back cover by sliding it down off the bottom of the BlackBerry.

Then insert the card with the metal contacts facing down and slide the card towards the top edge of the BlackBerry, with the printed label facing up.

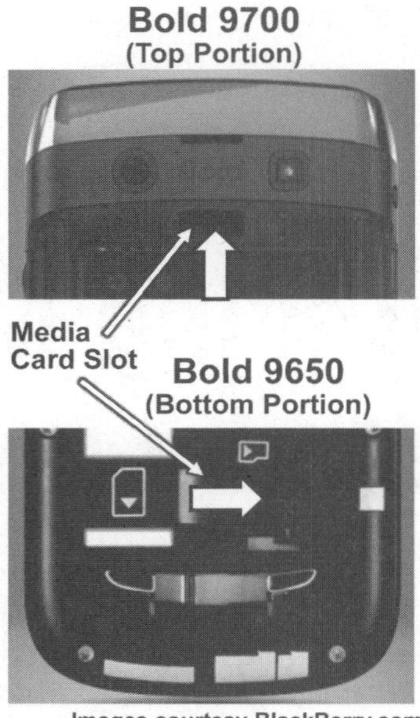

Images courtesy BlackBerry.com

You don't even need to power off the BlackBerry; you can insert the card with the BlackBerry on. When the card is correctly inserted, a **Media Card Inserted** message will appear on the screen.

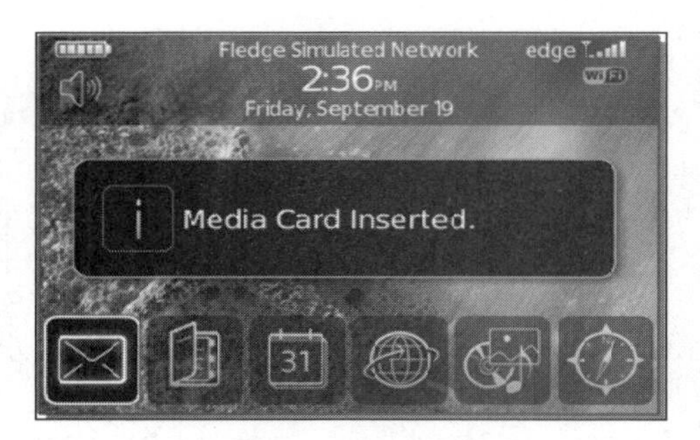

IMPORTANT: Removing the card

To remove the media card, press it into its slot further until it pops out a little. Then slide it down and out.

NOTE: You may damage the card if you simply pull it down without pushing it up first.

Verify the Media Card Installation and Free Memory

It is a good idea to double check that the card is installed correctly and how much free space is available.

1. Use the trackpad to glide to the **Options** icon and click on it. You may need to press the **Menu** key to the left of the trackpad to see this icon if it is not visible. You can also press the letter **O** if you have your **Hotkeys** turned on.
2. Scroll down and select **Memory** and click.

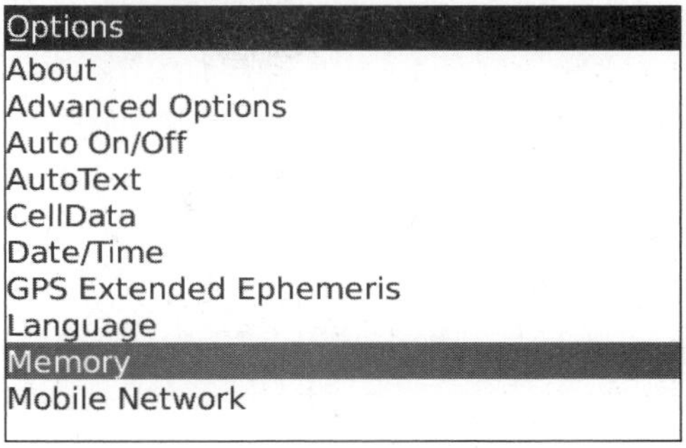

Options
About
Advanced Options
Auto On/Off
AutoText
CellData
Date/Time
GPS Extended Ephemeris
Language
Memory
Mobile Network

TIP: Pressing the first letter of an entry in the Options screen (or in menus) will jump to the first entry in the list starting with that letter. So pressing **M** will jump to the first entry starting with **M**, pressing it a second time will jump to the next **M** entry.

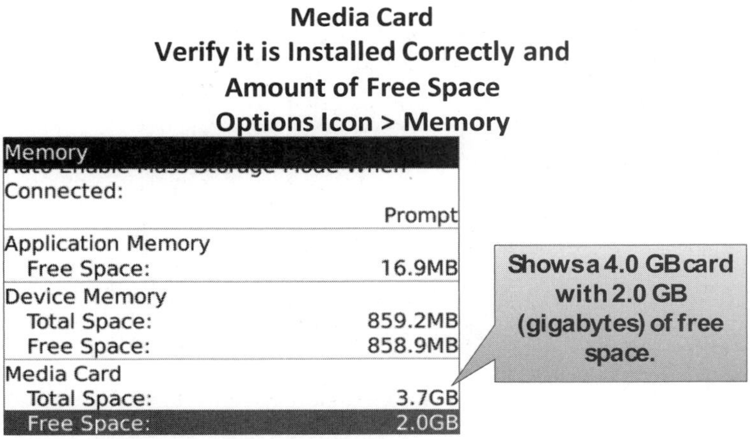

Media Card
Verify it is Installed Correctly and Amount of Free Space
Options Icon > Memory

Memory
Connected: Prompt
Application Memory
Free Space: 16.9MB
Device Memory
Total Space: 859.2MB
Free Space: 858.9MB
Media Card
Total Space: 3.7GB
Free Space: 2.0GB

Shows a 4.0 GB card with 2.0 GB (gigabytes) of free space.

Figure 20-1. *Free space available on media card*

3. Look at the total space figure at the bottom of the screen shown in Figure 20-1. A 4.0 GB card will read about 3.7MB (1GB equals about 1,000MB). If you see that, all is good.

Chapter 21

Your Music Player

Your BlackBerry is more than your personal digital assistant, your email machine, and your address book. Today's BlackBerrys are also very capable media players. With your BlackBerry, it is easy to carry your music, playlists, podcasts, and more. You can transfer playlists from iTunes and other media programs as well.

Listening to Your Music

You may be amazed at how great a music player your BlackBerry can be!

You can even listen to free streaming Internet Radio with software like Slacker and Pandora (see page 375).

With a good sized media card, the media capabilities of the BlackBerry might cause one to wonder, "Why do I need an iPhone or iPod? I've got a BlackBerry!"

> **NOTE:** Figure 21-1 shows some of the media player buttons and keys. See page 546 for a list of many important hotkeys, tips, and tricks.

Important Media Player Keys (Song and Video)

Figure 21-1. *Media player buttons and keys*

Getting Your Music and Playlists onto Your BlackBerry

The BlackBerry comes with internal memory but the OS and other pre-loaded programs take up some of that space, so what's left over is usually not enough to store all your music.

Step 1: Buy and Insert a Media Card

You can buy a media card (optional) to boost the memory available to store your favorite music (see page 359 for help).

> **NOTE:** Some BlackBerry smartphones come pre-installed with media cards.

Step 2: Transfer Your Music from Your Computer

If you are Windows user, please refer to page 105.

If you are an Apple Mac user, please refer to page 141.

Other ways to get music and videos on your computer are:

- Save the music or video from a web site to your BlackBerry
- Receive music or video using a Bluetooth file transfer

Downloading and Saving Music or Videos from the Web

Some web sites offer you videos or music to download to your BlackBerry. You are given the option to either **Open** or **Download**.

You may see this question You must select **OK** if you want to stream the video on your BlackBerry. Check the **Don't ask this again** to avoid seeing this question again.

After a little time, you may see a loading or buffering message and then the video will start playing.

> **NOTE:** When you choose to stream videos, you will not be able to save the video on your BlackBerry. If you want to save the video, you should select **Download** option instead of **Open** or **Play**.

Is that the Terminator?

No, it's the Governor.

Playing Your Music

Once your music is in the right place, you are ready to start enjoying the benefit of having your music with you at all times.

The fastest way to get to your music is by clicking on the **Music** icon.

You may first need to click on your **Media** folder icon then select **Music**.

Inside Music, you will see various preset options to find and play your favorite music.

- **All Songs** - Shows you every song on your BlackBerry.
- **Artists** - Shows you all artists, then you can click on an artist to see all their songs.
- **Albums** - Shows list of all albums.
- **Genres** - Shows list of all genres on your BlackBerry (Pop, Rock, Jazz, etc.)
- **Playlists** - Shows all playlists, or allows you to create new ones.

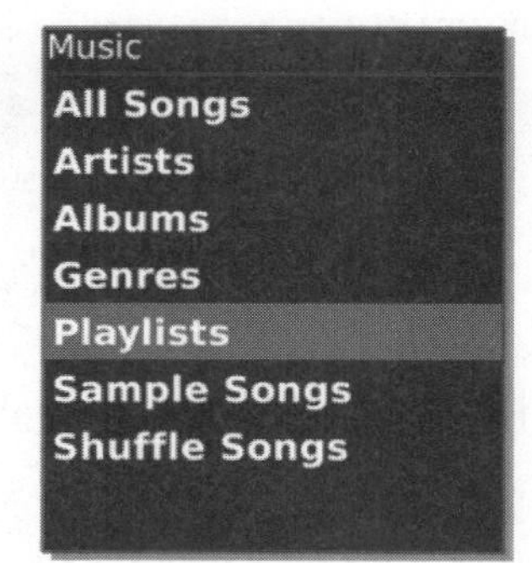

Sample Songs: Shows one or more sample songs preloaded on your device.

Shuffle Songs: Plays all your music in a shuffle mode or random order.

Finding and Playing an Individual Song

1. If you know the name of the song, then type a few letters of any word in the song's name in the Find field at the top to instantly locate all matching songs.
2. In this case, type "Love" and see all matching songs.

> **TIP:** To narrow the list, press the **Space** key and type a few more letters of another word. Notice that after typing "Love Per" only "Perfect Love… Gone Wrong" shows in the list.

3. Once you click on a song, the Music Player will open and your song will begin to play.
4. Clicking the trackpad will pause the player or continue playback.

TIP: Pressing the **Mute** key on the top of your BlackBerry will also pause or resume playback of songs and videos.

The volume keys on the side of the BlackBerry control the song volume.

Controlling Your Song or Video in the BlackBerry Media Player

There are a few ways to move around to a different section of song or video. First, if the song or video is playing, you need to click the trackpad once (or the **Mute** key on the top of the BlackBerry) to stop it.

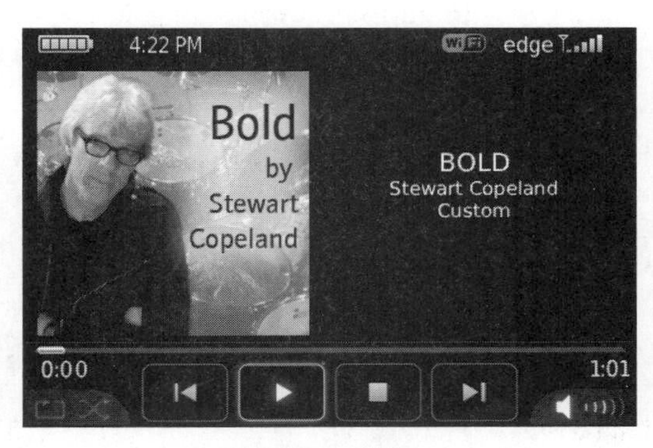

To stop the playback and go to the beginning of the song or video, click the **Stop** icon.

To go to the beginning of the song or video click this icon. To jump to the previous song/video in the list, click it again.

To go to next song/video in the list/playlist, click this icon. If you are at the last song/video, this will take you to the end of the current item.

To move around to a specific location in the song or video after the media is paused, glide the trackpad up, highlight the slider bar, and click on it. Then glide the trackpad left or right to move to a different location. When done, click the trackpad to complete the move. Click the **Play** button again to resume playback.

Click the slider to start moving it.

Glide the trackpad to move the slider.

Click the trackpad to complete the move.

Using Now Playing to Return to Your Video or Song

If you have pressed the **Escape** key a few times or the **Red Phone** key to leave your music playing or video paused, the next time you press the Menu key, you will now see a new menu item on most of your icons near the top—the **Now Playing** item. Just select this one to jump right back into the Media Player where you left off. Then you can continue playing your video or song see Figure 21-2.

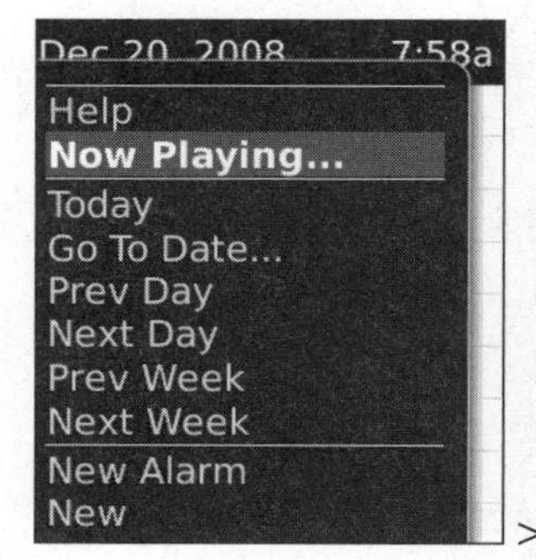

>>

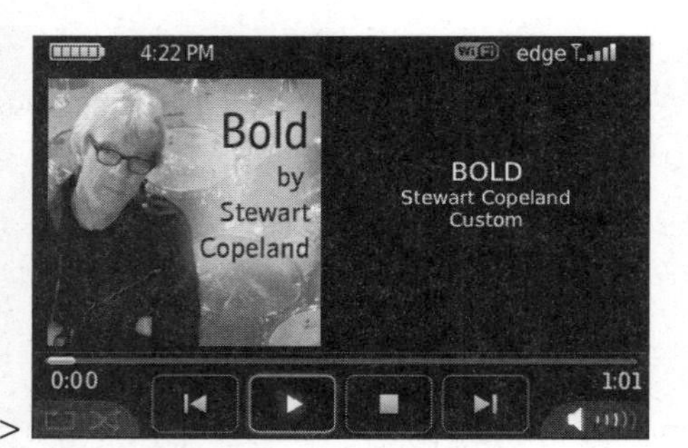

Figure 21-2. *Using the **Now Playing** option in the music menu*

Playing Your Music and the Shuffle Feature

Navigate to your music as shown above and highlight the first song in the folder or playlist you wish to play. If you have not set up individual folders or playlists, just highlight the first song. Then click the trackpad to begin playing the highlighted song and the music player will begin to play all the songs in that particular folder or playlist. (You can also select **Play** from the menu.)

To **Shuffle** your songs in that folder, navigate to the first song in your folder and press the **Menu** key. Click on **Shuffle**.

TIP: If there is a checkmark next to Shuffle in the menu, it's already on; clicking it again will turn Shuffle off.

The Shuffle icon: Two crossed Bold arrows 

To Repeat Playing a Song or Video

When a song or video is playing, press the **Menu** key and select **Repeat.**

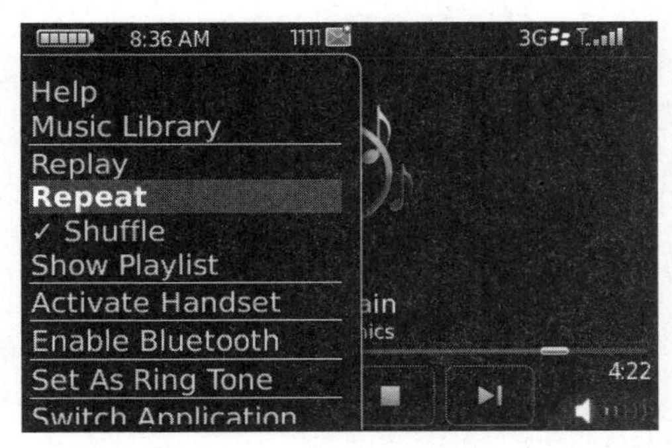

You will see this **Repeat** icon in the lower left corner.

To turn off the Repeat play, press the **Menu** key again and select **Repeat.** This will turn off the checkmark next to the menu item.

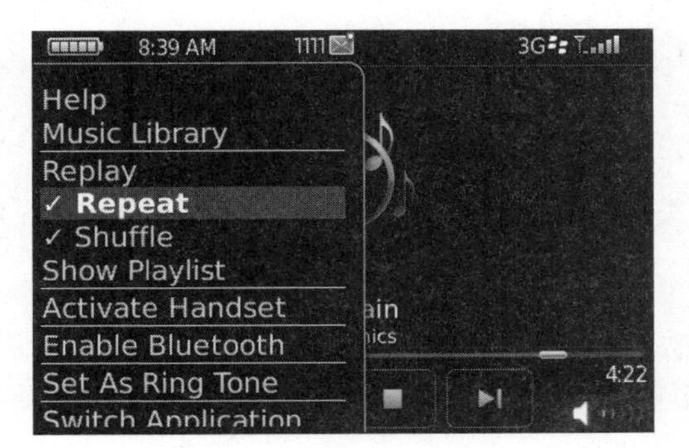

Finding Music Using a Memory Card

Assuming you have followed the steps above, your music is now on your Micro SD media card. Now, you want to play your music. What do you do?

From your Home Screen of icons, you have two ways to view and listen to music.

Option 1: Click on the **Music** icon

Option 2: Click on the **Media** folder, then click on the **Music** icon.

The available music folders are now displayed.

Click on the appropriate folder (if your music is on a Media Card, click that folder) and all of your music will now be displayed. Click on any song to start playing it.

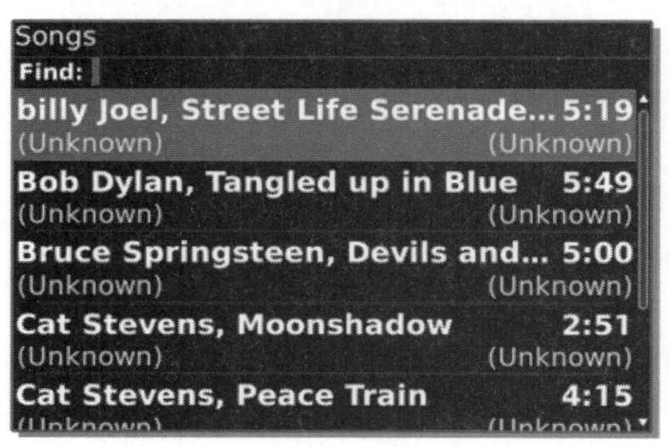

Playing a Playlist

1. To see the playlists on the BlackBerry, just glide over to the **Media Player** icon and click on it. Find the **Music** icon and click on that. Or, just click on the **Music** icon from the home page.
2. Then scroll down to **Playlists** and click the trackpad.

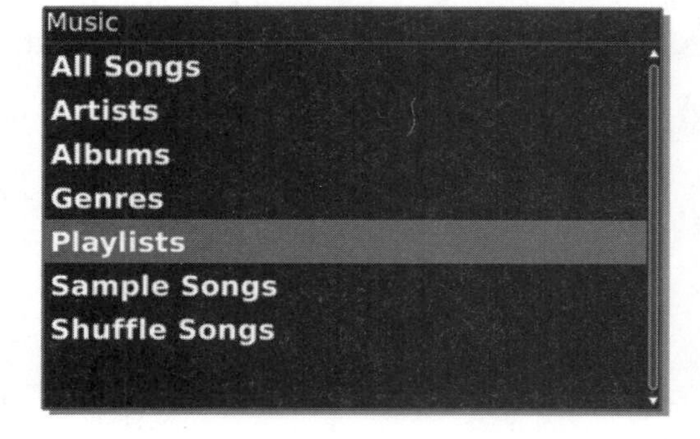

Playlists synced from iTunes are now listed right on the BlackBerry.

3. Click on the **Bike Riding** playlist to begin playing the songs.

4. To see the list of songs in a playlist, press the **Menu** key and select **Show Playlist.**

Many MP3 players utilize playlists to organize music and allow for a unique mix of songs. What types of music are supported on the BlackBerry? See page 373.

To Create Playlists from Your Computer

Use the supported computer software on your PC to create playlists and sync them to your computer.

Windows users, please refer to page 105.

If you are an Apple Mac user, please refer to page 141.

To Create Playlists on Your BlackBerry

1. Click on your **Music** icon or your **Media** icon then click on **Music**.

2. Click on **Playlists** to get into the Playlists section as shown.
3. You can either click on [**New Playlist**] at the top of the screen or press the **Menu** key and select **New Playlist** from the menu.

4. Now, you need to select Standard Playlist or Automatic Playlist
5. Select **Standard Playlist** to select and add any songs already stored on your BlackBerry.

6. Type your Playlist name next to Name at the top, press the **Menu** key, and select **Add Songs** to add new songs.

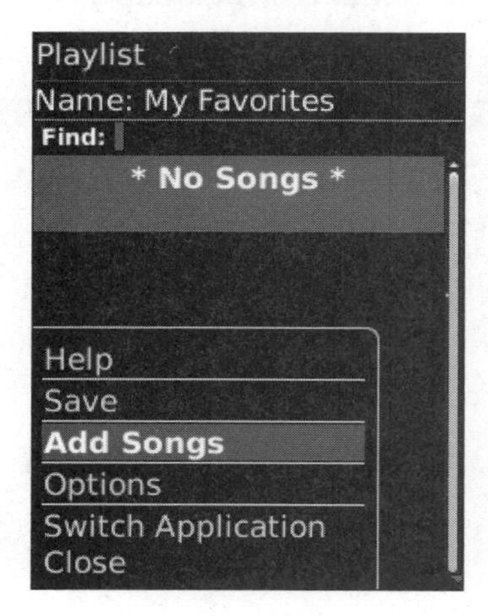

TIP: To find your songs, you can just scroll up down the list or type a few letters you know are in the title of the song like "love" and instantly see all matching songs.

7. When you find the song you want, just click on it to add it to your Playlist.

TIP: You can remove songs from the Playlist by selecting the song, pressing the **Menu** key, and selecting **Remove**.

Automatic Playlist

TIP: The **Automatic Playlist** feature allows you to create some general parameters for your playlists based on artists, songs, or genres.

Supported Music Types

The BlackBerry will play most types of music files. If you are an iPod user, all music except the music that you purchased on iTunes should be able to play on the BlackBerry.

However, if you burn your iTunes tracks to a CD (make a new playlist in iTunes, copy your iTunes tracks, then burn that playlist) Roxio Media Manager can convert these tracks to play on the BlackBerry.

The most common audio/music formats supported are:

- ACC - audio compression formats AAC,
- AAC+, and EAAC+ AMR - Adaptive Multi Rate-Narrow Band (AMR-NB) speech coder standard
- MIDI - Polyphonic MIDI
- MP3 - encoded using MPEG
- WAV - supports sample rates of 8 kHz, 16 kHz, 22.05 kHz, 32 kHz, 44.1 kHz, and 48 kHz with 8-bit and 16-bit depths in mono or stereo.

NOTE: Some WAV file formats may not be supported by your BlackBerry.

TIP: Mute Key for pause and resume

You can pause (and instantly silence) any song or video playing on your BlackBerry by pressing the **Mute** key on the top of your BlackBerry. Press **Mute** again to resume playback.

Media Player Options

To adjust media options, press the **Menu** key and select **Options** to see the options screen below. You can then turn on or off the equalizer for your headset. You can also enable Audio Boost which is disabled by default. You can turn on Auto Backlighting and adjust Closed Captions (enable, change style, change position) see Figure 21-3.

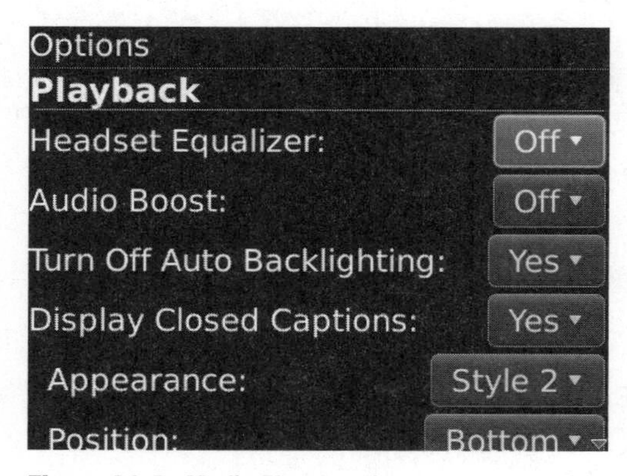

Figure 21-3. *Media Player options*

Music Player Tips and Tricks

TIP: Check out our Media Player hotkeys on page 546.

- To pause a song or video, press the **Mute** key on the top of the BlackBerry.
- To resume playing, press the **Mute** key again.
- To move to the next item, press **N** (also works in pictures).
- Also, pressing and holding the Volume Up key will move to the next track.
- To move to a previous item (in your playlist or video library,) press **P**.
- Also, pressing and holding the Volume Down key will move to the previous track.

Streaming Music

One of the amazing things about the BlackBerry is just how powerful it can be as a media center. Using a Bluetooth stereo in your car, you can stream your music via Bluetooth right through the car stereo.

To stream stereo music via Bluetooth, just pair your BlackBerry with a Bluetooth music source as described on Page 441.

Make sure that the Bluetooth device supports Bluetooth stereo uudio (sometimes called A2DP). Ensure that your BlackBerry is connected to the stereo music player or use the BlackBerry Remote Stereo Gateway mentioned on page 449. Then follow the steps for playing your music that were discussed earlier in this chapter.

Streaming Free Internet Radio

Internet Radio is a great way to listen to music on the go. There are several applications that allow you to setup and listen to streaming Internet radio. There are now several great free applications for the BlackBerry and all of them work well. The two most popular are **Pandora** and **Slacker Radio**, both found in the **BlackBerry App World**.

Just set up your account, specify the type of music you like, and you should be streaming Internet radio in minutes.

Pandora Internet Radio

Pandora is an outgrowth of the Music Genome project. Essentially, Pandora is streaming Internet radio where you control the radio stations. In other words, you build stations based around your favorite artists.

Download Pandora

Go to the BlackBerry App World (see page 431) and download the Pandora App. If the App is not part of the Featured Apps, you can find it in the Music section or by doing a simple search for Pandora.

You can also visit `www.pandora.com` and download the application from their web site right to your BlackBerry. Just agree to the terms and select **Download** when the download screen appears.

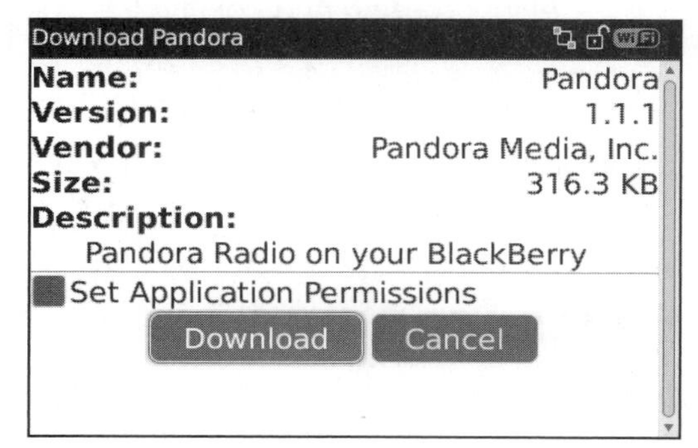

Starting Pandora for the First Time

When Pandora starts for the first time, you will be asked if you have an existing account or if you would like to create a new account.

Just type in your email and password (for an existing account) or input your email and create a password for a new account. The best thing about Pandora is that it is absolutely FREE!

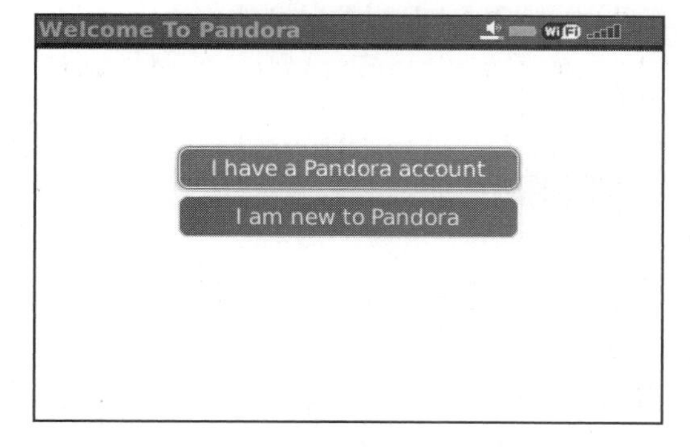

If this is your first time using Pandora, you will be asked to **Create a New Station**. If you have an existing Pandora account, you will see all your stations on the front page along with the **Create a New Station** option.

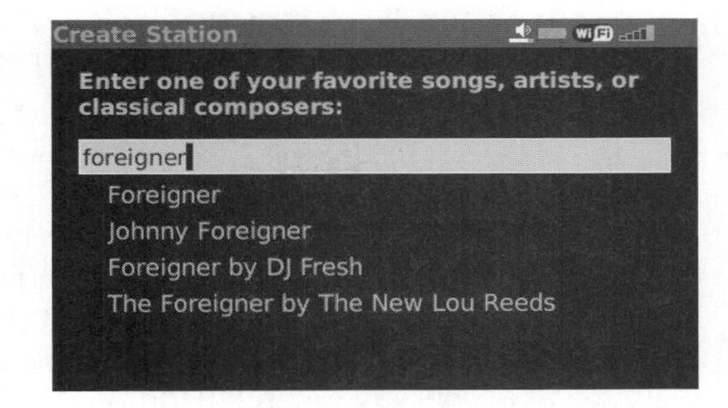

NOTE: Pandora also works on your PC or Mac. You can create your stations and listen on your computer. When you log in on the BlackBerry, all the stations created on your computer will appear.

Pandora Controls and Options

The interface of Pandora is very clean and intuitive.

To Play or Pause just press the Play arrow or Pause button.

To skip to the next song on the station, just click the Next arrow.

Using Thumbs Up and Thumbs Down

Two buttons that you can click are the Thumbs Up and Thumbs Down buttons.

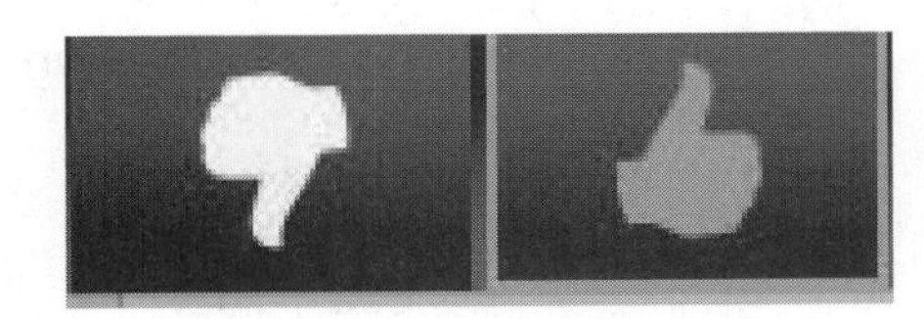

If you click the Thumbs Up button, a check mark is placed on the song letting Pandora know that you like this song and it can be used again.

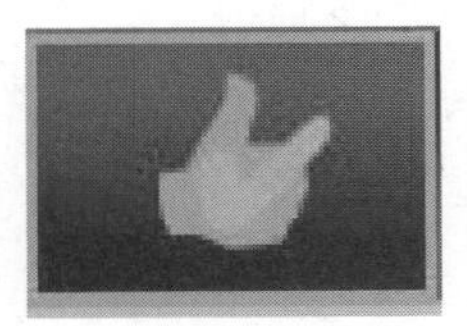

On the other hand, if you click on the Thumbs Down button, a very quick check mark appears and Pandora will skip to next song and remember to never play that song again.

Slacker Radio

Conceptually, Slacker is very similar to Pandora. You build stations around your favorite artists or choose from hundreds of existing stations for your listening pleasure. Slacker seems to have a few more pre-programmed stations than Pandora and has a cache feature that allows you to stream music to your computer then download the music to your BlackBerry media card. This allows you to listen to music without needing an internet c onnection—perfect for a long airplane trip or whenever you are out of radio coverage for a while.

Like Pandora, you can listen to Slacker Radio on your PC and Mac as well as your BlackBerry.

Download and Install Slacker Radio

As you did with Pandora, you have two options for downloading Slacker Radio. You can find Slacker in the BlackBerry App World, or you can also go to their website at `www.slacker.com` and download the latest build for BlackBerry. Follow the on screen instructions for download and installation. Accept the license agreement and you will be Slacking in no time!

Create or Log In to Slacker Account

As with Pandora, your first screen will ask if you have an existing Slacker Account or if you wish to create one. Either log in or create a new account by clicking the button at the bottom of the screen. As with Pandora, you can have Slacker running on your computer also!

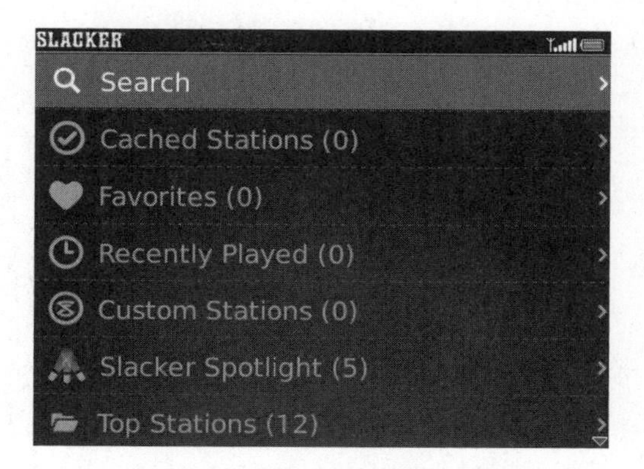

If you have an existing account, you will see a screen with your favorite stations displayed.

If you are signing on for the first time, you have two options:

Scroll down to check out stations by genre, or

click **Search** at the top and build a new station with music created from your favorite artist (as you did with Pandora).

In this example, we are creating a new station based on music by Peter Gabriel.

Choosing Your Station

Once you have some created stations, each time you log on to Slacker, you will be able to choose whether you wish to listen to one of your favorite stations or listen to other spotlight stations or genre in the Slacker library of music stations.

Slacker Controls

Slacker offers a few more controls and options than Pandora. You can access the controls via the dock of buttons at the bottom. The Home button takes you back to the Slacker Home screen where you can choose your station.

The Next arrow takes you to the next song in the station lineup. Notice that Slacker displays a picture of the next song or artist to the right of the album cover of the current song.

If you like the artist or song, just click the Heart button for Slacker to remember that.

If you don't want to hear this particular artist or song again, click the Do Not Play/Ban button.

Slacker Menu Commands and Shortcut Keys

Slacker offers some very cool commands in the Menu. Just press the **Menu** button on your BlackBerry and you can see your stations, lyrics to the current song, an artist bio, or an album review. There are also options to Ban this Artist, Filter Explicit Content, Enable Bluetooth (see page 449on A2DP Stereo Bluetooth), and Share Slacker with friends.

Slacker Shortcuts

- N - Skip to Next Song (with free Slacker, you are limited to six skips per hour)
- **Space** - Play/Pause Song
- H - Heart (Says you like the song so play more songs like it.)
- B - Ban Song (Says you don't like the song so don't ever play it again.)

Chapter 22

Snapping Pictures

Many BlackBerrys sold today come with a built-in camera—perfect for snapping a quick picture. (Note that some BlackBerrys issued by companies might not have a camera for security reasons.) Your camera and the photos you take can easily be shared with others via email or MMS messaging, or uploaded to Facebook and other social media programs.

Using the Camera

Your BlackBerry includes a feature-rich camera. This gives you the option to just snap a picture anywhere you are. You can then send the picture to friends and family to share the moment.

Camera Features and Buttons

You have plenty of options on your camera. Before we get to that, however, let's get familiar with the main buttons and features.

Starting the Camera Application

The camera can be started in one of two ways see Figure 22-1.

Option 1: The Right Side Convenience Key

Unless you have re-programmed your right convenience key (see page 187 for details), pressing the right side key (the one directly below the volume control buttons) will start your camera.

Figure 22-1. *Using Camera app on the BlackBerry*

Push this button once, and the camera should start.

Option 2: The Camera Icon

Press the **Menu** key to see all your icons.

Then, scroll to the **Camera** icon and click the trackpad.

Normal Viewfinder Mode

The BlackBerry has one standard Viewfinder mode.

You will see the picture in the background with the status bar at the bottom.

If you don't do anything, the status bar disappears.

As soon as you click a picture or touch any key, the status bar reappears.

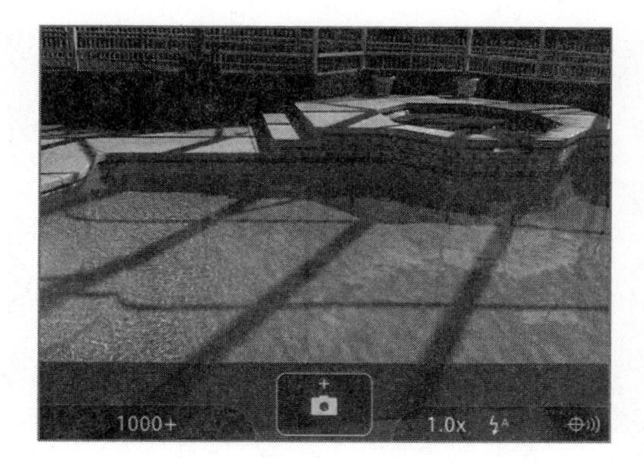

Taking Action Shots

Once you start the Camera and are in viewfinder mode. You can press the Right Convenience key in half-way to focus the camera. You will see the little box in the middle of the screen go from white to green. Green shows that the BlackBerry thinks it has focused.

This allows you to snap the picture immediately when you click the trackpad or press the Convenience key fully in.

If you do not use the Convenience key to focus the camera, when you click the trackpad, the camera will first focus, then take the picture. This may cause you to miss an action shot.

Using the Camera Zoom

As with many cameras, your BlackBerry camera gives you the opportunity to zoom in or out of your subject. Zooming could not be easier.

Frame your picture and gently glide up on the trackpad. The Camera will **zoom** in on the subject.

> **TIP:** You can also use the Volume Up/Down keys to zoom.

The digital Zoom level will be displayed to the right of the camera icon in the bottom status bar with a range of **1.0x – 2.0x** indicating the power of zoom chosen.

The image to the right shows **1.0x** zoom.

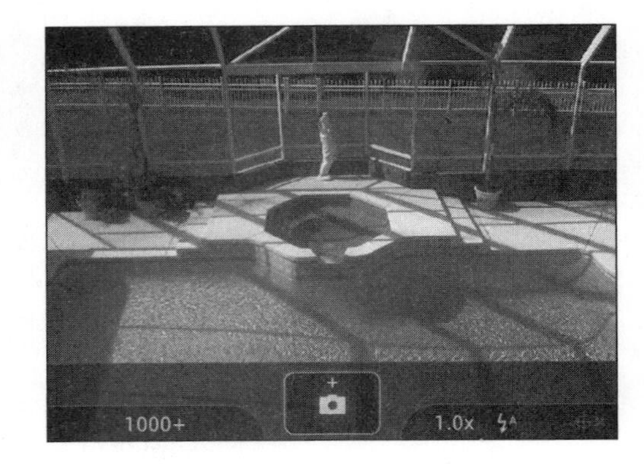

Here is the same shot at 2.0x zoom.

To Zoom back out, just glide the trackpad down.

Icons in the Camera Screen

Usually, when you open the Camera application either the last picture you took is in the window or the Camera is active. Underneath the picture window are five icons.

The Take a Picture icon

Click on this to take another picture.

The Trash Can icon

If you're not happy with the picture, simply scroll to the Trash Can and click. The last picture taken will be deleted.

The Rename/Folder icon

With the picture you desire to save on the screen, click on the folder to specify a new location or file name. We recommend renaming important pictures to a more meaningful name. For example, you might change the default name "IMG00008-20100314-1855" to "Martha Birthday."

Set As or Crop icon

You can use the picture in the main window as either home screen wallpaper or as a picture caller ID for someone in your Address Book by selecting the **Set As** button.

You can also set that picture as a background image for your Home Screen (Like the desktop background image on your computer)

1. Just open a picture from **My Pictures.**
2. Press the **Menu** key and select **Set as.**
3. Then select **Wallpaper.**

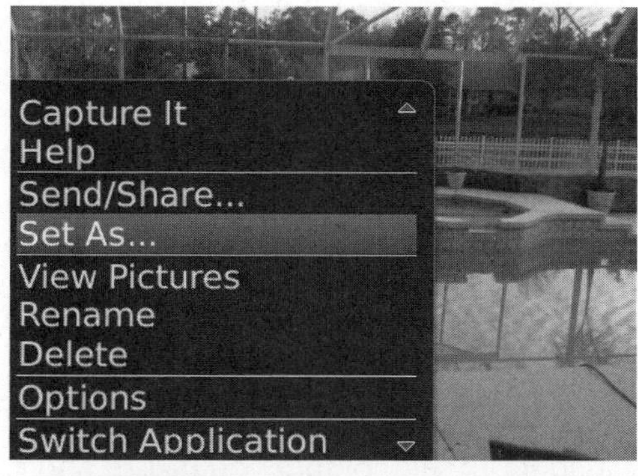

Sending Pictures with the Email Envelope Icon

There are several ways to send your pictures via email. The Camera application contains this handy Envelope icon that lets you email the picture on your screen quickly to one of your contacts.

Click the **Envelope** icon that brings up the **Send As** dialogue box.

Click one of these options:

- **MMS** – Multimedia message (as body of email); see page **335**.
- **Send to Messenger Contact** - This option may not be available if you have not yet setup BlackBerry Messenger; see page 345 to learn more.
- **Email** - Attach to an email as an image file.
- **Send to Facebook** - If you have this app installed.
- **Send to Flickr** -If you have this app installed.
- You may see additional options depending on which apps you have installed.

Navigate to the contact you desire and click the trackpad. Continue sending the message.

Fine-Tuning Your Camera Settings

You can fine tune your Curve camera to fit your needs with flash, zoom, white balance, picture quality and size.

Setting the Flash Mode and Zoom

One of the nice features of the BlackBerry camera is the inclusion of a flash. Just like with most digital cameras, you can adjust the properties of the flash (Figure 22-2).

Adjust Flash Mode: Tap the **Space** key.

Adjust Zoom: Glide the trackpad up/down or use the **Volume** keys on the right side.

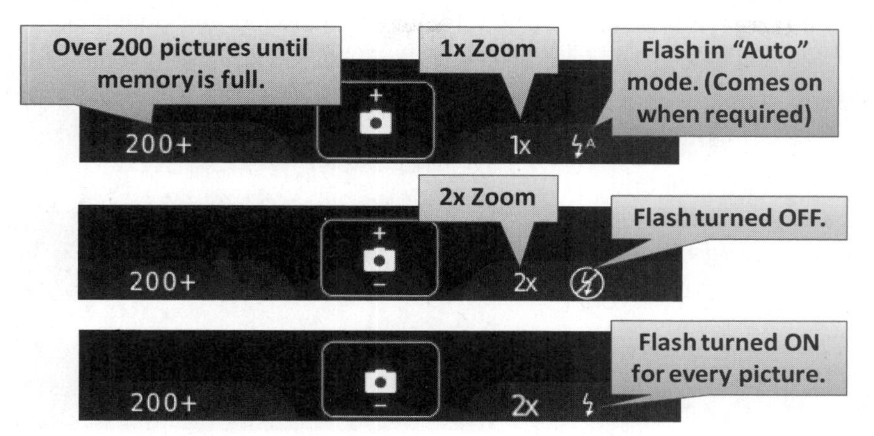

Figure 22-2. *Adjust flash and zoom settings*

With your subject framed in the camera window, press the **Space** key.

Look in the lower right hand corner of the screen to see the current flash mode displayed.

The three available flash modes are:

- **Automatic** indicated by the flash symbol with the A next to it.
- **On** indicated by the flash symbol (note that this mode will use more battery power).
- **Off** indicated by the no flash symbol.

Changing the Default Flash Settings

1. When in the **Camera**, press the **Menu** key and select **Options.**

2. Use the trackpad and highlight the **Default Flash Settings** in the upper right hand corner.
3. Select from **Off**, **On**, or **Automatic** (the default).
4. Press the **Menu** key and save your settings.

Adjusting the Size of the Picture

The size of your pictures corresponds to the number of pixels or dots used to render the image. If you tend to transfer your BlackBerry pictures to your desktop for printing or emailing, you might want a bigger or smaller picture to work with.

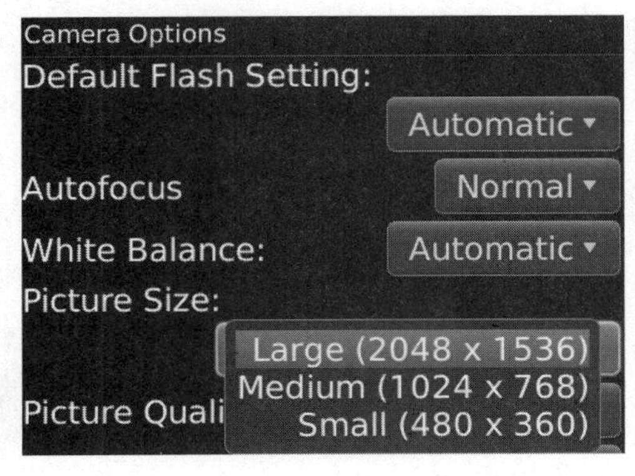

1. From the camera screen, press the **Menu** key and select **Options.**
2. Scroll down to **Picture Size** and select the size of your picture: **Small**, **Medium**, or **Large**.
3. Press the **Menu** key and save your settings.

TIP: Emailing your pictures will go faster if you set your picture size to Small.

Adjusting the White Balance

Usually the automatic white balance works fairly well. However, there may be times when you want to manually control it.

In this case, select an option from the White Balance menu (in the same Camera Options screen).

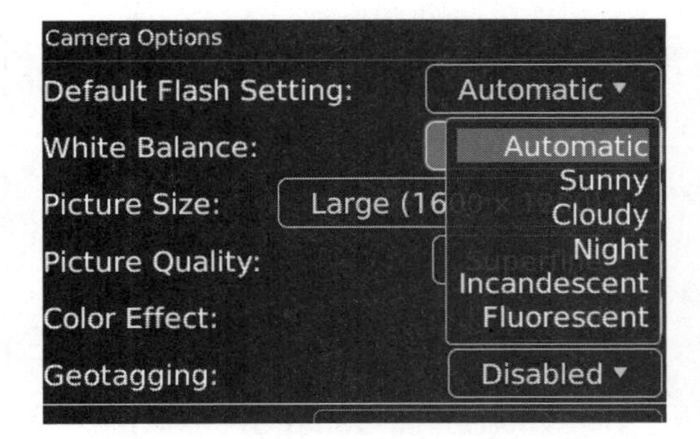

Adjusting the Picture Quality

There are times when you might want to change the picture quality. Fortunately, it is quite easy to adjust the quality of your photos. Realize, however, that **increasing the quality or the size will increase the memory requirements for that particular picture.**

In one non-scientific test, changing the picture quality resulted in the following changes to the file size of the picture at a fixed size setting of **Large (2048 × 1536)**:

- **Normal**: Image File Size: Approximately 80k (indoors picture)*
- **Fine**: Approximately 250% larger than Normal
- **Superfine**: Approximately 400% larger than Normal

NOTE: Pictures taken outdoors usually are larger in file size than those taken indoors.

1. Start your camera.
2. Press the **Menu** key, select **Options.**
3. Scroll to **Picture Quality**.
4. The choices **Normal**, **Fine**, and **Superfine** will be available. Click on the desired quality, press the **Menu** key, and save your settings.

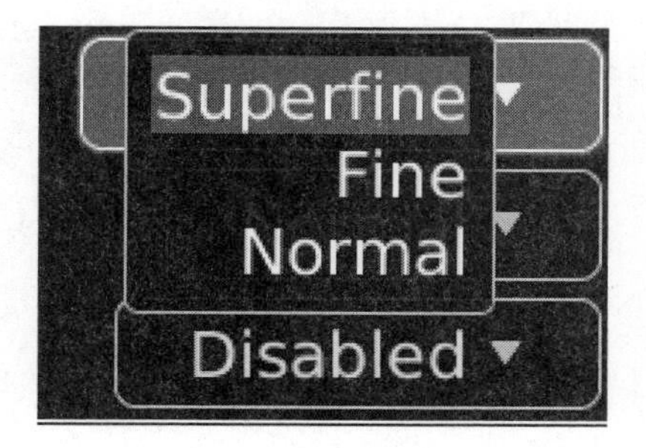

Geotagging: Adding GPS Location to New Pictures

What is **geotagging**? It is the assigning of the current GPS (Global Positioning System) longitude and latitude location (if available) to each picture taken on your BlackBerry camera.

TIP: To get **flickr** installed on your BlackBerry, go to `http://mobile.blackberry.com` from your browser and click on the **Social Networking** link.

Why would you want to enable geotagging?

Some online sites such as Flickr (photo sharing) and Google Earth (mapping) can put your geotagged photo on a map to show exactly where you took the picture.

Figure 22-3 shows a map from the `www.flickr.com` site showing what geotagging your photos can accomplish—essentially it will allow you to see exactly where you snapped your photos and help organize it.

Figure 22-3. *Geotagging a photo and displaying it on Flickr.*

Other programs can organize all your photos by showing their location on a map. To find such software, do a web search for "geotag photo software (Mac or Windows)".

To turn on or off geotagging, from the camera screen:

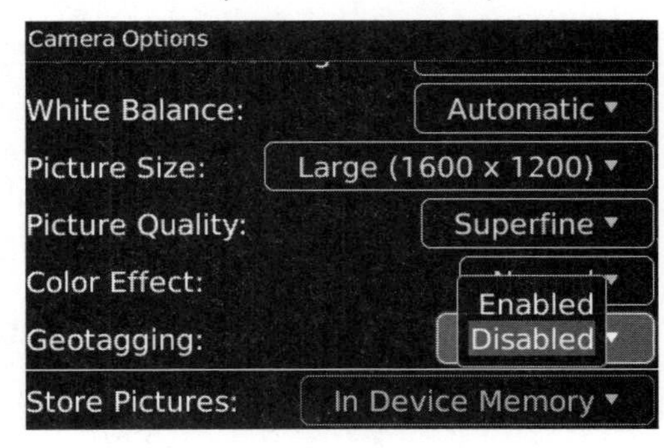

1. Press the **Menu** key and select **Options**.
2. Glide down to Geotagging and set it to **Enabled.**

3. When you see this warning message, make sure to check the **Don't ask this again** checkbox by clicking the trackpad or pressing the **Space** key.

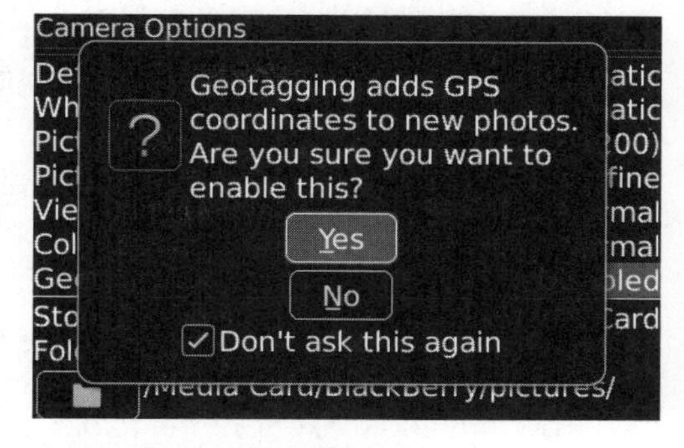

You know when geotagging is turned on if you see the white plus sign with circle around it and the three waves to the right of it in the lower right of your camera screen.

When geotagging is turned on, but your BlackBerry does not have a GPS signal to tag the pictures, you will see a red plus sign with circle around it and a red X in the lower right corner.

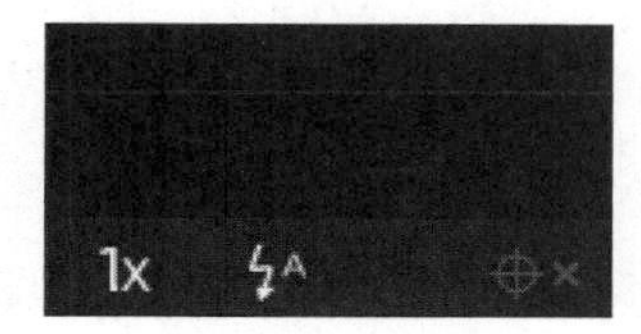

Managing Picture Storage

The authors strongly recommend buying a MicroSD media card for use in your BlackBerry. Prices are under US$20 for a 2GB card, which is very low compared to the price of your BlackBerry. For more information on inserting the media card, please see page 359. See below for help on storing pictures on the media card.

If you do not have a media card, then you will want to carefully manage the amount of your BlackBerry's main device memory that is used for pictures.

Selecting Where Pictures Are Stored

The default setting is for the BlackBerry to store pictures in Device Memory, but if you have a Media Card inserted, we recommend selecting that instead.

To confirm the default picture storage location:

1. Press the **Menu** key from the main camera screen and scroll to **Options** and click.
2. Scroll down to the **Store Pictures** and select **On Media Card**, if you have one, or **In Device Memory**, if you do not have a media card.
3. Look at the folder icon at the bottom and make sure the folder name ends in the word **/pictures**. This will help keep pictures together with pictures, videos with videos, music with music, and make it easier when you want to transfer pictures to and from your computer.

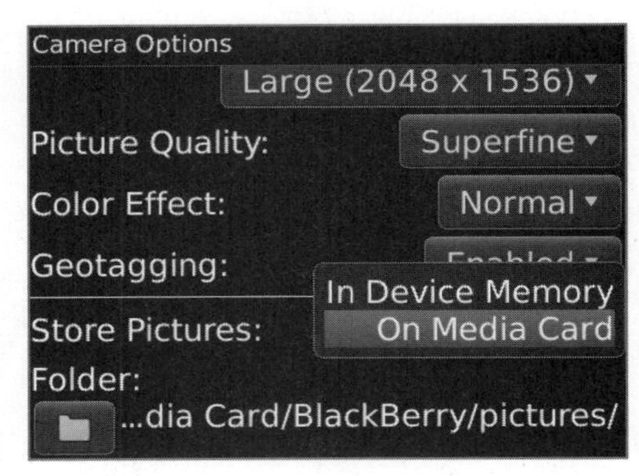

Using the Optional Media Card

At publishing time, your BlackBerry can support a 32GB sizable Micro SD media card. To give you some perspective, a 4GB card can store many times that of the BlackBerry internal device memory. This is equivalent to **several full-length feature films** and **thousands of songs**. Learn how to install a media card on page 359. Since program files can only be stored in main memory, we recommend putting as many of your media files on the Micro SD card as possible.

Viewing Pictures Stored in Memory (Media Player)

There are two primary ways to view stored pictures.

Option 1: Viewing from the Camera Program

1. Open up the Camera application and press the **Menu** key.
2. Scroll down to **View Pictures** (see Figure 22-4) and navigate to the appropriate folder to view your pictures.

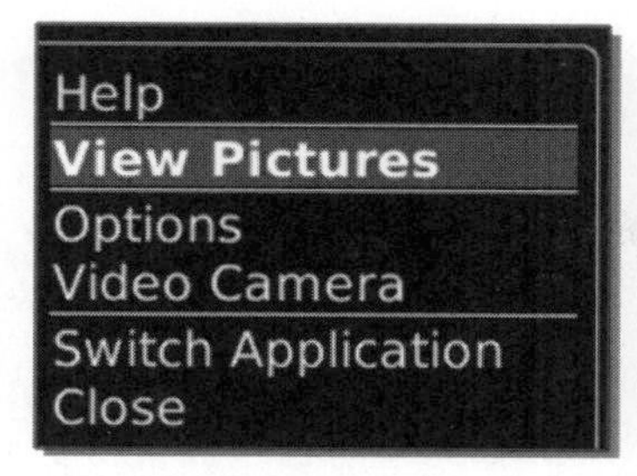

Figure 22-4. *Viewing pictures from the pictures folder*

Option 2: Viewing from the Media Player Menu

1. Navigate to the **Media** icon and click.
2. Scroll to **Pictures** and click. Your initial options will be **All Pictures**, or **Picture Folders** (Figure 22-5).
3. Click the appropriate folder and navigate to your pictures.

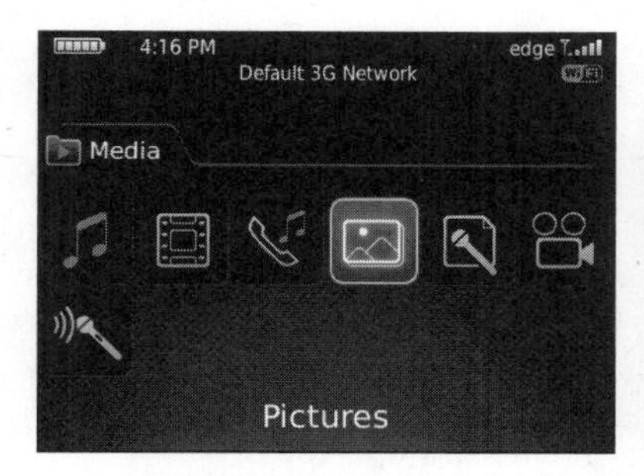

Figure 22-5. *Navigating to the picture folders*

To View a Slide Show

Follow the steps above and press the **Menu** key when you are in your picture directory.

Scroll to **View Slideshow** and click.

Take a Picture and Set as Caller ID or Home Screen Wallpaper

As discussed previously, you can assign a picture as a Caller ID for your contacts.

1. If you want to take a picture of someone and use it right away as caller ID, press the **Crop** button on the bottom of the camera screen:

2. Then select **Caller ID** or **Wallpaper.**

3. If the picture is the wrong size (too small/big) for the frame showing in the middle, then click the trackpad or **Menu** key and select **Zoom** to change the size.

4. You can also re-center the image by rolling the trackpad. Once you get it centered, click the trackpad and select **Crop and Save.**
5. If you were setting the **Wallpaper**, you are done.

6. If you selected **Caller ID**, then you select the contact in your Address Book (Contact list) to whom you are assigning this picture.
7. Type a few letters to find the contact and click the trackpad to select the person.

Find: Ma Trau
Trautschold (Florida), Martin
Trautschold, Martin
Trautschold, Martin
Trautschold, Martin
Cmt publications PP
Trautschold, Martin
Handheld Contact

Now, every time the person calls or you call them, you will see their picture in addition to their name on the phone screen.

Transferring Pictures to or from Your BlackBerry

There are a few ways to get pictures you have taken on your BlackBerry off it and to transfer pictures taken elsewhere onto the BlackBerry.

Method 1: Using Send/Share Command (Individually)

1. Take a picture and click on the **Envelope** or **Send/Share...** icon (see page 385).
2. Select where you want to send or share the picture. Depending on which apps you have installed you may see Facebook, Messenger Contact, Fliker, Email, MMS and more.
3. After choosing the method, follow the onscreen prompts to complete the send or share command.

You may also send pictures when you are viewing them in your **Pictures** application.

1. From **Pictures**, highlight the picture you want to send.
2. Press the **Menu** key and select **Send/Share...**
3. Follow the steps to send or share the image.

Method 2: Using Bluetooth

If you want to transfer pictures to/from your computer (assuming it has Bluetooth capabilities), you can. We explain exactly how to get this done in the Bluetooth chapter on page 441.

Method 3: Using Computer Software

Transferring pictures and other media to your computer is handled using the Media section of your desktop software. You would use BlackBerry Desktop Manager or BlackBerry Media Sync. For Windows users, see page 105. Mac users, see page 141.

Method 4: Using Mass Storage Mode

This method assumes you have stored your pictures on a media card. The first time you connected your BlackBerry to your computer, you probably saw a **Turn on Mass Storage Mode?** question.

> **NOTE:** If your BlackBerry was supplied to you by your workplace, this **Mass Storage Mode** feature may be disabled for security reasons by your BlackBerry Administrator. If this feature does not appear to work, contact your help desk for assistance.

If you answered **Yes**, then your media card looks just like another hard disk to your computer (just like a USB Flash Drive). Then you can drag and drop pictures to/from your BlackBerry and your computer. For more details, see page 144.

Chapter 23

Fun with Videos

We have already taught you how to get the most of music and pictures on your BlackBerry, but your BlackBerry can handle even more media that what we have discussed so far. Your BlackBerry can also handle videos—from short videos you shoot on the BlackBerry to watching a full length movie.

Having Fun with Videos

Your BlackBerry comes with a built-in video recorder to catch your world in full motion video and sound for those times when a simple picture just does not work. Your **Media** icon also can play all videos you record or transfer to your BlackBerry from your computer.

With an 8GB or 16GB media card, the media capabilities of the BlackBerry are greatly enhanced – you can store many videos and thousands of songs.

Adding Videos to Your BlackBerry

If you are a Windows user, check out the built-in media transfer and sync capabilities to BlackBerry Desktop Manger software on page 105. If you use a Mac, please refer to page 125.

Using Your Video Recorder

The video recorder is perfect for capturing parts of a business presentation or your child's soccer game. Videos can be emailed or stored on your PC for later use, just like pictures.

1. Push the **Menu** key and scroll down to the **Media** folder and click.
2. Glide to the **Video Camera** icon and click.

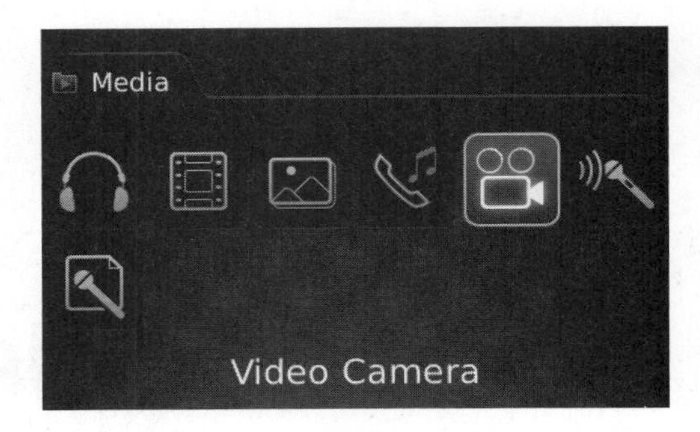

TIP: You can also start the camera by pressing the **Menu** key and selecting **Video Camera.**

3. If this is your first time using the video camera after installing a media card, you should be asked **Always save videos to media card?** We recommend answering **Yes** to this option.

4. Use the screen of the BlackBerry as your viewfinder to frame your video.
5. When you are ready to record, press the trackpad and the light in the center changes to a **pause** button as shown.

6. When you want to stop recording, click the trackpad again to see this screen. To continue recording, click the white circle in the left corner.
7. To stop recording, glide over and click on the white **Stop** square or press the **Escape** key.
8. To play what you just recorded, glide over and click on the **Play** triangle button.
9. To change the name of the video or location where it is saved, press the **File** icon.
10. To delete the video, click the **Trash Can** icon.

(Yes, this is Martin's fingertip.)

To send the video via email or MMS (if available), click the Envelope icon.

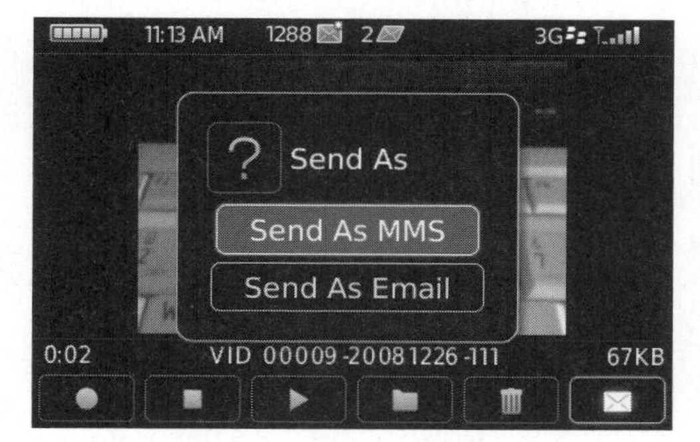

Changing Your Video Recording Options

From inside the video camera, press the **Menu** key and select **Options**.

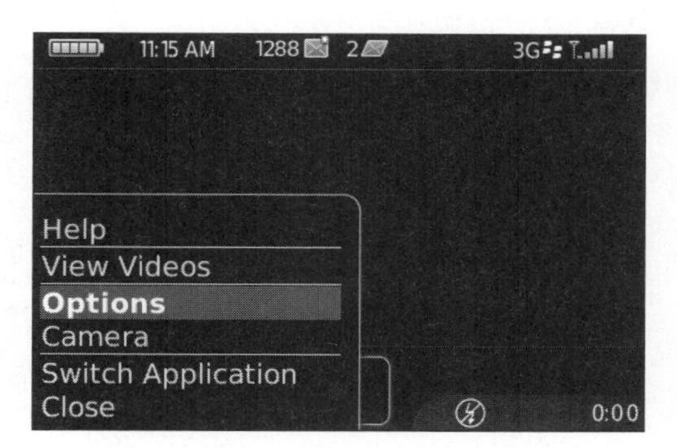

You can adjust the **Video Light**—either have a constant light from the Flash On or turn it Off.

You can also adjust the **Color Effect** choosing:

- **Normal** (Color)
- **Black and White**
- **Sepia** (Sort of a brownish old fashioned coloring)

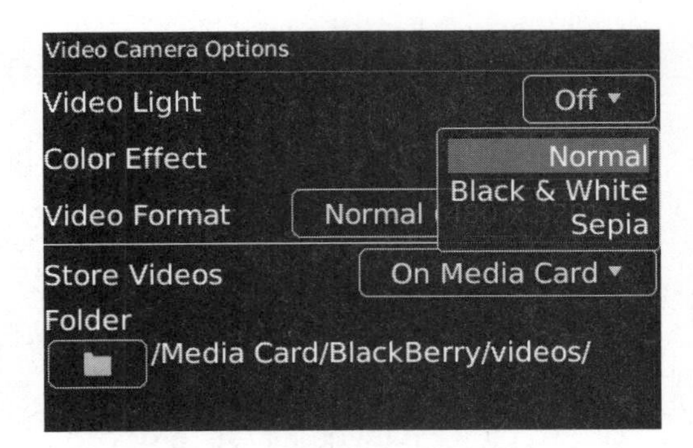

You can also adjust the **Video Format** for **Normal** video storage or **MMS Mode** for wireless transfer.

> **NOTE:** MMS mode is a much smaller screen and file size—suitable for sending wirelessly—but it will be of lower quality than Normal mode.

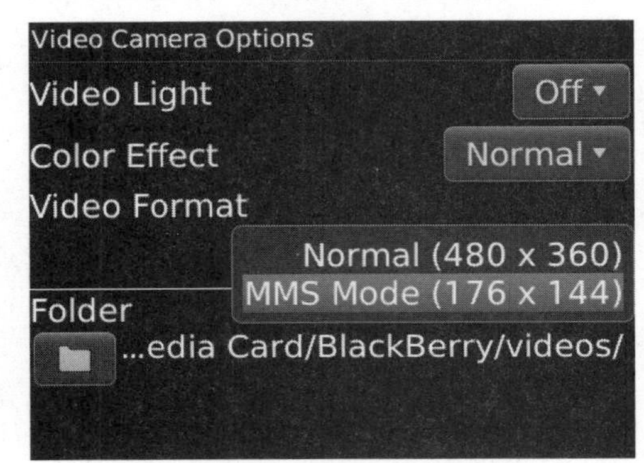

To choose where your videos are stored, click on the **Folder** icon and navigate to a new folder. Remember, we recommend storing your videos on your Media Card. This way you don't fill up your main BlackBerry device memory.

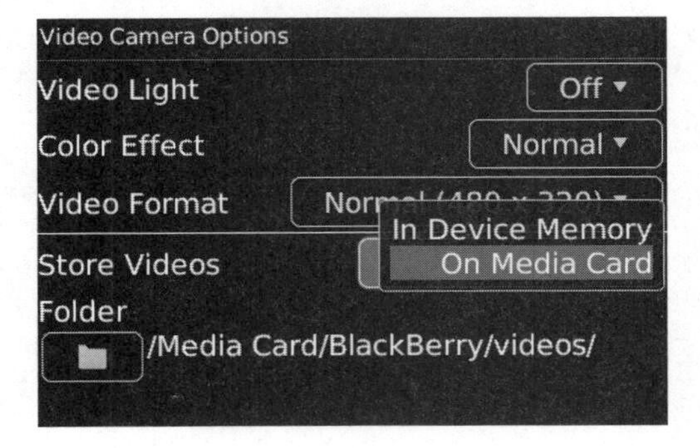

Supported Video Formats for the BlackBerry Bold Series Smartphone

Your BlackBerry can play many different types of video files. Check Table 23-1 for compatibility.

Table 23-1. *Supported Formats (Source:* *`www.blackberry.com`*)

File Format / Extension	Component	Codec	Notes	RTSP Streaming
MP4 M4A MOV 3GP	Video	H.264	Baseline Profile, 480×360 pixels, up to 1500 kbps, 24 frames per second	Supported
		MPEG4	Simple and Advance Simple Profile, 480×360 pixels, up to 1500 kbps, 24 frames per second	Supported
		H.263	Profile 0 and 3, Level 45	Supported
	Audio	AAC-LC, AAC+, eAAC+		Supported
		AMR-NB		Supported
AVI	Video	MPEG4	Simple and Advance Simple Profile, 480×360 pixels, up to 1500 kbps, 24 frames per second	Supported
	Audio	MP3		
ASF WMV WMA	Video	Windows Media Video 9	WMV3, Simple and Main Profile, 480×360 pixels, 24 frames per second	
	Audio	Windows Media Audio 9 Standard/Professional		
		Windows Media 10 Standard/Professional		
MP3	Audio	MP3		

Recommended video format for local playback

File Format / Extension	Component	Codec	Notes
MP4	Video	MPEG4	Advance Simple Profile, 480×360 pixels, up to 1500 kbps, 24 frames per second
	Audio	AAC-LC	

*** We recommend searching the online BlackBerry Knowledgebase for the most up-to-date listing. Go to `http://na.blackberry.com/eng/devices/blackberrybold9700/bold_specifications.jsp`

Viewing Videos on the BlackBerry

The BlackBerry contains a very sharp screen that is perfect for watching short videos. The video player is as easy to use as the audio player.

Playing a Video

1. Click on your **Media** folder.
2. Click on the **Video** icon and choose the folder where your video is stored.
3. The video player screen looks very similar to the Audio Player screen. Click the trackpad to pause or play and use the volume controls on the side of the BlackBerry.

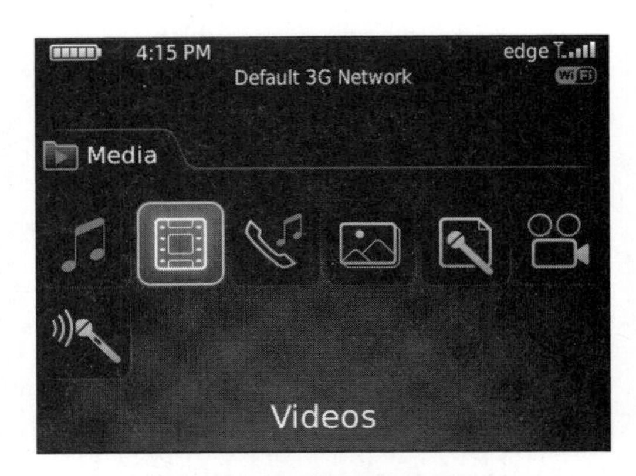

4. Just glide up and down to click on a video or you can type a few letters to quickly locate the video you want to watch. For example, typing the word **Address** would locate all videos with the word **Address** in the title.

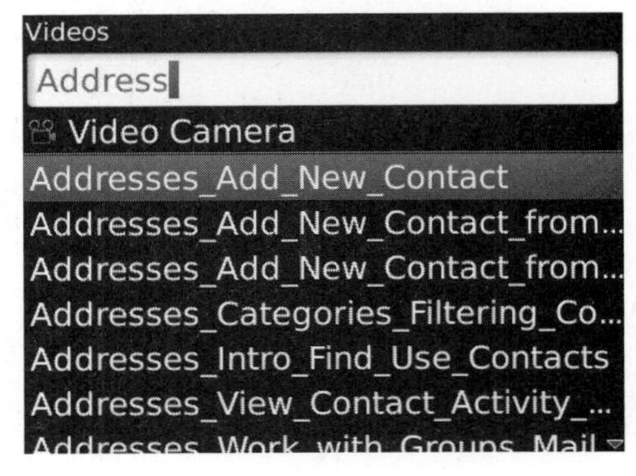

5. Click on the video to start playing it.

> **TIP:** Press the **Menu** key and select **Full Screen** to see the video without the top and bottom status bars.

6. Click the trackpad again or the **Mute** key on the top of the BlackBerry to pause the video.

You see the following menu items when you press the Menu key in your Video Player from List View:

- **Help** – Displays contextual help with Video Player
- **Now Playing**- Switches back to the currently playing song or video
- **Media Home** – Displays the Media folder
- **Play** – Plays selected item
- **Delete** – Erases selected item
- **Rename** – Changes the name of the selected item
- **Properties** – Displays properties such as location, file size, date created, and modified date.

Activate Handset
Activate XPLOD
Switch Application
Close

> **NOTE:** You see the **Activate XPLOD** option above because the BlackBerry used for the screen shot has an **XPLOD** Bluetooth device attached to it. You would see your Bluetooth devices (headsets) listed here, instead.

Learn more about Bluetooth in our chapter on it on page 441.

Other options you may see when you are viewing **Songs**:

- **Replay** – Replays the last viewed video
- **Repeat** – Repeats the last video
- **Show Playlist** – displays all items in the Playlist
- **Activate Handset** – Plays the audio of the video through your headset
- **Switch Application** – Allows you to go to any other open application
- **Close** – Closes the application

TIP: Select **Full Screen** and turn the phone sideways to watch a full screen video in Wide Screen Mode.

Chapter 24

Social Networking

BlackBerry smartphones a ren't j ust f or b usiness e xecutives a nymore—but you already know that. Your BlackBerry can keep you in touch in many ways beyond the messaging features, email, and the Web.

Some of the most popular places to connect these days social networking sites, places that allow you to create your own page and connect with friends and family to see what is going in their lives. Some of the most popular websites for social networking are Facebook, Twitter, and LinkedIn.

In this chapter, we will show you how to access these various sites. You will learn how to update your status, tweet, and keep track of those who of interest to you.

Downloading the Apps

To find social networking apps, type the name of the app(e.g. Facebook, LinkedIn or Twitter) in the **Search** feature in App World.

Type **Facebook**, **MySpace**, **Flickr LinkedIn**, or **Twitter**to quickly find each of these apps.

Follow the steps we show you in Chapter 26: "BlackBerry App World" to download and install each app.

At the time of this writing, BlackBerry had just released its own Twitter client. There are also more than half a dozen pretty good Twitter clients, and many of them are free. Try out UberTwitter or one of the others.

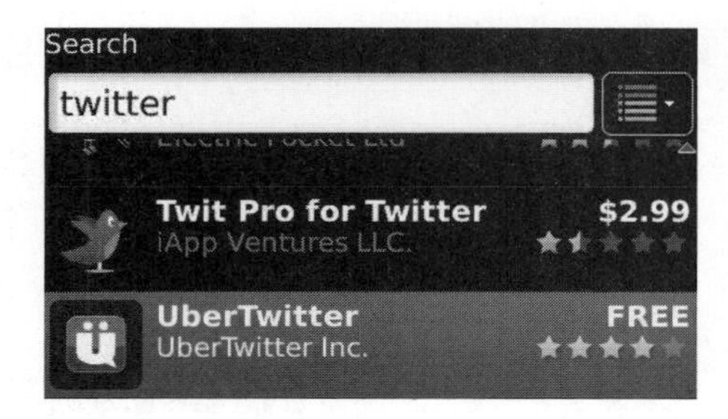

Logging into the Apps

You can log into the apps as follows:

To connect to your account on Facebook, Flickr, or MySpace, you will need click the icon you just installed. We use the example of Facebook here, but the process is very similar for the rest of the Apps.

Once Facebook is successfully downloaded, the icon in your **Downloads** folder should look something like this.

> **NOTE:** You might find your **Facebook** icon in your **Social Networking** folder, depending on your carrier.

There is a lot to like about having Facebook and other social networking apps on your Blackberry. You can always stay in touch. Just log in as you do on your computer, and you are ready to go—anytime, anywhere, from your BlackBerry.

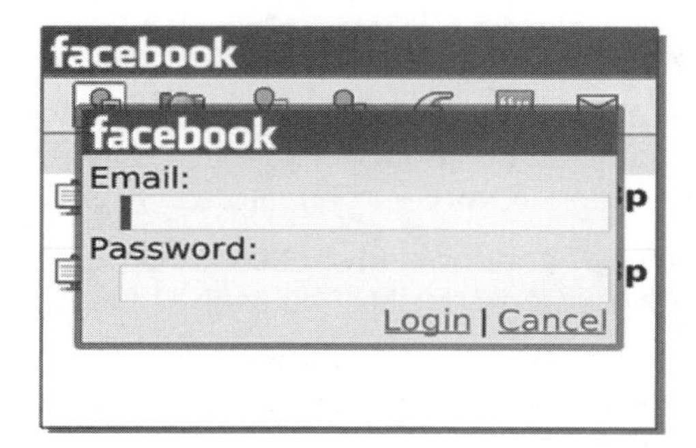

Facebook

Since Facebook was founded in February of 2004, it has served as the premier site for users to connect, reconnect, and share information with friends, coworkers and family. Today, over 400 million people use Facebook as their primary source of catching up with the people who matter to them.

Using Facebook for BlackBerry

Now that Facebook is a full stand-alone app for BlackBerry, you can do pretty much everything for Facebook account right on your BlackBerry.

Once Facebook is downloaded and installed, the first thing you will see is the login screen. Type your account information, that is, your email address and password.

Account Integration

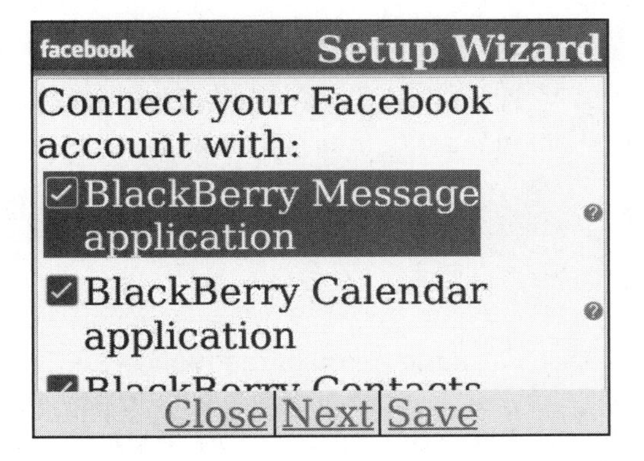

The next screens help you set up your Facebook account on your BlackBerry. You can integrate your Facebook friends with your BlackBerry contacts, as well as integrate any Facebook calendar events with your BlackBerry calendar. Just check the boxes for actions you wish to take.

You can also have every Facebook message sent to you, and any status update, Poke, or other action by your friends can show up as a Facebook message in your BlackBerry Message application.

Status Update and News Feed

Once you log on to Facebook for BlackBerry, you will have the option to write "What's on your mind" and see your News Feed from your friends.

Top Bar Icons

Along the top bar are icons for some of the most frequently used features of Facebook: **News Feed**, **Notifications**, **Upload a Photo**, **Friends**, **Add a Friend**, **Write on a Wall**, or **Send a Message**.

Communicating with Friends

Follow these steps to connect with your Facebook friends:

1. Click the **Friends** icon to display you list of friends. If you have contact information in your BlackBerry address book or phone numbers stored in your BlackBerry, the icons next to your friends' names will show that.

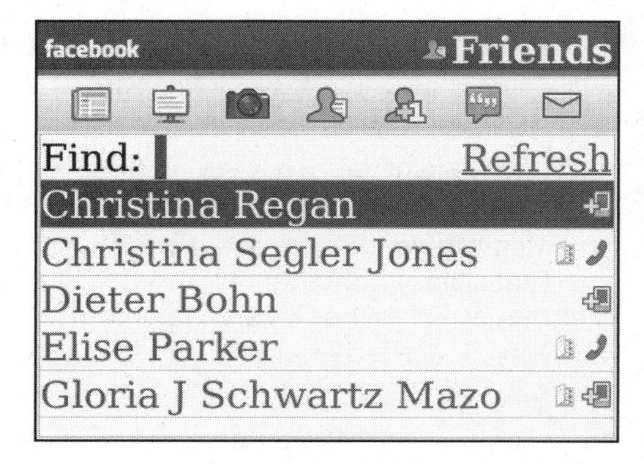

2. Click the name of the friend you wish to communicate with to see all the options available to you. In this example, Gloria Mazo's name was clicked, and now, we can Poke her, see her profile, write on her wall, or see her BlackBerry contact information.

Uploading Pictures

An easy and fun thing to do with Facebook is to upload pictures. You can upload pictures by either clicking the small **Camera** icon in the **Facebook** app. You can also upload any picture right after you take it or from your Pictures Application.

1. Just locate the picture in your photo album, and press the Menu key. Select **Upload to facebook**, and this screen will pop up allowing you to add a caption for the photo and select an album for the picture. We chose a picture of our book *CrackBerry: True Tales of BlackBerry Use and Abuse*, wrote a caption, and clicked **Upload**.

2. Now, from the Facebook account page, we click the **Menu** button, select **My Albums**, and navigate to the **Mobile Uploads** album to view the picture.

MySpace and Flickr

You will find the process of downloading and installing the icons for MySpace, Flickr, and other social networking sites to be virtually identical to the Facebook process described in the previous section. Go to `http://mobile.blackberry.com` on your BlackBerry browser for all the core BlackBerry applications, including these.

Twitter

Twitter, started in 2006, is essentially an SMS (text message)–based social networking site. It is often referred to as a micro-blogging site, where the famous and not so famous share what's on their minds. The catch is that you only have 140 characters to get your point across.

With Twitter, you subscribe to follow someone who tweets messages, and others can follow you. If you want to follow us, we are `@garymadesimple` on Twitter.

Making a Twitter Account

Making a Twitter account is very easy. Here's how:

1. We recommend that you first establish your account on the Twitter web site from your computer at `www.twitter.com`.

2. Click the **Join Today** button to see the account creation page shown to the right.

3. Type in the requested information to establish your account, you will be asked to choose a unique user name (we use `@garymadesimple`), email address, and a password.

4. You will then be sent an email confirmation. Click the link in your email to go back to the Twitter web site. You can choose people to follow or make tweets on the web site, as well as read tweets from your friends.

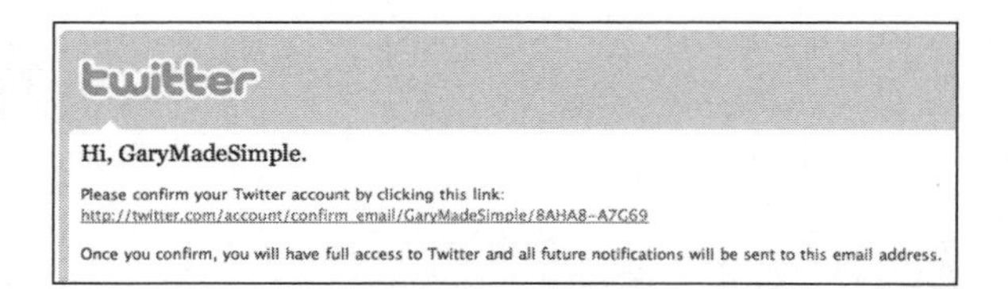

Download the Twitter App

There is a proprietary Twitter app for the Blackberry, just as there was for Facebook, and the best way to find it is to search BlackBerry App World for "Twitter."

Once App World lets you know that the installation is complete, you can click **OK** to close or **Run** to start the app.

NOTE: The Twitter app is often listed in the Featured Items at the top of the App World.

Using Twitter for BlackBerry

Here's how to log into and use the Twitter app:

1. The first time you use Twitter for BlackBerry, you will be asked to log into your twitter account with your twitter username and password.
2. Next, click the Login button, and the BlackBerry will sign into your Twitter account.

3. The initial view in Twitter has a small box at the top of the screen where you can tweet you message to those who are following you. Under that are the tweets from those you are following. Just scroll through the list of tweets to catch up on all the important news of the day!

Tweeting

At the very top of the Twitter home screen is the "What's happening?" box. That is your space to tweet (in 140 characters or less).

Type your message and click the **Update** button to do so.

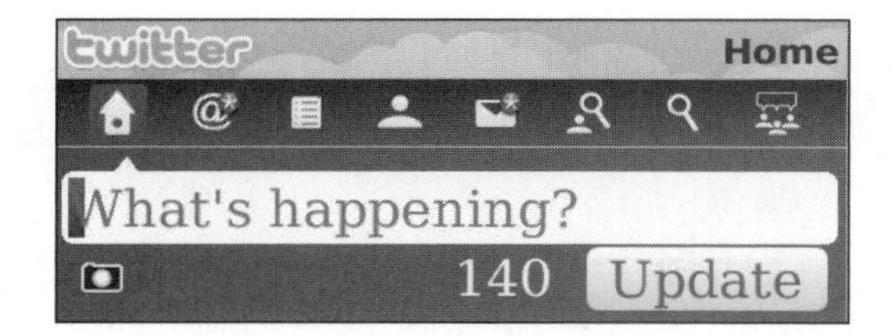

Twitter Icons

Along the top of the Twitter screen are eight icons shown in Figure 24-1; these are the quick links to the basic Twitter functions.

Figure 24-1. *Twitter icons at the top of the screen*

Home

The top left icon is your **Home** icon which takes you to the Twitter home screen where you can write your tweets.

Mentions

A mention, in Twitter speak, is a tweet that mentions your `@username` somewhere in the body. These are collected and stored in the **Mentions** section of your Twitter account.

My Lists

You can create a custom list so that your followers cannot only follow everything you say but can choose to follow a specific topic (a list) that you create. In this example, we created a list entitled Made Simple Announcements, so users can choose to only follow our special announcements.

My Profile

Just as it sounds, the **My Profile** icon will take you to your Twitter profile. You can see your followers, your tweets, and your biography. Just click any field in your profile, and a short menu comes us. To edit your information, just choose **Edit Profile** from the menu.

Direct Messages

Direct messages are messages between you and another Twitter user. Click the **Direct Messages** icon to see yours.

1. To compose a direct message, just click the **Compose Message** icon, fill in the to field with the Twitter user name, and type your 140-character message in the box.
2. When you are finished, click the **Send** button.

Find People

The Find People icon takes you to a search window where you can type in a user name, business name or last name to search for someone who might be on Twitter.

1. Type the name, and your query will be processed.
2. Scroll through the results to see if you find what you are looking for. If so, just click the entry that matches your search, and scroll down.

You will be able to see the statistics of how many followers you have and if they have particular lists to follow. You can also select **Follow** or **Block** to either follow this contact on Twitter or block all tweets.

Search

The **Search** icon takes you to another search window. Here, you can type any keyword to find opportunities to follow a topic or individual. For example, if you were to type in the word "golf," you might find popular golf courses, golfers, or driving ranges that use Twitter. You also might find thousands of individuals who just want to brag about their golf scores.

TIP: Be very specific in your search to help narrow down the search results.

Popular Topics

The **Popular Topics** icon will simply list those topics that are currently popular on Twitter. If scroll through, you might find everything from the current playoff series to the President's visit overseas—on Twitter, someone always has something to say about everything.

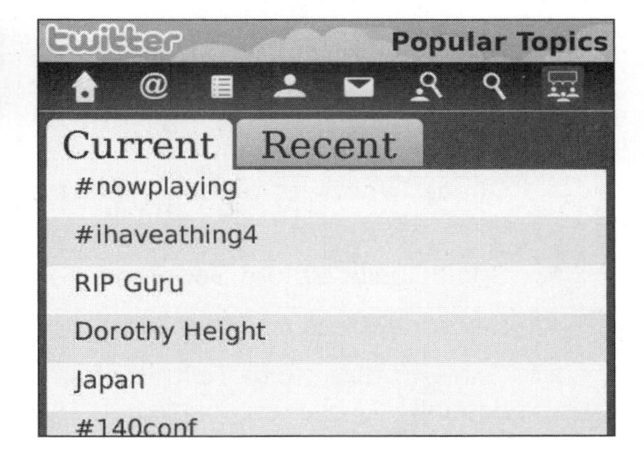

Customizing Twitter with Options

From your Twitter home screen, click the **Menu** button followed by **Options**. You will notice several check boxes for selecting options.

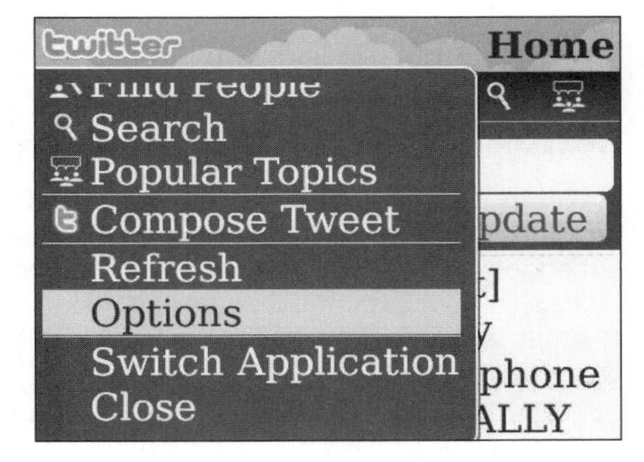

The available **Twitter** customization options follow:

- **Tweet refresh in background** (checked by default)
  - **Refresh tweets every** (5 min., 10 min – default, 15 min., 30 min, 1 hr.)
  - **Notify on new tweets** (unchecked by default)
  - **Notify on replies and mentions** (checked by default)
- **Number of tweets per refresh** (20 – default, 40, 60, 80, 100)
- **Integrate in Messages** (email) application (checked by default)
- **Show navigation bar** (checked by default)
- **Show tweet box on home screen** (checked by default)

- **Use system font settings** (checked by default) – uncheck to set your own fonts for Twitter
  - Font family
  - Font size (only available if you uncheck the **Use system font settings**)
- **Check tweets for personal information before sending** (checked by default)
- **Spell Check Before Sending** (checked by default)
- **Distance measured in** mi/km
- **Connect directly to Twitter when Wi-Fi is enabled** (checked by default)
- **Content cache – Clear** button (click to clear)
- **Version 1.0.0.40 – Check for Update** button (click to see if there is a newer version of the app available.)

Just check the box next to any option you would like to enable.

LinkedIn

LinkedIn has very similar core functionality to **Facebook**, but tends to be more business and career focused whereas Facebook is more personal friends and game focused. With **LinkedIn** you can connect and re-connect with past business associates, send messages, see what people are up to, have discussions and more.

The LinkedIn app is available as a free download from the **BlackBerry App World**.

Downloading the App

Once you find the LinkedIn app, choose to download it to your BlackBerry. Then, go to your **Downloads** folder to launch it.

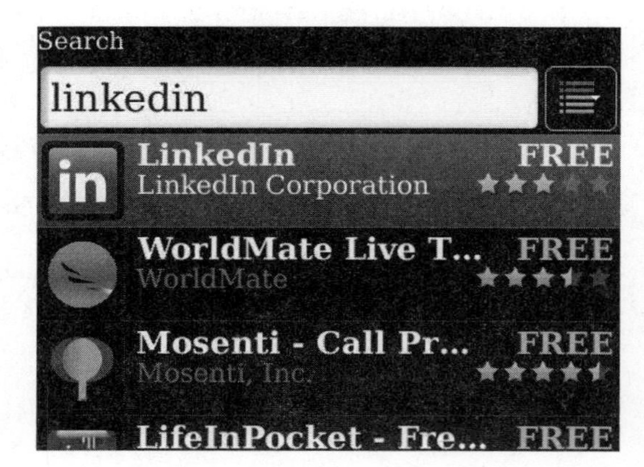

Logging In to LinkedIn

Once the App is installed, click the **LinkedIn** icon, and enter your login information.

> **NOTE**: We do recommend first setting up your LinkedIn account on your computer prior to installing the **LinkedIn** app as it is easier to type more quickly on your computer as you set up your LindedIn profile and add your picture.

Navigating the LinkedIn App

LinkedIn has a similar icon-based navigation as Facebook. The top of the home screen has a box for posting your profile update, which similar to a status update on Facebook.

Just type your message, and click the **Post** button.

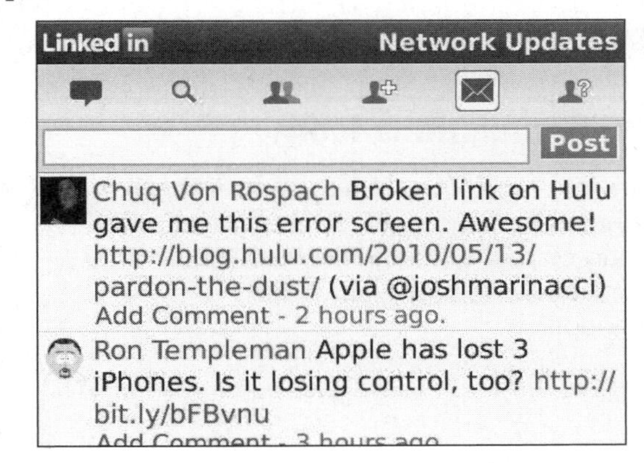

The LinkedIn Icons

There are six icons at the top of the LinkedIn home screen (Home Page, Search, Connections, Invitation, Messages, and Reconnect), each of which is described in the following sections.

The Home Page Icon

The **Home** Page icon will return you to the page with profile updates where you can update your own LinkedIn status.

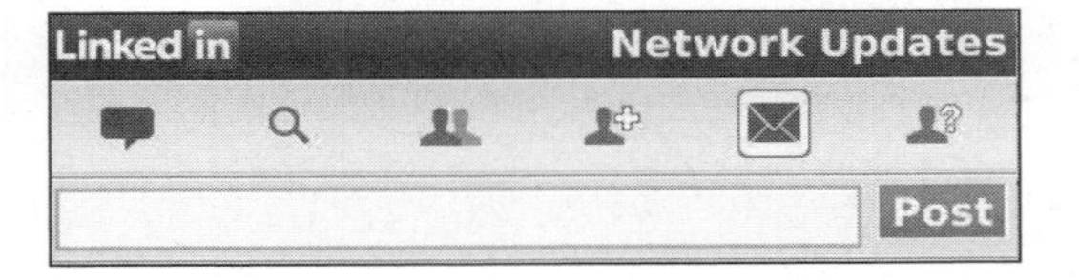

The Search Icon

Click the **Search** icon and type in the search box. **LinkedIn** will try to match your search with contacts in your **LinkedIn** directory, listed as 1st level connections.

Next, the search field will show individuals who match the search criteria who may be part of your network.

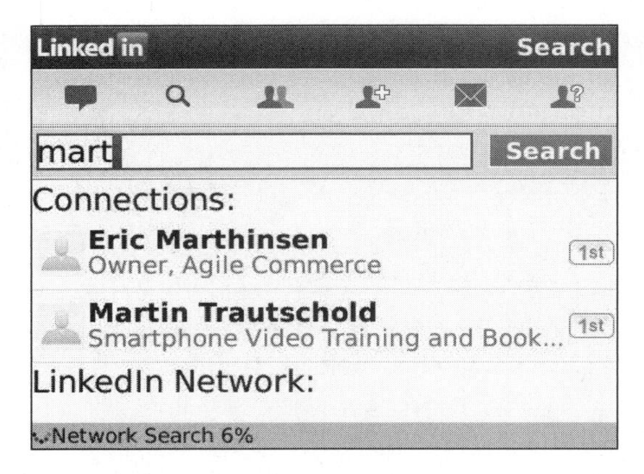

NOTE: In the LinkedIn network system, someone who is a direct contact is considered a first-degree connection in your network. Someone who is connected to one of your contacts (but not you) is considered a second-degree contact, and so on.

The Connections Icon

The next icon along the top is the **Connections** icon, which shows you a listing of your LinkedIn connections. Scroll down to find a particular connection.

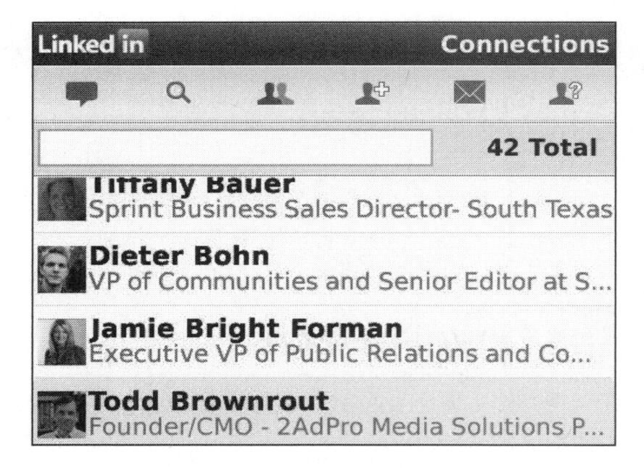

TIP: Press the **Menu** key on your BlackBerry when a connection is highlighted, scroll down, and click **Link to BlackBerry Contact** in the menu. This will attach all contact information (including the LinkedIn picture) to your BlackBerry contact for that individual.

The Invitation Icon

Click the **Invitations** icon to see if you have any pending invitations for connections. You can also press the **Menu** key and select **Compose New Invitation** to open a new invitation:

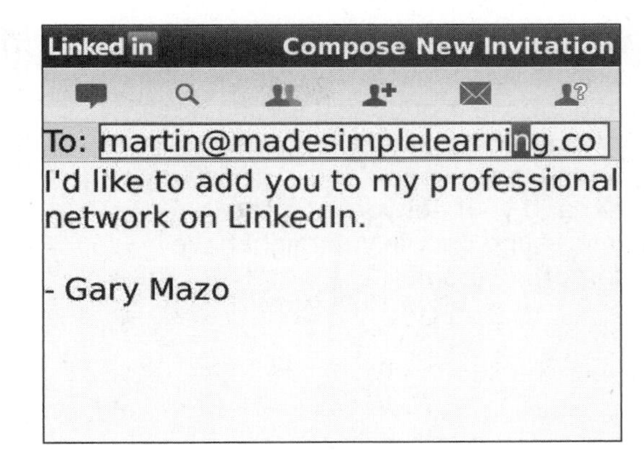

1. Type the email address of the individual you would like to add to your **LinkedIn** network.
2. Edit the text below the email field.
3. Press the **Menu** key, and select **Send**.

NOTE: Unlike the LinkedIn desktop application, you don't have to say how you know the individual with whom you would like to connect. This makes sending an invitation from the mobile application easier.

The Messages Icon

Click **Messages** icon for direct messages with your contacts. You can read messages that are sent to you here. You can also press the **Menu** key and select **Compose Message** to send a direct message to any of your LinkedIn contacts.

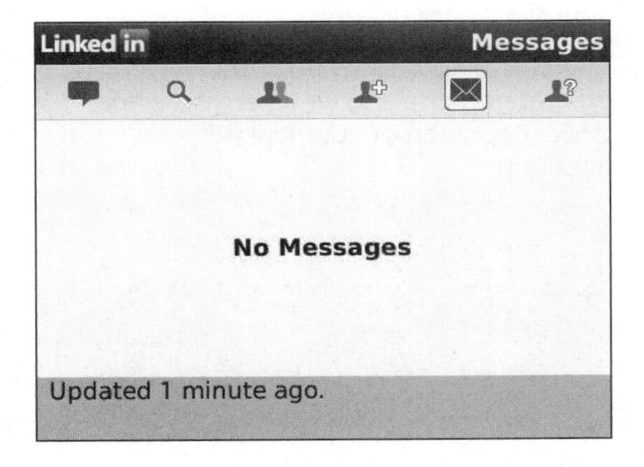

The Reconnect Icon

Click the **Reconnect** icon to see people you might know, people who might be in groups you belong to, or people who might have common connections to you. It is a good way to expand your network and see interesting people you might have forgotten about.

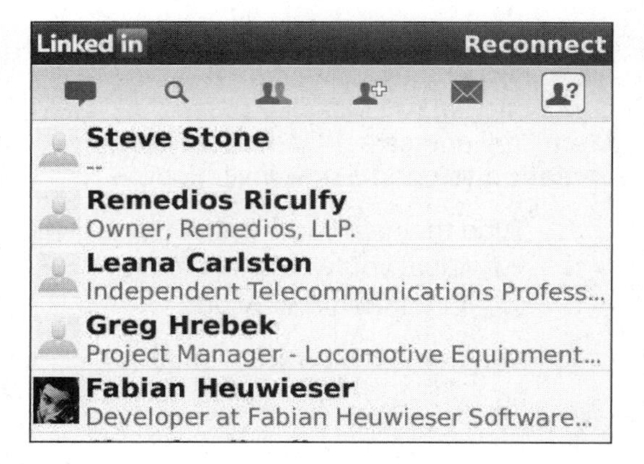

LinkedIn Options

Press the **Menu** key from any section of the LinkedIn app, and click **Options.** You will notice a selection of radio boxes that you can check or uncheck depending on your needs and desires.

Select **BlackBerry Mail** to have LinkedIn messages displayed in your BlackBerry message list. Select **BlackBerry Contacts** to have your LinkedIn contacts displayed as part of your **Contacts** directory.

The bottom boxes let you specify your type of network and whether you want BES integration of your LinkedIn account.

Gary Mazo's Options
BlackBerry Mail
BlackBerry Contacts
LinkedIn Network Updates
Show Connection Updates
Network Settings
Current Selection : BIS
Auto
BES (Not Available)
BIS
TCP

NOTE: Some options like BES (BlackBerry Enterprise Server), Wi-Fi, and WAP (Wireless Access Protocol – a type of communications protocol used over wireless networkds) are only available if your BlackBerry is connected to those servers or types of networks.

YouTube

One of the most fun sites to visit is YouTube, which shows short video clips on just about everything. Your new BlackBerry is able to view YouTube videos without doing anything special.

Just navigate to the YouTube website (m.youtube.com), and it will detect that you are on a mobile device.

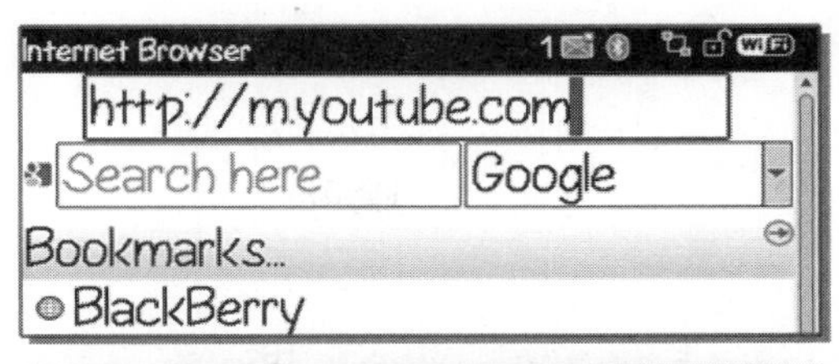

Search for videos just like you do on your computer. The video will load in your media player, and you can control it just like any other video.

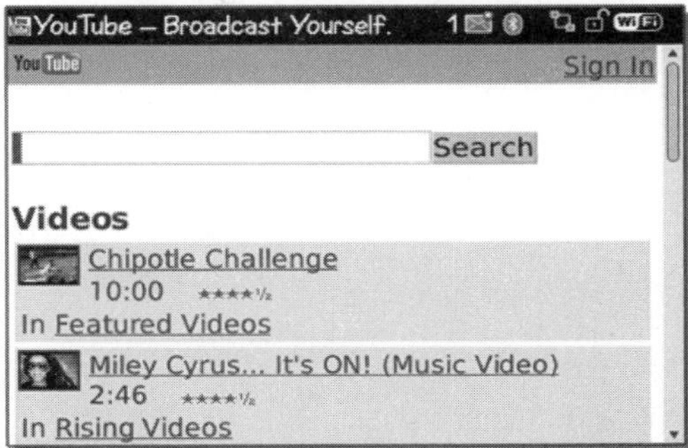

Chapter 25

Connecting with Wi-Fi

We live in a Wi-Fi world today. It is difficult to go anywhere and not hear about Wi-Fi. "Wi-Fi," according to some, stands for Wireless Fidelity (IEEE 802.11 wireless networking). Others say that the term is a wireless technology brand owned by the Wi-Fi Alliance.

In this chapter, we will show you how to connect your BlackBerry to available Wi-Fi networks, how to prioritize and organize your networks, and how to diagnose wireless problems.

Understanding Wi-Fi on Your BlackBerry

Your Internet signal can be transmitted wirelessly (using a wireless router) to computers, game consoles, printers, and now to your BlackBerry. If your BlackBerry is Wi-Fi equipped, you can take advantage of must faster web browsing and file downloading speeds via your home or office wireless network. You can also access millions of Wi-Fi hotspots in all sorts of places like coffee shops and hotels, some of which are free!

The Wi-Fi Advantage

Wi-Fi is a great advantage to BlackBerry users around the globe. The advantages to using a Wi-Fi connection as opposed to the carrier data connection (GPRS/EDGE/3G) are many:

- Web browsing speeds are much faster.
- You are not using up data from your data plan
- Most file downloads will be faster.
- You can get great Wi-Fi signals sometimes when you cannot get any regular cell coverage (GPRS/EDGE/3G), for example, on the bottom floors of a thick-walled building.

Setting Up Wi-Fi on Your BlackBerry

Before you can take advantage of the speed and convenience of using Wi-Fi on your BlackBerry, you will need to set up and configure your wireless connection.

1. Press the **Menu** key to see all your icons.
2. Scroll down, and click the **Set Up** or **Settings** folder.
3. There are two ways to get into Wi-Fi setup:
   a. (a) If you see a **Setup Wi-Fi** icon, click on it.
   b. (b) Otherwise, click on the **Manage Connections** icon, then click **Set Up Wi-Fi Network**.

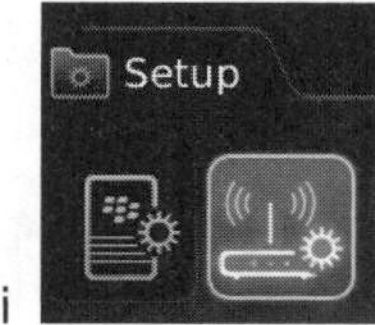

Setup Wi-Fi

Manage Connections

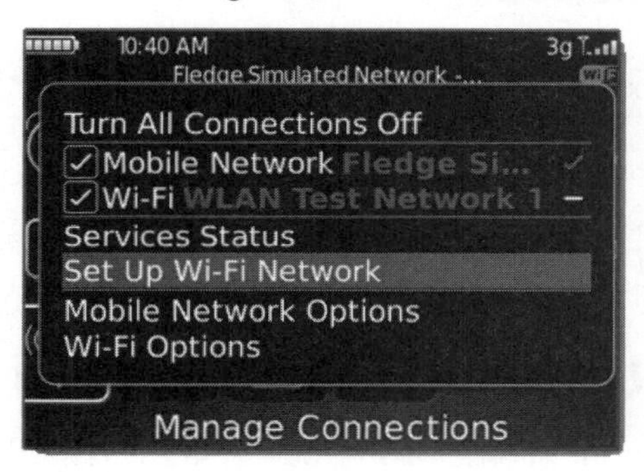

4. The "Welcome to Wi-Fi Setup screen" appears. Scroll down to read the introduction to Wi-Fi. When you are finished, click **Next**. (See Figure 25-1)

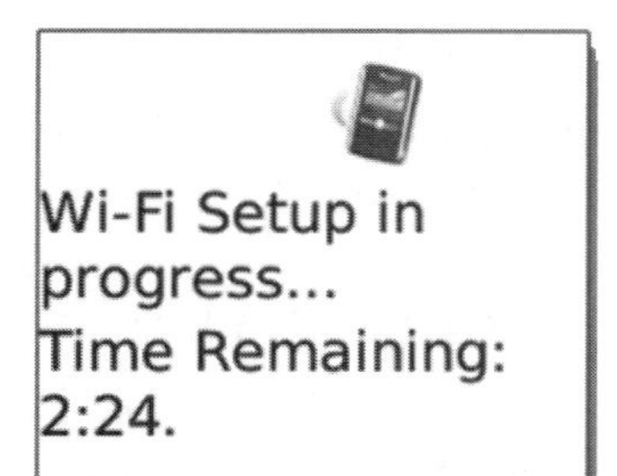

Figure 25-1. *Wi-Fi setup screen.*

5. You have two options available. We suggest first selecting **Scan for Networks**. In setting up our own BlackBerry smartphones, we've never had to choose **Manually Add Networks**. You should see the screens shown in Figure 25-2.

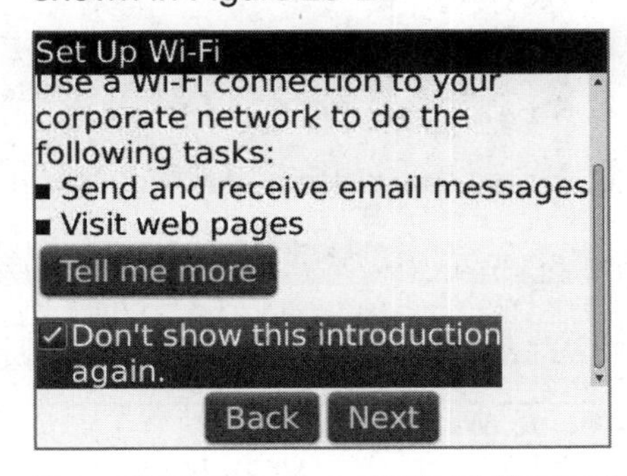

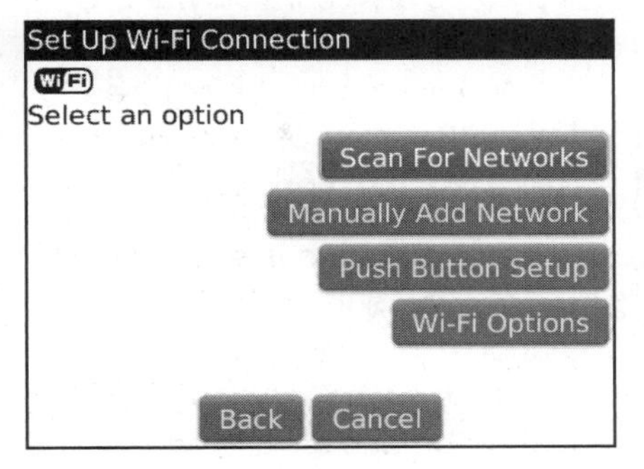

Figure 25-2. *Scan for available Wi-Fi networks.*

6. If Wi-Fi is not yet turned on, you will be prompted to do so; click the **Turn on Wi-Fi** option.

7. Next, you will see a status message telling you that your BlackBerry will automatically scan for available Wi-Fi networks.

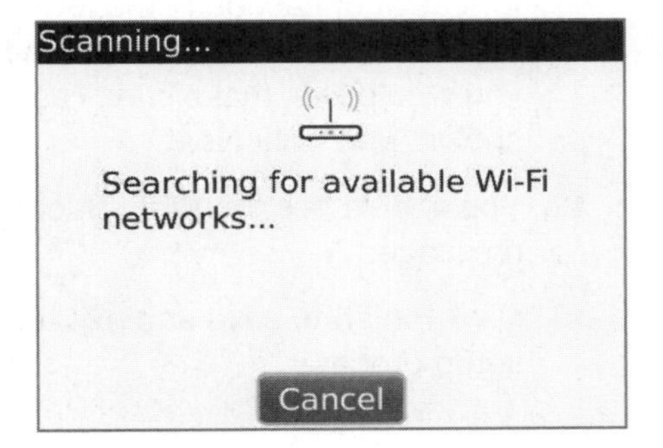

8. You will then see a list of available networks. When you see the correct network name, highlight it and click the trackpad.

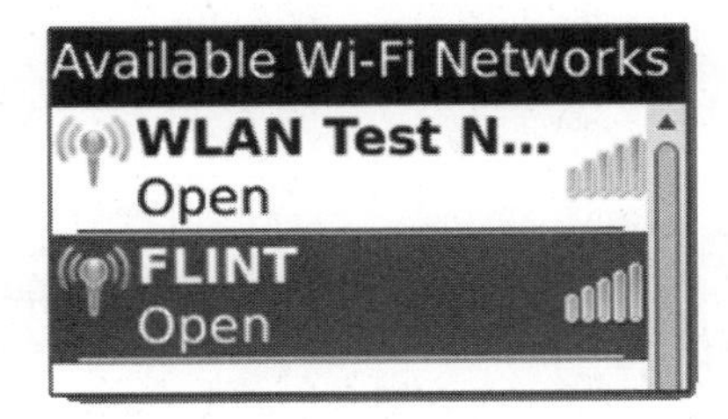

9. If your network uses a preshared key (PSK), you must type it at the prompt. Then click **Connect**. You might also be asked if you want to perform a WPS setup. (WPS is a type of Wi-Fi security.) Select **Yes** and follow the steps to connect to the WPS Wi-Fi network.

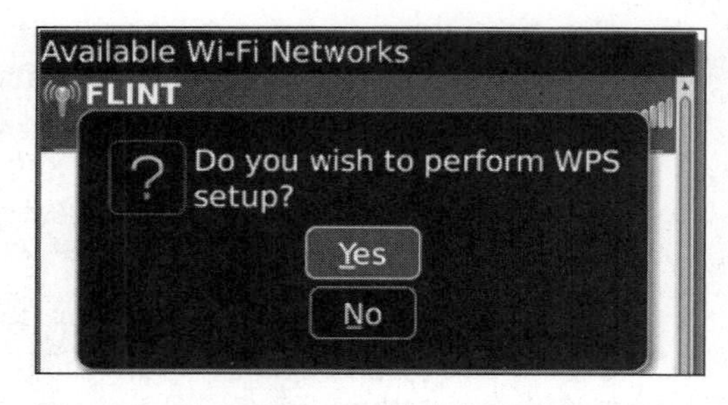

10. When the connection is successful you will see the screen to the right. If you choose **Yes**, your Wi-Fi profile will be saved, so you can connect automatically to this Wi-Fi network in the future.

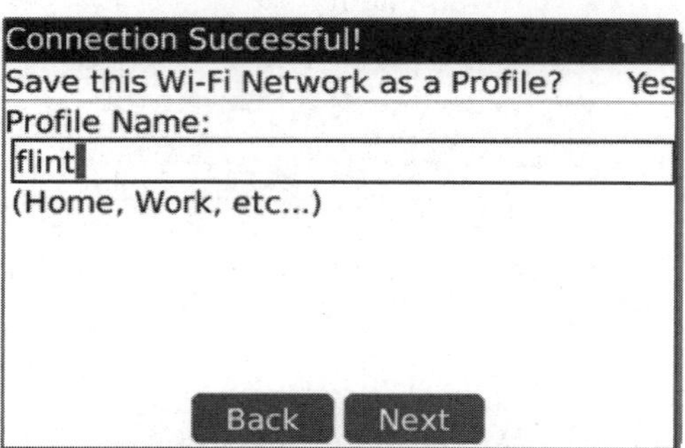

11. You will then be prompted to save your Wi-Fi connection as a profile for easy connection in the future. Type or adjust the **Profile Name** if you would like, make sure **Yes** is shown, and click **Next**.

12. You should see the Wi-Fi Success! message.

13. Click **Finish** to save and exit the setup process.

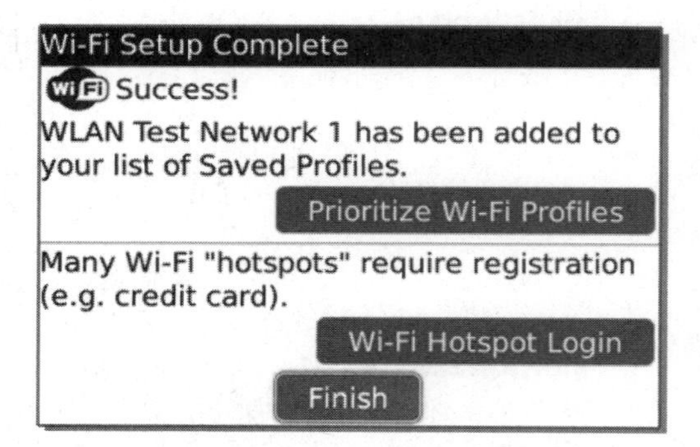

14. Once connected, you will see the Wi-Fi logo in the top right-hand corner of your screen.

Single Profile Scanning

New in version 5.0 of the BlackBerry operating system is the ability to give your Wi-Fi–enabled BlackBerry a single profile to scan or prompt you for a manual login.

A manual connection prompt is more secure, because your BlackBerry will never just connect to an available network (which could pose a security risk.)

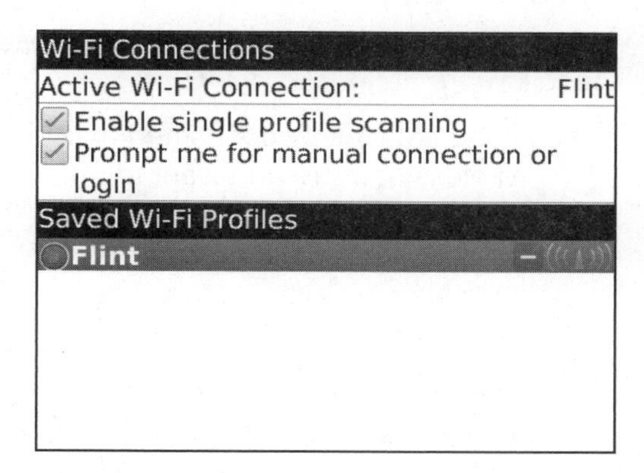

Checking **Enable single profile scanning** coupled and selecting a radio button in the **Saved WiFi Profiles** will allow your BlackBerry to connect automatically to preferred networks you select.

Connecting to a Wireless Hotspot

Connecting your BlackBerry to a wireless hotspot is also straightforward:

1. Navigate to your Wi-Fi setup screen, and press the **Menu** key.
2. Click the **Wi-Fi Hotspot Login** option shown in Figure 25-3.

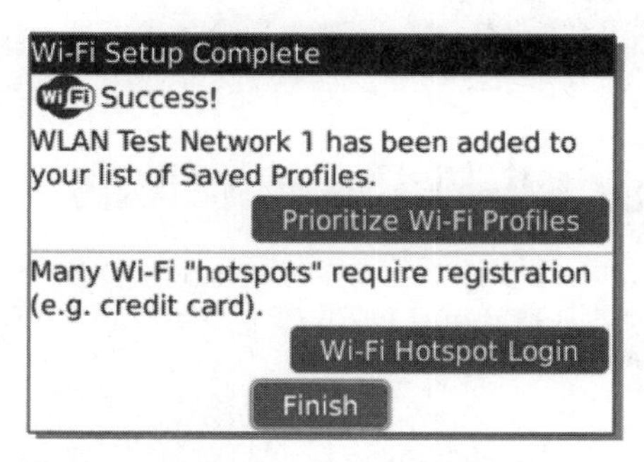

Figure 25-3. *Using Wi-Fi Hotspot login.*

3. You should be then taken to the logon screen for that particular hotspot. Type the required information, and a "You are Successfully Connected to the Internet" message should be displayed.

Changing your Wi-Fi Network

If you are already connected to one wireless network and want to change networks, do the following.

1. If you are already connected to one wireless network and want to change networks, from your **Home** Screen, pressing the letter **O** if you have hotkeys enabled (see page 541), or click the Options icon.
2. Press the letter **W** to jump down to **Wi-Fi**, and click it to see this screen.

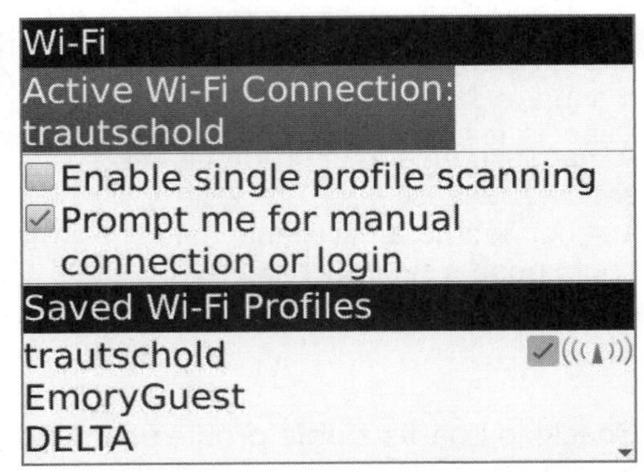

3. You will see the Active Wi-Fi connection highlighted.
4. Press the **Menu** key and select **New**.
5. Follow the instructions in the "Setting up Your Wi-Fi Connection" section.

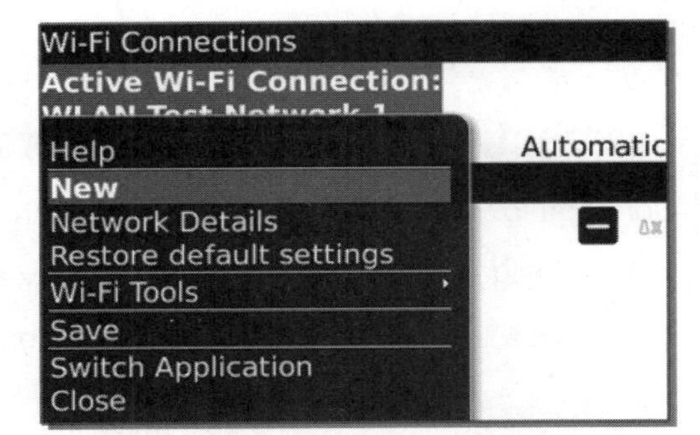

Changing or Prioritizing Your Wi-Fi Connections

One of the nice things about Wi-Fi is that you can connect to a wireless network just about anywhere. While you may have saved your home and work networks, there may be times you want to prioritize or change networks altogether.

Prioritizing Your Networks

Let's say that 90% of the time, the wireless network you connect to is your home network. You want to make sure that your home network is at the top of the list of networks your BlackBerry searches for.

To prioritize your networks, just do the following:

1. Follow the steps in the previous section to get into your Wi-Fi settings.
2. Scroll down to **Saved Wi-Fi Profiles** at the bottom of the screen, and highlight the top network.
3. Press the **Menu** key, and select **Move**.
4. The blue highlighted network becomes grey. Just scroll with the trackpad, and move the network into the position you desire.
5. Click the trackpad again to complete the move.

Using Wi-Fi Diagnostics

At times, your Wi-Fi connection may not seem to work for you. Thankfully, your BlackBerry has a very powerful diagnostics program built in to help you in those instances.

1. To launch the Wi-Fi Diagnostics, follow the steps in the previous section to get into your Wi-Fi settings, and again scroll to **Saved Wi-Fi Profiles**. This time, select the troublesome network.
2. Push the **Menu** button, click **Wi-Fi Tools** and then **Wi-Fi Diagnostics**.

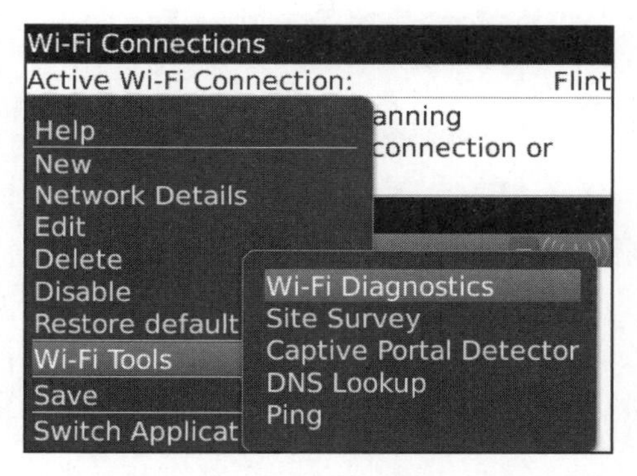

3. The available wireless networks and detailed connection information will now be displayed. If you are an advanced user with experience in wireless networking, click the Trackpad for access point details.
4. After the diagnosis, you can press the **Menu** key, and select **Email Report** to copy the details into an email and send it to your help desk for assistance.

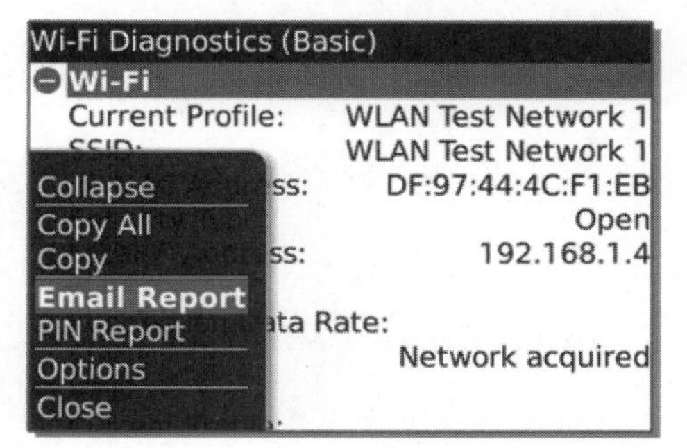

Chapter 26

BlackBerry App World

You can find software applications inside the App World, but since the App World is a relatively new addition to the BlackBerry environment, you will also find many applications outside of App World. Check out chapter 30 called "Adding and Removing Software" starting on page 475 for more information.

The App World Concept

Application stores are all the rage in the world of smartphones these days. One of the great things about a BlackBerry is that you can find applications in lots of places, not just the sanctioned application store. That said, BlackBerry App World is a new concept for BlackBerry and deserves some explanation. Remember that you can always go to the other locations mentioned on Page 475 to find additional applications for your BlackBerry.

Downloading the App World Program

BlackBerry App World is a free download for all BlackBerry users, and it works particularly well on the new BlackBerry Bold. If the App World icon was not already on your BlackBerry when you received it, you will need to download and install it.

1. Start your web browser by clicking the **Browser** icon or pressing the letter **B** if you enabled the home screen hotkeys (see page 541).
2. In the address bar of the browser, just type **mobile.blackberry.com**. The home page of your browser may already be set to the mobile BlackBerry site.

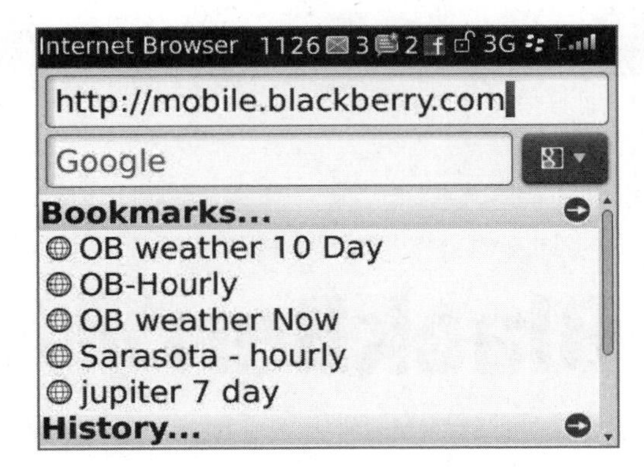

Locating the Link for App World

At the top (or bottom) of the Mobile BlackBerry page, you should see and link for the BlackBerry App World. Just glide to the link, and click the trackpad. You will then be taken to the Download page for the application. Accept any terms and conditions, and click the **Download** button shown in Figure 26-1.

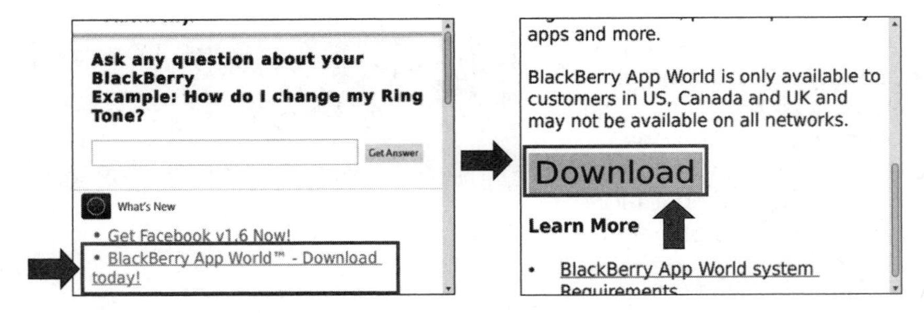

Figure 26-1. *Download BlackBerry App World.*

Starting App World for the First Time

The **App World** icon will probably be in your **Downloads** folder. If you want to move it to your home screen, just follow the directions on Page 169.

Find the **App World** icon, and click the Trackpad to start it. The first time you start the App World, it might take a little while to load; this is normal.

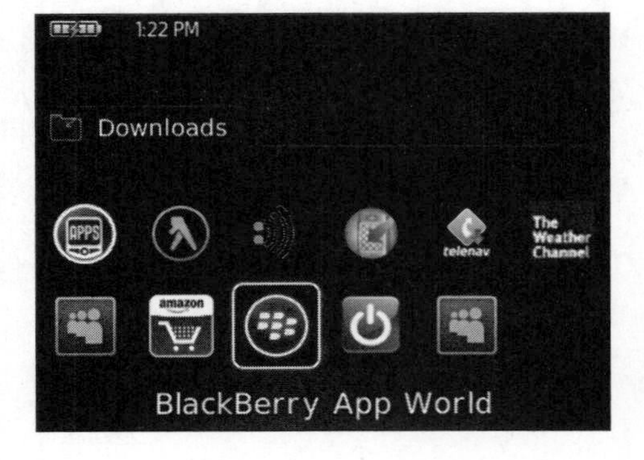

Downloading Themes from App World

We offer detailed instructions in our "Personalize Your BlackBerry" chapter on page 169.

Featured Programs

Highlighted in the App World program are large icons for the **Featured Programs.** Many of these are free; others need to be purchased (which we will discuss in the next section). Glide the trackpad through the featured programs to find one that interests you. To gain more information about the program, click the trackpad to be taken to the next screen.

Navigating around App World

Along the bottom of the App World screen you will see five small icons: **Categories**, **Top Free**, **Top Paid**, **Search**, and **My World**. Each gives you a different way to look for, download, or manage apps on your BlackBerry. To activate these soft keys, you need to glide the trackpad down.

Figure 26-2. *Layout of BlackBerry App World.*

Categories

Just like it sounds, the **Categories** icon gives you all the categories of applications in the new BlackBerry App World. At the time of this writing, there were over a dozen categories ranging from games and sports to finance and reference and much more.

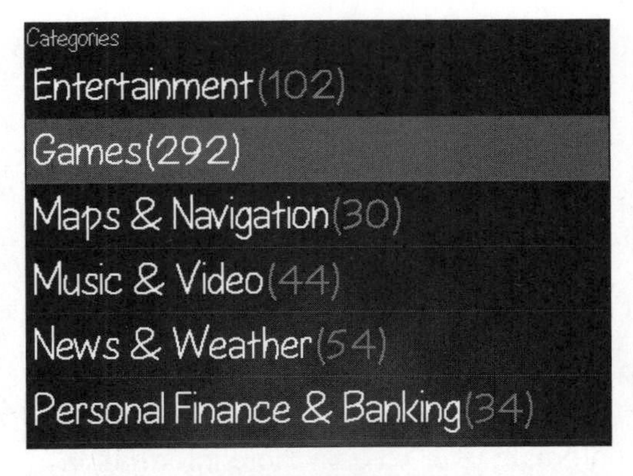

Just click a category to find more information about a particular application that may interest you. Once you click a category, the icons for the available programs will show on your screen. Glide your trackpad through the options and click. You can usually read reviews or see screen shots for most applications before you decide to download.

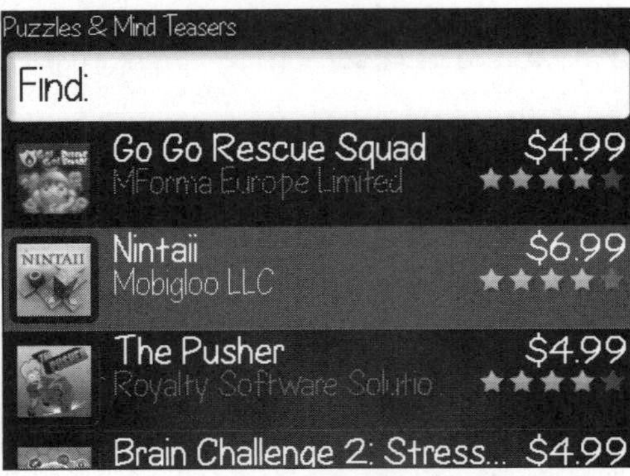

Top Free Downloads

As the name implies, the **Top Free Downloads** icon will show you the most frequently downloaded free applications from the App World. As you did previously, just click any program to see screen shots, read reviews, or download the App to your BlackBerry.

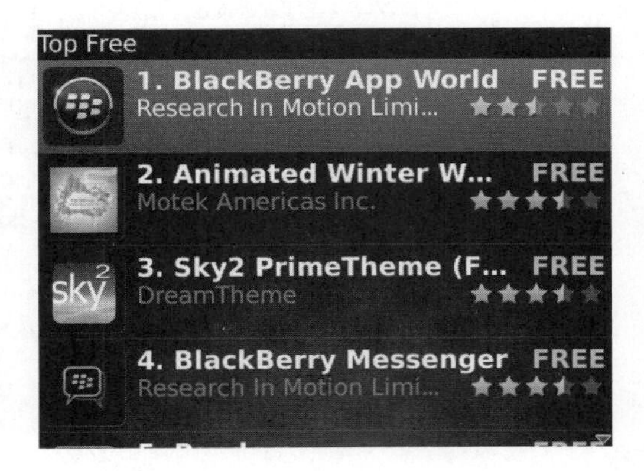

Top Paid Downloads

Top Paid Downloads

The **Top Paid Downloads** icon will show you the most often downloaded paid applications (which $0.99 or more each) from the App World. Here too, just click any program to see screen shots, read reviews, or download the app to your BlackBerry.

Search

The App World has a very good built-in Search tool . If you have an idea of what you might be looking for but don't know the exact name, just type a word into the search bar. In this example, we want to find music applications that are currently available, so we type **music** to see all the available music apps (see Figure 26-3).

Figure 26-3. *Search BlackBerry App World.*

You can sort the search results by clicking the **Sort By** icon in the upper-right corner (see Figure 26-4).

Figure 26-4. *Sort search results.*

After you select sort criteria, you can reverse the sort order (highest to lowest, A to Z, or Z to A) by clicking the same icon again.

Downloading Apps

At the time of this writing, the BlackBerry App World only accepts PayPal for purchase (this may change in the future). If you have a PayPal account, all you need to do is input your PayPal user name and password when you purchase an application. If you do not have a PayPal account, go to `www.paypal.com` from your computer and follow the instructions to set up your account.

Downloading and Purchasing an App

Follow these steps to download and purchase apps with your PayPal account:

1. Scroll through the applications as you until you find an app that you want to download. Click the **Download** button from the **Details** screen.

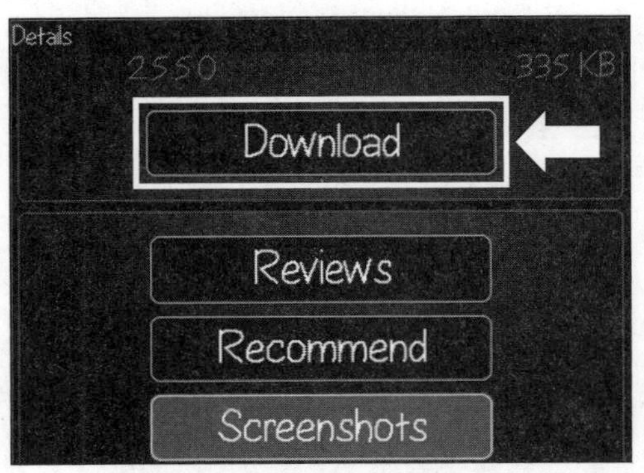

2. If the application is not a free download, the **Download** button will be replaced by a **Purchase** button. Just click it, and input your PayPal information, as shown in Figure 26-5.

NOTE: Your PayPal information will be stored to make purchases easier.

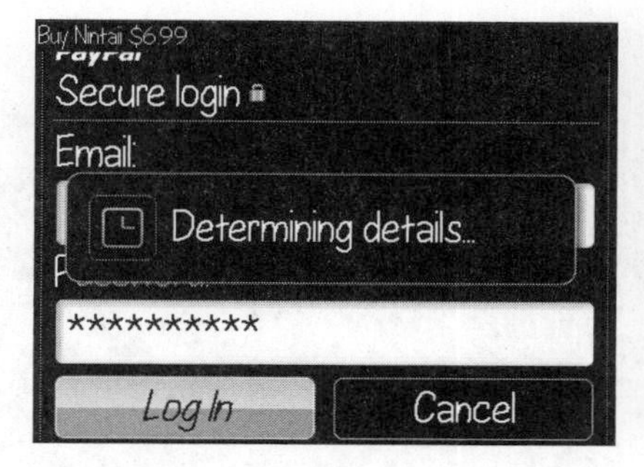

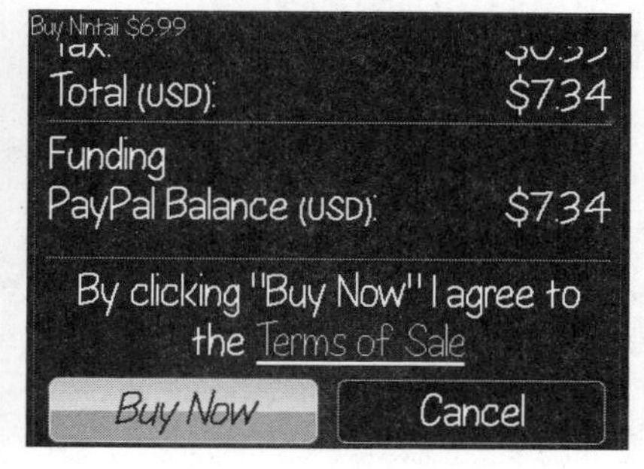

Figure 26-5. *Purchase an app using PayPal.*

3. You will see the progress bar of the application as it downloads, prepares to install, and installs on your BlackBerry.

Using the My World Feature

All the apps that you purchase or download for free are listed in your **My World** area of the App World.

Glide and click the **My World** icon (in the lower right-hand corner of the App World) to see all the programs you have downloaded or purchased recorded there.

NOTE: If you restart your BlackBerry or do a battery pull, you will be logged out of your account and you will be asked to reenter your PayPal password for your next app purchase.

My World Menu Commands

To access My World menu commands, highlight any app in your **My World** area, and press the **Menu** key.

From the **Menu**, you can log in to your account, view details about the app, run the selected app, review the app, recommend it, archive, delete or contact support.

> **NOTE:** If application updates are available, that will also be indicated in **My World**.

Removing or Uninstalling Programs

After you start to download lots of apps to your BlackBerry, you may decide that you no longer use some of them or simply want to free up space for new Apps. Follow these steps to remove apps?

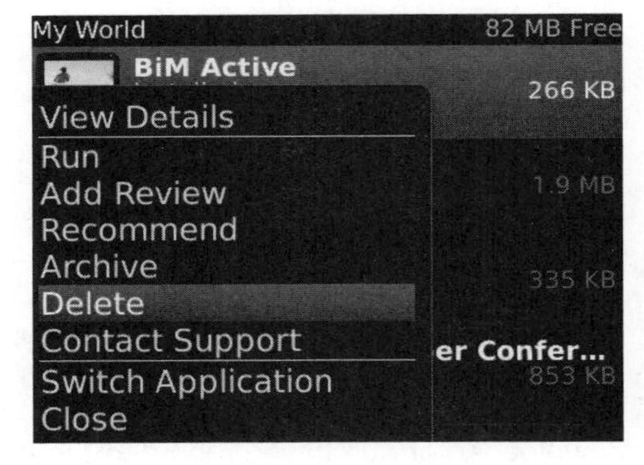

1. To delete an app, you need to be in the **My World section** of App World.
2. Highlight the app you wish to remove, and press the **Menu** key.

3. Select **Delete** to remove that program from your BlackBerry.
4. On the next screen you will be asked to confirm your selection to remove the program

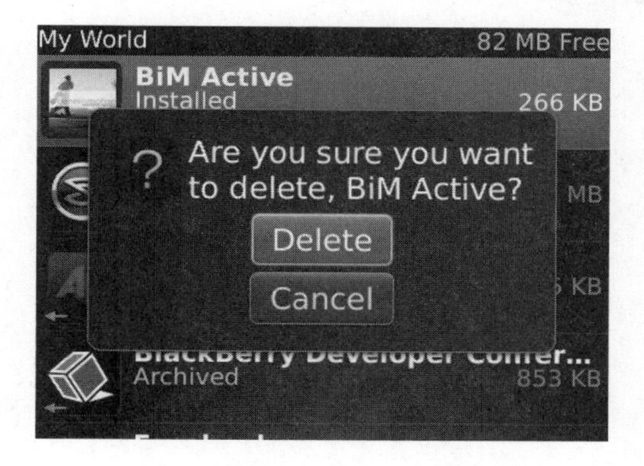

5. Most programs will require you to reboot (or reset) your BlackBerry to complete the uninstall process.

TIP: If you are removing more than one application, select **Reset Later** or **Reboot Later** until you have selected and uninstalled all the apps. At the end, select **Reset Now** to remove them all at once; this will save you a lot of waiting time!

Chapter 27

Connecting with Bluetooth

Bluetooth allows your BlackBerry to communicate with things like headsets, GPS devices, and other hands-free systems with the freedom of wireless. Bluetooth is a small radio that transmits from each device. The BlackBerry gets paired with (connected to) the peripheral. Many Bluetooth devices can be used up to 30 feet away from the BlackBerry.

In this chapter, we will show you how to connect (pair) to a Bluetooth headset, how to connect to a Bluetooth stereo device, and how to prioritize your Bluetooth connections.

History of Bluetooth

The BlackBerry ships with Bluetooth 2.0 technology. Think of Bluetooth as a short-range wireless technology that allows your BlackBerry to connect to various peripheral devices (headset, speakers, computer, and so on) wirelessly.

Bluetooth is believed to be named after a Danish Viking and king, Harald Blåtand (which has been translated as *Bluetooth* in English). King Blåtand lived in the tenth century and is famous for uniting Denmark and Norway. Similarly, Bluetooth technology unites computers and telecom. His name, according to legend, is from his very dark hair, which was unusual for Vikings—*Blåtand* means dark complexion. A more popular story states that the king loved to eat Blueberries, so much so his teeth became stained blue. Believe what you choose!

To learn more about Bluetooth, you can check out the following web sites:

```
http://cp.literature.agilent.com/litweb/pdf/5980-3032EN.pdf
http://www.cs.utk.edu/~dasgupta/bluetooth/history.htm
http://www.britannica.com/eb/topic-254809/Harald-I
```

Using Bluetooth on Your BlackBerry

To use Bluetooth on your BlackBerry, you must first turn it on. This is done through the **Setup** folder and the **Setup Bluetooth** icon.

TIP: You should also be able to get here from the **Manage Connections** icon.

Turning on Bluetooth

When Bluetooth is on, you will see this small Bluetooth icon in the upper part of your Home Screen. Depending on your theme, it may be on the left or right side.

If you don't see a **Turn On Bluetooth** or **Setup Bluetooth** icon, click the **Manage Connections** icon. Then click **Set Up Bluetooth**.

Configuring Bluetooth

Once Bluetooth is enabled, you will want to follow the steps in this section to take full advantage of the Bluetooth capabilities of the BlackBerry. There can sometimes be two or three ways to get into the Bluetooth setup and options screens, depending on your BlackBerry software version and BlackBerry carrier (cell phone company). We will show you the most common way.

1. Click the **Setup** folder.
2. Then click **Setup Bluetooth**.
3. If you see the window asking you to either search or be discovered, just press the **Escape** key.
4. Next, press the **Menu** key, and scroll down to **Options.**

5. If you have already paired your BlackBerry with Bluetooth devices, you will have seen those devices listed (we cover pairing in later in this chapter).

6. To change your device name (the way other Bluetooth devices will see your BlackBerry), click **Device Name**, and type a new name.

7. To make your BlackBerry discoverable to other devices click **Discoverable**, and select **Yes** (the default is **No**). As we note later, you should set this back to **No** after you finish pairing for increased security.

8. Make sure that **Always** or **If Unlocked** is visible after **Allow Outgoing Calls**.

9. Set **Address Book Transfer** to **Enable**. Depending on your software version and preferences, you may see the options of **All Entries** (the same as Enable), **Hotlist Only**, or **Selected Categories Only**. The **Address Book Transfer** option allows your address book data to be transferred to another device or computer using Bluetooth.

10. To see a blue flashing LED when connected to a Bluetooth device, make sure the **LED Connection Indicator** is set to **On**.

Bluetooth Security Tips

Here are a few security tips from a BlackBerry IT newsletter (`http://www.blackberry.com/newsletters/connection/it/jan-2007/managing-bluetooth-security.shtml?CPID=NLC-41`). These will help prevent hackers from getting access to your BlackBerry via Bluetooth:

- Never pair your BlackBerry when you are in a crowded public area.
- **Disable** the **Discoverable** setting after you are finished pairing your BlackBerry.
- Do not accept any pairing requests with unknown Bluetooth devices; only accept connections from devices with names you recognize.
- Change the name of your BlackBerry to something other than the default BlackBerry; this will help prevent hackers from easily finding your BlackBerry.

Supported Devices

Your BlackBerry should work with most Bluetooth headsets, car kits, hands-free kits, keyboards, and GPS receivers that are Bluetooth 2.0 and earlier compliant. At the time of this writing, Bluetooth 2.1 was just coming on the scene. You will need to check with the device manufacturer of newer devices to make sure they are compatible with your BlackBerry.

Pairing Your BlackBerry with a Bluetooth Device

Think of pairing as establishing a wireless connection between your BlackBerry and a peripheral device (headset, GPS device, external keyboard, Windows or Mac computer, and so on). Pairing is dependent on entering a required passkey, which locks your BlackBerry into a secure connection with the peripheral.

To pair your BlackBerry, put your Bluetooth device in pairing mode as recommended by the manufacturer. Also, have the passkey ready to enter.

There could be several ways to get into the Bluetooth setup screen to pair your BlackBerry and establish the connection. Navigate to the Bluetooth setup screen with one of the methods below:

- **Method 1:** Click the **Options** icon (which may be inside the **Settings** icon). Then click **Bluetooth**.
- **Method 2:** Click the **Set Up Bluetooth** icon.
- **Method 3:** Click the **Manage Connections** icon and then select **Setup Bluetooth**.

1. The BlackBerry will ask if you want to "Search for a device or Listen for another device to find me?" Click **Search** or **Listen** and then **OK**.

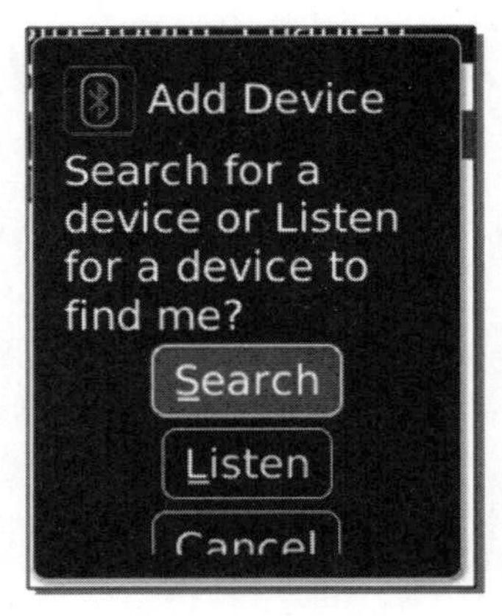

CAUTION: If you are pairing your BlackBerry with your computer, you need to make sure that both your BlackBerry and your computer are in **Discoverable** mode. Set this in the **Bluetooth Options** screen by setting **Discoverable** to **Yes** or **Ask**. The default setting is **No**, which will prevent you from pairing.

2. When the device is found, the BlackBerry will display the device name on the screen. Click the name to select it.

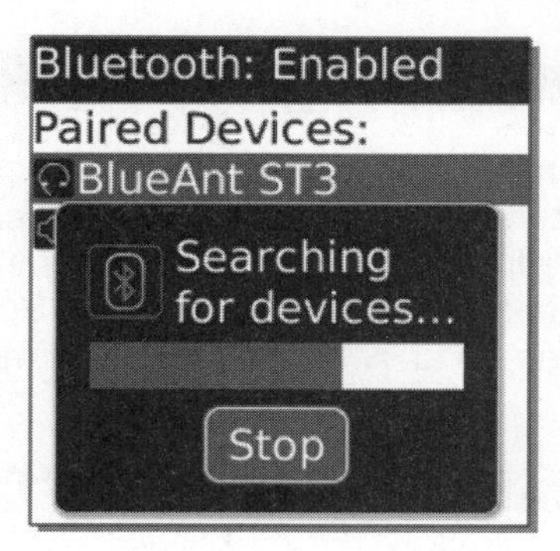

3. You will then be prompted to enter the four-digit passkey provided by the manufacturer of the Bluetooth peripheral. Once you enter in the passkey, click the trackpad. (Many default passkeys are just 0000 or 1234.)

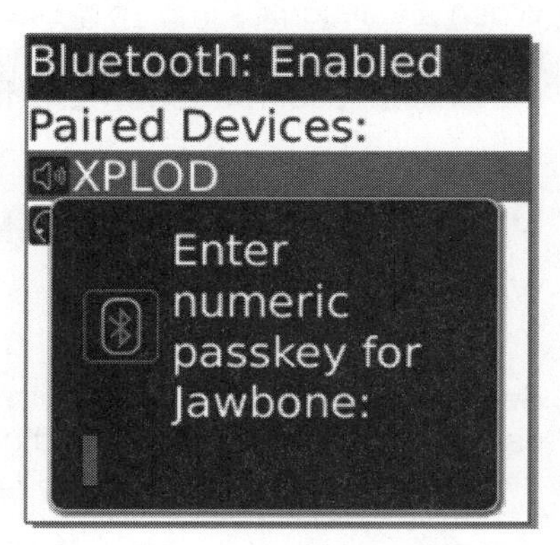

4. You will then be prompted to accept the connection from the new Device

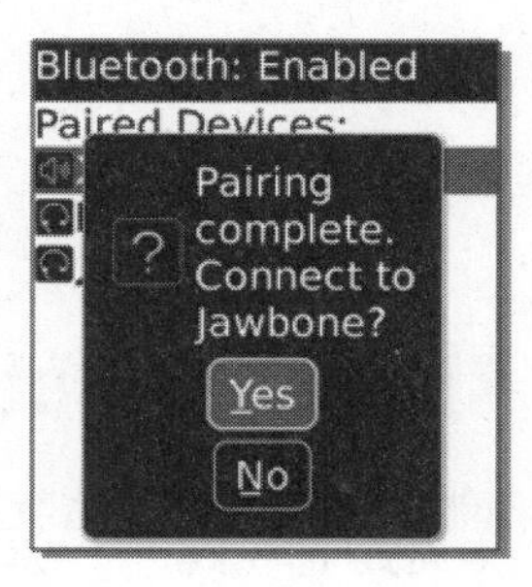

TIP: If you check the **Don't ask this again** box, you will only have to do this once.

5. Your device should now be paired and ready to use.

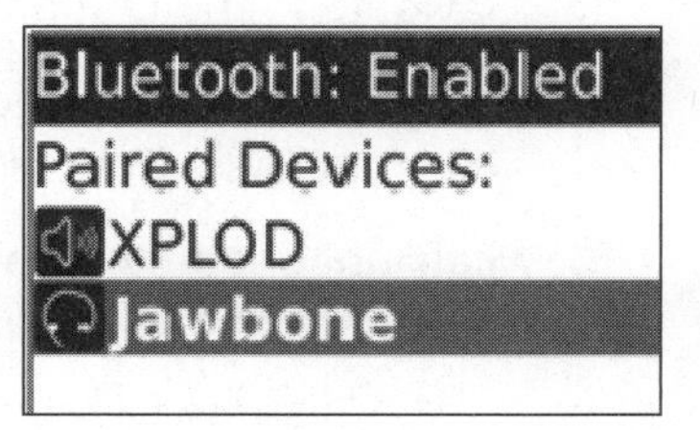

Answering and Making Calls with the Headset

Some Bluetooth headsets support an auto-answer protocol, which will, as it sounds, automatically answer incoming calls and send them right to the headset. This is very helpful when driving or in other situations where you should not be looking at your BlackBerry to answer the call. Sometimes, you will need to push a button—usually just one—to answer your call from the headset.

Option 1: Answering Directly from the Headset

When a call comes into your BlackBerry, you should hear an audible beep in the headset. Just press the multifunction button on your headset to answer the call. Press the multifunction button when the call ends to disconnect.

Option 2: Transferring the Caller to the Headset

When the phone call comes into your BlackBerry, press the **Menu** key.

Select **Activate (your Bluetooth headset name)**, and the call will be sent to the selected headset. In this image, the headset name is Jawbone.

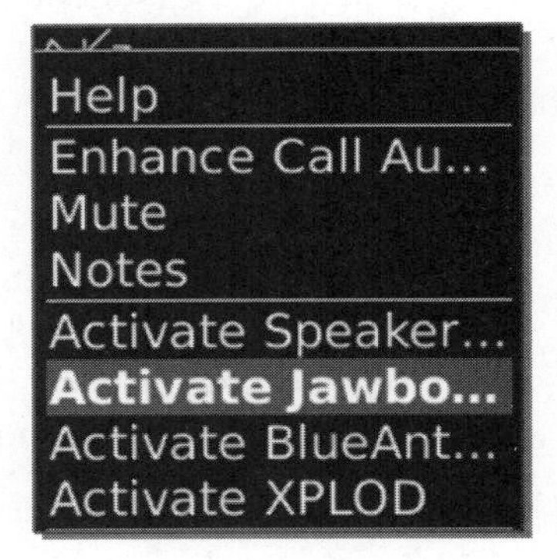

Bluetooth Setup Menu Commands

Several options are available to you from the Bluetooth menu. Learn these commands to be able to take full advantage of Bluetooth wireless technology on your BlackBerry.

1. Navigate to the **Options** icon, and click it.
2. Scroll to **Bluetooth and** click. You will now see the list of paired devices with your BlackBerry shown in Figure 27-1.
3. Highlight one of the devices listed and press the **Menu** key. The following options become available to you:
    - **Disable Bluetooth**: Another way to turn off the Bluetooth radio, this will help to save battery life if you don't need the Bluetooth active.
    - **Connect /Disconnect**: Clicking this will immediately connect or disconnect you to or from the highlighted Bluetooth device.

- **Add Device**: Use this to connect to a new Bluetooth peripheral.
- **Delete Device**: This one removes the highlighted device from the BlackBerry.
- **Device Properties**: To check whether the device is trusted or encrypted and if echo control is activated, use this option.
- **Transfer Contacts**: If you connected to a PC or another Bluetooth smartphone, you can send your address book via Bluetooth to that device.
- **Options**: This shows the Options screen (covered previously).

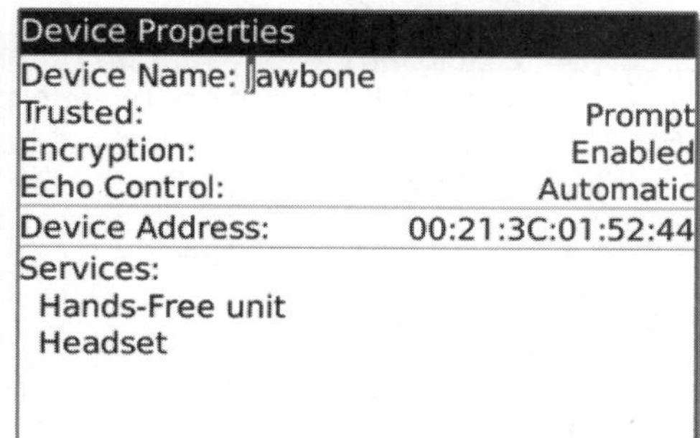

Figure 27-1. *Bluetooth device properties.*

You should change your **Device Name** to something other than BlackBerry Bold for security reasons. Use something that is not easily recognizable.

Also, make sure **Discoverable** is set to **No** after you are finished pairing.

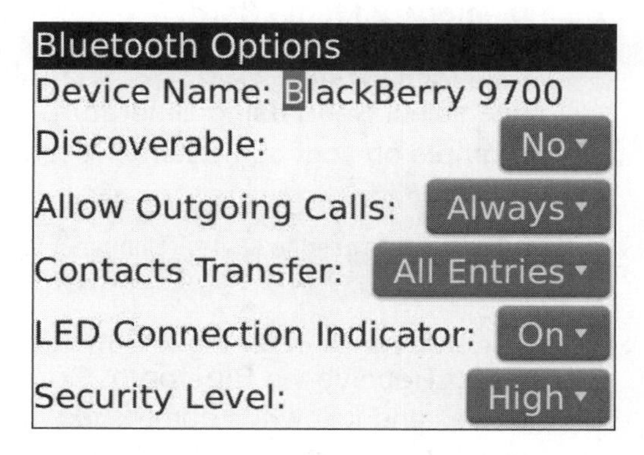

Contacts Transfer can be set to **Disabled**, **All Entries** (default), **Hotlist Only**, or **Selected Categories Only**.

Select the categories by pressing the **Menu** key and checking the categories you wish to be able to transfer.

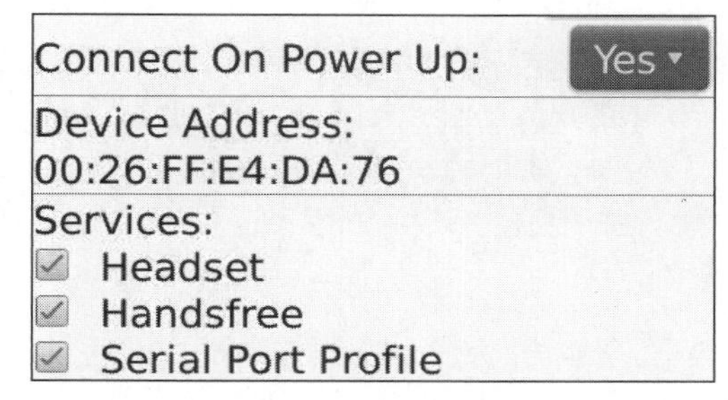

Services can be enabled or disabled for security purposes. Press the **Space** key to check or uncheck any services.

- Desktop Connectivity
- Wireless Bypass
- Data Transfer
- SIM Access Profile
- Dial-Up Networking
- Audio Source
- A/V Remote Control Target

Sending and Receiving Files with Bluetooth

Once you have paired your BlackBerry with your computer, you can use Bluetooth to send and receive files. At the time of this writing, these files were limited to media files (videos, music, and pictures) and address book entries, but we suspect that you will be able to transfer more types of files in the future.

To send or receive media files on your BlackBerry:

1. Click the **Media** icon.
2. Navigate to the type of file you want to send or receive: **Music**, **Video**, **Ringtones**, or **Pictures**.
3. Navigate to the folder where you want to send or receive the file: **Device Memory** or **Media Card**.
4. If you are *sending a file* to your computer, glide to and highlight the file and select **Send using Bluetooth**. Then, you will need to follow the prompts on your computer to receive the file.

> **NOTE:** You may need to set your computer to be able to **Receive via Bluetooth**.

5. 5. If you are *receiving a file* (or files) on your BlackBerry, you need to select **Receive via Bluetooth**. Go to your computer, and select the file or files and follow the commands to send via Bluetooth. You may be asked to confirm the folder receiving the files on your BlackBerry.

> **NOTE:** You can send (transfer) only media files that you have put onto your BlackBerry yourself. The preloaded media files cannot be transferred via Bluetooth.

Streaming Bluetooth Stereo

Devices, like the BlackBerry Remote Stereo Gateway, allow you to stream any music source via Bluetooth or stream music from your BlackBerry to your home stereo. Just pair the Bluetooth Device with your BlackBerry and begin playing music. The music from your BlackBerry will then be sent to your home or car stereo.

Bluetooth Troubleshooting

Bluetooth is still an emergent technology, and sometimes, it doesn't work as well as we might hope. If you are having difficulty, perhaps one of these suggestions will help.

My Passkey Is Not Being Accepted by the Device

You may have the incorrect passkey. Most Bluetooth devices use either 0000 or 1234, but some have unique passkeys.

If you lost your manual for the Bluetooth device, many times you can use a web search engine such as Google or Yahoo to find the manufacturer's web site and locate the product manual.

I Have the Right Passkey, But I Still Cannot Pair the Device

The device may not be compatible with the BlackBerry. However, one thing you can try is to turn off encryption.

Click **Options** and then **Bluetooth**, and then highlight the problem device and click it. In **Device Properties**, disable encryption for that device, and try to connect again.

I Can't Share My Address Book

Inside the Bluetooth setup screen, press the **Menu** key, and select **Options**. Make sure that you have enabled the **Address Book Transfer** field.

> **NOTE:** Many Bluetooth headsets and car kits do not fully support address book transfers; double-check the documentation that came with your Bluetooth device.

Chapter 28

Connect as a Tethered Modem

Tethering is the process of connecting your BlackBerry to your computer and using the BlackBerry as a modem to access the Internet. This is particularly useful if you are in an airport or a hotel with no Internet connection on your notebook and need more than the capabilities of your BlackBerry.

Connecting Your Laptop to the Internet Using Your BlackBerry

Depending on your wireless carrier (phone company) and the type of software you use, you should be able to use your BlackBerry to connect your laptop (PC or Mac) to the Internet. This is called tethered modem or tethering, or sometimes an IP modem. You need the USB cable for PCs. With Macs you can usually tether with Bluetooth, but if that does not work, you can use the USB cable.

> **NOTE:** Not every BlackBerry wireless carrier supports using your BlackBerry as a modem. Please check with your carrier if you have trouble connecting.

Tethering Your BlackBerry Usually Costs Extra

Most, but not all, wireless carriers charge an extra fee to allow you to use your BlackBerry as a tethered modem, or for dial-up networking or phone-as-modem. Also, you may not be able to connect using your BlackBerry as a modem unless you have specifically signed up for the "BlackBerry as Modem" or similar data plan. We have heard of users getting a surprise phone bill in the hundreds of dollars, *even with an "unlimited BlackBerry data plan"* because their particular carrier did not include BlackBerry modem data in the unlimited plan.

TIP: Some carriers allow you to turn the "BlackBerry as Modem" extra service on and off. Check with your particular carrier. Turn it off when you don't need it if you can. Also, beware that changing turning this modem service on and off might extend or renew your two-year commitment period. Ask your phone company for its policies. Assuming there are no extra hidden costs or commitments, you could just enable it for a scheduled trip and then turn it off when you return home.

We also want to thank Research In Motion, Ltd. and `BlackBerry.com` for valuable information contained in their extensive BlackBerry Technical Solution Center. We strongly encourage you to visit this site for the latest information on using your BlackBerry as a modem and anything else! Visit `http://www.blackberry.com/btsc/supportcentral/supportcentral.do?id=m1`, and search for "modem how to."

TIP: When searching the BlackBerry Technical Solution Center knowledgebase, do not enter your specific BlackBerry model; use the series. For example, if you have a Bold 9700, enter **Bold** or **9700 series**, or just leave that out of the search.

To locate the modem instructions for your particular BlackBerry, you will need to know the network—3G, EDGE, GPRS, CDMA, EVDO—on which your BlackBerry operates. The Bold 9700 series runs on the 3G/EDGE/GRPS networks.

The steps and software to use your BlackBerry as a modem have changed and continue to change frequently, almost on a monthly basis. Therefore, some of the software and information contained in this chapter may be slightly different when you are reading it than when the book was published.

So how can you find the latest information?

Please do a web search using terms like "BlackBerry as Modem," "BlackBerry Tethered Modem," or "BlackBerry IP Modem."

Understanding the Options for Tethering

You have a few options to use your BlackBerry as a tethered modem for your laptop:

- **Option 1:** Purchase third-party software.
- **Option 2:** Use your wireless carrier's software; contact your carried to ask if this is available.

- **Option 3:** Use the BlackBerry Desktop Manager Software. This is usually only an option for you if carrier-specific software is not available.

Purchase Third-Party Software

Tether is a popular option (see Figure 28-1) and now works for both Windows and Mac computers. Check out its site at www.tether.com. The bonus claimed by Tether is that you do not need to pay for a separate modem plan from your carrier, and you may be able to save a lot on monthly connection charges. Cost for Tether was about $30 at the time of this writing, but the company offers a seven-day free trial, so you can give it a try before you buy.

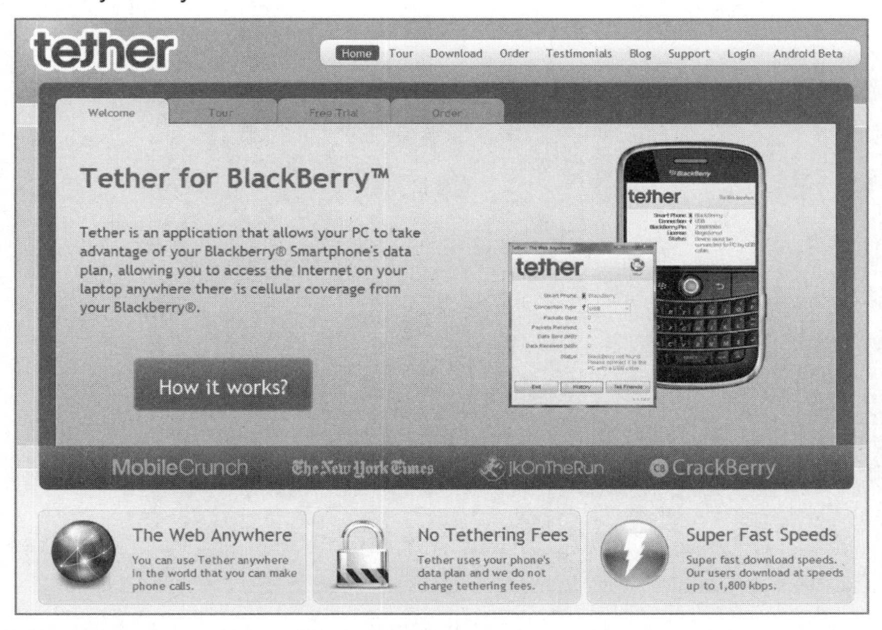

Figure 28-1. *Tether third-party tethering software*

Use Your Wireless Carrier's Software

Contact your wireless carrier's technical support to find out if they offer software and service plans for this option. Tell them you want to use your BlackBerry as a modem for your laptop. In the United States, AT&T, Sprint/Nextel, and Verizon have simple software (e.g. Communication Manager or Verizon Access Manager) that you can download and install.

Use Desktop Manager to Connect to the Internet

This works for both Windows and Mac computers. Windows computer users read on below. Mac users see page 457.

Using Desktop Manager for Windows

If you have confirmed with your carrier that you can use the IP Modem feature in BlackBerry Desktop Manager, follow the steps shown here. (Learn how to download and install Desktop Manager for Windows on page 65.)

1. Start Desktop Manager by going to **Start ➤ BlackBerry ➤ Desktop Manager** (Figure 28-2).

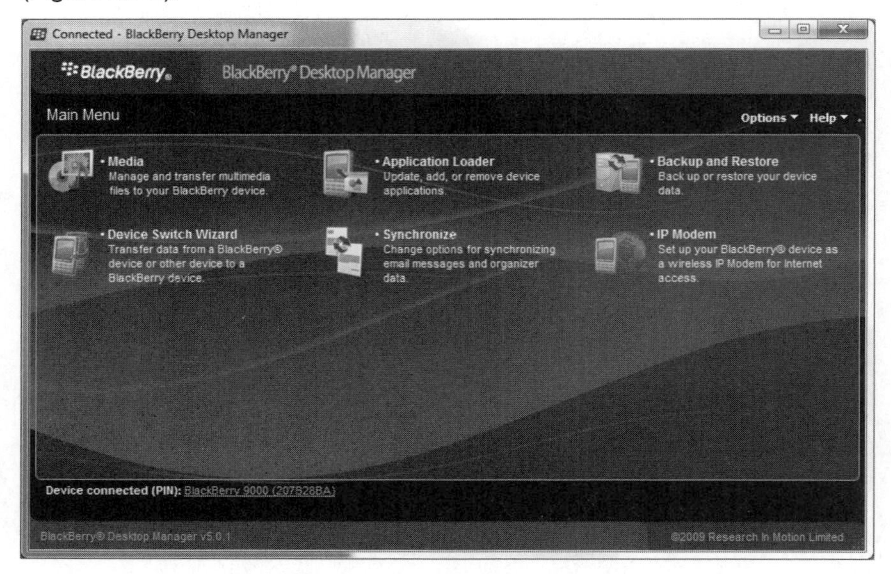

Figure 28-2. *Configue IP Modem in Desktop Manager for Windows.*

2. Click the **IP Modem** icon. If you don't see this icon, you will need to confirm with your carrier that you can use it. Sometimes, you can reenable this icon by editing a specific file.

3. Configure the IP Modem by clicking the **Configure...** button as shown in Figure 28-3.

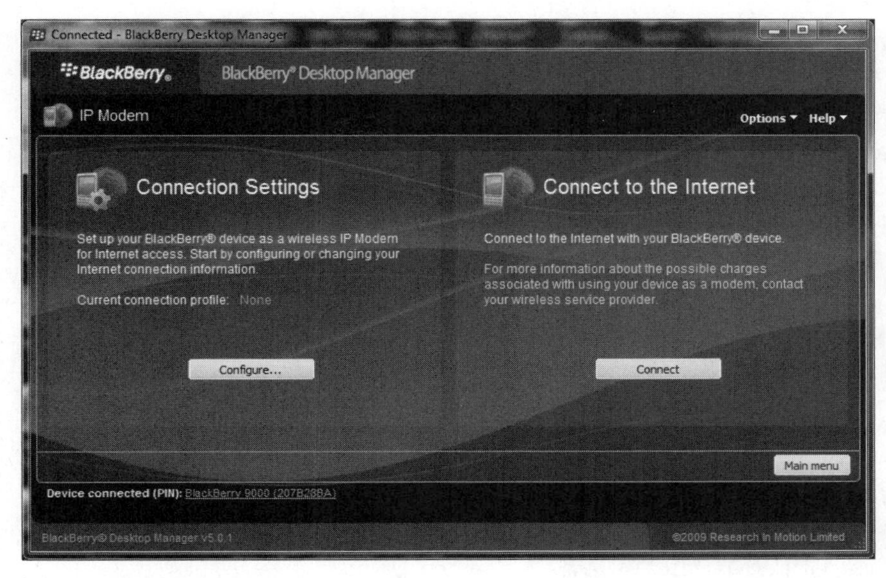

Figure 28-3. *Click Configure in connection settings.*

4. Now, see if your carrier is in the drop-down list next to **Connection Profile** (see Figure 28-4).

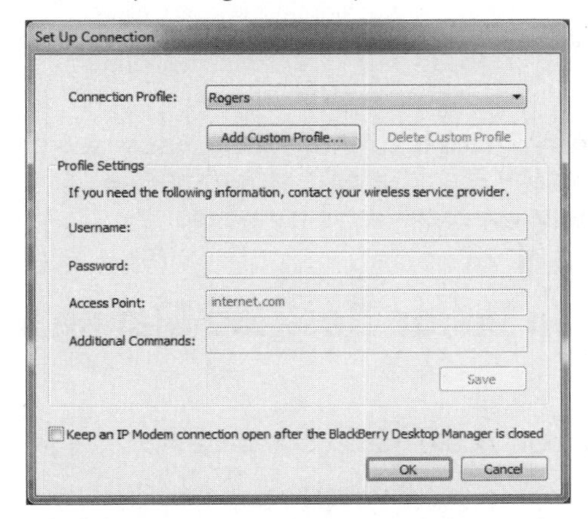

Figure 28-4. *Look for yoru carrier in the dropdown menu.*

5. If not, select **Add Custom Profile**, and follow the steps to enter the information. Most likely, you will need to contact your wireless carrier for the setup information.

Add Custom Profile

Profile Name:

Add Profile Cancel

6. Make sure to check the box at the bottom of the **Set Up Connection** screen if you want your Internet connection to remain active even after you close out of Desktop Manager. Click **OK**, and save your settings.

7. Make sure your BlackBerry is connected to your computer and recognized by Desktop Manager. You can tell if Desktop Manager sees your BlackBerry by looking in the lower-left corner:

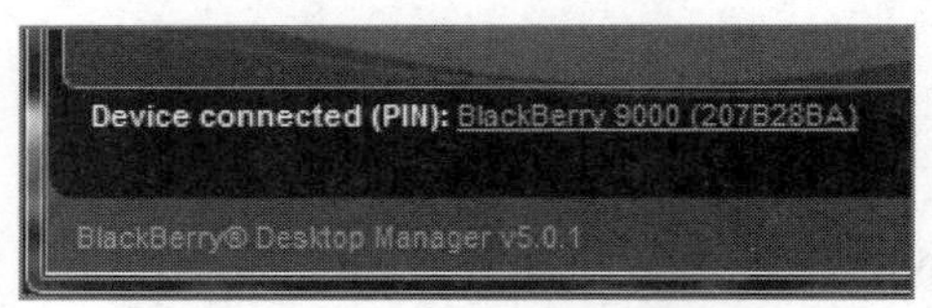

8. Now, you are ready to connect. Click the **Connect** button. If you see any error messages, follow the directions, or contact your wireless carrier for setup assistance.

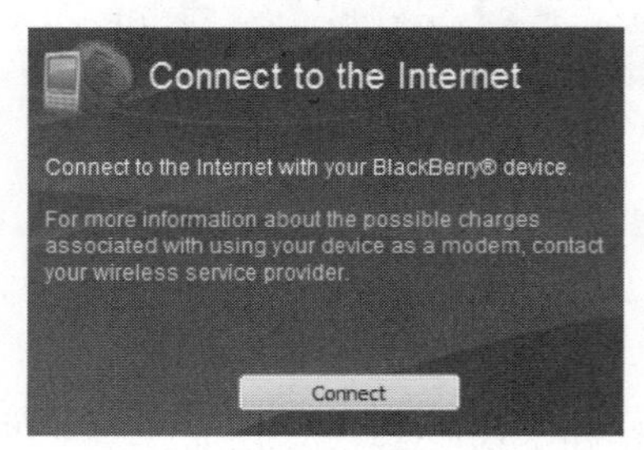

For help, also consider visiting the **BlackBerry Technical Knowledge Base** (see page 535). Or, you can ask your question at one of the BlackBerry user forums like `www.crackberry.com`, `www.pinstack.com`, and `www.blackberryforums.com`.

Re-Enabling the IP Modem Icon inside Desktop Manager for Windows

What if you're IP Modem icon disappeared; how could you I get it back?

First, you should be sure to contact your carrier to see if you can use Desktop Manager's IP Modem feature. If their answer is yes, follow these steps to edit the `ip_modem_configuration.xml` file. If your carrier says no, following these steps will likely not work; please use the software recommended by your carrier.

1. Locate the `ip_modem_configuration.xml` file. It is probably best to use the Windows search feature or Google Desktop to find this file, but you may find it in one of these places:

   - **Vista**: `C:\ProgramData\Research In Motion\BlackBerry`
   - **Windows XP**: `C:\Documents and Settings\All Users\Application Data\Research in Motion\BlackBerry`

2. Open the file using a text-only editor such as Windows Notepad or Wordpad.

3. Locate your wireless carrier's name. Take note that carrier names with ampersands like AT&T will be listed as "AT&T".

4. Change **enabled="false"** to **enabled="true"** as shown in Figure 28-5.

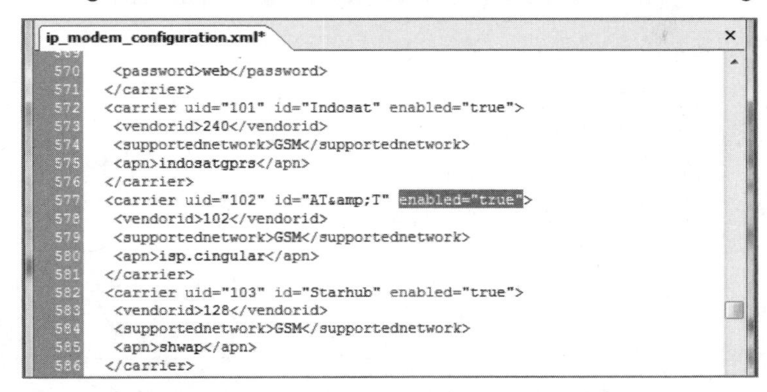

Figure 28-5. *Use text based editor to edit connection string.*

5. Save your changes, and restart Desktop Manager. The Icon should reappear. If it does not, restart your computer, and try again.

Using Desktop Manager for Mac

First, download and install the latest version of Desktop Manager for Mac from `www.blackberry.com`. Learn how on page 125.

Now, connect your BlackBerry to your Mac. Go to **Settings ➤ Network** on your Mac to see the window shown in Figure 28-6.

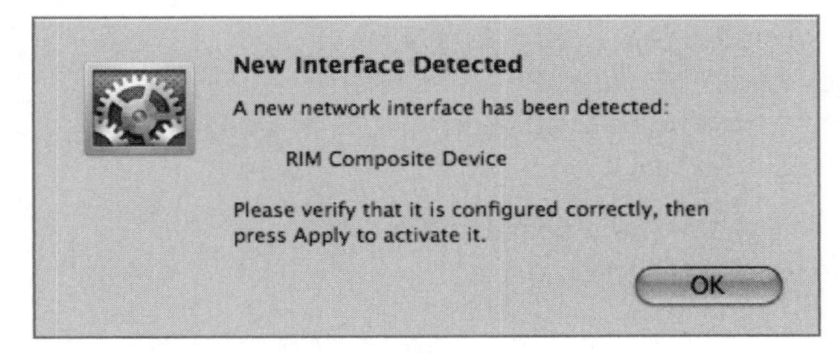

Figure 28-6. *Configure BlackBerry as modem on a Mac.*

Your Mac now sees your BlackBerry as a new network interface and will attempt to configure it as a dial up modem for tethering.

If you have a tethering or BlackBerry as a modem service plan with your carrier, you can continue with the setup and choose **Network Preferences...** as shown in Figure 28-7.

> **NOTE:** Most, but not all, carriers charge extra for using your BlackBerry as a modem for your Mac to connect to the Internet. Contact your carrier for pricing.

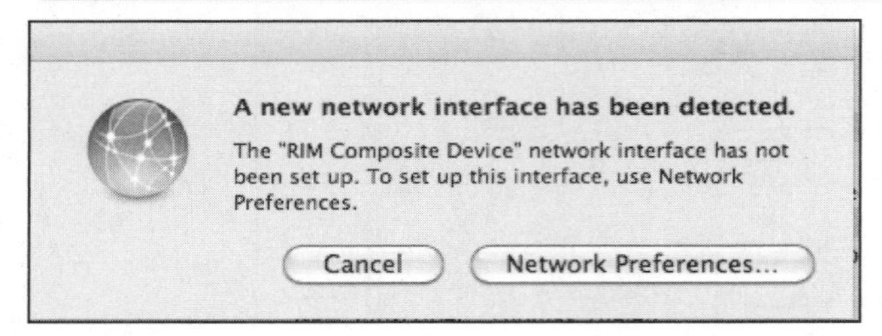

Figure 28-7. *Pop up window advising you to use Newtwork Preferences to configure BlackBerry as modem.*

> **TIP:** Many carriers will allow you to turn on and off this extra service. Say you were taking a trip and needed to use the modem feature for three weeks. Your carrier may allow you to do and then turn it off again when you return home.

Setting Network Preferences

This section explains how to configure your network preferences:

> **NOTE:** Only proceed with these steps if have all the necessary carrier setup strings provided by your wireless carrier.

1. Click **Default** next to the **Configuration** field, and select **New Connection**. Type a connection name; in Figure 28-8, we used BlackBerry 8900, but you might want to type **BlackBerry 9700** or **Bold 9700**.
2. Fill in your phone number, account name, and password in the appropriate boxes to begin the configuration process.

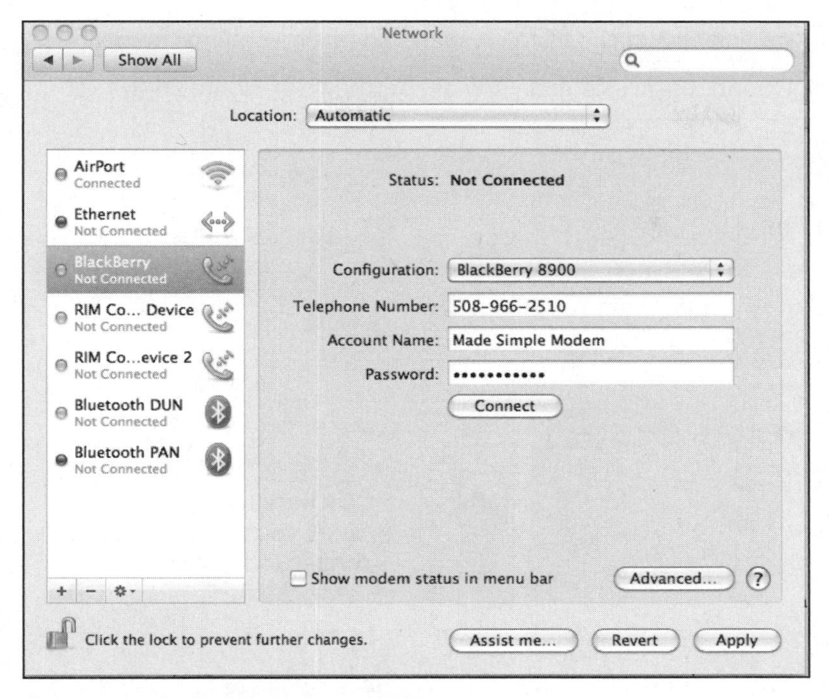

Figure 28-8. *Configuration settings on a Mac.*

3. If you need to access more settings, click the **Advanced...** button in the lower-right corner. The advanced options screen (see Figure 28-9) will have tabs along the top for Modem, DNS, WINS, Proxies, and PPP settings.

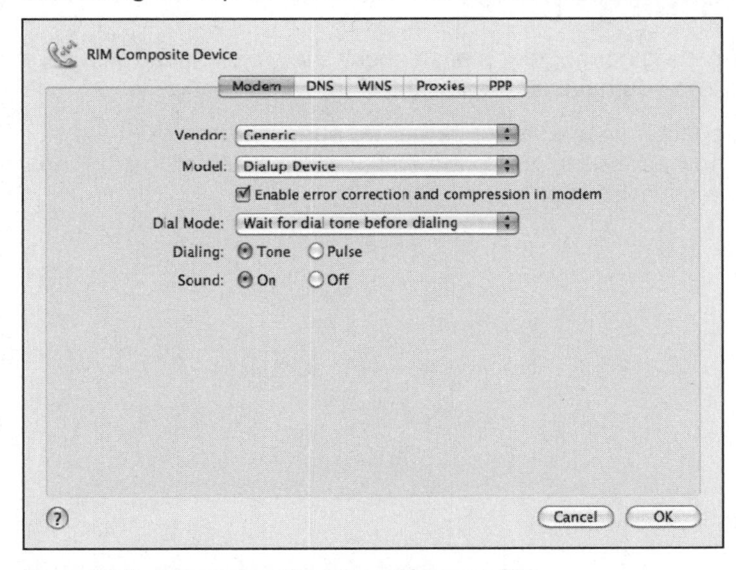

Figure 28-9. *Advanced configuration settings on a Mac.*

4. When you are finished, click **OK** to close the advanced settings screen.
5. Next, click **Connect** from the main screen, as shown in Figure 28-10.

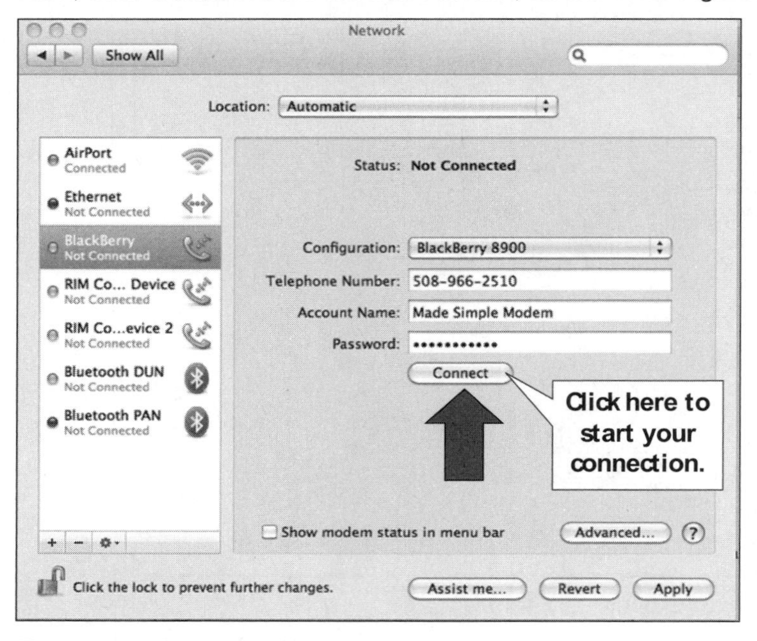

Figure 28-10. *Click Connect to go online.*

Having Trouble Getting Online?

If you have any trouble getting connected, please check that your BlackBerry is connected to your Mac and you have entered all information correctly.

If everything looks OK, contact your wireless carrier for support, check out the BlackBerry Technical Knowledgebase (see page 535), or visit some of the on-line BlackBerry forums as we have listed on page 537.

Chapter 29

Surfing the Web

A prerequisite for any smartphone today is the ability to get online and browse the Web. Now, smartphone web browsing will never be a substitute for desktop browsing, but you will be surprised by all the features available on your BlackBerry.

In this chapter, we will show you how to get online, how to set browser bookmarks, how to use your browsing history, and how to use some hotkeys to get around the browser quickly.

Web Browsing on Your BlackBerry

One of the amazing features of smartphones like the BlackBerry is the ability to browse the web right from your handheld. More and more web sites are now supporting mobile browser formatting. These sites sense you are viewing them from a small mobile browser and automatically reconfigure themselves for your BlackBerry so they load quickly—some even quicker than their desktop counterparts.

Using Alternative Web Browsers

The native web browser may seem slow to some people, so vendors are creating alternative web browsers for your BlackBerry. RIM, BlackBerry's maker, has recently purchased a web browser development company so we may see an overhaul of the BlackBerry browser in the coming months.

We have seen several alternatives to the native BlackBerry web browser appear on the market. A couple of the more notable ones recently have been Opera Mini (`www.opera.com/mini` or `m.opera.com`) and Bolt Browser (`www.boltbrowser.com`). We have not extensively tested them, but they do keep improving the look, feel, and performance of the overall web browsing experience. You can locate these browsers and other alternatives by performing a web search for "BlackBerry Bold web browser" or "BlackBerry web browser."

NOTE: The rest of this chapter focuses on the primary, or native, web browser that is included with every BlackBerry.

Locating the Web Browser from the Home Screen

Web browsing can actually start with a few of the icons on your home screen. The easiest way to get started is to press the hotkey **B** for the Internet Browser or **W** for the WAP Browser. (For hotkeys help, see page 541) Or, you can find the **Browser** icon; it looks like a globe.

Most of our screen shots in this book are from the Internet Browser, not the WAP Browser.

1. Use the trackpad to navigate to the **Browser** icon and click it. It might say **Browser**, **BlackBerry Browser**, or **Internet Browser** or even something different like **mLife** or **T-Zone**.

2. You will either be taken directly to the home screen of your particular carrier or to your list of bookmarks.

The start page is nice because on a single screen, it allows you to

- Type a web address (URL).
- Search the Web.
- See your popular bookmarks.
- See your web history.

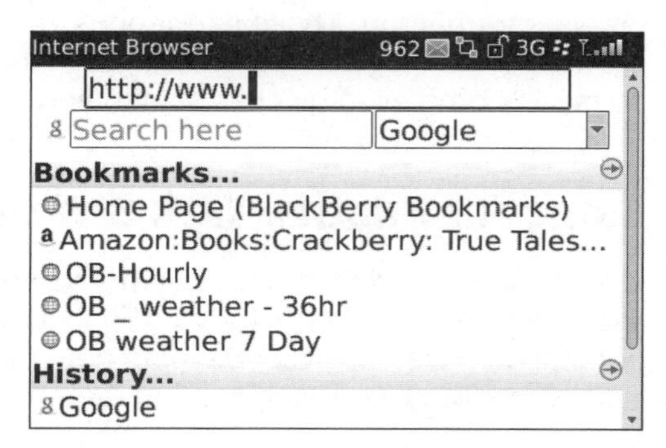

Learn how to add bookmarks to the **Bookmarks** list on page 467.

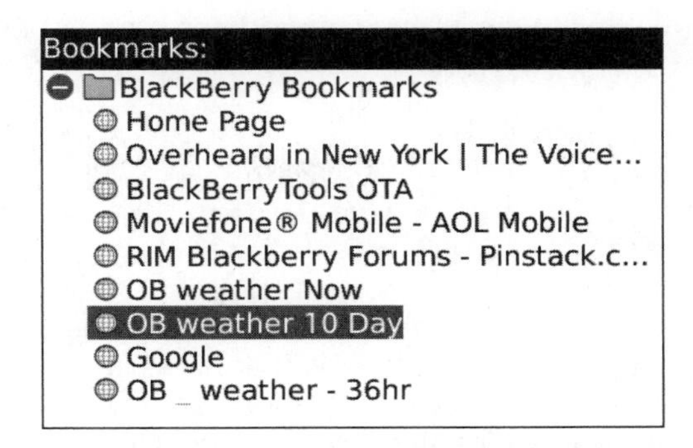

TIP: Type a few letters of the Bookmark's name to instantly Find it. (Just like finding names in your Contact list.)

You can set your customized home page to anything you want (e.g. `www.google.com` as it is here); see page 469.

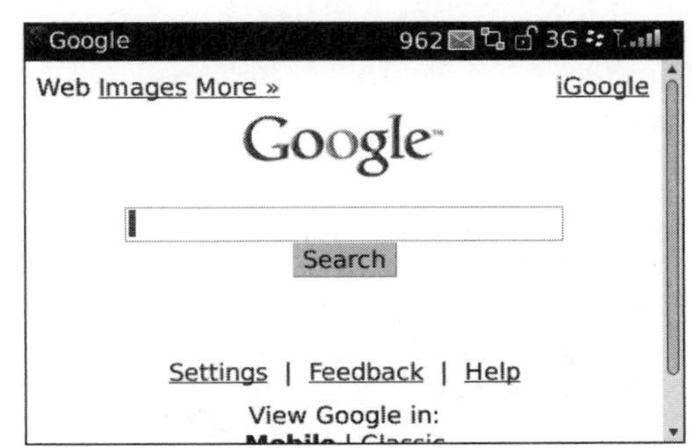

Speed Up Your Browser by Clearing the Browser Cache

You can really speed up your web browser by periodically clearing out the cache. See page 523.

Web Browser Shortcut Keys

We put many of the hotkeys and shortcuts in the beginning of this book for easy access and to keep them all together. Please go to page 544 to see the complete list of web browser hotkeys.

Going to a Web Page

To get to any web page, you can either type it on your start page or press the **Menu** key and select **Go To...** from the menu.

The address bar comes up with the `http://www.` in place, waiting for you to type the rest of the address.

TIP: Pressing the **Delete** key will quickly erase the `www.` but leave the `http://`.

Simply, type in the web address (remember, pushing the **Space key** will put in the dot (.).

> **TIP:** Pressing Shift + Space is the shortcut to type a slash (/), as in `www.google.com/gmm` (the address to download Google Maps application at the time of this writing).

The first thing you see on the top of the start page will be the web address line. Then you will see your Bookmarks and history of recently visited pages. Click any **Bookmark** or **History** item to instantly jump to it.

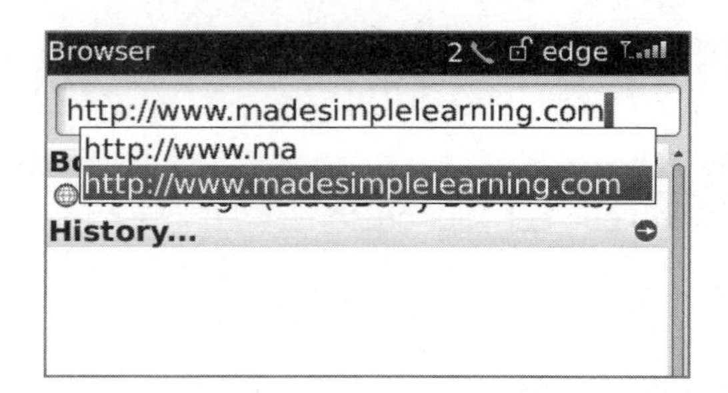

> **TIP:** As you start typing a web address, any similar addresses instantly are shown below.

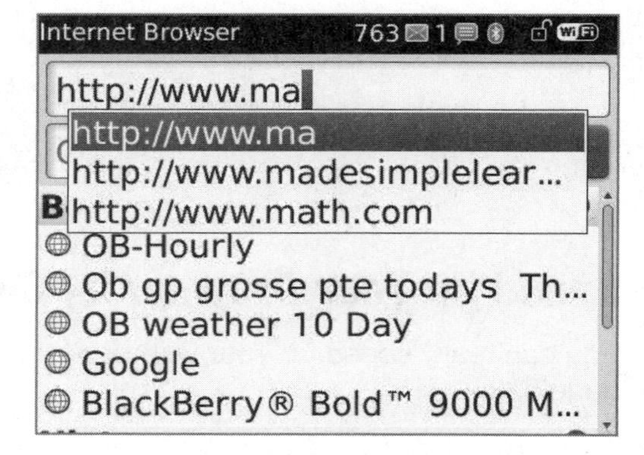

Gide down and click the address you want as soon as you see it in the list.

If necessary, type a few more letters to narrow the list of web addresses shown.

> **TIP:** To edit the address in the drop-down list, glide down to it and glide to the right with the trackpad to change it in the top window.

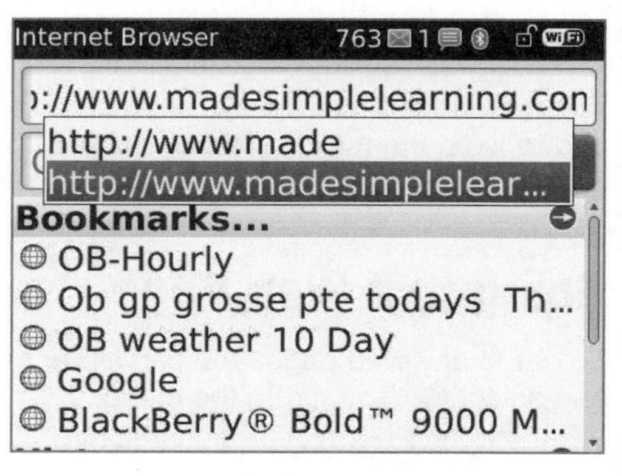

Press the Enter key or click the trackpad to go to the selected web page or history item.

Browser Menu Options

While in the Browser application, press the **Menu** key, and the following options are available to you:

- **Help**: See on-screen text help for the Web Browser, useful when you forget something and need quick help.
- **Column View/Page View:** Toggle between the two views (hotkey: **Z**). Column view offers are more zoomed in view. Page view gives you a view of the entire page, more like a PC web browser .
- **Zoom In:** Zoom in multiple times to see text more clearly (hotkey: **I**).
- **Zoom Out:** Zoom out to see more of the page (hotkey: **O**).
- **Find:** Search for text on a web page (hotkey: **F** / **V**).
- **Find Next:** Search for the next occurrence of the last **Find**.
- **Home**: Go to the browser home page (hotkey: **H**). You can set or change the browser home page inside the **Browser ➤ Options** screen.
- **Go To…**: Type a specific web address for browsing. See page 463 for details (hotkey: **G**).
- **Recent Pages**: View the most recent web pages browsed.
- **History**: Show your entire web browsing history (hotkey: **Y**).
- **Refresh**: Update the current web page (hotkey: **R**).
- **Set Encoding**: Change character encoding of web browsing with this advanced feature. (You probably won't need to use this.)
- **Add Bookmark**: Set the current page as a favorite or bookmark (Hotkey: A). This one is *extremely* useful. See page 467 for details.

- **Bookmarks**: Lists all your bookmarks (hotkey: **K**). This one is also Extremely useful. See page 467 for details on using bookmarks and page 467 for details on organizing bookmarks with folders.
- **Page Address**: Show you the full web address of the current page (hotkey: **P**).
- **Send Address:** Send the current page address to a contact.
- **Options**: Set the browser configuration, properties, and cache settings (hotkey: **S**).
- **Save Page**: Save the page as a file accessible from your **Messages** icon (your email inbox).
- **Switch Application:** Jump or multitask over to other applications while leaving the current web page open (hotkey: **D** jumps to home screen).
- **Close**: Close the web browser and exit to the home screen (hotkey: Press and hold **Escape** key).

NOTE: These hotkeys work only while you are viewing web page and only if the cursor is not in an input (typing) field. If the cursor is in the Google search field, for example, typing these hotkey letters will only show you the letters, not perform the command.

Copying or Sending the Web Page You Are Viewing

Follow these steps to copy or send a web page:

1. Open the **Browser** icon, and press the **Menu** key.
2. Click Page Address.
3. The web address is displayed in the window. Scroll down with the trackpad for options.
4. Click **Copy Address** to copy the web address to the clipboard, so it can easily be pasted into a contact, an email, a memo, or your calendar.

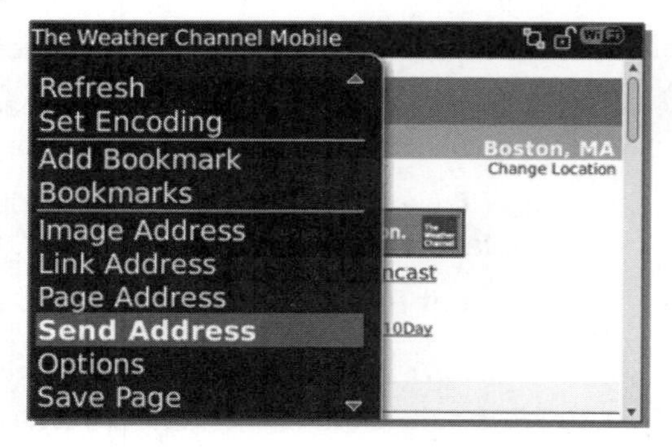

5. Alternatively, click **Send Address**. This will allow you to send the particular web address information via email, MMS, SMS, or PIN messaging. Just select the form and the contact.

Setting and Naming Bookmarks

TIP: You can instantly find bookmarks by typing a few letters of the bookmark name, just like you lookup contacts in your address book. Keep this in mind as you add new bookmarks.

One of the keys to great web browsing on your BlackBerry is the liberal use of bookmarks. Your BlackBerry will come with a couple of bookmarks already set. It is very easy to customize your bookmarks to include all your web favorites for easy browsing.

Adding and Naming Bookmarks to Easily Find Them

Let's set up a bookmark to find our local weather instantly.

1. Open the Browser and use the **Go To...** command, or the **.** (period) hotkey, to input a favorite web page. In this example, we will type **www.weather.com**.

2. Type your zip code or city name to see your current weather.

3. Once the page loads with your own local weather, press the **Menu** key, and select **Add Bookmark** (or use the **A** hotkey).

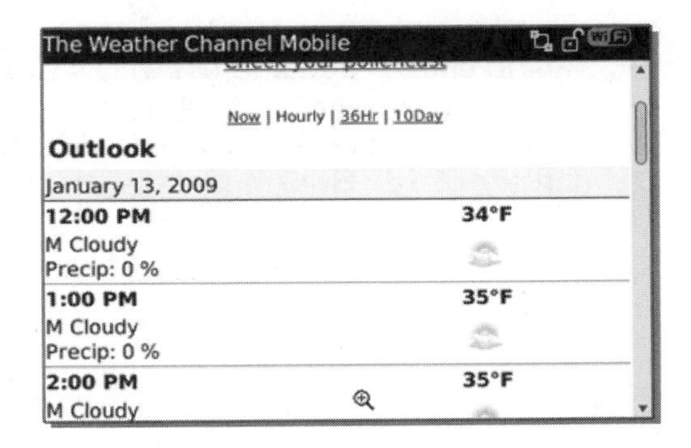

4. The full name of the web address is displayed. In this case, you will probably see TWC Weather. In most cases, we recommend changing the bookmark name to something short and unique.

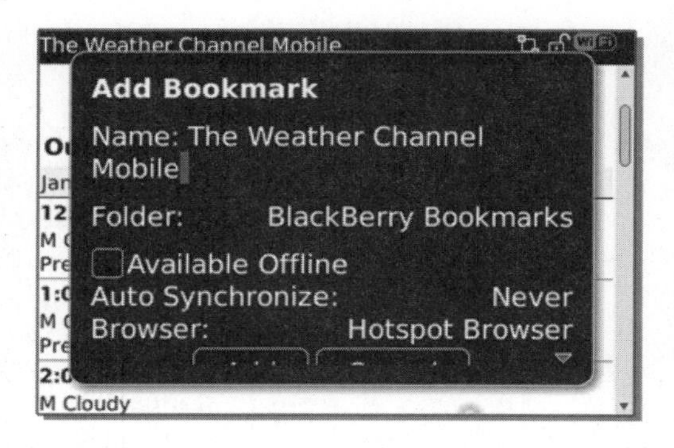

NOTE: if you bookmarked four different weather forecasts, the default bookmark names would all show up as The Weather Channel—sort of useless if you want to get right to the ten-day forecast.

Keep these things in mind as you name your bookmarks:

- Make all bookmark names fairly short. You will only see the first 10–15 characters of the name in your list (because the screen is small).
- Make all bookmark names similar but unique. For example, if you were adding four bookmarks for the weather in New York or your area, you might want to name them as follows:
  - NY Now
  - NY 10-day
  - NY 36-hour
  - NY Hourly

This way, you can instantly locate all your forecasts by typing the letters **NY** in your bookmark list. Only those bookmarks with the letters "NY" will show up.

You can save time by editing a bookmark. If you want to enter a web address that is similar to a bookmark, you should highlight the previously entered address, press the **Menu** key, and select **Edit Bookmark**.

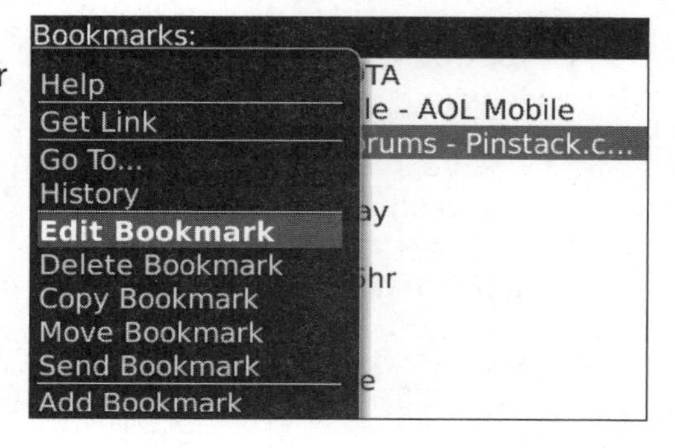

Then type your changes and glide to the bottom to click **Accept** to save your changes or **Cancel** to stop editing and not save any changes.

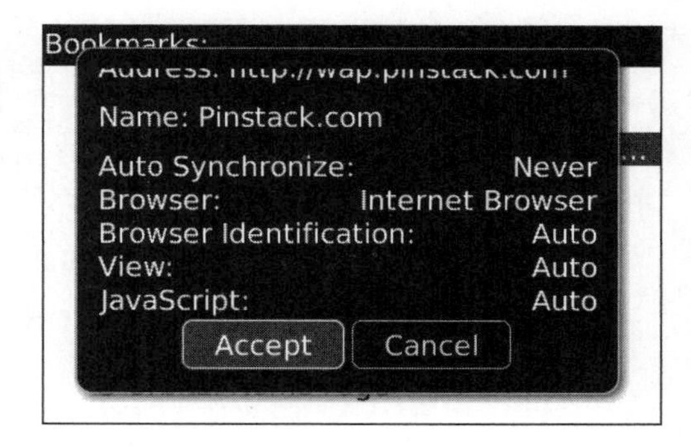

Setting Your Browser Start Page

When you open the browser, you might prefer to see the bookmark list rather than the start page or you might prefer to see your home page.

One reason to set your bookmark list as your start page is that it gives you a very fast way to get to your favorite web pages.

If you have home screen hotkeys turned on (see page 541), and your browser bookmarks as the start up page, you could type the letters **BNY** to instantly see all your weather bookmarks for New York City.

Pressing the home screen hotkey **B** opens the browser, showing you your bookmarks (assuming you have set your bookmarks to be your startup page in browser options).

Bookmarks:
BlackBerry Bookmarks
Home Page
Overheard in New York | The Voice...
BlackBerryTools OTA
Moviefone® Mobile - AOL Mobile
RIM Blackberry Forums - Pinstack.c...
OB weather Now
OB weather 10 Day
Google
OB _ weather - 36hr

Typing **NY** shows you all matching entries in your bookmarks with the letters "NY" in the name.

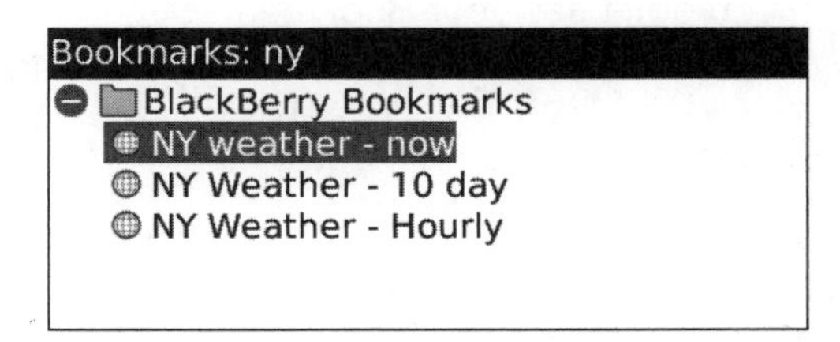

The reason is simple: typing letters in the bookmark list allows you to instantly locate the bookmarks with matching letters (like in the address book).

The benefit is that it's much faster to get to favorite web pages that are bookmarked like local weather or your favorite search engine.

Your BlackBerry may automatically open to your bookmarks list, but you may prefer to see a selected home page instead. You can use these instructions to make that change as well.

Here's how to set your bookmarks (or something else) to appear when you start your browser:

1. Click your **Browser** icon or press the hotkey **B**.
2. Press the **Menu** key, and select **Options** (or press the hotkey **S**).
3. Click Browser Configuration.
4. Click **Start Page**, all the way down near the bottom. You will most likely see three options **Bookmarks Page** (list of bookmarks), **Home Page** (the web site currently listed as your home page, which you can change on this screen), or **Last Page Loaded** (the last web page in memory can be brings up when you reenter your browser).

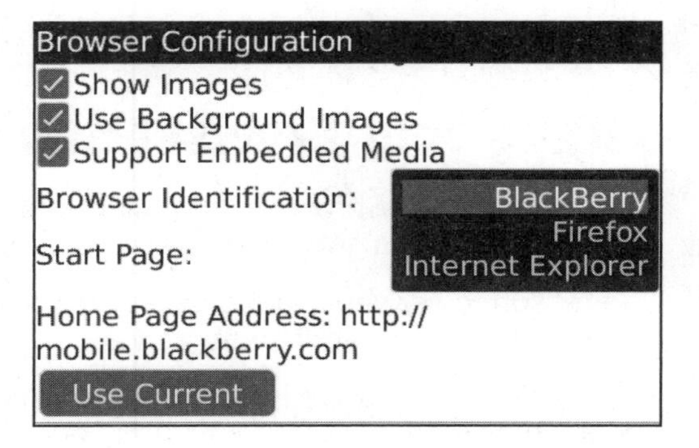

5. To select bookmark list, choose **Bookmarks Page**, and make sure to save your settings.

Using Your Bookmarks to Browse the Web

If you'd like to surf the Web on your BlackBerry, follow these steps.

Click your **Browser** icon, or press the hotkey **B** to start it.

If you don't see your Bookmark list automatically when you start your browser, press the **Menu** key, and click **Bookmarks**.

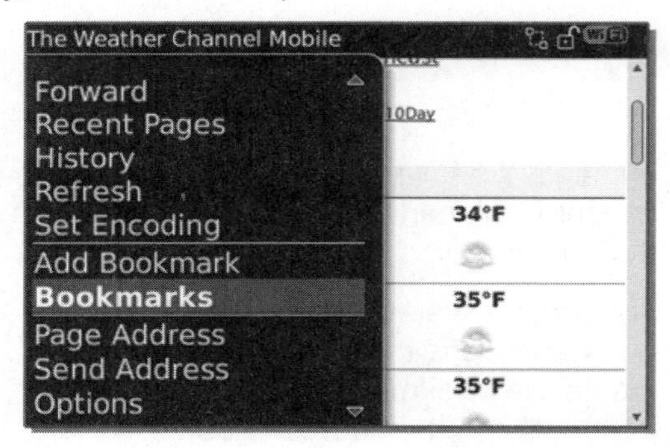

All of your Bookmarks will be listed, including any default bookmarks that were put there automatically by your phone company.

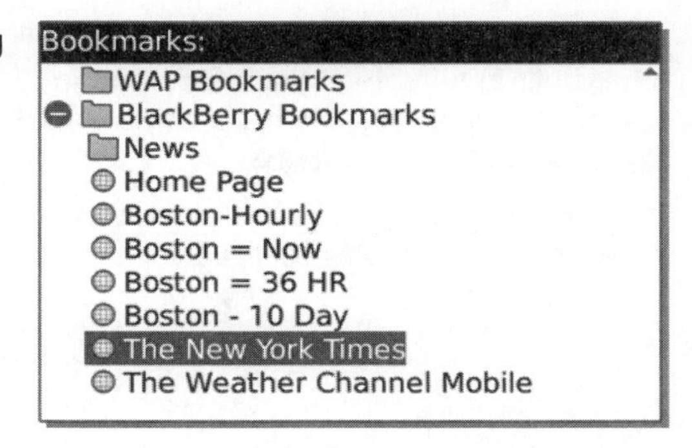

You might want to click a particular folder to open all the bookmarks contained within, or if you see the bookmark you need, just click it.

However, if you have a lot of bookmarks, you should use the **Find** feature—type a few letters matching the bookmark you want to find.

In the preceding image, typing the letters **NY** will immediately find all bookmarks with "NY" in the bookmark name.

Searching with Google

Google also has a mobile version that loads quickly and is quite useful on your BlackBerry. To get there, just go to `www.google.com` in your BlackBerry web browser.

As an example, to find all the pizza places in the zip code 32174, just type **pizza 32174** in the Google search field.

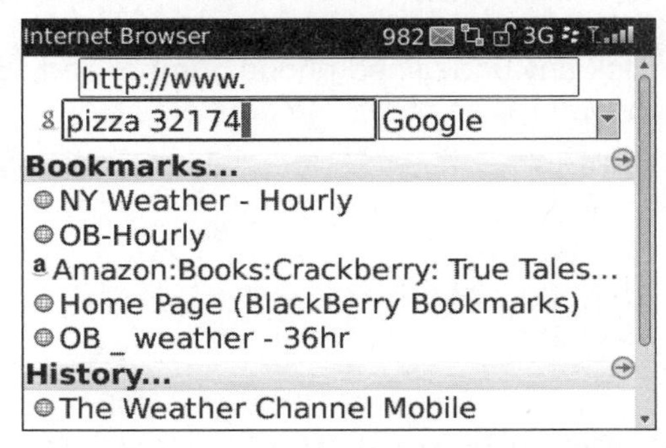

You can set this search bar for Google or other built-in search engines. To change search engines, click the drop-down arrow to the right of the search bar.

You can change the default search provider in the **Options ➤ Browser Configuration** screen.

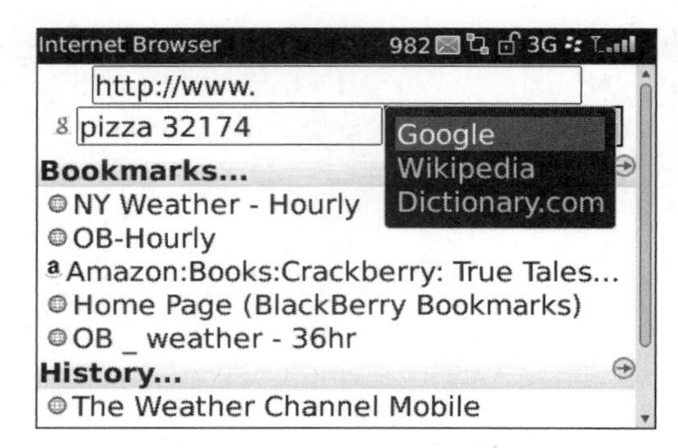

Once you press the **Enter** key or trackpad, your web search is started. Here is the results screen from Google.

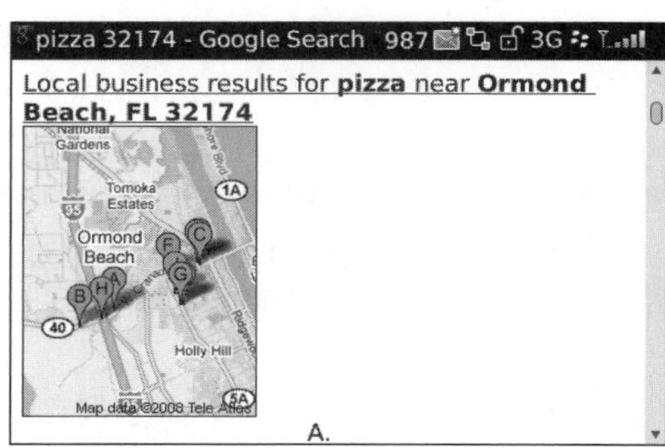

Scroll down a little to see all the local pizza establishments. The great part is that, since your BlackBerry is a phone, you can just click any underlined phone number and call for your pizza order immediately.

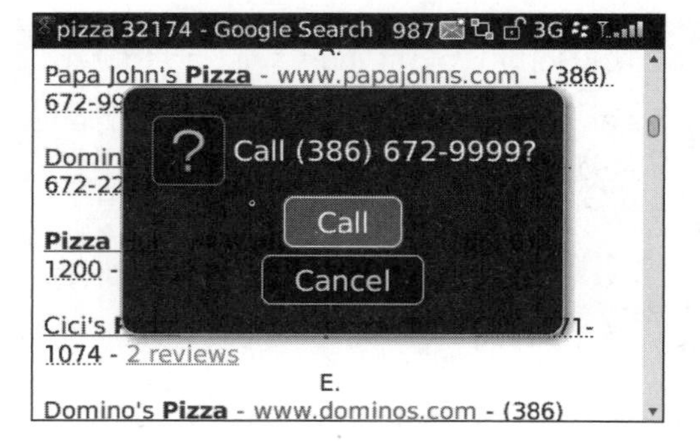

Finding a Quick Map of Your Search

You can see the small map shown on the regular Google search results screen, but we highly recommend using the Google Maps application for all your mapping and directions needs. See page 491 for more information.

You can even get driving directions using Google Maps to find the quickest path from my current location to the restaurant—all on your BlackBerry!

Finding Things Using Google Maps

We describe in detail how to obtain and use Google Maps on your BlackBerry on page 491.

Searching with Yahoo Mobile

Yahoo, like most of the other big search engines, has a mobile version. The web site senses you are viewing it from your small BlackBerry browser adjusts to a scaled-down version that gives you most of the functionality on a smaller screen.

How can you get access? Go to `www.yahoomobile.com` from your BlackBerry web browser. It's free!

Web Browser Tips and Tricks

There are some helpful shortcuts to help you navigate the Web faster and easier. We have included a few for you here:

- To insert a period in the web address in the **Go To...** dialog box, press the **Space** key.
- To insert a forward slash (/) in the **Go To...** dialog box, hold the **Shift** key and press the **Space** key.
- To open the bookmark list from a web page, press **K**.
- To add a bookmark from a web page, press the **A** Key.
- To stop loading a web page, press the **Escape** key.
- To go to a specific web page, press the **G** key.
- To close a browser, press and hold the **Escape** key.

Visiting YouTube

One of the most fun sites to visit on the computer is YouTube for viewing short video clips on just about everything. Your new BlackBerry is able to view most YouTube videos without doing anything special. We show you how to visit YouTube (`m.youtube.com`), on page 420.

Chapter 30

Add or Remove Apps

Your BlackBerry comes with most of the major apps you need already installed. However, there are literally thousands of third-party apps available in virtually any category you can think of to help you get the most out of your BlackBerry.

There are apps for productivity, reference, music and fun—like lots of great games. One nice thing about the BlackBerry platform is that there are lots of ways to find and download apps beyond the official App World. (See chapter 26 on page 431 to learn about BlackBerry App World.) In this chapter, we will show you how to download, install and remove apps outside of App World.

Downloading and Adding New Software

One of the very cool things about your BlackBerry is that, just like on your computer, you can go on the Web and find software to download. You can download everything from ring tones and games to content that is, like your email, pushed to your BlackBerry on a regular basis.

Today, there are hundreds of software applications and services to help extend the capabilities of your BlackBerry. We have used or are currently using the programs or applications in Table 30-1. Find many more programs using BlackBerry App World on page 431.

Table 30-1. *Selected Apps Used By the Authors*

App Name	What It Does	Where To Get It
AP Mobile News	Associated Press reader	http://ap.mwap.at/
CrackBerry & Shop CrackBerry	CrackBerry Blogs CrackBerry Store	http://wap.crackberry.com
ESPN	ESPN website	www.espn.com
Facebook Mobile	Social networking	http://mobile.blackberry.com
Flycast	Internet streaming radio	www.flycast.fm
Google Maps	Mapping and search software	http://google.com/gmm
Google Mobile	Voice-activated search and access to all Google Apps	http://m.google.com
Kindle	Allows you to read your kindle books on your BlackBerry	www.amazon.com/kindlebb
Newsweek Mobile	*Newsweek*'s website	http://mobile.blackberry.com
NY Times	The *New York Times* website	www.nytimes.com
Pinstack	Pinstack BlackBerry forum	www.pinstack.com
Slacker	Internet streaming radio	www.slacker.com
Viigo	RSS news reader	www.viigo.com
Wall Street Journal Mobile	The *Wall Street Journal*'s website	http://mobile.blackberry.com
YouTube Mobile	YouTube website	http://m.youtube.com

Installing New Software

There are several ways to install software on your BlackBerry.

The first option is to browse mobile.blackberry.com or another web site to locate software designed for your BlackBerry. Then click a link and download wirelessly or over the air (OTA). You sometimes have the link emailed to you after entering your address into a web form on the software vendor's or web store's site. Another way is to type a web address directly into your BlackBerry web browser.

> **TIP:** You may need to use the **Go To…** menu command in your Browser to type a web address.

Some sites ask you to enter your mobile phone number (your BlackBerry phone number) to send you an SMS text message with the download link.

The second option is to download a file to your computer and connect your BlackBerry to your computer to install it via USB cable or Bluetooth connection. Mac users, see the chapter starting on page 125 and Windows PC users see the chapter starting on page 65.

You can use the third option only if your BlackBerry is connected to a BlackBerry enterprise server. Your BlackBerry server administrator can push new software directly to your BlackBerry device. Using this option is both wireless and automatic from the user's perspective.

We will describe options 1 and 2 in this book.

Wirelessly Installing Software Directly from Your BlackBerry Over the Air

The beauty of the wireless over-the-air (OTA) software installation process is that you do not need a computer, a CD, or even a USB cable to install new software. It is really quite easy to install software OTA on the BlackBerry.

> **NOTE:** Software installation used to be a much more painful process in older BlackBerry models—thank goodness for Research In Motion's continuous technical advances!

To install software from your BlackBerry, as we described above, you need to either click a link you received in your email or SMS text message or start the Web Browser and **Go To…** a web address where the software download files are located. In this example, we will use `www.mobile.blackberry.com`:

1. Find and click the link for the software title you want to install.
2. Usually, a license agreement screen pops up. You must click **Accept** to continue with the download.

3. The next screen to appear is the standard Download screen. This screen shows the name, version, the vendor (software company), and size of the program. Below all this information, you see a **Download** button to click.

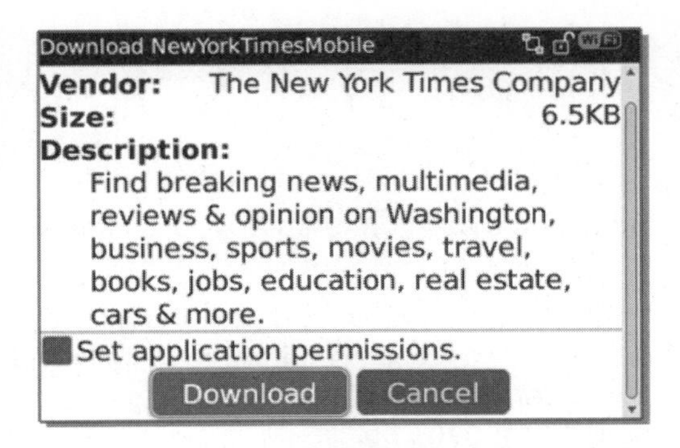

4. The progress of the download will be displayed in the center of the screen.

5. You may see a warning message that says something like "Warning: Application is not digitally signed. Do you want to continue? Yes or No". You must click **Yes** to continue, but make sure this is software from a vendor that you recognize or have confidence that it is not malicious.

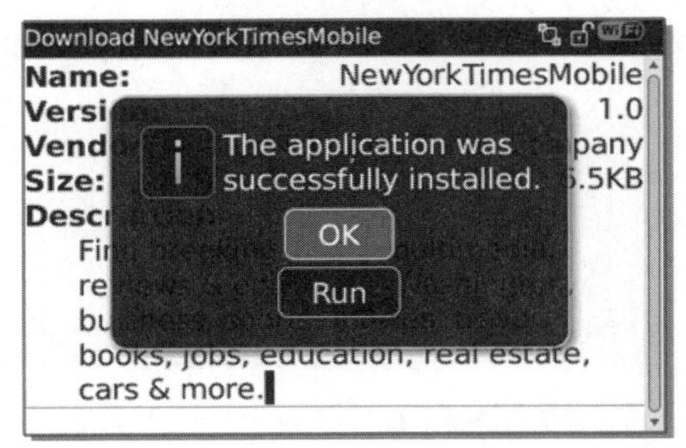

6. Once the application is completely downloaded, a dialogue box pops up notifying you the download was successful. You may be given an option to run the program.

7. If you click **Run**, you start the newly installed software; clicking **OK** brings you back to the original web page where you downloaded the file. To exit to your home screen, press the **Escape** key a few times.

8. You will then see the new icon either on your home screen of icons, in the **Applications** subfolder, or after you press the **Menu** key to see all your icons.

Downloading and Installing Games and Other Apps on Your BlackBerry

The BlackBerry can truly be a multimedia entertainment device. Sometimes, you might want to play a new game on your BlackBerry. While your BlackBerry may have come with a few games, there are many places on the Web where you can find others.

The **BlackBerry App World** (see page 431) is a great place to find new games!

One place to start looking for additional games is the mobile BlackBerry site at `http://mobile.blackberry.com`. Another good site is `www.shopcrackberry.com`, which also has an applications store that you can access from your device.

See the web browser section on page 463 to learn how to get to these web pages on your BlackBerry.

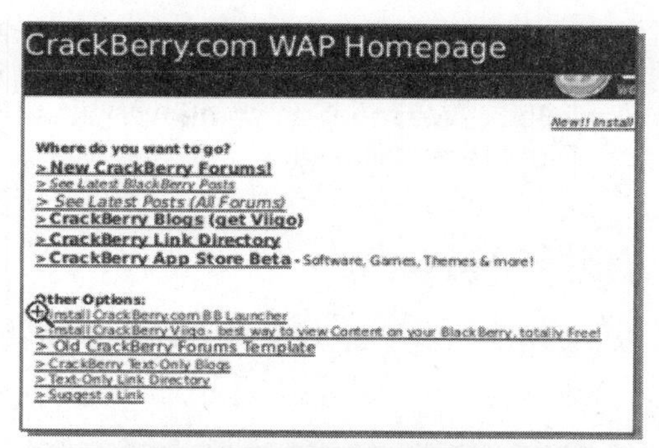

Scroll through the choices until you find a game that you want to try, and then just select the **Download** link.

Another site to look at is `www.bplay.com` where you can also find some of the most popular games available for BlackBerry entertainment.

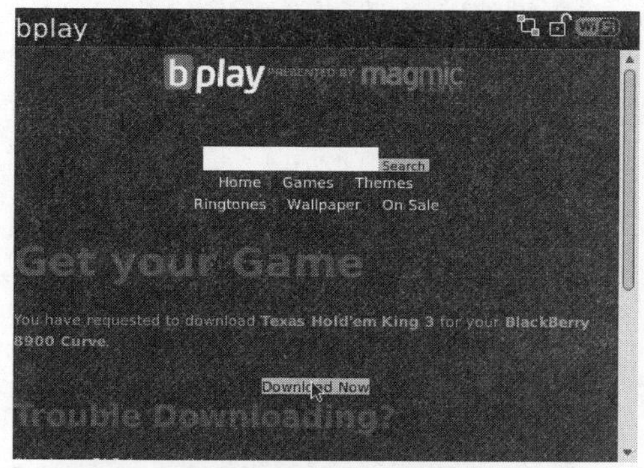

Usually, you can just add the game to a shopping cart and have it billed to your mobile number or pay on line with a credit card or PayPal.

Then, just download and install exactly as you did when installing the web push icon. Initially, the program file will go into your **Downloads** folder as stated previously. You can always move the icon for the game into your **Games** folder; just take a look back at page 169 to see how to do that.

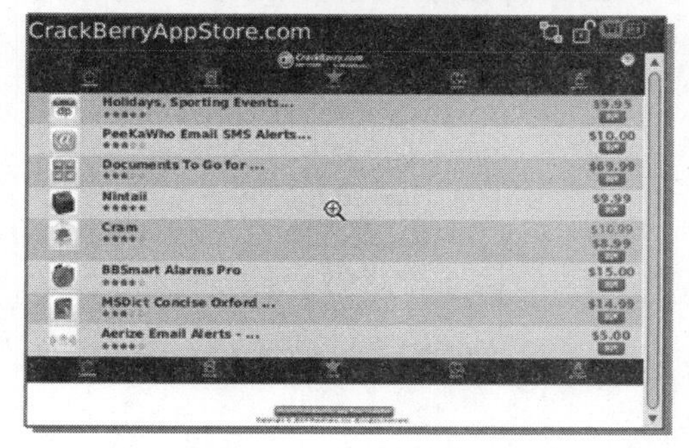

Visiting the BlackBerry App World Store

Earlier in 2009, BlackBerry launched the on-the-BlackBerry mobile software store called BlackBerry App World to compete with the Apple iPhone AppStore. We have an entire chapter dedicated to getting the most out of the BlackBerry App World, see page 431.

Getting BlackBerry Software Other Places

Since your BlackBerry is not locked to only use the BlackBerry App World, you really do have a myriad of possibilities as to where you can find quality software for your BlackBerry. We mention some websites below, but you can also do a simple web search for "BlackBerry Software" and see if other options come up.

Web Stores

You can usually purchase software from these stores:

- www.crackberry.com
- www.shopcrackbery.com
- www.bberry.com
- www.handango.com
- www.mobihand.com

Online reviews of software, services, ring tones, themes, wallpaper, accessories, and other BlackBerry-related news and Technical Support can be found here:

- www.allblackberry.com
- www.bbhub.com
- www.berryreview.com
- www.blackberrycool.com
- www.blackberryforums.com
- www.boygeniusreport.com
- www.howardforums.com (The RIM-Research In Motion Section)
- www.pinstack.com
- www.RIMarkable.com

BlackBerry Solutions Catalog

The *BlackBerry Solutions Catalog* is available at http://www.blackberrysolutionscatalog.com/

Software and services you will find in the Solutions Catalog will be focused toward business users more than individuals.

BlackBerry App World on the Web

The App World is also available for you to browse from your computer's web browser at: `http://appworld.blackberry.com/webstore/`.

Removing Apps Directly from your BlackBerry

There will be times when you wish to remove a software icon from the BlackBerry and you are not connected to your computer. Fortunately, it is very easy and intuitive to remove programs from the BlackBerry itself.

> **TIP:** If you have installed software using the BlackBerry App World, you can also remove it from the My World section; please see page 438 for help.

Deleting an Icon from the Home Screen

Probably the easiest way to delete an icon is to highlight it, press the **Menu** key, and select **Delete**.

> **NOTE:** The icons for some core apps like Messages, Calendar, and Contacts cannot be deleted, so you won't see the **Delete** menu item.

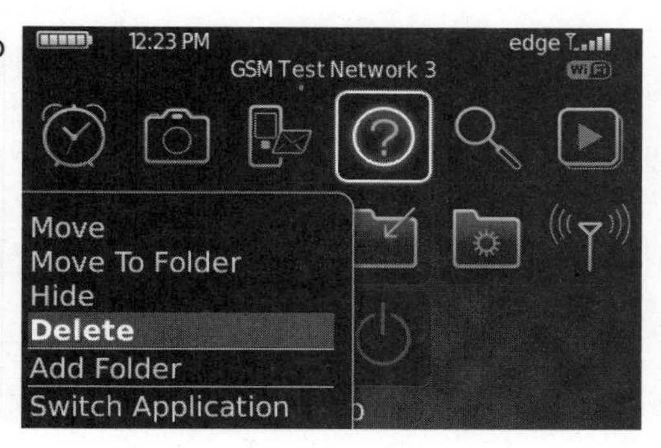

Using the Options Icon to Remove Software

To remove software directly from the BlackBerry, just do the following:

1. Press the **Menu** key, and click the Options icon.

2. Click Advanced Options.
3. Scroll to the top of the list, and click **Applications**.

Options
About
Advanced Options
Auto On/Off
AutoText
CellData
Date/Time
GPS Extended Ephemeris
Language

4. The screen will now display all the programs installed on your BlackBerry. Scroll to the application you want to delete to highlight it, press the **Menu** key, and select **Delete**.
5. Then, you will need to confirm that you want to delete the program and it will be removed.

NOTE: Sometimes you BlackBerry will need to reboot itself to complete the program deletion process. It will tell you if it needs to reboot and ask you to confirm.

Troubleshooting Software Removal

If the software is not completely removed from your BlackBerry after following the above steps, then try this:

1. Go back into the **Options** icon.
2. Select **Advanced Options**.
3. Select **Applications**.
4. Press the **Menu** key, and select **Modules**.
5. Now scroll down the list of modules and make sure to delete every module with a name that is related to the software you are trying to remove. Highlight a module, press the **Menu** key, and select **Delete** from the menu.

NOTE: If you have a Windows computer, you can also use the **Application Loader** icon built into BlackBerry Desktop Manager software to remove icons (see page 89).

Chapter 31

Traveling: Maps and More

If you are traveling with your BlackBerry, there are a few things you should know how to do. There are plenty of mapping applications that use the global positioning system (GPS) location in the BlackBerry to help you find your destination and provide you with turn-by-turn directions.

If you are going to another country, there are some definite things to do before you take off so you do not get surprised with a huge data roaming phone bill.

Traveling Internationally

Some BlackBerry smartphones are equipped for international travel, and some are more of a challenge. Networks in the United States like Verizon and Sprint use CDMA technology— widely available in North America, but not in other places. BlackBerry smartphones that use SIM cards (AT&T and T-Mobile) in the United States are easier to use throughout the world.

We always recommend that you call your cell provider well in advance of your trip to see if there is an international feature you can turn on.

Avoiding a Surprisingly Large Roaming Phone Bill

We have heard of people who traveled to another country being surprised with $300 or $400 monthly data or voice roaming charges after their trips. You can avoid these costs by taking a few easy steps before and during your trip.

Checking to See If You Have a SIM Card inserted

If you are traveling to a country where you will need to connect to a GSM network, you will need to have a SIM Card in your BlackBerry. Most times, with a BlackBerry Bold, you will already have a SIM card pre-inserted by the phone company.

You can check that you have a SIM card by clicking on the **Options** icon followed by **Advanced Options**. Tap the letter **S** a couple times to jump down to **SIM Card** and click it. If you see a screen that says "SIM Card: No Valid SIM Card," then you do not have a SIM Card or your SIM card is not inserted correctly.

You can also quickly check if you have a SIM card by removing the back cover of your BlackBerry and taking a look. See page 5 for a picture of where the SIM Card Slot is located.

Calling Your BlackBerry Phone Company to Turn on a Temporary International Rate Plan

Check with your wireless carrier about any voice and data roaming charges. You can try searching on your phone company's web site, but usually, you will have to call the help desk and specifically ask what the *voice and data roaming* charges are for the country or countries you are visiting. If you use email, SMS text, MMS messaging, web browsing, and any other data services, you will want to specifically ask about whether or not any of these services are charged separately.

Some phone companies offer an international rate plan that you will need to activate. In some cases, you must activate such a plan to use your BlackBerry at all; in other cases, activating such a plan will allow you to save some money on the standard data and voice roaming charges. Check out these plans to see if they can save you some money while you are on your trip, especially if you need to have access to data while you are away.

Unlocking Your BlackBerry to Use a Foreign SIM Card

In some cases, your BlackBerry phone company does not offer special deals on international data roaming plans or their rates are unreasonably high. In these cases, you may want to ask your phone company to unlock your BlackBerry so that you can insert a SIM card you purchase in the foreign country. Many times, inserting a local SIM Card will eliminate or greatly reduce data and voice roaming charges. However, you should check carefully the cost of placing and receiving international calls on that foreign SIM card. Using a foreign SIM card may save you hundreds of dollars, but it's best to do some web research or try to talk to someone who has recently traveled to the same country for advice.

Getting into Airplane Mode

See page 16 to learn how to turn off the wireless radio and other connections on your BlackBerry. Airlines force you to completely power down your electronic devices during take-off and landing, but then allow use of "approved electronic devices" (i.e., BlackBerry with the radio turned off) while in flight.

Things to Do While You Are Abroad

Once you do everything above and prepare for your trip, there are still some things to do once you get to your destination.

After you get there, adjust your time zone to the local time zone.

See page 175 for help with setting the time zone.

If Data Roaming Charges are unknown or too high, turn off data roaming. If you were unable to find out about data roaming charges from your local phone company, try to contact a phone company in the country where you are traveling to find out about any data roaming or voice roaming charges.

In the worst case, if you are worried about the data roaming charges and can do without your email and web while away, you should disable data services when you are roaming:

1. Click the Options or Manage Connections icon.
2. Click **Mobile Network.**
3. Your data services will most likely be set to **On**; there is a secondary tab for **While Roaming.** We recommend setting the switch to **Off** in the **While Roaming** field so you can continue to receive data in your home network.
4. Just press the **Escape** key; your settings are saved automatically.

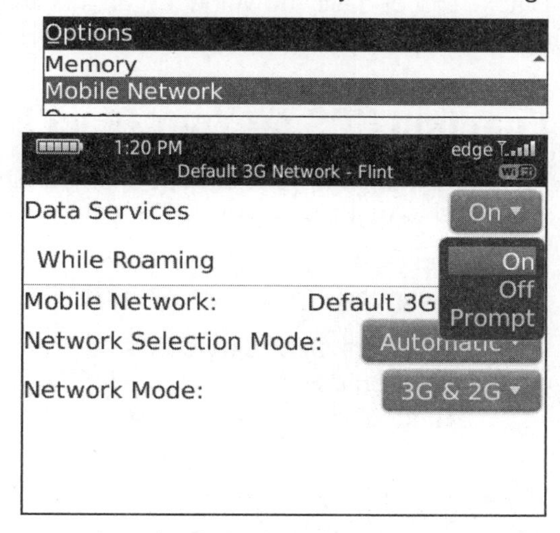

Setting **While Roaming** to **Off** should help you avoid any potentially exorbitant data roaming charges. You still need to worry about voice roaming charges, but at least you can control those by watching how much you talk on your phone.

Remember, with the **Off While Roaming** setting, when you return to your home network, all data services (email, web, etc.) will work automatically.

You may also want to register on a local network. If you are having trouble connecting to the local wireless network, you may need to register your BlackBerry on the local network. See the Host Routing Table (Register Now) on page 549.

Returning Home

Once you arrive at home again, make sure to adjust your time zone back to your local time zone. See page 175 for help with setting the time zone.

If you have turned Data Services to Off, you will need to make sure to turn them back **On** so you can receive data when you return home.

You may need to register your BlackBerry on the local network again. See the Host Routing Table (Register Now) on page 549.

You may want turn off special international plans, if they're not needed. If you have activated some sort of special international roaming rate plan with your BlackBerry phone company and do not need it any more, turn it off to save some money.

BlackBerry Maps, Google Maps, Bluetooth GPS

In addition to the myriad of possibilities in which your BlackBerry can manage your life, it can also literally take you places. With the aid of software that is either preloaded on the BlackBerry or easily downloaded on the Web, you can find just about any location, business, or attractions using your BlackBerry.

Enabling GPS on Your BlackBerry

To get the maximum benefit out of any of any mapping software on your BlackBerry, you need to make sure your GPS receiver is enabled or turned on.

1. Click your **Options** icon. Select **Advanced Options** and then **GPS**.
2. Make sure **GPS Services** is set to **Location ON** as shown, press the **Menu** key, and select **Save**.

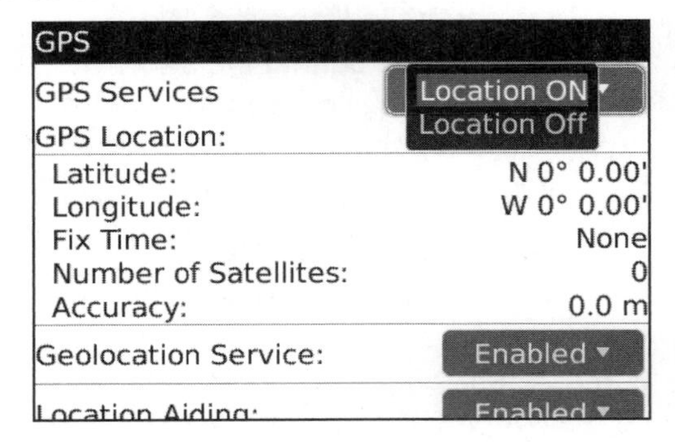

Using BlackBerry Maps

The BlackBerry ships with the BlackBerry Maps software, a very good application for determining your current location and tracking your progress via the built-in GPS receiver on your BlackBerry.

> **NOTE:** For GPS to work, you first have to turn on or enable GPS on your BlackBerry; see the section above.

To Enable GPS use on BlackBerry Maps (using either Bluetooth GPS receiver or built-in GPS)

1. Click the **Maps** icon in the applications menu, and then press the **Menu** key.
2. Click **Options**.
3. Under **GPS Source**, select either **Internal GPS** or an external GPS unit if you have one.
4. Press the **Escape** key, and save your choices.

Viewing a Particular Map from a Contact

You can view a map as follows:

1. From the **Map** icon, press the **Menu** key from the main map screen and select **Find Location.**
2. In the Mapping screen, press the **Menu** key and select **Start GPS Navigation**.

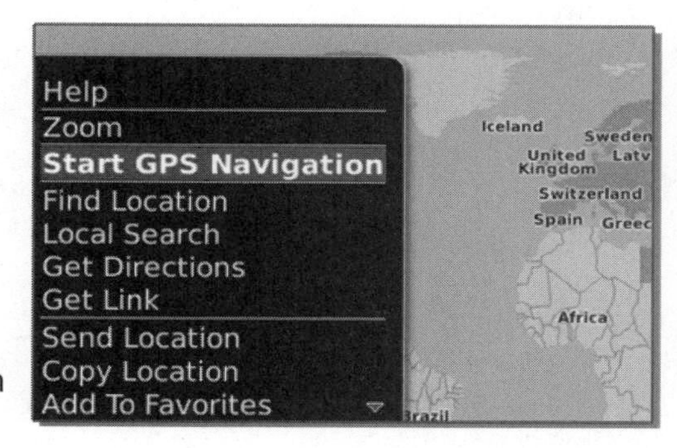

The BlackBerry Map software will now search for your current location.

Alternatively, you can follow these steps to view a map of a contact's location:

1. Click the **Map** icon, press the **Menu** key from the main map screen, and select **Find Location**.

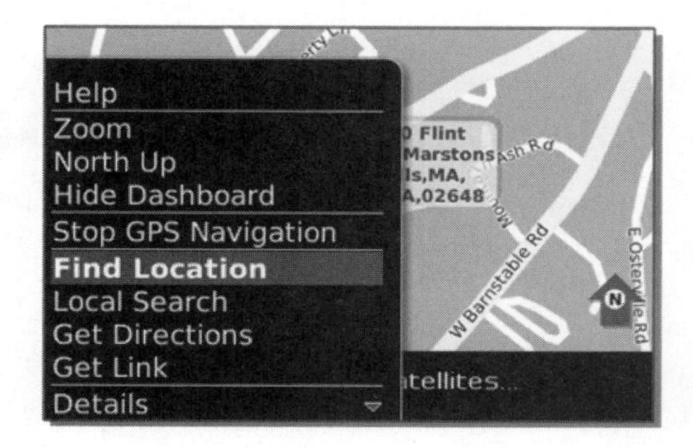

2. You can search based on your current location or you can enter an address, search from contacts, or look at recent searches. Click **From Contacts**.

3. On the next screen scroll down to and click the contact you want.

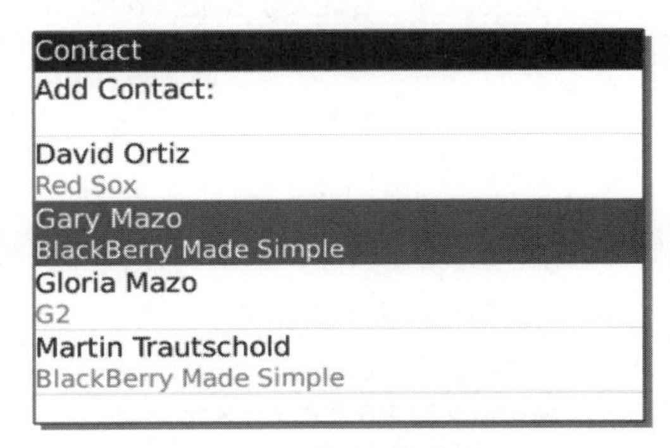

4. A map of that contact's location will appear on the screen.

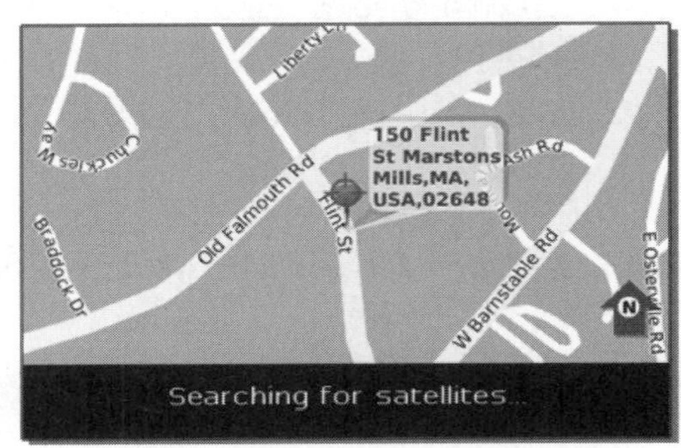

Getting Directions with BlackBerry Maps

Follow these steps to get directions from the native Maps app:

1. Press the **Menu** key from the map screen, and select **Get Directions**.

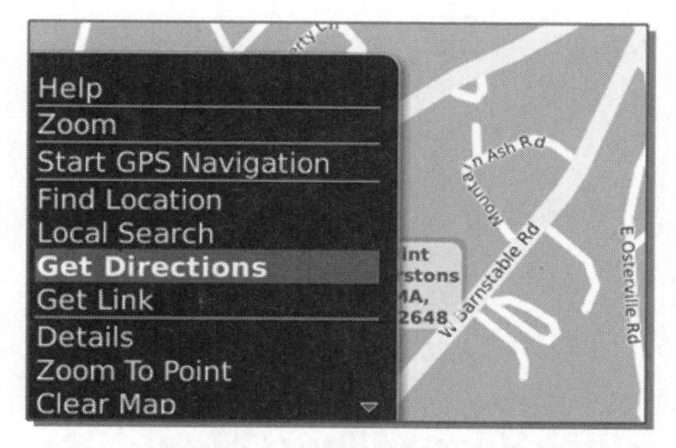

2. Select a start location by scrolling down to **Where I Am**, **Enter Address**, **From Contacts**, **From Map**, **Recent**, or **Favorites**.
3. Repeat the steps now for the end location.

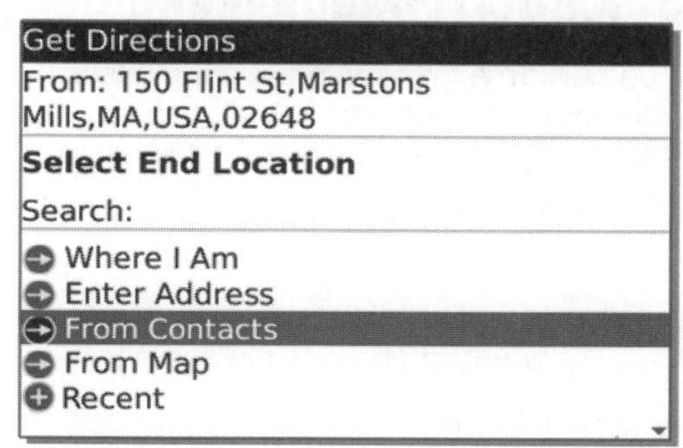

4. When you're finished, click the trackpad, and the BlackBerry Map program will create a route for your trip.

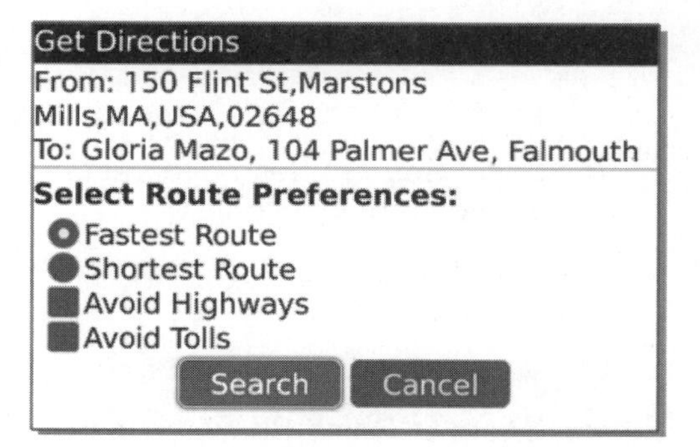

5. You can choose whether you want the fastest or shortest route and whether you want to avoid highways or tolls. Then, just click **Search** to have your route displayed for you. If you click **View on Map** you will see a map view of the route.

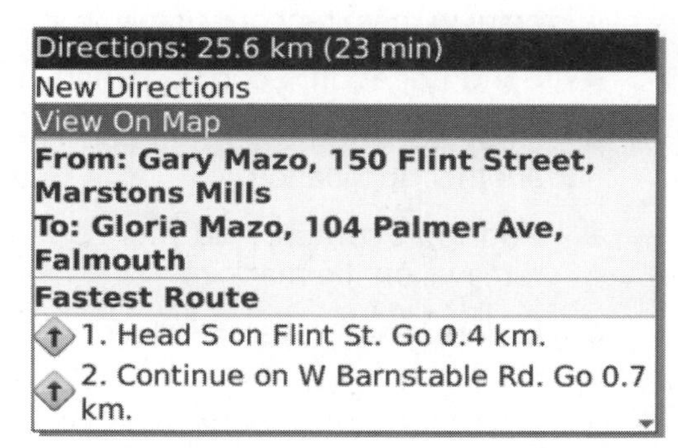

6. If you have enabled GPS use, the GPS will track you along the route but it will not give you voice prompts or turn-by- turn voice directions.

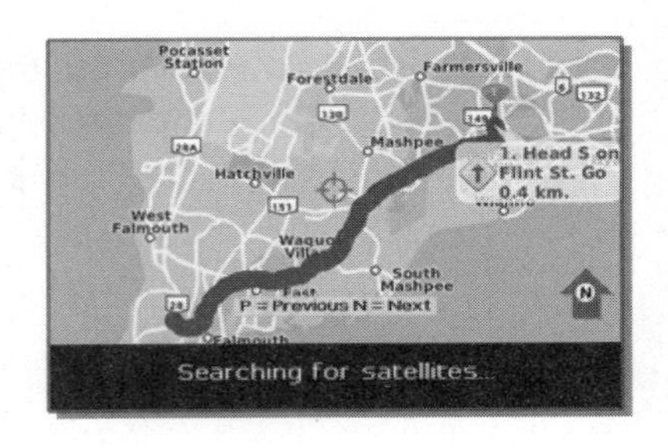

BlackBerry Maps Menu Commands

You can do a great deal with the BlackBerry Maps application. Pressing the **Menu** key offers you many options with just a scroll and a click. From the menu, you can do the following:

- **Zoom**: Go from street level to the stratosphere (hotkeys: **L** for zooming in and **O** for zooming out).
- **North Up**: Place the north arrow straight up and keeps the map oriented so up is always north (only with GPS enabled).
- **Hide Dashboard:** Put the map in full-screen mode, without the bottom border.
- **Stop GPS Navigation:** End GPS tracking (only with GPS enabled).
- **Find Location**: Type in an address to jump to that address.
- **Local Search:** Search for restaurants, stores, or places of interest near your current location.
- **New Directions**: Find directions using your location history or typing in new addresses.
- **View Directions**: Switch to text, if you are in map mode.
- **Zoom to Point**: Show the map detail around the currently selected point in the directions. (Only when viewing directions.)
- **Send Location**: Send your map location via email.
- **Copy Location**: Add your current location to your address book or another application.
- **Add to Favorites**: Add your current location as a favorite for easy retrieval on the map.
- **Layers**: Show layers of recent searches, favorites, or links on the map.

- **Options**: Change GPS Bluetooth devices, disable backlight timeout settings, change units from metric (kilometers) to imperial (miles), enable or disable tracking with GPS, and show or hide title bar when starting.

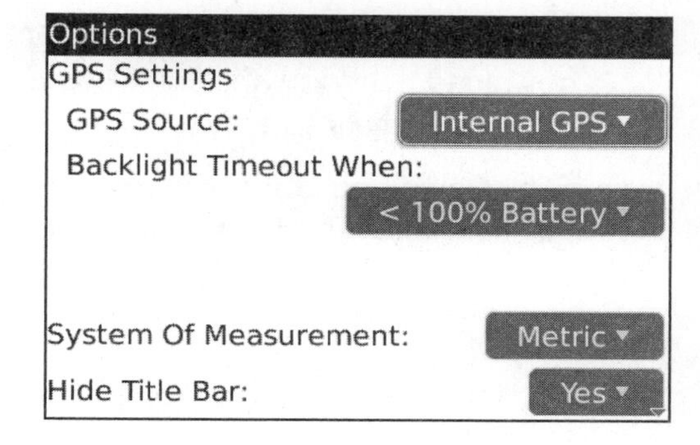

- **About:** Show information about the current provider of the mapping data and software.
- **Cell Data:** Control whether you can be tracked or not
- **Switch Application:** Jump over to any other application (multitask).

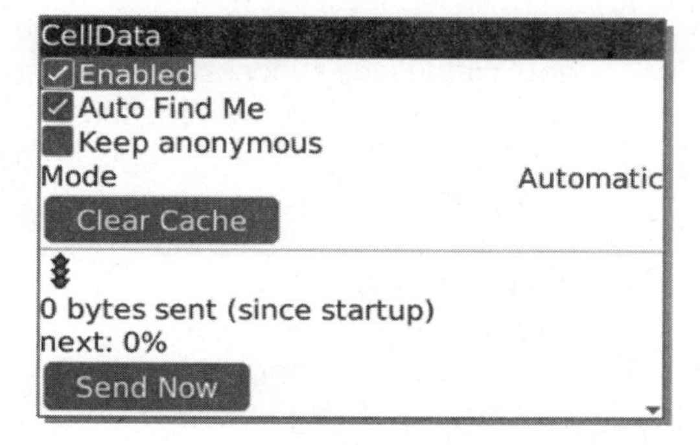

Using Google Maps

If you have ever used Google Earth, you have seen the power of satellite technology in mapping and rendering terrain. Google Maps Mobile brings that same technology to handheld devices including the BlackBerry.

Downloading and Installing Google Maps

With Google Maps you can view 3-D rendered satellite shots of any address, anywhere in the world. To get started with this amazing application, you need to first download it onto your BlackBerry.

1. Click the **Browser** icon from the application menu.
2. Press the **Menu** key, and click the **Go To...** option.

> **TIP:** Use the shortcut hotkey of . (period) for **Go To...**.

3. Enter the address to perform an OTA download right onto the BlackBerry: `http://www.google.com/gmm`.
4. Click **Download Google Maps**, and the installation program will begin.

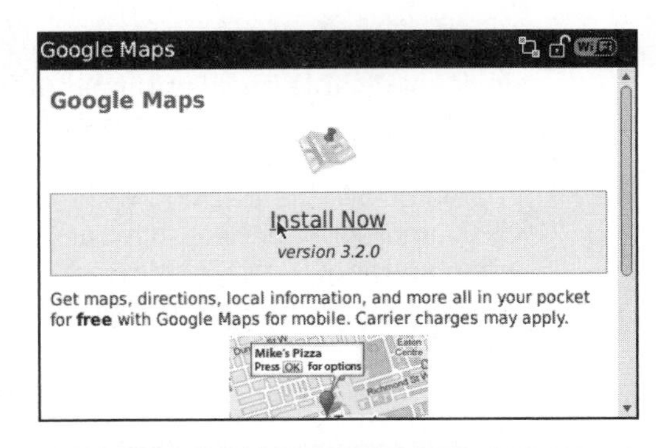

5. Click the **Download** button on the next screen.
6. Finally, you will see a screen that the application was successfully installed. Select **OK** to close the window, or **Run** to start Google Maps right away. Or click on the icon on the BlackBerry, which should be in the **Downloads** folder.

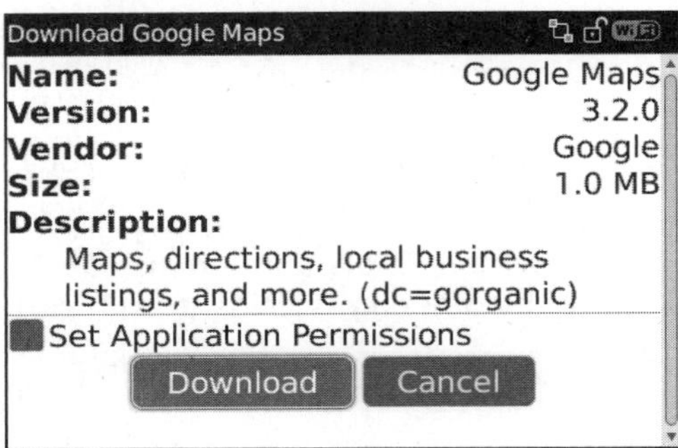

7. You might get prompted to reboot your device; if so, just click **Reboot** and then try to run the program.

8. The first time you start up Google Maps Mobile, just click the icon (usually in your **Downloads** folder. Read the terms and conditions, push in the trackpad, and select **Accept**.

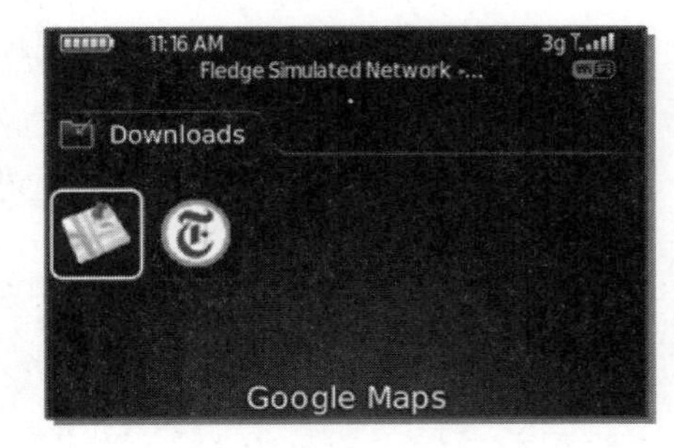

Google Map Menu Commands

Google Maps is full of great features, most of which are accessed right from the menu. Press the **Menu** key to see them.

One of the very cool new option is **Street View**, which shows you actual photos of your location on the map (where available).

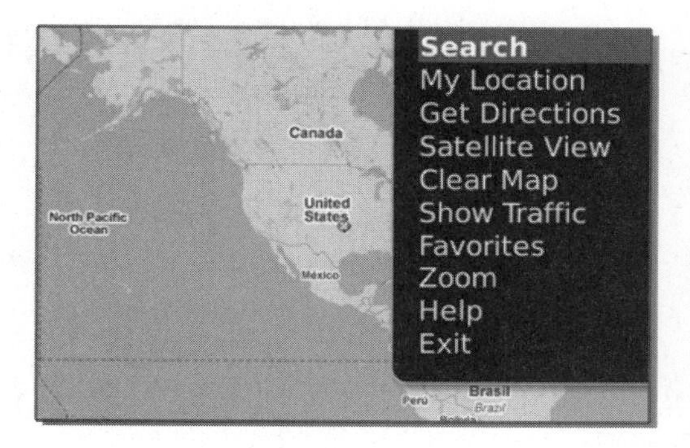

Searching for an Address (Location) or Business

Finding an address or business is very easy with Google Maps.

1. Start the **Google Maps** application by clicking on the icon.
2. Press the **Menu** key and choose **Search**,
3. On this screen you can type a new search or glide down to click on a recent search below.
4. You may type about anything in the search s tring—an address, city name, type of business and zip code, business name and city/state, and so on. If we wanted to find bike stores in Ormond Beach, Florida, we would enter "bike stores Ormond Beach, FL" or "bike stores 32174" (if you know the zip code).

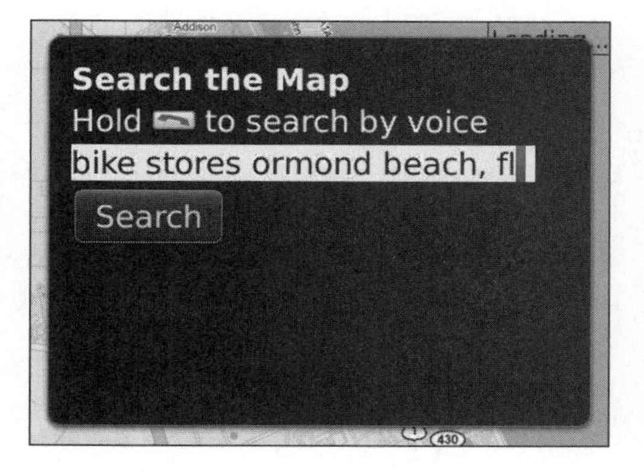

5. You will see a number of pins on the map, each with a letter, showing your search results.

6. From here you can do a number of things:
   Press **6** to see the next search result
   Press **#** to see the list of results
   Press the **Enter** key to see details on this location.

7. The list of search results looks like this image. From here you can glide the trackpad up/down to select an entry and click the trackpad to see details.

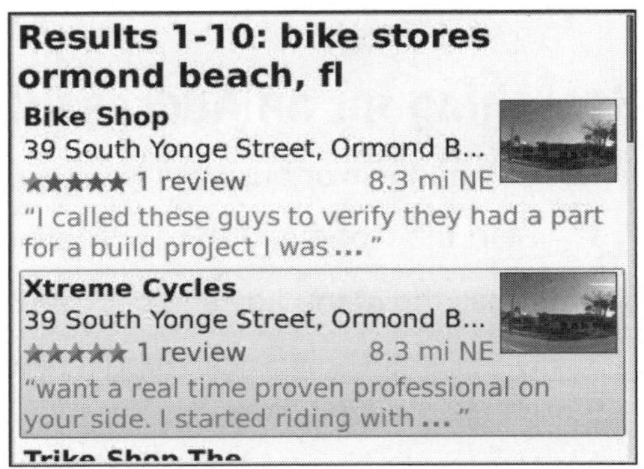

8. You will see this details screen if you click on one of the items from the search list, press the **Enter** key when the pin is highlighted or click the trackpad on the pin. From here you can use the icons in the middle of the page to:

   map the location,

   get directions (see page 499 for more on directions),

   call the location, or,

   see the street view of the front of the location.

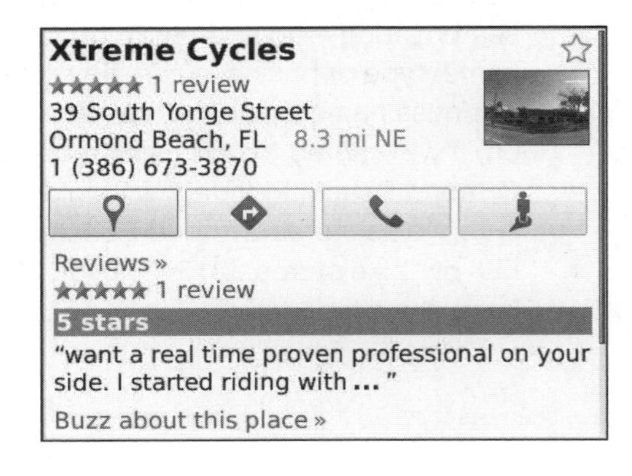

9. Pressing the street view icon will show you the front of the location. Roll around with the trackpad to see the other side of the street or up and down the street. Press the **Escape** key to get back to map view.

> **NOTE:** If you have enabled GPS use on your BlackBerry, the GPS will track you along the route but it will not give you voice prompts or turn-by-turn voice directions.

Google Map Hotkeys

Use these keys in Table 31-1 on your BlackBerry keyboard to control Google Maps.

> **NOTE:** Just press the key with the number without holding ALT (e.g., to press the 4 in Google Maps, you just press the S (4) key by itself).

Table 31-1. *Google Hotkeys*

4	Previous search results
6	Next search results
#	Toggle between the map view and search results list
2	Toggle between satellite and map views
1 / O	Zoom out
3 / I	Zoom in
*	Favorites list / add a new favorite location
9	More search options and search tips
0	Show or hide location (if available)

Switching Views in Google Maps

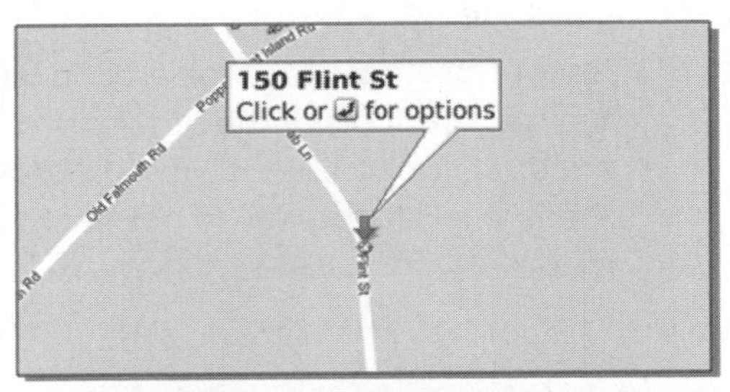

Google Maps gives you the option of looking at a conventional map grid or real satellite images.

To switch between views, use the hotkey = **2**, or press the **Menu** key and select **Satellite View** (or 2).

To switch back, press the **Menu** key, and select **Map View** (or press **2**).

Seeing Your Current Location—The Little Blue Dot

Versions 3.0 and higher of Google Maps allow you to show your location as a little blue dot, which is pinpointed by the GPS receiver built into your BlackBerry.

> **NOTE:** When your GPS cannot get a satellite lock, or for your friends without GPS built-in to their BlackBerry devices, the approximate location is determined by triangulation using cell phone towers. When cell towers are used, a shaded circle around the blue dot shows that the location is approximate.

This is a great feature and very easy to access using the shortcut key of zero **0**. Pressing **0** or selecting **My Location** from the menu will show you the blue dot.

Another new feature is Google Latitude, which allows you to enable your Google contacts to see where you are in real time. You can also see them once you accept invitations.

Just press the **Menu** key, and select **Latitude** to get started. Once set up, your location on the map is sent to whoever you choose.

You will see all your friends and family on the map if they also share their location using Google Lattitude.

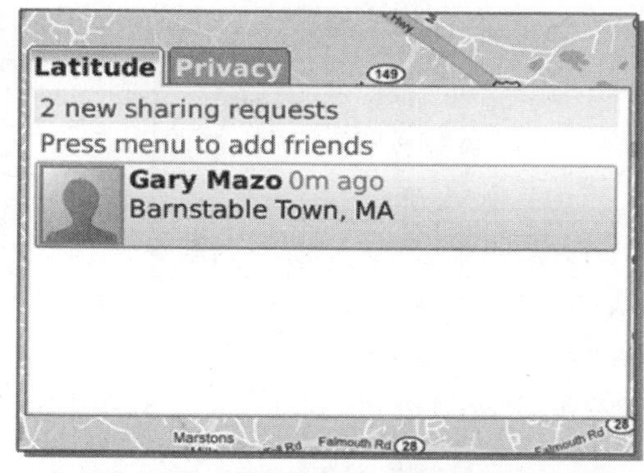

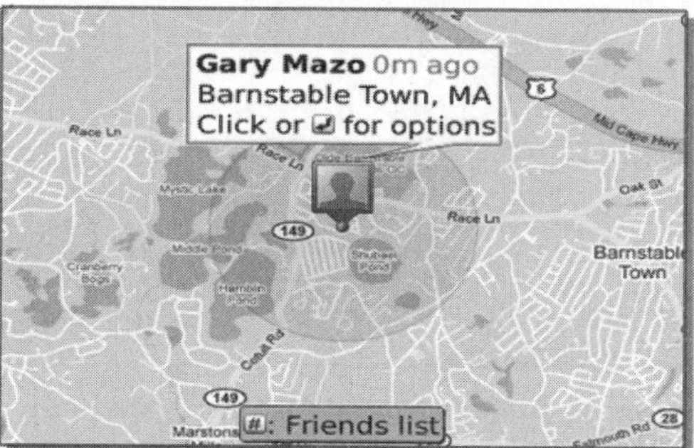

> **NOTE:** In some places, the software cannot find enough information to show your location; it will tell you if that's the case.

Finding More Things Nearby with Layers

In the newest version of Google Maps (version 4.02 at the time of this writing,) there is a new feature called Layers. Essentially, Layers will show you Wikipedia entries for things nearby, traffic conditions, transit lines, or personal maps. The data will be overlaid on the current map.

In the example shown in Figure 30-1, we have used Latitude to locate Martin in Florida and then used Layers to find things nearby.

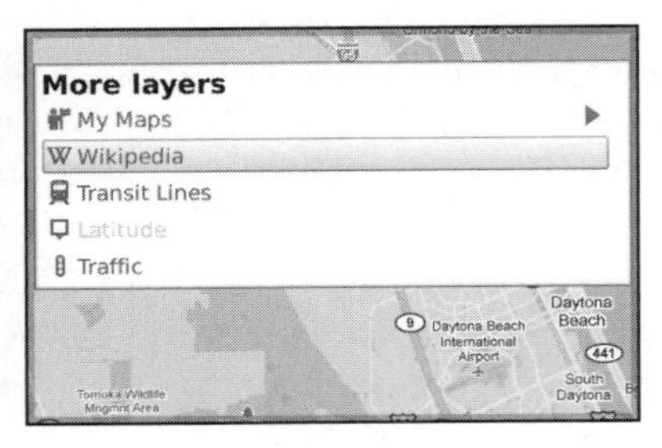

Figure 30-1. *Using Layers in Google Maps.*

The Wikipedia entries (in Figure 31-2) show that the Granada Bridge is near him, so we click it. We can now find more details and even see a picture of the Bridge.

Figure 31-2. *Viewing the stree view of a Wikipedia entry using Layers.*

Seeing Current Traffic in Google Maps

Another useful thing you can do in major metropolitan areas or on major highways (this does not work everywhere) is to view current traffic with Google Maps. First, map the location you want to view traffic.

Press the **Menu** key and select **Show Traffic** to display the traffic as shown in Figure 31-3.

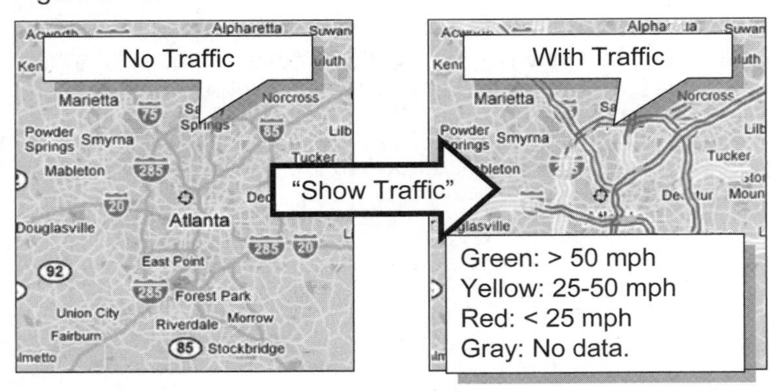

Figure 31-3. *Using the Show Traffic feature of Google Maps.*

Getting Directions with Google Maps

Here's how you can obtain driving directions with Google Maps:

1. Press the **Menu** key, and select **Get Directions**.

2. If you want to change your starting point, click **Start Point** (the default will be **My Location** (with the blue dot). In this case we will leave it alone.

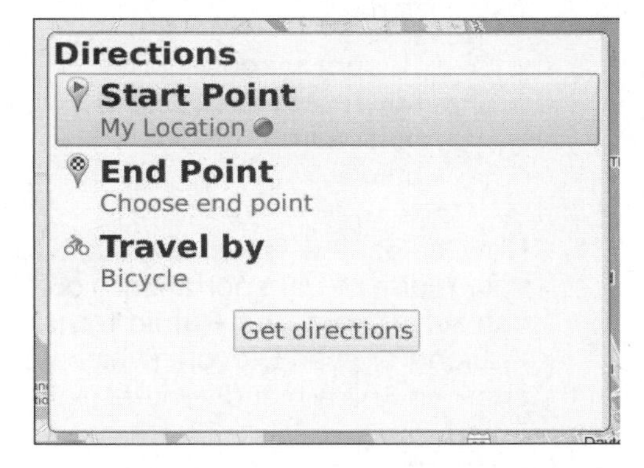

3. To change or set your destination, click **End Point**.

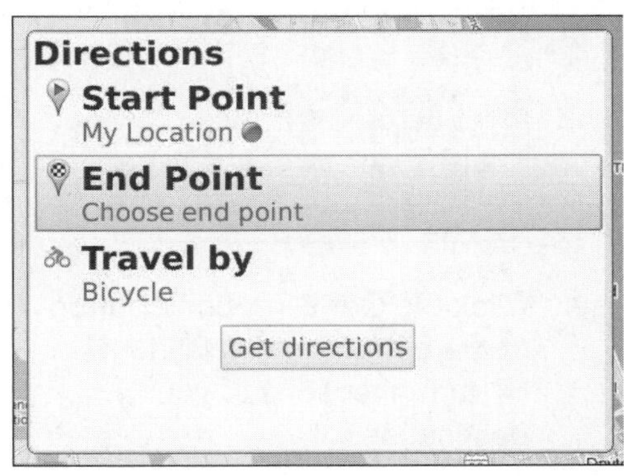

4. Now you have several ways to choose your End Point, just like selecting your starting point: (We clicked on a recent place: **Bantam Chef**)

   - **Enter an address** – type in a street address
   - **Select a point on map** – click on the map with the crosshairs in the center of the screen
   - **Starred items** – use one of your saved starred items
   - **Recent places** – click any recent searches or places from the list below.

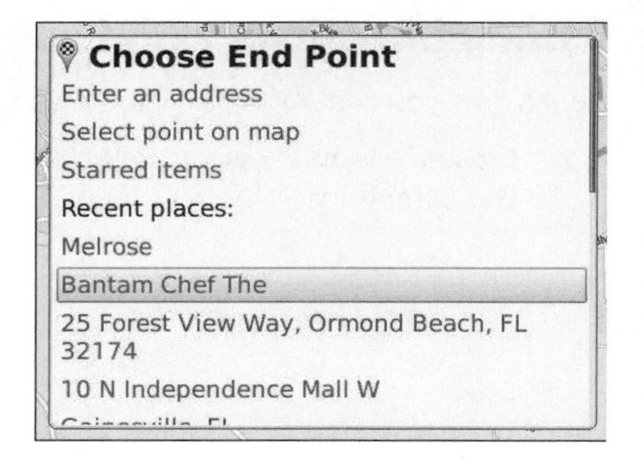

5. Now, select the **Travel By** to change your mode of transportation. You can select from **Car**, **Public transit**, **Walking** or even **Bicycle** (which will show you bike-friendly routes).

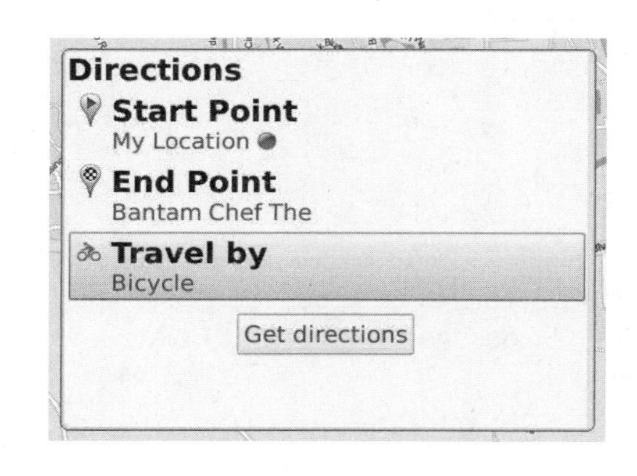

6. Click the **Get Directions** button at the bottom to display turn-by-turn directions to your destination.

7. Click **Show on map** to see the map with your directions.

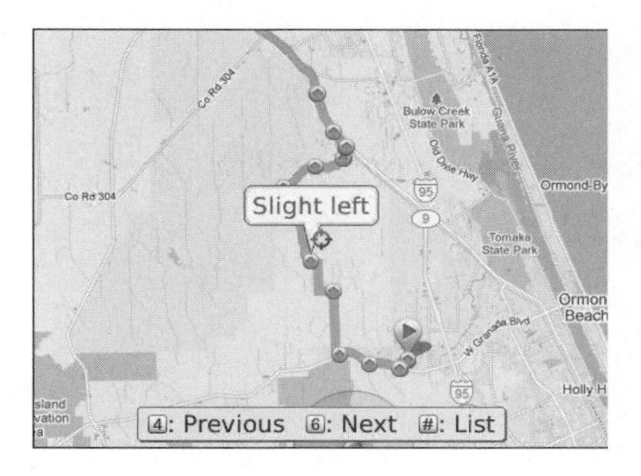

Enabling GPS Use in Google Maps

To use GPS with Google Maps, start Google Maps, and press the **Menu** key. Then select **Bluetooth GPS** or **Enable GPS** depending on how you are accessing the GPS signal in your BlackBerry.

In the next screen, select the GPS device that you paired with your BlackBerry and the GPS commands will now be available to you.

Once GPS is enabled, your location will be shown with a little blue dot. If you press the zero (0) number key (to the left of your **Space** key) then the map will be re-centerd on the blue dot – your location.

NOTE: For GPS to work, you first have to turn on or enable GPS on your BlackBerry see the section on page 486.

Chapter 32

Other Applications

As if your BlackBerry doesn't do enough for you, RIM was very thoughtful and included even more utilities and programs to help keep your organized and to help manage your busy life. Most of these additional programs will be found in the **Applications** folder.

Calculator

There are many times when having a Calculator nearby is handy. Gary usually likes to have his 15-year-old "math genius" daughter nearby when he has a math problem, but sometimes, she's in school and not available to help.

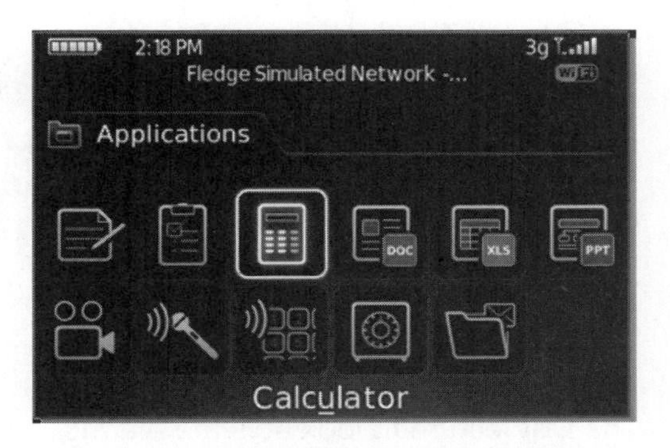

To start the Calculator program, just click the **Applications** folder. One of the icons should say **Calculator** (hotkey U).

Just input your equation as you would on any calculator program.

One handy tool in the Calculator program is that it can easily convert amounts to metric. Just press the **Menu** key, and scroll to **Convert to Metric.**

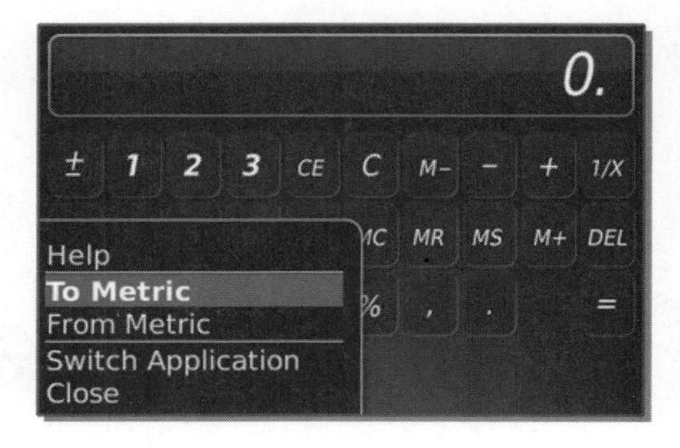

Clock and Alarm Clock

While having a clock is nothing new for a BlackBerry, the features and the ability to customize the clock, especially when the BlackBerry is on your bedside table, is new and appreciated.

To start the Clock application, just go once again to the **Applications** folder and find the icon for the **Clock**. On some devices, the clock icon might just be an icon in the menu, not in a folder. By default, the clock is initially set to an analog face, but that be changed.

The first feature to look at is the Alarm. The easiest way to bring up the settings for the Alarm is to simply click the trackpad. The **Alarm** menu will appear on the bottom.

Just glide the trackpad to the desired field and glide up or down in that highlighted field to change the option. Highlight the farthest field to the right, and glide to select **Off, On**, or **Weekdays** for the alarm setting. Press the **Escape** key to save your changes.

Also built into the Clock application are a stopwatch and a timer, which can be started by selecting them from the menu.

> **TIP:** The clock can be changed to a digital clock, a flip clock or an LCD digital face by selecting the clock face from the **Options** menu.

Want to turn off your clock being displayed when you connect your BlackBerry to your computer?

By default, you will see your clock every time you connect your BlackBerry to your computer (when it is plugged in) or when you connect your BlackBerry to the power charger.

You can turn off this feature in the **Clock Options**. Inside the Clock app, press the **Menu** key, and select **Options**. Glide down, and set **When Charging** to **Do Nothing**.

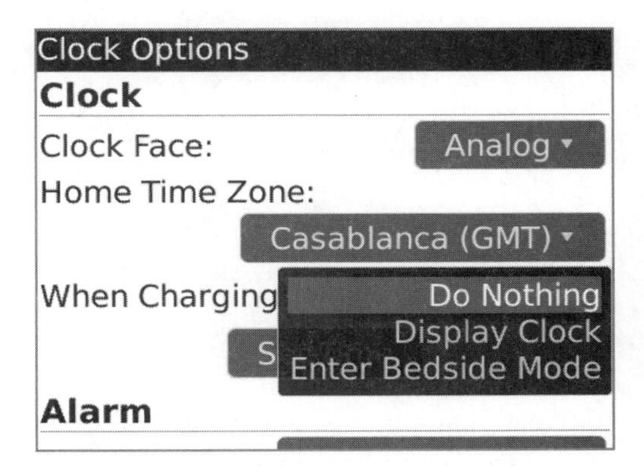

Using the Bedside Mode Setting

One of the nice new features of the Clock application is the Bedside Mode setting. Since many of us keep our BlackBerry devices by the sides of our beds, we now have the option of telling our BlackBerry not to flash, ring, or buzz in the middle of the night. We just have to activate Bedside Mode by pressing the **Menu** key, and selecting **Enter Bedside Mode.**

Press the **Menu** key in the Clock app, and then scroll down to **Options** to configure the **Bedside Mode** options, which are toward the bottom of the options menu. You can disable the LED and the radio and dim the screen when with these settings.

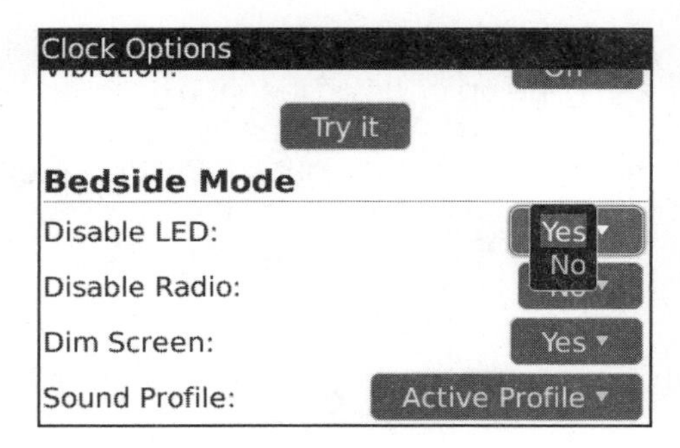

Voice Notes Recorder

Another useful program in your **Media** folder is the Voice Notes Recorder. Say you need to dictate something to recall at a later time. Or perhaps you would rather speak a note instead of composing an email. Your BlackBerry makes that very easy for you.

This program is a very simple program to use. Just click the trackpad when you are ready to record your note, and click it again when you are finished.

You will then see options along the lower part of the program that will allow you to: continue recording, stop, play, resume, delete, or email the voice note.

Just click the corresponding icon to perform the desired action.

Password Keeper

In today's Web world, keeping track of different passwords and password rules on many sites can be difficult. Fortunately, your BlackBerry has a very effective and safe way for you to manage all your passwords—the Password Keeper program.

Just go back to your **Applications** folder, and click the **Password Keeper** icon. The first thing you will need to do is set a program p assword— pick something you will remember!

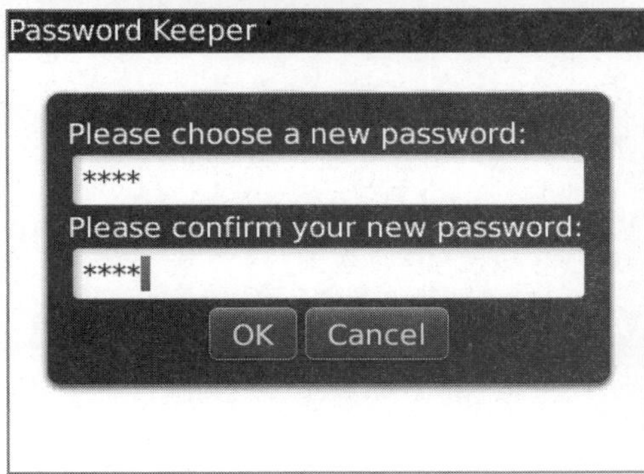

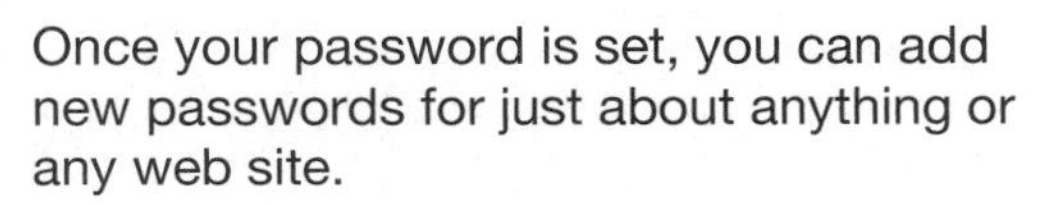

Once your password is set, you can add new passwords for just about anything or any web site.

Press the **Menu** key, and select **New** to add passwords, user names, and web addresses to help you remember all your various login information.

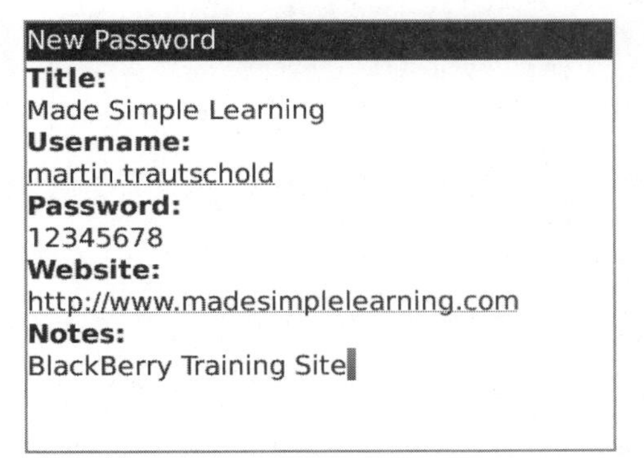

Chapter 33

Searching for Lost Stuff

Your BlackBerry can hold so much information that it can sometimes be difficult to keep track of where everything is. Thankfully, BlackBerry comes with a very comprehensive search program to help you find exactly what you are looking for. For example, with the Search icon, you can search Calendar events to answer questions like, "When is my next meeting with Sarah?" and "When did I last meet with her?" (For this to work, you need to type people's names in your events, like "Meet with Sarah.")

Understanding How Search Works

Once you get used to your BlackBerry, you will begin to rely on it more and more. The more you use it, the more information you will store within it. It is truly amazing how much information you can place in this little device.

At some point, you will want to retrieve something—a name or a word or phrase—but you may not be exactly sure of where you placed that particular piece of information. This is where the **Search** icon can be invaluable.

Search Built into Messages (Email)

A nice search program is built right into your Messages (email) icon that allows you to search for names and subjects and has many options. Learn how to use this built-in Messages Search on page 252.

Finding the Search Icon

In your applications menu, there is a **Search** Icon. If you have enabled hotkeys (see page 541), pressing the hotkey **S** will start the **Search** icon. From your home screen of icons, use the trackpad to find the **Search** icon. It may be located within the **Applications** Folder.

Searching Several Apps at Once

It is possible that the desired text or name could be in one or several different places on your BlackBerry. The Search tool is quite powerful and flexible. It allows you to narrow down or expand the icons you want to search. If you are sure your information is in the Calendar, just check that box, if not, you can easily check all the boxes using **Select All** from the menu.

1. Click the **Search** Icon (or press the hotkey **S**), and the main search screen is visible.
2. By default, only the **Messages** field is checked, as shown in Figure 33-1.

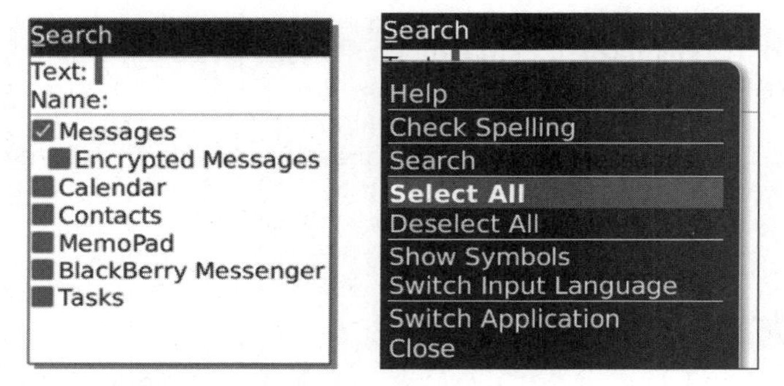

Figure 33-1. *Search several apps at once.*

3. If you would like to search all apps at once, then Press the **Menu** key and click **Select All.**
4. Type the text or name you'd like to search for.
5. Click the trackpad and select **Search**.
6. Now the search results will be shown separated by app. (e.g. **Messages**, **Calendar**, **Contacts**, etc.)

Searching for Names or Text

In the **Name** field, you can search for a name or email address. The **Text** field allows you to search for any other text that might be found in the body of an email, inside a calendar event, address book entry, memo, or task.

If you decide to search for a name, you can either type a few letters of a name, or press the **Menu** key and choose **Select Name.**

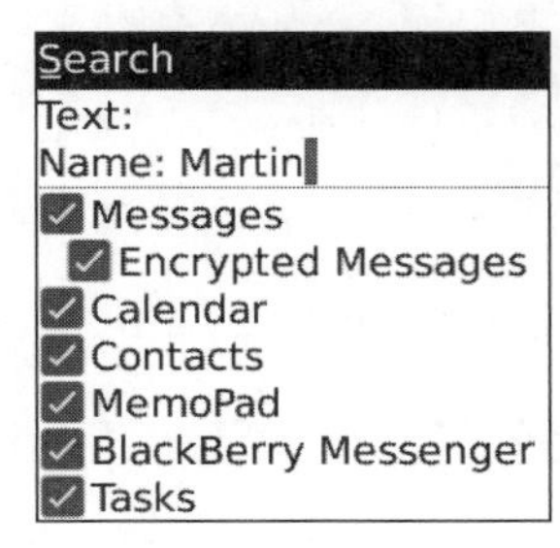

Type the name.

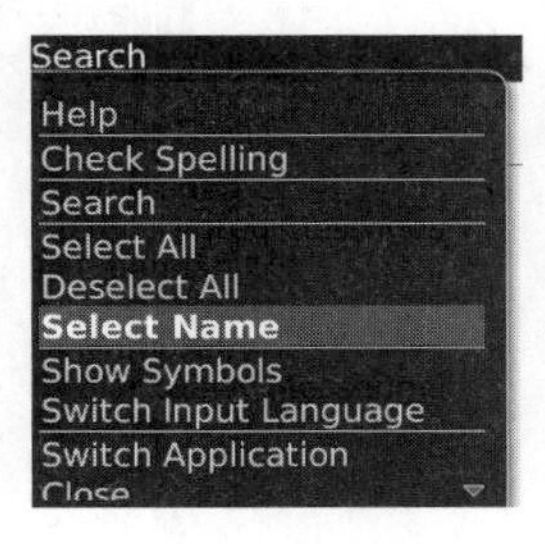

Or press the **Menu** key and click **Select Name**.

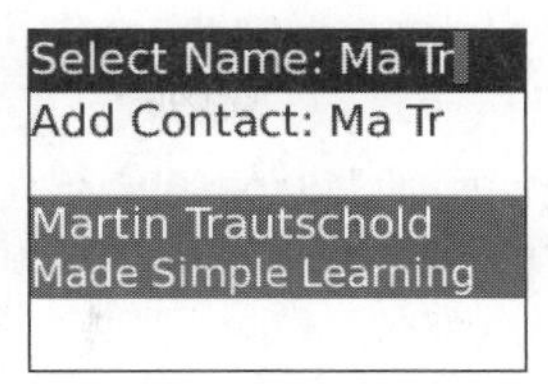

Then type a few letters to select the name from your contact list and click it.

If you are looking for specific text, like a word, phrase, or even phone number that is not in an email address field, type it into the **Text** field.

When you are ready to start the search, click the trackpad, and select **Search**.

The results of the search are displayed with the number of found entries. In this image, there were four total matches found, three in Messages (email) and one in Contacts.

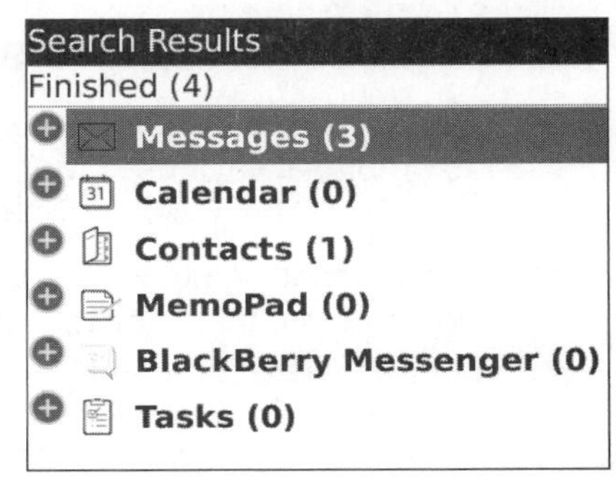

To expand your search results, just click one of the items that is shown to see more details. Click again to contract the details.

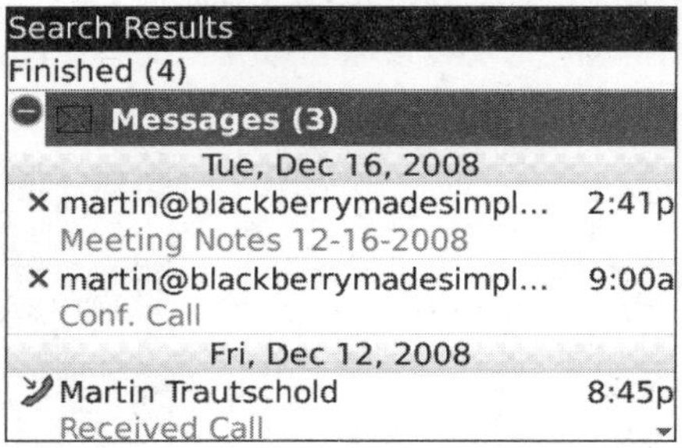

NOTE: Pressing the **Escape** key will clear your search results and bring you back to the search screen.

Search Tips and Tricks

You can see that the more information you enter on your BlackBerry (or enter on your desktop computer and sync to your BlackBerry), the more useful the device becomes. When you combine a great deal of useful information with this search tool, you truly have a very powerful handheld computer.

As your BlackBerry fills up, the number of possible places where your information is stored increases. Also, the search might not turn up the exact information you are looking for due to inconsistencies in the way you store the information.

TIP: You can even search notes added to Calendar events. Remember to add notes to your calendar events and the notes field at the bottom of your contacts. You can do this right on your BlackBerry or on your computer and sync them. Use the Search app to find key notes on your BlackBerry right when you need them.

Here are some tips for successful BlackBerry searching:

- Try to be consistent in the way you type someone's name. For example, always use "Martin" instead of "Marty" or "M" or any other variation. This way, the Search will always find what you need.
- Occasionally, check your address book for duplicates of contact information. It is easy to wind up with two or three entries for one contact if you add an email one time, a phone number another, and an address a third. Try to keep one entry per contact. It is usually easier to do this clean up work on your computer and then sync the changes or deletions back to your BlackBerry.
- If you are not sure whether you are looking for "Mark" or "Martin," just type in "Mar" and then search. This way, you will find both names.
- Remember, if you want to find an exact name, glide to the **Name** field, press the **Menu** key, and choose **Select Name** to select a name from your address book.

- Do your best to put consistent information into Calendar events. For example, if you wanted to find when the next dentist appointment for Gary was, you could search for "Gary Dentist" in your Calendar and find it. But only if you made sure to put the full words "Gary" and "dentist" in your calendar entry. It would be better to just search for "dentist."
- If you want to find a phone number and remembered only the area code, type that area code into the **Text** field and search the address book.
- If you want to find when the name "Gary" was in the body of an email, not an email address field (To, Cc, From, or Bcc), enter the name in the **Text** field, not the **Name** field, on the **Search** screen.

TIP: Traveling and want to find local contacts? Let's say you are traveling to New York City and want to find every entry in your Blackberry address book with a 212 area code. Type "212" in the **Text** field, check **Contacts**, and click **Search** to immediately find everyone who has a number with a 212 area code.

Chapter 34

Securing Your Data

In this chapter, we talk all about security and securing your BlackBerry data. Most of us keep very important information on our devices and would want to be prepared in case anything happens to the BlackBerry.

What If Your BlackBerry Was Lost or Stolen?

Would you be uncomfortable if someone found and easily accessed all the information on your device? For most of us the answer would be yes!

On our BlackBerry smartphones, many of us store friends' and colleagues' names, addresses, phone numbers, confidential email, and notes. Some of our devices may even contain Social Security numbers, passwords, and other important information in your Contact notes or your MemoPad.

In that case, you will want to enable or turn on the password security feature. When you turn this on, you will need to enter your own password to access and use the BlackBerry. In many large organizations, you do not have an option to turn off your password security; it is automatically turned on by your BlackBerry enterprise server administrator.

Preparing for the Worst Case Scenario

There are a few things you can do to make this worst case scenario less painful.

1. Back up your BlackBerry to your computer.
2. Turn on (enable) BlackBerry password security.
3. Set your Owner Information with an incentive for returning your BlackBerry.

Backing Up or Syncing Your Basic BlackBerry Data

For Windows users, check out our full description of how to back up and restore your BlackBerry data using Desktop Manager software on page 95. Mac users, use the backup feature in Desktop Manager for Mac shown on page 125.

CAUTION: Do you have important media (personal pictures, videos, music, and so on) stored on your media card (See page 359 for more on media cards)? If so, you will have to copy your media card information separately from the sync or backup process. See our chapter 3 for Windows users and chapter 5 for Mac users.

Turning On Password Security

Make sure you turn on enable password security. This is really your best backup plan to safeguard all your data should you happen to lose your BlackBerry.

1. Click your **Options** icon on your home screen. Press the letter **P** to jump down to **Password**, and click it.

2. Select **General Settings** to see this screen.
3. Click the **Disabled** setting next to **Password**, and change it to **Enabled.**
4. Then click **Set Password** to create your password.
5. You can adjust the other settings unless they are locked out by your BlackBerry Administrator.
6. Save your settings.

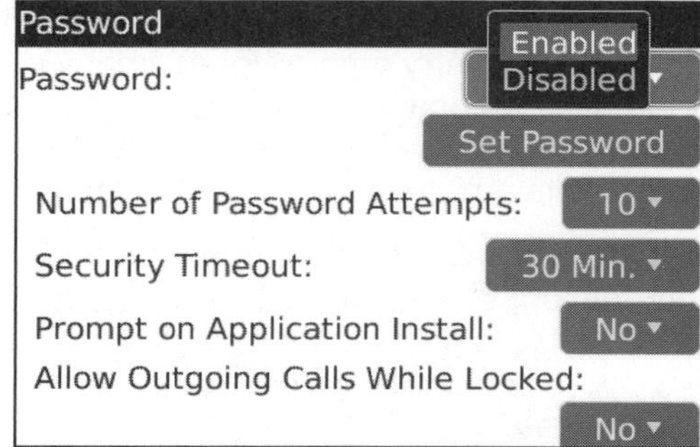

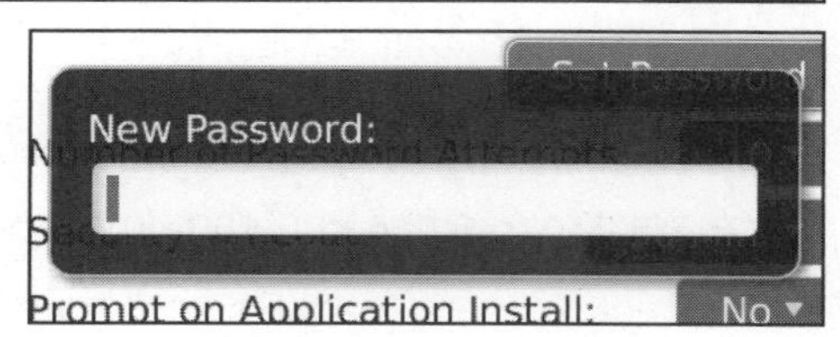

CAUTION: If you cannot remember your password, you will lose all the data (email, addresses, events, tasks—everything) from your BlackBerry. Once you make more than the set number of password attempts (the default is ten attempts), the BlackBerry will automatically wipe or erase all data.

Setting Your Owner Information

We recommend including something to the effect of "reward offered for safe return" in your owner information, which would show up when someone found your BlackBerry in a locked mode. You may have already set your owner information in the Setup Wizard, but you can also via the **Options** icon.

1. Click your **Options** icon on your home screen. Press the letter **O** to jump down to **Owner** in the list of options settings, and click the trackpad to open it.

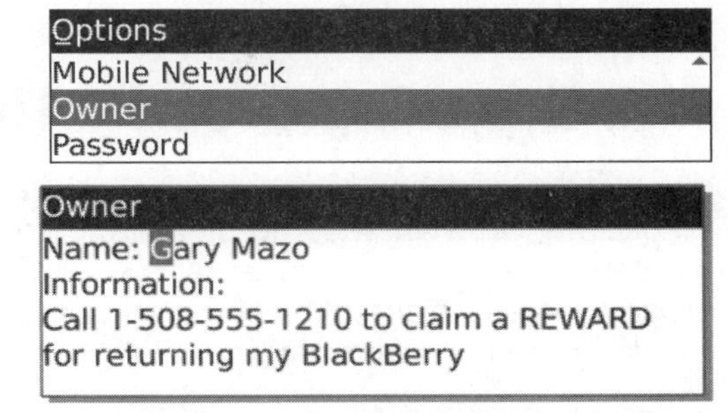

2. In the **Owner Information** settings screen, put yourself into the mind of the person who might find your BlackBerry and give both the information necessary and an incentive to return your BlackBerry safely to you. Press the **Menu** key, and ave your settings.

> **TIP:** Shortcuts to Lock
>
> From your Home Screen, quickly tapping the **K** key or pressing & holding the **A** key will **Lock** your BlackBerry. (Assumes you have turned on Home Screen Hotkeys – see page 541). You can also press the **Lock** button on the top of your device.

3. Test what you've entered by locking your device. Tap the **Lock** button on the top of your BlackBerry, or click on the **Lock** icon. Now, you will see your owner information.

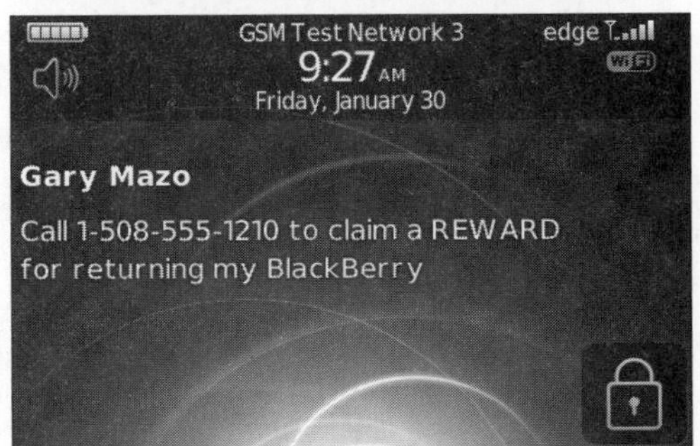

Email Security Tips

Never send personal information via email. Examples of information not to send are credit card numbers, Social Security numbers, date of birth, mother's maiden name, sensitive passwords or PIN numbers (e.g., bank account ATM card PIN). If you have to transmit this information, it's best to call the trusted source or, if possible, visit in person.

Web Browsing Security Tips

If you are in an organization with a BlackBerry enterprise server, when you use the BlackBerry Browser secured by the encryption from the server, your communications will be secure, just like your email, as long as you are browsing sites within your organization's intranet. Once you go outside to the Internet and are not using an HTTPS connection, your web traffic is no longer secure.

Anyone that visits a site with HTTPS (or a secure socket layer) connection will also have a secure connection from your BlackBerry to the web site, like with your computer's web browser.

Whenever you are browsing a regular HTTP web site on the Internet (not your organization's secure Intranet), be aware that your connection is not secure even if are using a BlackBerry enterprise server. In this case, please make sure not to type or enter any confidential, financial, or personal information.

If You Lose Your BlackBerry

If you work at an organization with a BlackBerry enterprise server or use a hosted BlackBerry enterprise server, immediately call your help desk and let them know what has happened. Most help desks can send an immediate command to wipe or erase all data stored on your BlackBerry device. You or the help desk should also contact the cell phone company to disable the BlackBerry phone.

If you are not at an organization that has a BlackBerry enterprise server, you should immediately contact the cell phone company that supplied your BlackBerry and let them know what happened. Hopefully, a Good Samaritan will find your BlackBerry and return it to you.

Disabling Password Security

To turn off password security, go into the **Options** icon as above. Then click **Password** and change it to **Disabled**. Save your settings; you will be required to enter your password one last time to turn it off, for security purposes.

NOTE: If you work at an organization with a BlackBerry enterprise server, you may not be allowed to turn off your password.

SIM Card Security Options

With any GSM phone, like your BlackBerry, if your phone was every lost or stolen, your SIM card could be removed and used to activate another phone. To prevent that, follow these steps:

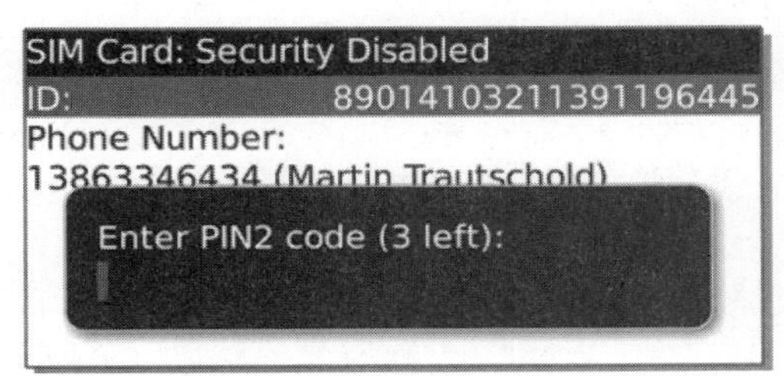

1. Click Options and Advanced Options.
2. Click **SIM card**.
3. Press the **Menu** key and select **Enable Security** The PIN code locks your SIM card to your BlackBerry.
4. Also, change the **PIN2 code** for added security.

Chapter 35

Fixing Problems

Your BlackBerry is virtually a complete computer in the palm of your hand. Like your desktop computer, sometimes it needs a little tweaking or maybe even a re-boot to keep it in top running order. This chapter has some of the most valuable tips and tricks to fix problems and keep your BlackBerry running smoothly.

BlackBerry Running Slowly?

If your BlackBerry is running slowly, try the following troubleshooting tips and tricks.

Clearing Out the Event Log

Your BlackBerry tracks absolutely everything it does to help with debugging and troubleshooting in what is called an event log. Your BlackBerry will run smoother and faster if you periodically clear out this log.

1. From your home screen of icons, press and hold the **ALT** key and type LGLG (*do not* press the **Shift** key at the same time) to bring up the Event Log screen as shown.

2. Press the **Menu** key, and select **Clear Log**.

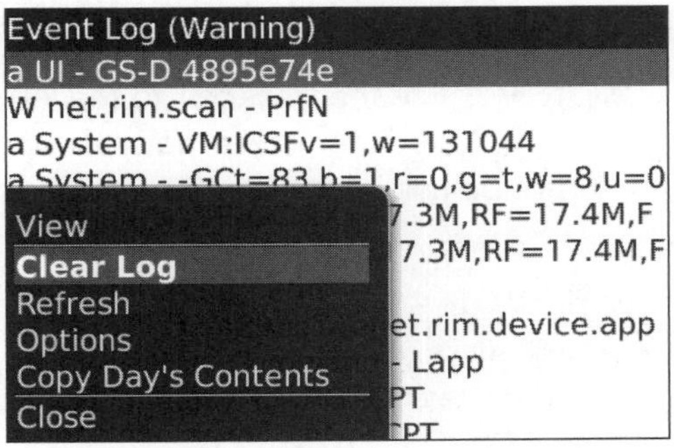

3. Select **Delete** when it asks you to confirm, and finally, press the **Escape** key to get back to your home screen.

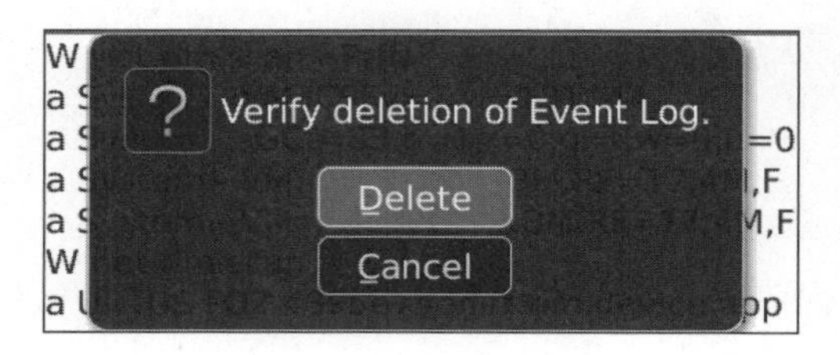

Closing Unused Icons Running in Background

If you have been using the **Switch Application** menu item, the **Red Phone** key, or the **ALT+Escape** trick to multitask (see page 322), you may have many icons running in the background and slowing down your BlackBerry.

Periodically, you should verify that you have closed out all unnecessary application icons.

1. First, bring up the **Switch Applications** pop-up window to see what is running. Press and hold the **ALT** key, and tap the **Escape** key.

NOTE: The following icons cannot be closed: **BlackBerry Messenger**, **Messages**, **Browser**, **Phone Call Logs**, **Home** (home screen). You won't see the BlackBerry Messenger icon unless the app is installed.

2. Now, glide the trackpad to the left or right to look for icons that you may not be using, like BlackBerry Maps.

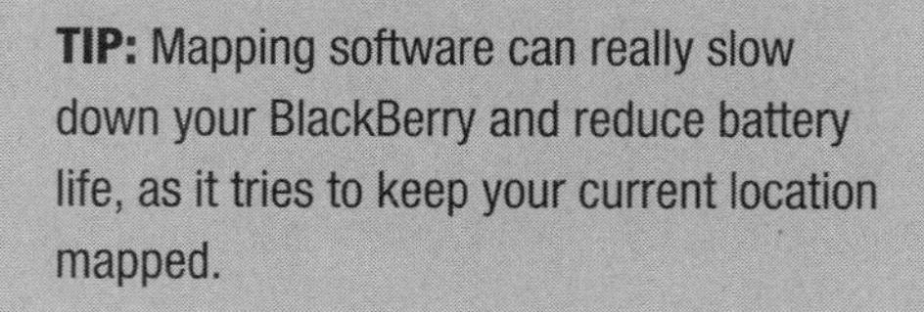

TIP: Mapping software can really slow down your BlackBerry and reduce battery life, as it tries to keep your current location mapped.

3. Press the **Menu** key, and select **Close** or **Exit**, usually the bottommost menu item.

4. Repeat this procedure for every unneeded icon that is running to help speed up your BlackBerry.

> **TIP:** In the future, when you are done using an icon, press the **Menu** key, and select **Close** or **Exit** to make sure that icon is actually exiting out and freeing up memory and other resources on your BlackBerry.

Web Browser Running Slowly?

If your web browser is slow, you can make a dramatic increase in the speed by clearing out the cache.

1. Start your browser by clicking its icon.
2. Inside the Browser, click the **Menu** key, and select **Options**.

3. Select Cache Operations.

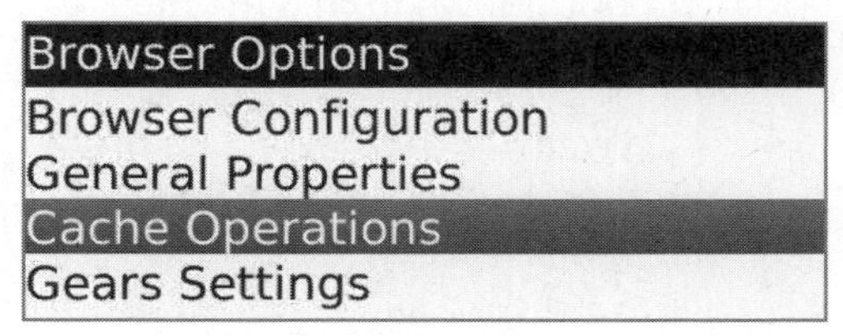

4. Click **Clear History**, and click all the **Clear** buttons below it.
5. Answer **Yes** when it asked about clearing pushed content.
6. Press the **Escape** key a few times to return to the browser. Clearing the cache will usually speed up your browsing speed significantly.

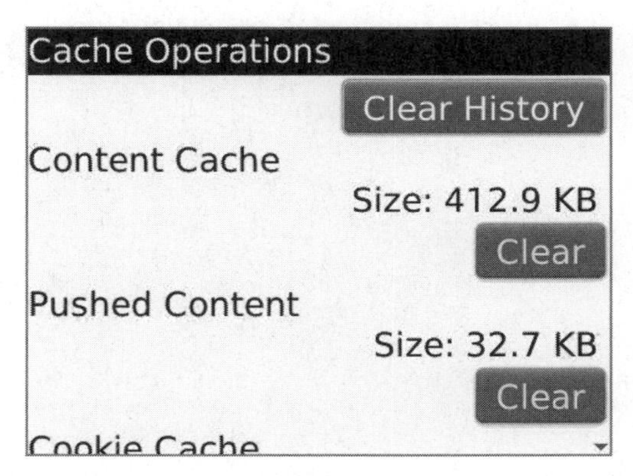

Automatic Memory Cleaning to Keep Running Smoothly

One thing that may help keep your BlackBerry running smoothly is to utilize the automatic memory cleaning feature. Although your BlackBerry has lots of built-in memory, all smartphones are prone to memory leaks, which can cause the device to really slow down and programs to hang or not work properly.

1. Click your **Options** icon, and then **Security Options**.
2. Glide down, and click **Memory Cleaning**.

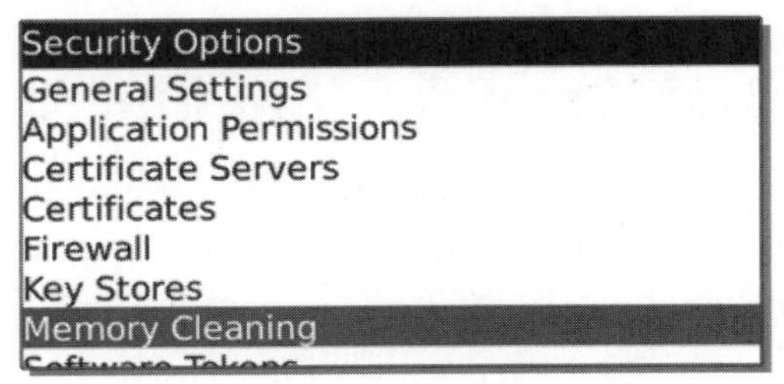

3. Change the Status to **Enabled**, and you will see more options. We recommend leaving the first two set to **Yes** and adjusting the **Idle Timeout** down from 5 minutes to 1 or 2 minutes.
4. Finally, if you want, you can **Show Icon on Home Screen**. Save your changes.

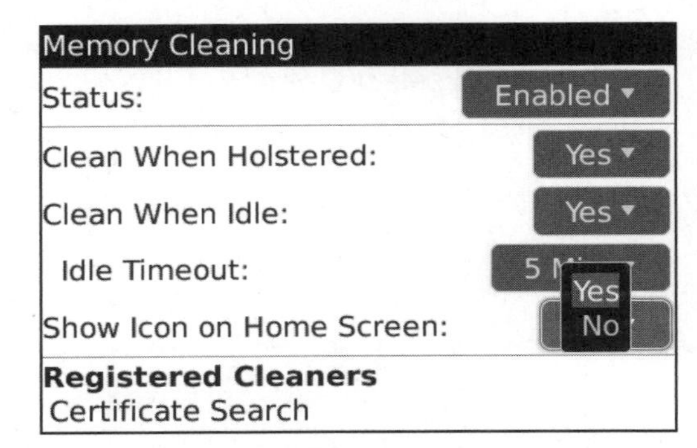

5. If you selected to show your icon, you will see it in the **Applications** folder as shown.

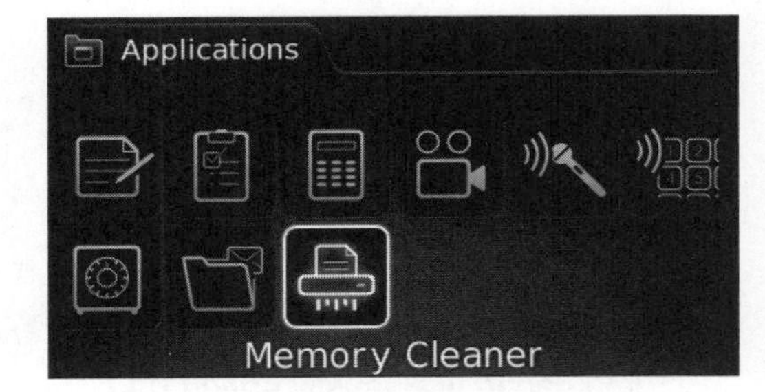

6. When you click the **Memory Cleaner** icon, you will see the cleaning happen.

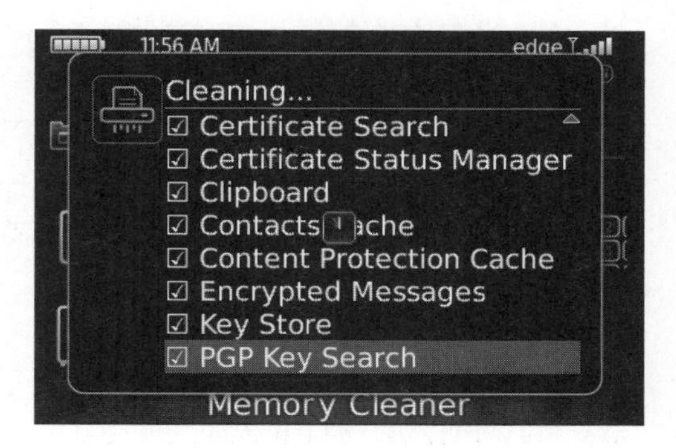

Understanding Your Wireless Data Connection

Normally, you don't have to think about your wireless signal; it just works. However, sometimes things happen to cause you troubles. You may try the Browser but realize that you have no data connectivity. If you see an hourglass or X instead of a check mark when you try to send an email, you know something is wrong. The BlackBerry makers were nice enough to give you a wireless signal status meter at the top of your BlackBerry screen (or it might be on the bottom, depending on your particular BlackBerry wireless carrier or theme).

There are two components to your wireless status meter: The **Signal Strength Meter** and the **Data Connection Letters**, which are usually right next to the **Signal Strength Meter**.

Data Connection Letters (Vary by Network Type)

Your data connection allows you to send and receive email and data and browse the Web. The confusing thing is that you may have very strong signal strength (e.g., four or five bars) but still not be able to send or receive email. The three of four letters or numbers shown next to the **Signal Strength Meter** will vary based on the type of wireless network to which you are connected.

- **If you are connected to a CDMA Network** (e.g. Verizon, Sprint, Telus, or Bell-Canada), you will see 1XEV, 1X, 1xev, or 1x.
- **If you are connected to a GSM Network** (e.g., AT&T, T-Mobile, or Rogers), you will see 3G (with or without a logo), EDGE, GPRS, edge, gprs, or GSM.

Take a look at Table 35-1 to understand whether or not you have a data connection.

Table 35-1. *Interpreting Data Connection Information*

If you see. . .	It means that. . .	You should. . .
OFF	Your radio is turned off.	Turn your radio back on: Click **Manage Connections** **Restore Connections.**
GSM (no logo)	There is no data connection, only phone and SMS work.	Try the "Turn Off and On Your Radio" section below and other sections in this chapter.
1XEV, 3G with logo	The highest-speed data, phone, and SMS text are available and working.	If email and web are not working, try the "Register Now" section on page 528, and other sections in this chapter.
1X, EDGE	High-speed data, phone, and SMS are working.	If email and web are not working, try the Register Now section on page 528 and other sections in this chapter.
GPRS	Low-speed data, phone, and SMS are working.	If email and web are not working, try the "Register Now" section on page 528, and other sections in this chapter.
3G (no logo), 1xev, 1x, edge, gprs	There is no data connection. Phone and SMS only work.	Try the "Turn Off and On Your Radio" section below, and other sections in this chapter.

Trouble with Email, Web, or Phone?

If you're having difficulties with email, web browsers, or making calls, try the tips in the following sections.

Turning Off and On Your Radio

If you're having difficulty with your email, web browser, or phone, many times the simple act of turning your radio off and back on will restore your wireless connectivity.

To do this, first go to your home screen and click the **Manage Connections** icon. In the Manage Connections screen shown in Figure 35-1, click **Turn All Connections Off.**

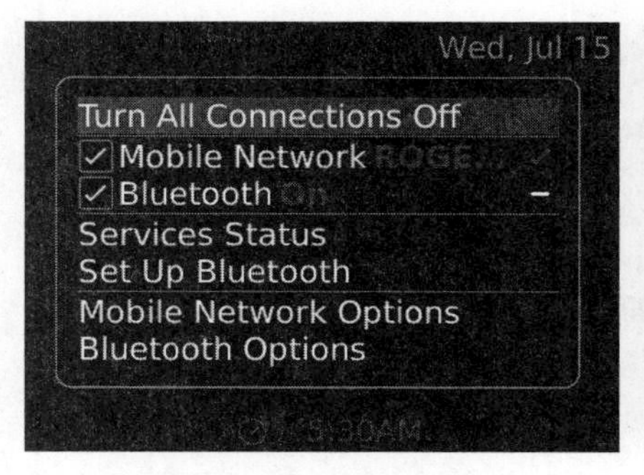

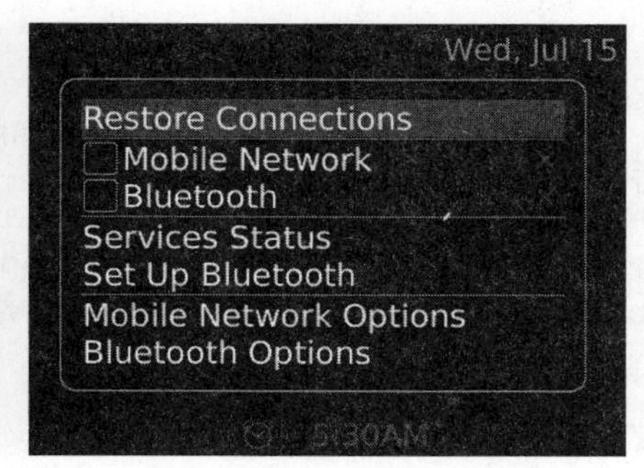

Figure 35-1. ***Manage Connections*** *screen.*

Wait until you see the word "OFF" next to the wireless signal strength icon on your home screen, and then click **Restore Connections** to turn on your radio again.

Look for your **Wireless Signal Meter** and uppercase 1XEV, 1X, 3G (with logo), EDGE or GPRS. Check to see if your email and web apps are working. If not, try some more of the troubleshooting tips that follow.

Registering for Host Routing

Another possible solution to connectivity issues is to select register for host routing:

1. From your home screen, click the **Options** icon, or press the hotkey **O**.
2. Click Advanced Options.
3. Now, click **Host Routing Table** (use hotkey **H** to jump down to this option); see Figure 35-2.

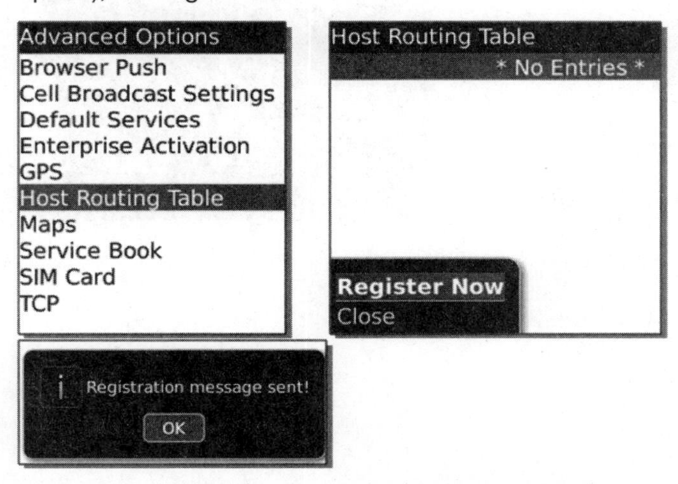

Figure 35-2. *Register BlackBerry with the* ***Host Routing Table.***

4. In the Host Routing Table screen, you will see many entries related to your BlackBerry phone company.

5. Press the **Menu** key, and select **Register Now.** If you see "Registration message sent!", you should be OK. If you see a message like "Request queued and will be send when data connection is established", try some more troubleshooting steps in the following sections.

Sending Service Books

This trick for solving connectivity issues will work for you only if you have set up Internet or personal email. (For more on Internet and personal email, see page 231)

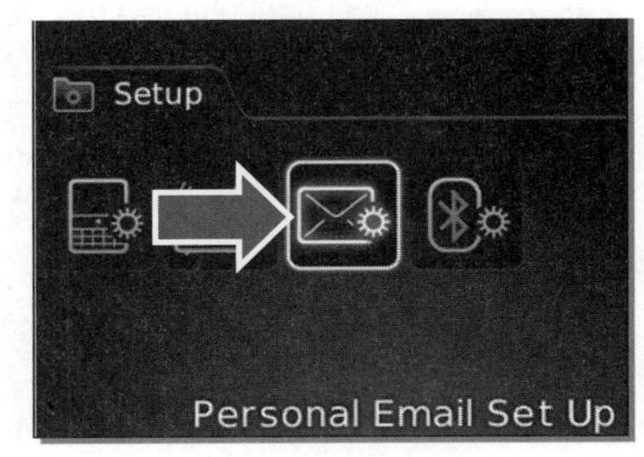

1. From your home screen, click the **Setup** or **Settings** folder and then the **Manage Personal Email** or **Set Up Internet Mail** icon (each carrier names these a bit differently).

2. If this is your first time setting up these email accounts, you may be asked to create an account or log in to get to personal email setup.

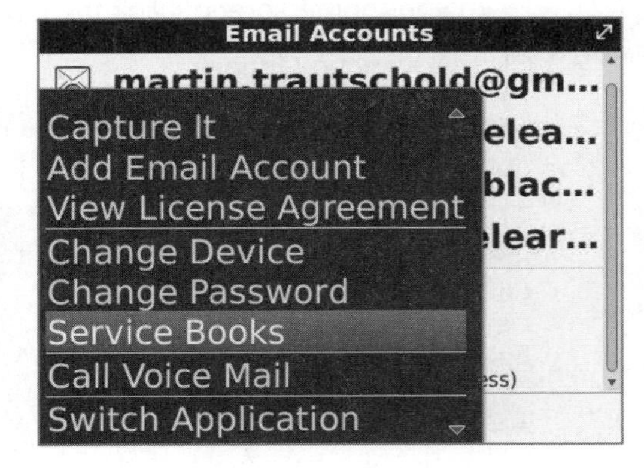

3. After you have logged in to your email setup screen, press the **Menu** key, and select the **Service Books** item.

4. Roll to the bottom of the screen, and click **Send Service Books**.

You should see a "successfully sent" message and, in your **Messages** icon, a number of ''Activation" m essages—one per email account, as shown in Figure 35-3.

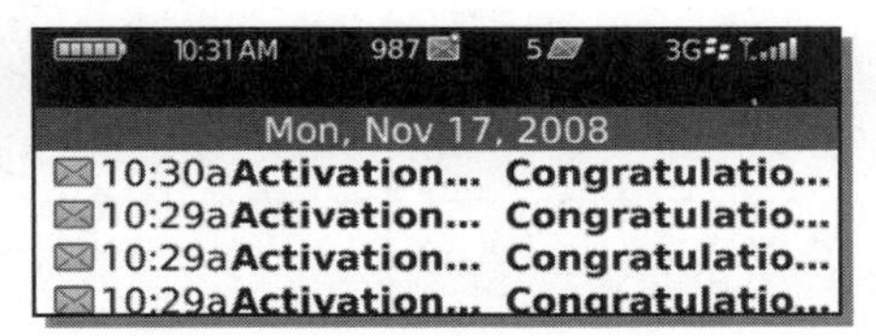

Figure 35-3. *Service Books successfully sent.*

If you see any other message indicating the message was not sent, you may want to verify that you have good coverage at your current location. You may also want to repeat some of other the troubleshooting steps in this chapter or contact your help desk or wireless carrier support.

Trying a Soft Reset

If you're having trouble with general slowness in email or web applications, sometimes, like on your computer, you simply need to try a reset or reboot (like pressing **Ctrl** + **Alt** + **Del** on a Windows computer). You can do a similar soft reset on your BlackBerry by pressing and holding three keys simultaneously **ALT** + **CAP** + **DEL** (Figure 35-4).

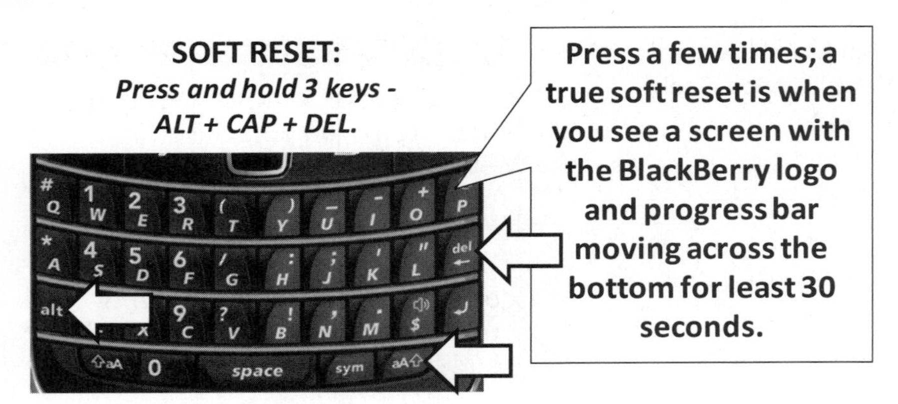

Figure 35-4. *Performing a soft reset*

Using a Hard Reset (Removing the Battery)

When you really need to start over, you can try a hard reset. Check out more images and help with this on page 5.

1. Turn off your BlackBerry by pressing and holding the power button (the **Red Phone** key). In some cases, if your BlackBerry is locked, it may not turn off, just remove the battery as shown in the next steps.

2. Remove the battery cover. Slide the back cover down and off the back of your BlackBerry.
3. Gently pry out and remove the battery. You will have to put your fingernail in the top right or left edge next to the printed gray or black smooth semicircle and pry the battery out from the bottom edge. It has a few small tabs holding it in at the top of the BlackBerry.

Image courtesy of `BlackBerry.com`

4. Wait about 15 seconds.
5. Replace the battery, making sure to slide in the top of the battery first and press down from the bottom.
6. Replace the battery door by sliding it up until it latches.

7. Power on your BlackBerry (it may come on automatically). Then, you will need to wait a little for the timer to go away (up to several minutes).

8. If you see **OFF** next to your radio tower indicator in the upper-left or upper-right corner, you will need to turn on your wireless radio again. To do this, tap the **Menu** key, and click the radio tower icon, which says **Manage Connections**, and then click **Restore Connections**.

Updating the BlackBerry Operating System

BlackBerry is continually updating the operating system (OS) software for its devices. Sometimes, if you are having problems, they might actually already have been addressed by a newer version of the software. It is generally a good idea to check for software updates on a regular basis for just this reason.

How Can I Tell What OS Version My BlackBerry Is Running?

To see which OS you have, you can click your **Options** icon, and select **About**. An even faster way is to bring up the **Help Me!** screen by simultaneously pressing and holding **ALT** + CAP + H.

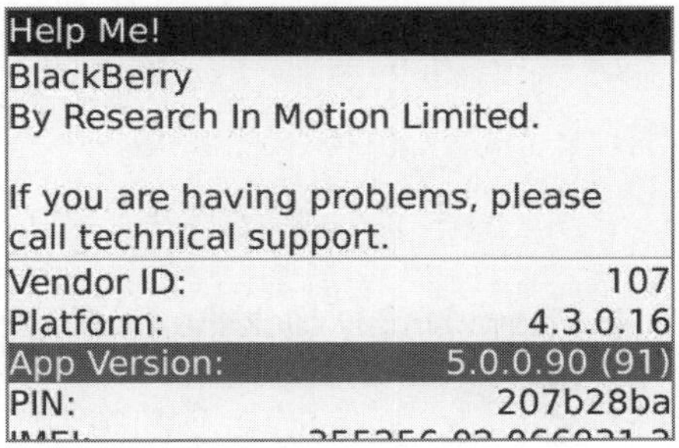

> **TIP:** You can also view things like free memory and battery strength if you scroll down to the bottom of the **Help Me!** screen.

The first two digits next to **App Version** are your OS version; the preceding image shows this BlackBerry is running version 5.0.

How Can I Update My BlackBerry OS?

You can update your OS (also called firmware), using Desktop Manager or with a wireless update from your BlackBerry.

Using Desktop Manager

If you use Desktop Manager for Windows (see page 65) or Mac (see page 125), you will be notified of OS updates and be able to update using either software.

Using Wireless Update

Most BlackBerry smartphones also now offer the ability to update your OS directly from the device itself in the **Options** icon.

1. Click the **Options** icon by locating, or use the hotkey **O** if you have turned on home screen hotkeys. (See page 541)
2. Click **Advanced Options** and press **W** to jump down to **Wireless Update**; click it.

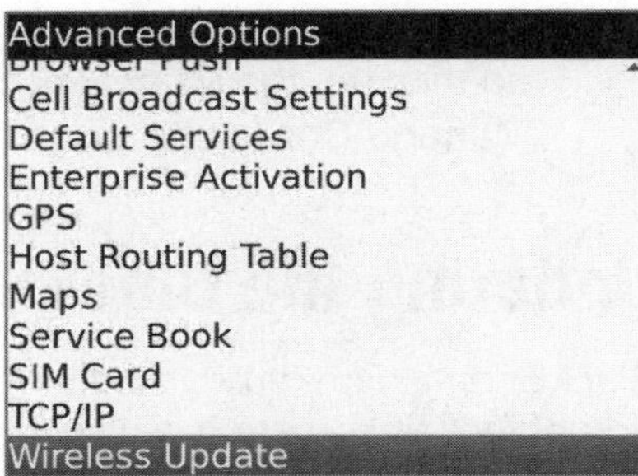

3. Click **Next** at the bottom to get to the **Updates** screen.

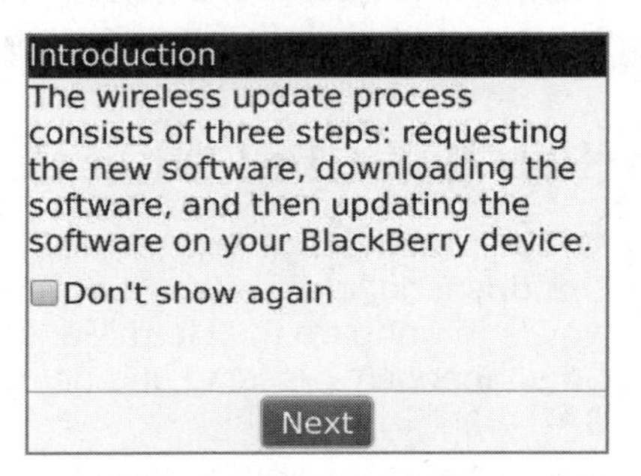

4. Now, simply click the **Check for Updates** button at the bottom.
5. If updates are found, you will see the status on the screen. Follow the on-screen instructions to install the update.

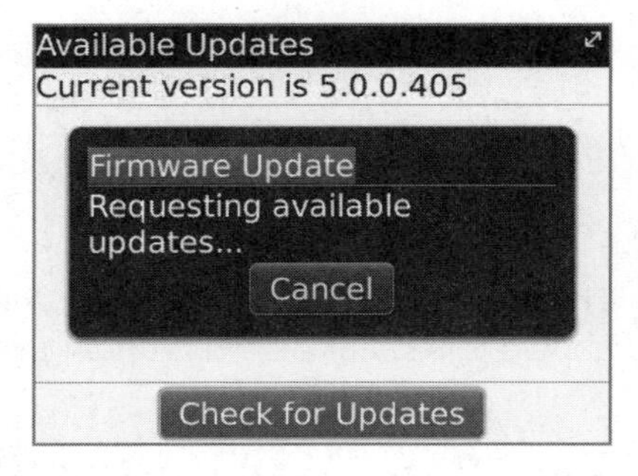

NOTE: A full update can take an hour or more, so you may want to complete the update only at night.

6. Once the update is complete, you will see a status message as shown to the right. If there are no updates, you will see a message similar to the one following.

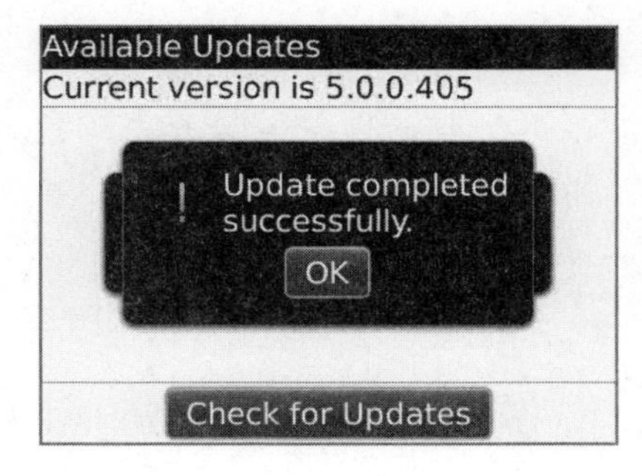

Still Can't Fix Your Problems?

Check out all the resources found in the "More Resources" chapter on page 535.

Chapter 36

More Resources

One of the great things about owning a BlackBerry is that you immediately become a part of a large, worldwide community of BlackBerry owners.

Many BlackBerry owners could easily be classified as enthusiasts and are part of a number of BlackBerry user groups. These user groups, along with various forums and web sites, serve as a great resource for BlackBerry users.

Many of these resources are available right from your BlackBerry, and others are websites that you might want to visit on your computer.

Sometimes, you may just want to connect with other BlackBerry enthusiasts, ask a technical question, or keep up with the latest and greatest rumors, apps, and accessories. This chapter gives you some great resources for doing so.

Visiting the BlackBerry Technical Solution Center

Go to the BlackBerry Technical Solution Center (see Figure 36-1) for help with anything to do with your BlackBerry:

1. Do a web search for "BlackBerry Technical Knowledge Base." One of the top links will usually get you to this page on `www.blackberry.com`. You may also be able to get directly to the Technical Solution Center by typing in the web address `www.blackberry.com/btsc/`.
2. Just type your question in as few words as possible in the **Search** box in the middle of the screen.

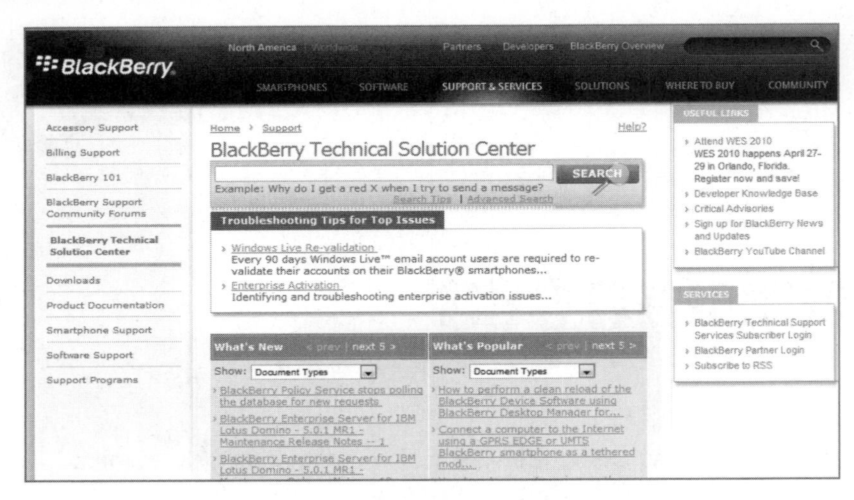

Figure 36-1. *BlackBerry Technical Solution Center web page.*

Accessing Resources from Your BlackBerry

The first place to start is the **Help** item found on most of the menus. Learn all about how to use BlackBerry built-in help on page 159.

Some of the resources actually have software icons to install right on your BlackBerry home screen. Others, you will want to set as bookmarks in your BlackBerry web browser (learn about bookmarks on page 467)

1. The official BlackBerry mobile website is `mobile.blackberry.com`
2. Start your BlackBerry Web Browser.
3. If you don't see a place to type a web address, then press the **Menu** key, and select **Go To…**.
4. Type mobile.blackberry.com.
5. The page is organized with a help directory, a **What's New** section, as well as pages for **Fun and Games**, **Great Sites**, **Messaging**, **Maps & GPS**, and **BlackBerry**.
6. Click any link to download software or visit the linked website.

Finding BlackBerry Forums and Discussion Groups

There are many of discussion groups such as CrackBerry, BerryReview, BlackBerry Rocks, and BlackBerry Forums. A forum is an organized discussion about anything and everything BlackBerry. On each of these sites, you can discuss your BlackBerry and find tips and tricks, as well as post questions and read answers to other users' questions.

At the time of this writing, these are a few of these more popular forums:

- `www.boygeniusreport.com`
- `www.blackberrycool.com`
- `www.blackberryforums.com`
- `www.crackberry.com`
- `www.berryreview.com`
- `www.blackberryrocks.com`

Some sites even provide you a way to install an icon in your BlackBerry application menu for one-click access to participate in the discussion right from your BlackBerry.

Navigate through the forums or use the Search tool to find a particular topic of interest. One of the forums will be specific to your particular BlackBerry, so look for forums that match your BlackBerry model or series, for example, BlackBerry Bold 9700 or BlackBerry Bold 9650.

Various discussion threads on topics related to the BlackBerry Bold 9650 and Bold 9700 will then be listed like those from the "Made Simple Learning" website (`http://www.madesimplelearning.com`.

> **TIP:** Try your best to enter your question under the correct forum topic; you are much more likely to find people answering you if you are in the correct topic area.

There are many, many other web sites to visit for more information and helpful resources that offer news, information and discussion forums.

Made Simple Learning Free BlackBerry Tips

We host a website at `www.madesimplelearning.com` that provides free BlackBerry tips via email and free sample video tutorials. We also have a Contact Us page, so please drop us a line and let us know what is on your mind.

Thanks!

The authors sincerely thank you for purchasing this book and hope it has helped you learn how to get every last drop of productivity and fun out of your BlackBerry!

Hotkey Shortcuts

Hotkeys – Open Almost Any App with One Key!

You have learned the basics and can now navigate around your BlackBerry, but sometimes, you just want to get somewhere even more quickly. You may not use all the following information on a daily basis, but if you take the time to learn the hotkeys – you will save time doing thing things you do most often.

Throughout this book, we have shown you how to get the most out of your BlackBerry, where to find key information, and even suggested lots of tips and tricks to help along the way.

BlackBerry's built-in **hotkeys** to quickly take you to an app or through a menu without needless scrolling and searching.

For example, let's say you want to go to your **Messages** app – once **hotkeys** are enabled, just hit the letter **M**. No scrolling, searching, or fumbling. These hotkeys can be very useful.

Home Screen Hotkeys

Hotkeys can get you around your BlackBerry home screen, but there are also hotkeys inside of each of the core apps. We will show you how to not only launch programs quickly, but how to jump around inside of your apps and menus.

Once you enable your **Home Screen Hotkeys**, you will be able to launch most of the built-in applications with just one letter. This is extremely useful when you have lots of programs and you don't exactly remember where the launch icons are located.

NOTE: Depending on the version of the operating system software installed on your BlackBerry smartphone, some of these Home screen hotkeys or key combinations may not work.

Enable Home Screen Hotkeys

Use these steps to turn on the Blackberry **Home Screen** Hotkeys (listed in the following table).

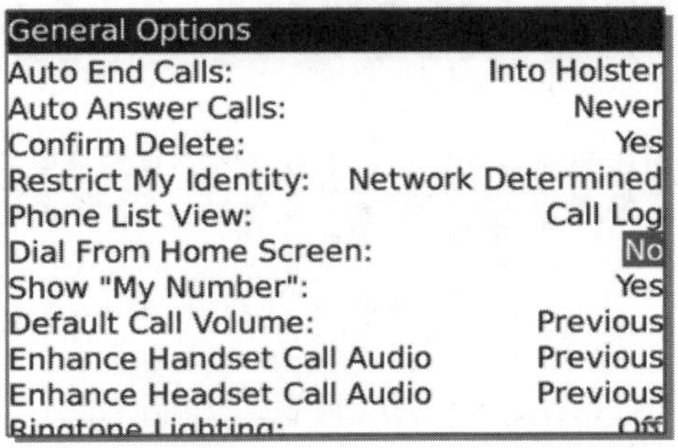

1. Access your **Phone call logs** by tapping the **Green Phone** key once.
2. Press the **Menu** key.
3. Select **Options** from the **Phone Logs** menu.
4. Click **General Options**, as shown in the screen to the right.
5. Glide to **Dial from Home Screen** and change it to **No** by pressing the **Space** key.
6. Press the **Menu** key and save your **Options** settings.

General Options	
Auto End Calls:	Into Holster
Auto Answer Calls:	Never
Confirm Delete:	Yes
Restrict My Identity:	Network Determined
Phone List View:	Call Log
Dial From Home Screen:	No
Show "My Number":	Yes
Default Call Volume:	Previous
Enhance Handset Call Audio	Previous
Enhance Headset Call Audio	Previous
Ringtone Lighting:	Off

List of Home Screen Hotkeys

Following are the list of one key shortcuts to start most of your major icons.

M	Messages (Email)	H	Help
L	Calendar	K	Keyboard Lock
C	Contact or A = Address Book	O	Options Icon
V	Saved Messages	U	Calculator
N	BlackBerry Messenger	S	Search Icon
T	Tasks	D	MemoPad
B	Web Browser	P	Phone (Call Logs)
W	WAP Browser		

Once turned on, the hotkeys are denoted by the underlined letter. For example, highlight the **Options** icon and see that the hotkey O is underlined, as shown on the right.

CAUTION: Don't press and hold any of these hotkeys, otherwise you will start Speed Dialing (page **216**)

Email Messages Hotkeys

Not only are hotkeys helpful to launch apps, they are also very useful from inside an app to take you to a specific menu command. The following are the hotkeys that can get you around your **BlackBerry Mail** app.

NOTE: There is nothing you need to do to activate the hotkeys in the **Mail** app.

T Top of Inbox B Bottom of Inbox U Go to next newest Unread Message	**Space** key - Page Down **ALT + Glide trackpad** - Page Up/Down ENTER - Open Item (or compose new message when on date row separator)	C Compose new email R Reply to selected message L Reply All to selected message F Forward selected
N Next day P Previous day ALT + U Toggle read / unread message	E Find Delivery Error(s) V Go to Saved Messages Folder Q (When highlighting email/name) Show/Hide Email Address Or Friendly Name	K Search for next message in thread (Replies, Forward, etc.) S Search
Shift + Glidetrackpad - Select messages DEL Delete selected message(s) **CAUTION:** If wireless email synchronization is on, this will also delete email from your inbox.	ALT + O – Filter for Outgoing Messages ALT + I – Filter for Incoming Messages	ALT + M – Filter for MMS Messages ALT + S – Filter for SMS Messages

Web Browser Hotkeys

Just like in your **Email** app, you can quickly launch certain commands or jump to specific places in your **Browser** menu by using certain hotkeys, such as the ones in the following table.

> **TIP:** When typing a web address, try these handy shortcuts:
>
> Use **Space** key for the "." (dots)iln the web address.
>
> Press **Shift** + **Space** key to get the "/" (slash) (for example, in www.google.com/gmm).
>
> Press the **Escape** key to stop loading a web page and back up one level; or press and hold the **Escape** key to close the Browser.

T/X Top of Web Page	**Enter** – Select (click-on) highlighted link	I Zoom into web page
B Bottom of Web Page	Y View history of web pages	O Zoom out of web page
Space key – Page Down	R Refresh the current web page	U Show/hide top status bar
Shift + Space key – Page Up	P View the address for the page you are viewing with an option to copy or send the address (via email, PIN, or SMS).	**F/V** Search for text on the current web page
G Go to your Start page where you can type in a web address	L View the web page address the clickable link that is currently highlighted	Z Switch between column and page view
H Go to your Home page (Set in **Browser menu → Options → Browser Configuration**)	J Turn on JavaScript support	A Add new bookmark
Escape key/**DEL** key Go back 1 page		K View your Bookmark List
D Jump out of Browser, go to Home screen of icons		S View Browser Options screens
C View connection information (bytes sent/received, security)		

> **NOTE:** These hotkeys do not work in your Bookmark List or **Start** page; only when you are actually viewing a web page. In the Bookmark List, typing letters will find your bookmarks that match the typed letters.

Calendar Hotkeys

You can turn on the **Calendar** hotkeys listed in the following table on your BlackBerry using these steps:

1. Start your **Calendar**.
2. Press the **Menu** key.
3. Select **Options** from the **Calendar** Menu.
4. In the **General Calendar Options** screen, glide to **Enable Quick Entry** and change it to **No** by pressing the **Space** key.
5. Save your **Options** settings.

General Calendar Options

Initial View:	Day
Show Free Time in Agenda View:	Yes
Show End Time in Agenda View:	Yes
Actions	
Snooze:	5 Min.
Default Reminder:	15 Min.
Enable Quick Entry:	Yes / No
Confirm Delete:	
Keep Appointments:	60 Days
Show Tasks:	No

Key	Action
D	Day View
W	Week View
M	Month View
A	Agenda View
N	Next day
P	Previous day
G	Go To Date
T	Jump to Today (Now)
C	Schedule new event (detail view)
Space key	Next Day (day view)
DEL	Delete selected event

Shift + glide trackpad – select several hours – example if you wanted to quickly schedule a 3 hour meeting you would highlight the 3 hours and press **Enter**.

Enter – (in Day View – not on a scheduled event) Start Quick Scheduling

Enter – (in Day View – on a scheduled event) Opens it.

Quick Scheduling New Events in Day View:

Enter key Begin Quick Scheduling, then type Subject of Appointment on Day View screen.

4:00p Quick scheduling
4:15p

ALT + Glide trackpad down – Change **Start** time

4:30p Quick scheduling
4:45p

Glidetrackpad up/down – Change **Ending** time

4:30p Quick scheduling
5:30p

Enter or click **trackpad** – Save new event

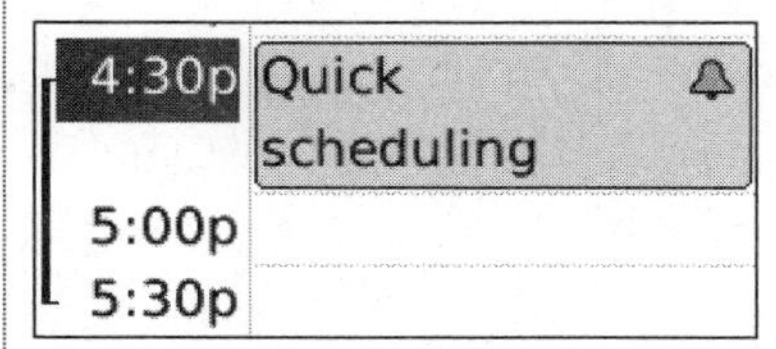

Enter or click **trackpad** – Open the event you just scheduled to change details such as Recurring, Notes, Alarms, and so forth.

Moving Around: Glide **trackpad** left/right to move a day-at-a-time (Day or Week View)

Media Player Hotkeys

The hotkeys and tips below will help you enjoy media on your Bold by performing many of the common functions with a single keystroke or special key combinations.

You also have a **Mute** key on the top right of your BlackBerry that will play/pause any song or video.

These keys work only inside the Media Player App	These keys work anytime (Whether inside the Media App or not)
N Next track/video P Previous track/video **Space key** Play / pause **Enter** key Selects highlighted button (same as clicking trackpad / clicking trackball) **DEL** Delete selected media item (song, video, and so forth, when in list or icon view)	Press & hold **Volume Up** key – Next track/video Press & hold **Volume Down** key – Previous track/video **Mute** Key on top – Play/pause **$/Speaker** key – Switch between speaker and handset/headset (including Bluetooth headset)
Jumping Out of the Media Player: Press the **Red Phone** key to jump back to the Home screen.	Jumping Back into the Media Player and currently playing song/video: Tap the **Menu** key and select **Now Playing…** from the top of the menu. Press and hold the **Menu** key, then select the **Media** icon from the pop-up.

Index

C

E

F

G

H

I

J

K

L

M

N

O

P

Q

R

S

T

U

V

W

Y

Z